AVIAN ECOLOGY AND CONSERVATION IN AN URBANIZING WORLD

AVIAN ECOLOGY AND CONSERVATION IN AN URBANIZING WORLD

edited by

John M. Marzluff
College of Forest Resources
University of Washington

Reed Bowman
Archbold Biological Station

Roarke Donnelly
College of Forest Resources
University of Washington

KLUWER ACADEMIC PUBLISHERS
Boston / Dordrecht / London

Distributors for North, Central and South America:
Kluwer Academic Publishers
101 Philip Drive
Assinippi Park
Norwell, Massachusetts 02061 USA
Telephone (781) 871-6600
Fax (781) 681-9045
E-Mail <kluwer@wkap.com>

Distributors for all other countries:
Kluwer Academic Publishers Group
Distribution Centre
Post Office Box 322
3300 AH Dordrecht, THE NETHERLANDS
Telephone 31 78 6392 392
Fax 31 78 6546 474
E-Mail <services@wkap.nl>

Electronic Services <http://www.wkap.nl>

Library of Congress Cataloging-in-Publication Data

Avian ecology and conservation in an urbanizing world / edited by John Marzluff, Reed Bowman, Roarke Donnelly.
p. cm.
Includes bibliographical references (p.).
ISBN 0-7923-7458-4
1. Birds—Ecology. 2. Urbanization—Environmental aspects. 3. Birds, Protection of. I. Marzluff, John M. II. Bowman, Reed. III. Donnelly, Roarke.

QL698.95 .A84 2001
598.17—dc21

2001042818

Printed on acid-free paper. Printed in the United States of America

TABLE OF CONTENTS

PREFACE

Humans dominate earth's ecosystems in a variety of ways. One of the most striking and persistent is our ability to change land-cover as we settle a region. Settlement modifies water and sediment fluxes; reduces, and fragments natural habitats; introduces exotic organisms; severely alters energy flow and nutrient cycles; and overstresses entire ecosystems. Despite these important effects, the ecological implications of settlement and resulting urbanization are just beginning to be understood. Much of our understanding derives from studies of birds, yet the existing literature is scattered, mostly decades old, and rarely synthesized or standardized. In addition, a new paradigm for urban ecological research is emerging, one based solidly on traditional ecological theory. Increasingly, studies of birds in urban environments consider the mechanisms by which human changes affect birds, they consider these effects at multiple levels of ecological organization, including the community, population, and individuals, and they consider these effects across a gradient of different levels of urbanization. Therefore, we felt it timely to summarize existing studies, suggest standard approaches, and showcase ongoing studies of birds in urban environments.

We investigate how birds respond to human settlement in 27 chapters authored by leaders in this expanding field. Our objectives were to: 1) review what is known about birds in urbanizing settings, 2) suggest important ways to standardize our study of birds and urbanization, 3) investigate connections between urbanization and individual birds, the demography of bird populations and the structure and composition of bird communities, 4) probe the evolutionary importance of urbanization to birds, 5) understand how best to conserve birds in the face of increasing urbanization, 6) investigate the policy and management implications of urban bird studies, and 7) suggest important areas where research is needed to better understand these topics. We introduce these objectives and summarize the authors' progress at fulfilling them in an introductory and summary chapter.

This volume stands alone as a summary of current research on birds in settled environments ranging from wildlands to exurban/rural to urban. As such, we feel it would be suitable for the ecologist, land manager, wildlife manager, evolutionary ecologist, urban planner, landscape architect, and conservation biologist. This wide array of professionals will find our information useful because we address the conservation and evolutionary implications of urban life from an ecological and planning perspective. Graduate students in these fields also will find the volume to be a useful

Graduate students in these fields also will find the volume to be a useful summary and synthesis of current research, extant literature, and prescriptions for future work. All interested in human-driven land-cover changes will benefit from a perusal of this book because we present high altitude photographs of each study area. Simply looking at these images shows the variety of ways in which human settlement affects earth--from the tropics to the arctic and from intensive urban centers to small native settlements.

We organized this volume into five sections. Section I introduces the study of birds in urban environments. Section II investigates the important anthropogenic processes that affect birds in urban environments. Sections III and IV provide case studies of bird populations and communities in urban settings. These are all new empirical works, often summarizing decades of research, that clearly relate the processes discussed in Section II to demographic and community-level processes. Section V is a single synthetic chapter summarizing the book and highlighting its implications for future studies.

In Section I, the reader learns: how we have studied birds in urban environments (Chp. 1), standard definitions of settlement (Chp. 1), how humans have settled Earth (Chp. 2), how birds respond to settlement (Chps. 2, 3), the degree to which North American birds are synanthropic (Chp. 3), how human perception and appreciation of birds is affected by urbanization (Chp. 4), how to accurately quantify urban settlement patterns so they can rigorously be related to birds and other organisms (Chp. 5), and how to consider multiple scales in our study of birds (Chps. 6, 7,). These latter chapters emphasize the need to consider many scales in ecological study as well as policy implementation. We need to understand how individual animals and landscapes respond to urbanization and then use this understanding to motivate individual landowners and increasingly complex governments to plan urbanization in ways compatible with biodiversity.

Armed with this background, we discuss important mechanisms that affect urban birds in Section II. The importance of "top down" mechanisms such as predation and competition are contrasted with more subtle, less studied "bottom up" mechanisms such as arthropod availability (Chp. 8). Changes in vegetation caused by settlement are reviewed in detail (Chp. 9) and specifically related to bird changes in the midwestern United States (Chp. 10). The role birds play in aiding the spread of invasive plants is also discussed (Chp. 9). The well-studied influence of urbanization on avian nest predators is discussed in a case study from the American South (Chp. 11) and related to fragmentation of the American West (Chp. 8). Humans visiting parks disturb native birds, but susceptibility depends on park vegetation, amount of visitation, and bird size (Chp. 12). These results come from Europe, but have broad applicability to the design of trails in urban parks. European Starlings are one of the world's most successful urban birds. Part of their success stems from their ability to utilize lawns and anthropogenic foods in urban and suburban settings (Chp. 13). Foods provided to birds in suburban settings produce complex physiological responses that are now amenable to study. For example, suburban Florida Scrub-jays have elevated plasma protein levels prior to egg laying suggesting that anthropogenic foods may provide essential nutrients that, in turn, cue breeders to lay eggs (Chp. 14). This may allow suburban jays to breed earlier than wildland jays, but does not guarantee them greater fledgling success (see below). The general mechanisms we cover in Section II (exotic vegetation, food availability, exotic competitors, predation, and human disturbance) apply worldwide and are the primary reasons why bird populations and communities change in response to urbanization.

We illustrate many of the mechanisms discussed in Section II with specific examples of how they affect bird populations in Section III. Western Gulls decline in response to urbanization because of changes in nesting habitat and available food (Chp. 15). This is striking because gulls are usually thought of as being strongly synanthropic (Chp. 3). But, long-term observations and detailed experiments on Western Gulls indicate that relying on human refuse for food reduces reproductive performance (Chp. 15). On the other hand, corvids appear to more effectively use human foods and their populations increase in response to urbanization. American Crows (Chps. 16 and 17) appear to be most successful in suburban and rural environments, but this success may fuel dramatic increases in urban settings as well. Florida scrub-jays (Chp. 18) also use anthropogenic foods, but differences in food and predators between wildland and suburban areas appear to have a complex and interactive effect on nest success. European magpies are increasing rapidly in suburban and urban areas of the Palearctic (Chp. 19) because their reproduction and survivorship is increased as they utilize anthropogenic foods and flourish in response to predator reductions. Human settlement of the Amazon and Arctic is low at present, but chapters 20 and 21 anticipate the potential effects on native birds should settlement in these regions increase. Parrots are likely to decline in the Amazon in response to direct harvest, changes in fruit abundance, and habitat modification. Reserves that protect these species will likely need to be far from even small human settlements. Waterbirds in the Arctic are likely to have mixed responses to human settlement; many may benefit from man-made water bodes while others may decline in response to wetland draining and human use of wetlands. At least in Alaska, current policies (Clean Water Act) can be used to conserve these sensitive areas.

The demographic responses of birds to urbanization detailed in Section III eventually lead to restructuring of avian communities. Approaches and introductions to such community-level responses were investigated earlier (Chps. 2, 6, 7, 8). In Section IV a series of empirical studies of avian communities in urbanizing areas is presented. Urban bird communities worldwide are characterized by ecologically similar species. Often, they even host identical species. This homogenization of avifaunas is explored by contrasting western and mid-western sites in the United States (Chp. 22), but its implication for worldwide avian diversity is sobering. Many recent investigators of bird communities relate changes in land-cover, land-use, and resulting landscape patterns to community assembly. New empirical work on avian communities in St. Louis (Missouri), Chicago (Illinois), Cincinnati (Ohio), and Toronto are detailed in Chapters 23-26. These studies quantify the large diversity of birds living in some of the most urbanized regions of North America. They suggest that despite extensive land-cover change in these areas, many native birds persist and rely on these areas for breeding (Chps. 23, 26), during migration (Chp. 24), and over winter (Chp. 25). Challenges abound to conserve diverse, native bird communities in highly urban areas, but these studies suggest that a variety of policy instruments (e.g., greenspace reserves and private-public partnerships) are feasible and effective. The extensive group working to conserve native areas in the Chicago region is exemplary and is being used as a model for international efforts (Chp. 24).

We conclude the book in a final section with a single chapter summarizing the main points made by contributors and reflecting on the type of research that is still needed to increase our ability to provide for birds in the face of global urbanization. Current research suggests that the effect of urbanization on birds is immense, yet our understanding of these effects is rudimentary. We identify three broad approaches that are necessary to improve this understanding: (1) long-term research that utilizes an experimental and observational approach, (2) research that explores the influence of multiple urban spatial patterns, and (3) research that examines urbanization effects on individuals, populations and communities across gradients of variable urbanization intensity. Many effects of urbanization are obvious (loss and degradation of native habitat, introduction of exotics, changes in predator communities). However, it is important to remember the less obvious and often indirect effects of urbanization (climate change, human disturbance, ecosystem disruption, physiological stress, toxins, food supplementation) and determine how these affect birds. We end by suggesting that effective conservation of birds in urban settings will require involvement of all relevant stakeholders and interdisciplinary approaches. Our hope is that future urban avifaunas will retain a large and self-sustaining component of that which existed before settlement. The thoughts presented in this volume should help us attain that goal.

We could not have completed this book without the patience and support of many. Each chapter was peer reviewed by two of our colleagues. We thank David Aborn, Curtis Adkisson, Marina Alberti, Keith Bildstein, David Bird, Robert Blair, Carl Bock, Jeff Bradley, Jeff Brawn, Dan Brooks, Joanna Burger, John Russell Butler, Carolee Caffrey, Dave Craig, Robert Curry, Richard DeGraaf, John Faaborg, Fred Gehlbach, Jon Greenlaw, John Hadidian, Harlo Hadow, Lori Henning, Mark Hostetler, Richard Johnston, Jukka Jokimaki, Matthias Leu, Dave Manuwal, Don McIvor, James Miller, Michael Molton, Ronald Mumme, Charles Nilon, Michael North, Peter Paton, Raymond Pierotti, Thane Pratt, Marco Restani, James Rodgers, Steven Rottenborn, Tom Scott, Dawn Sherry, Navjot Sodhi, Harrison B. Tordoff, Ian Warkington and Dave Willard for providing these reviews. John Marzluff and Roarke Donnelly were supported by The National Science Foundation (DEB 9875041), The Washington Department of Fish and Wildlife, and The University of Washington, particularly its Rachael Wood's Graduate Program and the Tool's for Transformation Program. Reed Bowman was supported by Archbold Biological Station, the Florida Fish and Wildlife Conservation Commission, the United States Fish and Wildlife Service, and the National Science Foundation (IBN-0077469).

John Marzluff, Seattle, WA
Reed Bowman, Lake Placid, FL
Roarke Donnelly, Seattle, WA

Chapter 1

A historical perspective on urban bird research: trends, terms, and approaches

John M. Marzluff[1], Reed Bowman[2], and Roarke Donnelly[1]
[1]*College of Forest Resources, University of Washington, Box 352100, Seattle, WA 98195 USA, e-mail: corvid@u.washington.edu*

[2]*Archold Biological Station, P.O. Box 2047, Lake Placid, FL 33862 USA*

Key words: Community ecology, exurban, human population growth, landscape ecology, population ecology, study design, study duration, suburban, urban, urban metrics, urbanization, wildland

Abstract: As Earth's human population continues to increase and urbanize, it is likely to increasingly affect biodiversity. Avian ecologists have been studying these effects for over a century. Here, we review these studies to: 1) characterize the type of research approaches that have been used, 2) suggest strengths and weaknesses of these approaches, 3) offer a standardized nomenclature for the degree of settlement that will be used throughout this volume, and 4) suggest how our approach can be strengthened to better inform public policy. The majority of urban bird studies were conducted since 1980. The typical study is a one- or two-year correlational investigation of breeding bird relative abundance in the forests of the United States or northern Europe. Experimental studies are rare despite the frequent and replicated land transformations conducted by developers. Studies of birds in tropical, urban settings are especially rare. This is problematic because human populations are expected to rapidly grow and urbanize in such regions and biodiversity there is rich. Recognizing trade-offs among study duration, spatial extent, and mechanistic understanding, we suggest that researchers use short-term, correlational studies of the entire gradient of urbanization to inform long-term, mechanistic studies of bird populations. We define five points along the gradient of urbanization for consistent use throughout this volume (wildland, exurban or rural, suburban, and urban; Table 1.1). These are useful for categorizing study areas at the landscape scale ($>1 km^2$). Briefly, wildlands are unsettled lands that may occasionally include dwellings. Rural and exurban lands are sparsely settled by individual homesteads, recreational development, and small towns (they are distinguished by the surrounding matrix: rural = agricultural, exurban = native habitat). Suburban lands are characterized by moderate- to high-density, single-family housing with lot sizes of 0.1-1.0 ha. Urban lands are primarily covered by multi-family and/or multi-storied buildings. Urban bird studies would more effectively inform policy, planning, and management if they were more relevant, rigorous, compelling, and visible. We suggest that this can be accomplished by 1) understanding how development pattern affects birds, 2) increasing study of birds in tropical urban areas, 3) increasing the use of experimental, mechanistic investigations, 4) using landscape ecological metrics to quantify the urban settings we study, and 5) distilling clear graphs or photographs of important results.

1. INTRODUCTION

Human populations continue to grow and dominate ecosystems around the world (Horiuchi 1992; Vitousek et al. 1997). Wild portions of Earth are urbanizing and humans are increasingly on the move in many parts of the world, shifting from rural regions to burgeoning urban centers. As a result, the world's urban population multiplied tenfold last century. The United Nations estimates that by 2050, the global urban population will equal today's total population (~6.5 billion; United Nations 1996). Most urban growth is occurring in developing countries, where human populations are increasing at exponential rates (Fig. 1.1; WRI 1996). By 2025 the urban population in developing countries will reach 4 billion (3x the expected urban population of developed countries; United Nations 1996). In those countries urban growth will occur so rapidly that it will strain the ability of local governments to provide adequate housing, infrastructure, sanitation, public safety and other essential services. The result will be an increased human impact on ecosystems.

In the developed world, however, a greater proportion of the land is urbanized and populations are moving away from traditional city centers. Emigrants tend to redistribute themselves in a suburban or exurban pattern, causing cities to grow in area much faster than in population and resulting in strong gradients of human density. These gradients of urbanization become increasingly complex and multi-modal as suburbs take on increasingly urban characteristics (Alberti et al. 2001). These processes, collectively known as urbanization, had a staggering effect on native flora and fauna. Urbanization is likely to be the single most important driver of extinction during this century. Already, urbanization is the second most frequently cited cause of species endangerment in the United States (Czech and Krausman 1997).

This book aims to document and discuss the effects of human settlement on birds. To put this current discussion into perspective, we have reviewed the published literature on the effects of urbanization on birds. In this chapter, we use this review to: 1) characterize the type of research approaches that have been used, 2) suggest strengths and weaknesses of these approaches, 3) offer a standardized nomenclature for different levels of human settlement along the urban gradient that will be used throughout this volume, and 4) suggest how ecological research on birds along this can be strengthened to better inform public policy.

2. METHODS

We reviewed the modern literature (1900-2000) using electronic databases and manual search. The databases *Biosis, Wildlife Worldwide, and Current Contents* were queried using *urban, birds, human effects on birds, fragmentation, settlement, suburbs, and development* as key words. The literature cited section of each article found by electronic search was reviewed for additional published works on birds in urbanized settings. In total, we found 101 articles that reported new empirical research on birds and urbanization. We excluded review and theoretical articles.

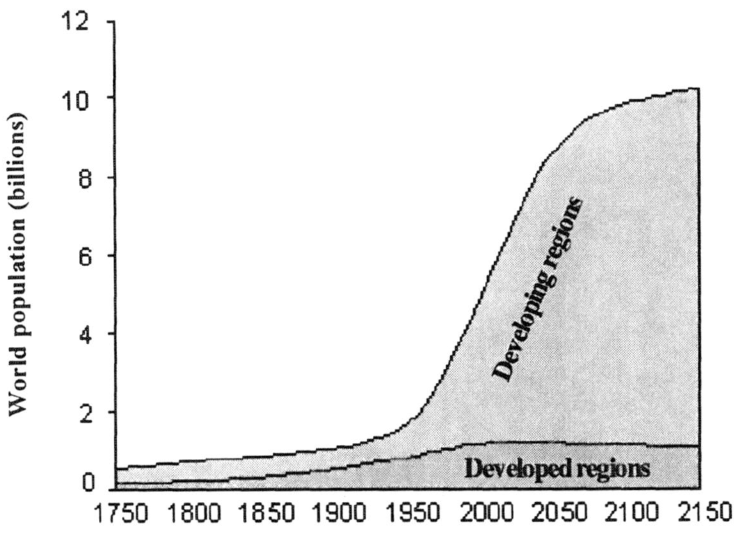

Figure 1.1. Past and projected growth of human populations in developed and developing regions of Earth compiled by the World Resources Institute (WRI 1996). Growth is expected to increase most dramatically in developing regions.

3. RESULTS AND DISCUSSION OF LITERATURE REVIEW

The study of urban effects on birds has increased steadily (Fig. 1.2) since Pitelka (1942) characterized how the coastal avifauna of California responded to suburbanization and recreational development. Nearly 75% of the studies we reviewed were conducted during the last 20 years.

The typical study of the impact of urbanization on birds is one- or two-year (Fig. 1.3) correlational investigation of the relative abundance of breeding birds in the forests of the United States or northern Europe (Fig. 1.4). Most studies of urbanization (63%) simultaneously compared avifaunas at two to four points along a portion of the gradient from urban areas to wildlands. Rarely (8%) the study areas spanned the entire gradient. Often (29%), only one point on the gradient was studied. Longer-term studies were not uncommon; 39% of studies lasted >5 years and four (Batten 1972, Walcott 1974, Bezzel 1985, and Luniak and Mulsow 1988) documented avifaunal change at a site that became increasingly urbanized over more than a century. These long-term studies often employed a time

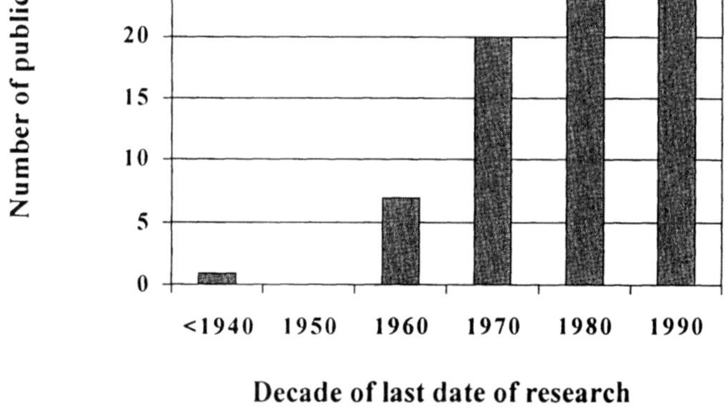

Figure 1.2. Occurrence of published studies (n = 101) on urban effects on birds during the past 60 years. Numbers of studies that conducted research in each decade is plotted.

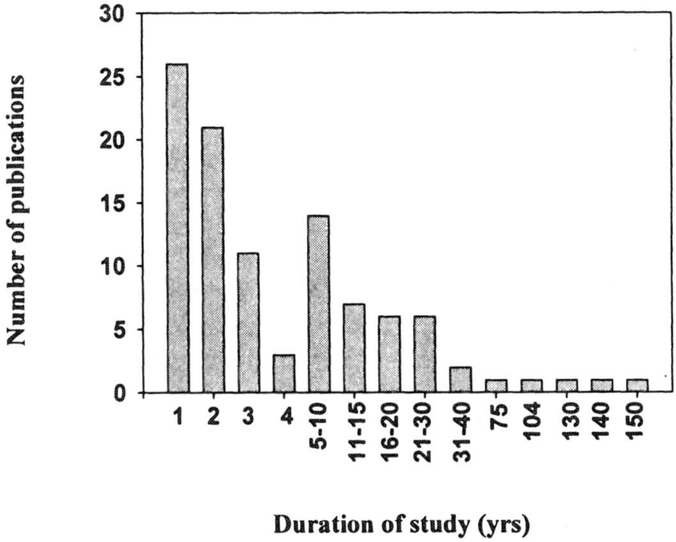

Figure 1.3. Duration (years) of 101 studies on the effects of urban development on birds.

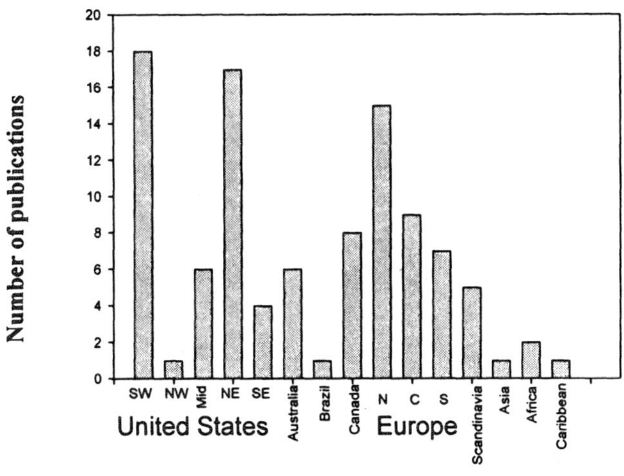

Figure 1.4. Locations where 101 studies of the effects of urbanization on birds were conducted. N Europe = Germany, Poland, Russia, and Switzerland. C Europe = Estonia and the United Kingdom. S Europe = Spain and Italy. The two African studies were conducted in South Africa. The Caribbean study was done on Trinidad.

series method of analysis and hence could be considered quasi-experimental. However, only five studies used a true experimental approach with replicates and controls. These five were all investigations of predation rates using experimental nests deployed at varying distances from human settlement. Despite the rapid pace of urbanization, its proximity to major research institutions, and its obvious effects on ecosystems, we found no studies that used actual human settlement patterns as replicated factors in a controlled, experimental study.

The preponderance of one- and two-year studies is not surprising given the usual tenure of funds and graduate students, but the occurrence of truly long-term studies is notable. Ecological studies in general (Kareiva and Anderson 1989), and studies of birds in particular (Marzluff and Sallabanks 1998) are of notoriously short duration. Yet long-term records of birds in areas settled by humans have allowed us to relate avian community composition to increasing human settlement over decades and even centuries. The study of Blackbirds (*Turdus merula*) over a period of 150 years by Luniak and Mulsow (1988) is especially notable, because they documented increasing density of Blackbirds in response to urbanization and identified increased fecundity and survival as the mechanisms for the density increase.

Our inferences about the response of birds to urbanization may be based on solid temporal studies, but this literature is woefully limited to breeding birds of temperate, upland forests. Over 60% of the studies (n = 65) were conducted only during the breeding season and over 70% were conducted in temperate forests (n = 71). Studies of other ecoregions do exist, but in small numbers. We found 10 studies of shrublands, that while modest in size, were distributed across a broad geographical area (America, Europe, Australia, and South Africa). Five or fewer studies were conducted in each of the following ecoregions: aquatic (n = 2), coastal or island (n = 4), desert (n = 3), grassland (n = 5), and riparian forest (n = 5).

Studies of birds in tropical urban settings are especially lacking. We found only six such studies. Two were from the neotropics (Caribbean: Fonaroff 1974; Brazil: Ruszczyk et al. 1987) and four were from the Australian tropics (Jones 1981, Green et al. 1989, Munyenyembe et al. 1989, Kentish et al. 1995). Given the expected increase in human populations and urbanization in the tropics (Fig. 1.1) and the rich biodiversity that characterizes this region, more studies are desperately needed to inform public policy so that the negative consequences of human development are mitigated.

Most studies of the effects of urbanization on bird community structure and composition have quantified the relative abundance of each species (Fig. 1.5). These studies are essential to understanding how community structure and composition change with urbanization, but they provide limited, and potentially misleading, information about the mechanisms that drive community changes with increasing urbanization. Such studies cannot distinguish among source, stable, and sink populations (Pulliam 1988) which is essential to understand if we are to manage birds effectively (Van Horne 1983). They do not inform us about the relative movement patterns of individuals along the urban gradient or about the potential metapopulation structure that may exist among populations along that gradient. Nor do they inform us about the ecological changes along that gradient and how these changes influence demographic patterns. Therefore, to improve our understanding of how birds respond to urbanization we need community-level studies, but we also need studies examining demography, including fecundity and survivorship and studies examining population dynamics, including immigration and emigration, at a variety of spatial scales. This requires long-term studies that focus on populations and demography rather than entire communities. Such studies are not uncommon in urbanizing environments. Thirty-seven percent (n = 37) of studies focused on populations, 19 measured annual breeding success, and 12 measured

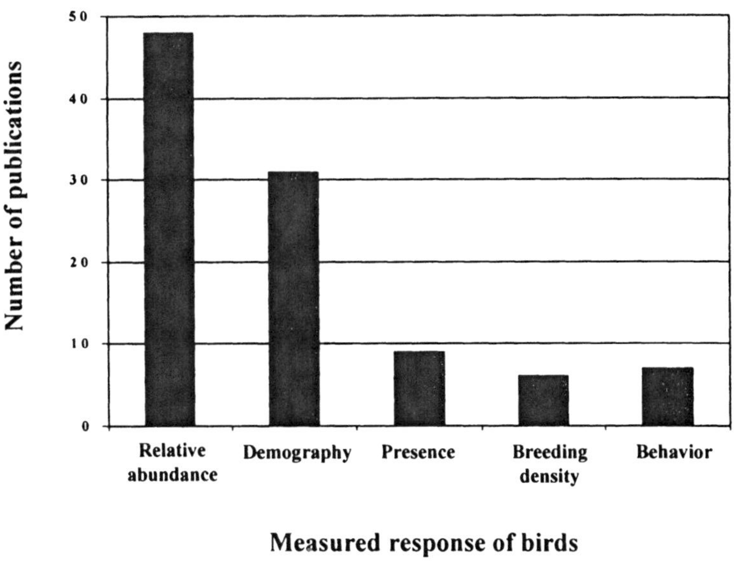

Figure 1.5. Measured responses of birds to urbanization in the published (n = 101 articles) literature. Demography includes 19 studies of annual breeding success and 12 studies of fecundity or survivorship. Presence documents studies that recorded presence or absence of a particular species. Behavior includes 4 studies of diet, 1 of time budget, and 2 of other behaviors.

fecundity and survivorship. Sixteen studies discussed the source-sink structure of populations in urbanizing environments. However, few of these studies examined emigration or immigration rates and none examined the large scale structuring of populations along an urban gradient.

Urban bird studies are distinctly population- (n = 37) OR community-level (n = 64) investigations. We found no studies that explicitly discussed the connection between population-level dynamics and community structuring. Yet, a full understanding of ecological communities requires understanding of individual, population, and higher order biotic effects, as well as abiotic effects (Wiens 1989, Marzluff and Ewing 2001, Bolger 2001).

In accord with the large spatial extent of urbanization, urban bird studies often (n = 42 studies) related landscape features to urban bird communities and populations. However, measures of urban landscape features are crude or nonexistent (Fig. 1.6). Forty percent (n = 41) of the studies we reviewed did not quantify the urban setting in which their study occurred! Those that did quantify the degree of urbanization used relatively simple measures such as the percentage of the landscape that was urban or the density of houses surrounding the study area (n = 38). Distance to human settlement was used to quantify urban landscapes in five studies. Sophisticated measurement of urban patterns using typical landscape ecology metrics, as proposed by Alberti et al. (2001), has not been utilized in past avian studies. Likewise, multi-scale analyses that recognize the varying spatial sensitivity of individuals and populations as recommended by Hostetler (2001) are lacking.

Our review suggests that studies of birds in urbanizing landscapes must often make tradeoffs among increasing duration, increasing spatial resolution, and increasing mechanistic understanding. Understanding the mechanisms of population responses to urbanization does not appear to reduce a study's duration. On average, population-level studies lasted 12.7 years (SE = 4.0; n = 37) and community-level studies lasted 12.0 years (SE = 3.5; n = 64). However, mechanistic understanding appears to carry a cost of reducing spatial extent of study (Fig. 1.7). Studies that included landscapes ranging from urban or suburban to wildlands were twice as likely to employ a community-level approach, than a population-level approach (Fig. 1.7). Some community-level studies also were done at only a single point along the gradient. Increasing spatial extent appears to reduce study duration. The average duration of studies conducted at one point on the gradient was 12.8 years (SE = 5.1, n = 29). In contrast, studies spanning the entire urban to wildland gradient lasted an average of 10.9 years (SE = 4.3, n = 8). Studies spanning intermediate portions of the gradient lasted 12.1 years (SE = 4.3, n = 64).

What are the advantages and disadvantages of trading spatial extent for mechanistic understanding and study longevity? Short duration, community-level studies at many points along a gradient of urbanization can reveal patterns of community assembly quickly and relatively cheaply. However, such patterns will remain unexplained until longer-term, population-level studies are done. Therefore, a prudent study of birds in urbanizing environments would first conduct a short-term (perhaps even only a single year) correlational study designed to document the presence or absence of species along the full extent of the gradient. This study would simply be used to inform future long-term, population-level studies with hypotheses suggesting what points along the gradient affect birds differently and what species are most responsive to urbanization. Populations of responsive species could then be studied at points along the gradient where they should differ in demography and/or density. These latter studies would be designed to reveal mechanisms responsible for bird changes and amenable to policy adjustment (Marzluff et al. 2000).

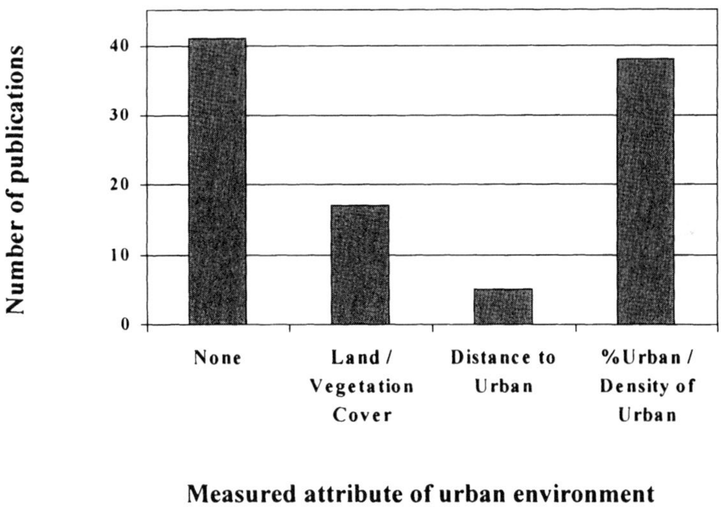

Figure 1.6. Measured attributes of the urban environments where birds studies (n = 101 articles) were conducted. Often no attributes of the environment are quantified (none).

4. STANDARDIZING TERMS

In our view, urbanization is the process of human settlement that gradually transforms uninhabited wildlands into lands including some degree of relatively permanent human presence. It is a continuous process that produces a range of settlement densities and patterns from widely spaced agricultural and recreational homesteads to the concrete and steel heart of a large metropolitan area. For a variety of social, economic, political, behavioral, and temporal reasons, much of this range in settlement patterns can often be found in proximity to large cities around the world (Alberti et al. 2001, Marzluff 2001). This range in settlement density and pattern is often referred to collectively as a gradient of urbanization (McDonnell et al. 1993) in recognition of its continuous, as opposed to discrete, nature. Certainly urbanization is continuous, even if complexly multinuclear with

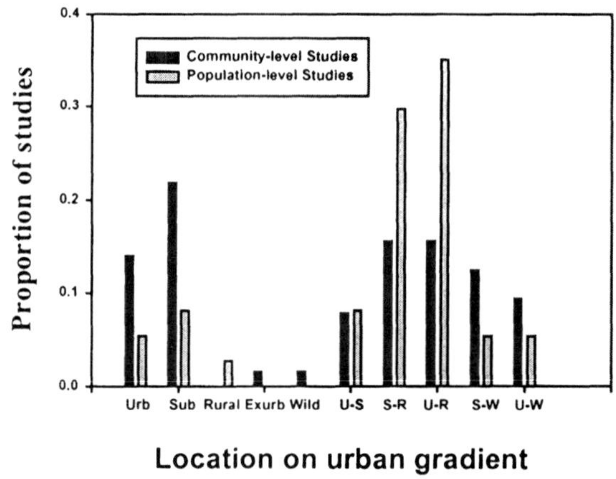

Figure 1.7. The proportion of studies reviewed (n = 101) that were conducted at single (Urb, Sub, Rural, Exurb, Wild) versus multiple (all other bars) points along the gradient of urbanization. Only the bar to the extreme right hand side of the graph (U-W) represents studies along the entire gradient of urban to wildland settings. Proportions are calculated separately for community- and population-level studies to emphasize tradeoffs in mechanistic understanding and spatial extent. Abbreviations are as follows (Urb=urban; Sub=suburban; Rural=rural; Exurb=exurban; Wild=wildland; U-S=sites in urban and suburban areas; S-R=sites in suburban and rural; U-R-sites in at least 2 locales spanning urban to rural areas; S-W=sites in at least 2 locales spanning suburban to wildland areas; U-W-sites in at least 2 locales spanning urban to wildland areas.

respect to distance from a city center (Alberti et al. 2001). However, researchers often characterize their study areas with discrete terms that describe their position along this continuum. To this end, a variety of terms have been used in the published urban ecology literature (e.g., *wildland, hinterland, greenbelt, rural, countryside, rurban, outskirts, exurban, extraurban, urban fringe, suburban, dwelling quarters, residential area, periurban, urban, city center*). The diverse terminology is especially confusing because terms often have different regional and cultural meanings. For example, to some "rural" means a relatively wild area far from humans (Yaukey 1996); to others this is the "extra-urban" (Plesník 1990), "rurban" (Hohtola 1978), "hinterland" (Fonaroff 1974), "outskirts" (Lo Valvo et al. 1985), or "greenbelt" (Erz 1966), because "rural" implies some degree of pastoralism (Nuorteva 1971).

To avoid confusion created by diverse, colloquial terminology we have defined a few, often studied points along the urban gradient (Table 1.1). Use of these terms is convenient for the author and reader, but even these terms are relatively broad and can still describe a range of human settlement patterns. Therefore, many of the studies included here report quantitative measures describing the urban characteristics most relevant to the particular study. Measures such as housing density, human density, and percentage of undeveloped land are useful. Additionally, because such measures are scale-dependent, the extent of the classified area must be defined. For most of these studies, we have included a large-scale aerial image of the study area and surrounding landscape.

We use five terms to describe the landscape of settlement (Table 1.1). In order of increasing settlement they are: wildland, rural or exurban, suburban, and urban. Building density and the proportion of built area increases monotonically in this terminology. However, resident human density is not necessarily greater in urban than suburban areas because urban areas often

Table 1.1. Standardized terms that describe major points along the gradient of urbanization. These terms are most meaningful at relatively large (>1km^2), landscape scales. Other attributes of each class and suggestions for subdividing classes are provided in the text. Residential human density refers to the density of humans occupying living quarters (dwellings) in the area.

Term	Percent built	Building density	Residential human density
Wildland	0-2	0	<1 / ha
Rural / exurban	5-20	<2.5 / ha	1-10 / ha
Suburban	30-50	2.5-10 / ha	>10 / ha
Urban	>50	> 10 / ha	>10 / ha

include large amounts of industrial or commercial development characterized by low resident human, but high building density. In general, ***wildlands*** are unsettled lands that may occasionally (especially at large scales) include dwellings. ***Rural and exurban lands*** are sparsely settled by individual homesteads, recreational developments, small towns, and villages. Unsettled land is much more abundant than settled land, but the actual pattern of settlement can vary widely (see below). Rural lands are distinguished from exurban lands by the matrix surrounding settlement. Settlements in Exurban areas are surrounded by a natural matrix; settlements in rural areas are surrounded by an agricultural matrix. ***Suburban lands*** are characterized by moderate- to high-density, single-family housing with lot sizes of 0.1 to 1.0 ha. Lawns and gardens are common. Basic services, light industry, and multi-family housing are interspersed with the typical single-family dwellings. Most buildings are single- or double-storied. ***Urban lands*** are areas where the majority of the land is covered by buildings. Building density is high and many buildings are for commerce, service, or industry. Single-family homes are rare and typically densely packed with little garden or lawn space. Multi-family housing and multi-storied buildings characterize urban areas.

The continuous nature of the urban gradient dictates that each of our discrete categories can be infinitely subdivided. Rural and exurban settlement could profitably be categorized based on a combination of landscape metrics such as contagion, interspersion, or patch size. As a starting point, Nilon et al. (1995) classified exurban lands as "dispersed" if lots were ≥ 1 ha or "clustered" if they were <1 ha. Suburban lands could be classified based on development age and retention of original vegetation. For example, Guthrie (1974) distinguished new, mature, and old suburbs based on their age (1, 25, and 75 years, respectively) and DeGraaf and Wentworth (1986) distinguished 15-year-old subdivisions in native forest from those initiated on denuded areas. Within suburban and urban landscapes, classifications might also include characterization of land use, such as single-family dwelling, multi-family dwelling, industrial, commercial, parkland, woodland, garden, etc. (Blair 2001, Mirabella et al. 1996, Lancaster and Rees 1979).

Descriptive classifications, such as the wildland to urban terminology described above, and quantitative measures of urban characteristics may differ in scale. The urban to wildland scale (Table 1.1), the complete urban gradient, is a landscape scale and is often broader than many urban ecological studies, which may be conducted in just a small portion of the overall gradient. Quantitative measures of urban characteristics within a study area describe a more local scale. Typically we consider areas of 1 to many km^2 when we characterize study areas into these categories. The finer subdivision various authors use to describe their actual point of data collection is usually at a local plot scale that is <1 km^2. At this scale land use can be consistent and categorized as suggested above. It would reduce confusion and perhaps more accurately describe areas of study if these differing scales were considered before characterization. Throughout this volume, we have consistently used the coarse, landscape-scale classification scheme (Table 1.1). We urge others to do likewise and to provide quantitative measures of the urban characteristics in their study area so that we might improve our knowledge of animal ecology in urban landscapes.

5. POLICY AND MANAGEMENT IMPLICATIONS

Science is but one of many sources of information that urban planners, managers, and policy makers use in their decision making process. The decisions they make will determine in large part how much and what type of biological diversity survives in the face of a growing and increasingly urban human population. Much of the current urban planning and policy-making process concerns the impacts of urbanization on human society, rather than on biodiversity. For biodiversity to become a major consideration in the planning and implementation process, the available science must be relevant, rigorous, compelling, and visible. Our review of the literature suggests that we can improve in each of these areas.

Urban ecological research is relevant if it explicitly compares biodiversity and ecological and biotic patterns that contribute to biodiversity at a variety of spatial and temporal scales, including the range of human settlement patterns. Population growth, urbanization, and suburbanization are inevitable for the foreseeable future. Therefore, our research should serve to direct inevitable development so that threats to biodiversity are minimized.

To effectively direct development from a biological conservation perspective, we need to understand: 1) where development will have the least effect on organisms, 2) how biodiversity responds to settlement density, and 3) if certain patterns of settlement minimize expected effects of density. Our review suggests that, for birds, we have addressed (2); avian diversity is routinely compared at various points along a gradient of urbanization. However, we have not explicitly considered (1). Are there certain areas within urbanizing landscapes where development would be especially damaging to birds? We could make educated guesses based on general relationships between habitat features and bird species diversity, but this needs more explicit consideration. We have rarely addressed (3). Nilon et al. (1995) compared the relative abundance of bird species to dispersed and clustered recreational development, but such study is sorely needed in rapidly urbanizing areas, especially at the urban/suburban and the suburban/exurban or suburban/rural interfaces. Developments with similar human densities, but different patterns of settlement may have different ecological impacts; thus policy might be crafted to favor urban development with ecologically benign settlement patterns. The relevance of urban avian ecology and of urban ecology in general to the urban planning process can be increased by identifying those processes or patterns that have important implications to both humans and to biodiversity. In addition, focusing research on those areas where relatively little is known, but predictably large impacts of future urbanization are likely to occur, will make data relevant to future planning decisions. Developing, tropical countries have the largest human growth rates, most of the Earth's biodiversity, and rapid expected rates of urbanization, yet studies of the impacts of urbanization in the tropics are exceedingly rare. Current investigations are just beginning address the complex effects of human settlement in the tropics (Brooks and Begazo 2001), but future research should address all three areas of investigation mentioned previously.

Only rigorous science is likely to produce well-informed, effective public policy. In addition to the surveys and correlational investigations that typify past studies of urban avian ecology, we need to incorporate the combination of observational and experimental studies common to more traditional ecological investigations (see Bolger 2001). These studies can identify mechanisms and causal relationships underlying patterns identified in the former studies (Marzluff and Sallabanks 1998, Marzluff et al. 2000, Bolger 2001). Past studies rarely connect population-level mechanisms (reproduction, survivorship, dispersal) to community-level patterns. Nor have most studies searched for ecological correlates to population patterns and experimental research will be needed to establish cause and effect. The lack of experimental studies in urbanizing areas is puzzling. Developers conduct large-scale, replicated land transformations on a regular basis in urbanizing areas. We could learn much about the causal link between urbanization and diversity by comparing population and community measures among, before, and after such disturbances. Increasing the rigor with which we quantify urban landscapes also will strengthen our science. Past studies often only subjectively categorize their study areas. Alberti et al. (2001) and Hostetler (2001) provide examples and advice for quantifying the urban environment.

A challenge facing scientists working on urban ecology issues is how to make their complex, and to some degree uncertain, science accessible to policy makers. Meeting with planners, managers, and policy makers is critical, as is joint development of research goals and publishing in non-traditional outlets (Kochert and Collopy 1998; Young and Varland 1998). Tailoring the format of results to a more diverse audience is also important. For example, maps or photographs that communicate the association of land cover with avian population viability would be especially compelling to planners, managers, and policy makers. Such summaries are not to be overly simplified or gross extrapolations. Rather, they should be distillations based on our rigorous studies that have been peer reviewed and published.

ACKNOWLEDGEMENTS

John Marzluff and Roarke Donnelly were supported by The National Science Foundation (DEB 9875041), The Washington Department of Fish and Wildlife, and The University of Washington, particularly its Rachael Wood's Graduate Program and the Tool's for Transformation Program. Reed Bowman was supported by Archbold Biological Station, the Florida Fish and Wildlife Conservation Commission, the United States Fish and Wildlife Service, and the National Science Foundation (IBN-0077469).

REFERENCES

Alberti, M., E. Botsford, and A. Cohen. 2001. Quantifying the urban gradient: linking urban planning and ecology, p. 89-115. *In* J. M. Marzluff, R. Bowman, and R. Donnelly [EDS.], Avian ecology and conservation in an urbanizing world. Kluwer Academic, Norwell, MA.

Batten, L. A. 1972. Breeding bird species diversity in relation to increasing urbanisation. Bird Study 19:157-166.

Bezzel, E. 1985. Birdlife in intensively used rural and urban environments. Ornis Fennica 62:90-95.

Blair, R. 2001. Creating a homogeneous avifauna, p. 461-488. *In* J. M. Marzluff, R. Bowman, and R. Donnelly [EDS.], Avian ecology and conservation in an urbanizing world. Kluwer Academic, Norwell, MA.

Bolger, D. 2001. Urban birds: population, community, and landscape approaches, p. 155-177. *In* J. M. Marzluff, R. Bowman, and R. Donnelly [EDS.], Avian ecology and conservation in an urbanizing world. Kluwer Academic, New York, NY.

Brooks, D., and A. Begazo. 2001. Mccaw abundance in relation to human population density, p. 429-439. *In* J. M. Marzluff, R. Bowman, and R. Donnelly [EDS.], Avian ecology and conservation in an urbanizing world. Kluwer Academic, Norwell, MA.

Czech, B., and P. R. Krausman. 1997. Distribution and causation of species endangerment in the United States. Science 277:1116-1117.

DeGraaf, R. M. and J. M. Wentworth 1986. Avian guild structure and habitat associations in suburban bird communities. Urban Ecol. 9:399-412.

Erz, W. 1966. Ecological principles in the urbanization of birds. Ostrich Sup. 6:357-363.

Fonaroff, L. S. 1974. Urbanization, birds, and ecological change in Northwestern Trinidad. Biological Conservation 64:258-262.

Green, R. J., C. P. Catterall, and D. N. Jones. 1989. Foraging and other behaviour of birds in subtropical and temperate suburban habitats. Emu 89:216-222.

Guthrie, D. A. 1974. Suburban bird populations in Southern California. American Midland Naturalist 92:461-466.

Hohtola, E. 1978. Diferential changes in bird community structure with urbanisation: a study in Central Finland. Ornis Scand. 9:94-100.

Horiuchi, S. 1992. Stagnation in the decline of the world population growth rate during the 1980s. Science 257:761-765.

Hostetler, M. 2001. The importance of multi-scale analyses in avian habitat selection studies in urban environments, p. 139-154. *In* J. M. Marzluff, R. Bowman, and R. Donnelly [EDS.], Avian ecology and conservation in an urbanizing world. Kluwer Academic. Norwell, MA.

Jones, D. N. 1981. Temporal changes in the suburban avifauna of an inland city. Aust. Wildl. Res. 8:109-119.

Kareiva, P., and M. Anderson. 1989. Spatial aspects of species interactions: the wedding of models and experiments, p.35-50. *In* A. Hastings [ED.], Community ecology. Springer-Verlag, New York, NY.

Kentish, B. J., P. Dann, and K. W. Lowe. 1995. Breeding biology of the common blackbird *Turdus merula* in Australia. Emu 95: 233-244.

Kochert, M. N., and M. W. Collopy. 1998. Relevance of research to resource managers and policy makers, p.423-430. *In* J. M. Marzluff and R. Sallabanks [EDS.], Avian conservation: Research and management. Island Press, Washington, DC.

Lancaster, R. K. and W. E. Rees 1979. Bird communities and the structure of urban habitats. Can. J. Zool. 57:2358-2368.

Lo Valvo, M., La Mantia, T., and B. Massa. 1985. Bird population of Palermo's urban and suburban areas. Boll. Zool. 52:347-354.

Luniak, M. and R. Mulsow. 1988. Ecological parameters in urbanization of the European blackbird, p.1787-1793. *In* L. Tomialojc and F. R. Gehlbach [EDS.], Avian population responses to man-made environments. Acta XIX Congressus Internationalis Ornithologica, Ottawa, Canada.

Marzluff, J. M. 2001. Worldwide increase in urbanization and its effects on birds, p. 19-47. *In* J. M. Marzluff, R. Bowman, and R. Donnelly [EDS.], Avian ecology and conservation in an urbanizing world. Kluwer Academic, Norwell, MA.

Marzluff, J. M. and K. Ewing. 2001. Restoration of fragmented landscapes for the conservation of birds: a general framework and specific recommendations for urbanizing landscapes. Restor. Ecol.

Marzluff, J. M., M. G. Raphael, and R. Sallabanks. 2000. Understanding the effects of forest management on avian species. Wildl. Soc. Bull. 28:1144-1155.

Marzluff, J. M., and R. Sallabanks. 1998. Past approaches and future directions for avian conservation biology, p.5-14. *In* J.M. Marzluff and R. Sallabanks [EDS.], Avian conservation: research and management. Island Press, Washington, DC.

McDonnell, M. J. and S. T. A. Pickett. [EDS.]. 1993. Humans as components of ecosystems. Springer-Verlag. New York, NY.

McDonnell, M. J., Pickett, S. T. A., and R. V. Pouyat. 1993. The application of the ecological gradient paradigm to the study of urban effects, p.175-189. *In* M. J. McDonnell and S. T. A. Pickett [EDS.], Humans as components of ecosystems. Springer-Verlag. New York, NY.

Mirabella, P., M. Fraissinet, and M. Malone. 1996. Breeding birds and territorial heterogeneity in Naples city (Italy). Acta Ornithologica 3:25-31.

Munyenyembe, F., J. Harris, and J. Hone. 1989. Determinants of bird populations in an urban area. Australian Journal of Ecology 14:549-557.

Nilon, C. H., C. L. Long, and W. C. Zipperer. 1995. Effects of wildland development on forest bird communities. Landscape Urban Plann. 32:81-92.

Nuorteva, P. 1971. The synanthropy of birds as an expression of the ecological cycle disorder caused by urbanization. Ann. Zool. Fennici 8:547-553.

Pitelka, F. A. 1942. High population of breeding birds within an artifical habitat. Condor 44:172-174.

Plesník, J. 1990. Long-term study of some urban and extra-urban populations of the Kestrel (*Falco tinnunculus* L.), p.453-458. *In* K. Stastiny and V. Bejcek [EDS.], Bird census and atlas studies. Proceedings XIth International Conference on bird census and atlas work, Prague.

Pulliam, H. R. 1988. Sources, sinks, and population regulation. The American Naturalist 132: 652-661.

Ruszczyk, A., J. J. S. Rodrigues, Roberts, T.M.T., Bendati, M.M.A., del Pino, R.S., Marques, J. C. V, and M. T. Q. Melo. 1987. Distribution patterns of eight bird species in the urbanization gradient of Porto Alegre, Brazil. Ciéncia e Cultura 39:14-19.

United Nations. 1996. World Urbanization Prospects: The 1996 Revision. United Nations, New York, NY.

Wiens, J. 1989. The ecology of bird communities. Cambridge University Press. Cambridge, UK.

Van Horne, B. 1983. Density as a misleading indicator of habitat quality. J. Wildl. Manage. 47:893-901.

Vitousek, P. M., H. A. Mooney, et al. 1997. Human domination of Earth's ecosystems. Science 277:494-499.

Walcott, C. F. 1974. Changes in bird life in Cambridge, Massachussetts from 1860 to 1964. Auk 91:151-160.

WRI (World Resources Institute). 1996. World resources 1996-97. Oxford University Press, Oxford, UK. 400p.

Yaukey, P. H. 1996. Patterns of avian population density, habitat use, and flocking behavior in urban and rural habitats during winter. Prof. Geogr. 48: 70-81.

Young, L. S., and D. E. Varland. Making research meaningful to the manager, p.415-422. *In* J. M. Marzluff and R. Sallabanks [EDS.], Avian conservation: Research and management. Island Press, Washington, DC.

Chapter 2

Worldwide urbanization and its effects on birds

John M. Marzluff
College of Forest Resources, University of Washington, Box 352100, Seattle, WA 98195 USA, e-mail: corvid@u.washington.edu

Key words: Conservation, density, demography, evenness, food supplementation, habitat fragmentation and loss, human population, land cover, land use, nest predation, richness, urbanization, urban sprawl

Abstract: Human populations are increasing and becoming predominantly urban. Resulting land cover changes reduce, perforate, isolate, and degrade bird habitat on local and global scales. I review: 1) urbanization of the Earth, and 2) published studies of bird responses to human settlement, and then: 3) suggest how and why birds respond to settlement. In a slight majority of studies, bird density increased, but richness and evenness decreased in response to urbanization. The most consistent effects of increasing settlement were increases in non-native species of birds, increases in birds able to nest on buildings (esp. swifts and swallows), increases in nest predation, and decreases in interior- and ground-nesting species. Effects of urbanization on hawks, owls, and cavity nesters were less consistent, in part being dependent on the surrounding habitat. The factors favoring species in urbanizing areas appear simpler than those reducing species. Increased availability of food was primary among factors benefiting species; predator reduction, reduced human persecution, and habitat enhancement were less important. Decreased habitat availability, reduced patch size, increased edge, increased non-native vegetation, decreased vegetative complexity, and increased nest predation were commonly associated with bird declines in response to human settlement. Urban planners and policy makers can profoundly affect how and where cities grow. Avian ecologists can help inform these important decisions by: 1) quantifying how the pattern of settlement affects birds and 2) understanding how bird populations and resulting communities change along entire gradients of urbanization.

1. INTRODUCTION

Humans are a rapidly increasing, social species whose industrial abilities allow us to colonize virtual wilderness, settle small towns, and build sprawling metropolises. We live virtually everywhere on Earth and wherever we settle we significantly transform the natural habitats (Berry 1990, Meyer and Turner 1992, Houghton 1994, Marzluff and Hamel 2001). Near the end of last century, human dwellings occupied 1-6% of the Earth's surface; human agriculture covered another 12% (Meyer and Turner 1992). Virtually all lands have been impacted by human settlement or agriculture or have been used to provide the natural resources or recreational opportunities needed to sustain the burgeoning human population. Models suggest that over the last three centuries forests have declined by 19%, grasslands have declined 8%, and cropland has increased over 400% (Matthews 1983, Richards 1990, Meyer and Turner 1992, Marzluff and Hamel 2001). Our domination of Earth is evidenced by our use of 40% of all terrestrial net primary productivity (Vitousek et al. 1986) and radiation from our lights that is visible from space at night (Fig. 2.1; Elvidge et al. 1997).

Human populations are becoming increasingly urban. In 1700, only 14 cities (all in Eurasia) with populations of more than 200,000 people existed (Berry 1990). By 1900, 42 cites on four continents (Eurasia, North America, South America and Africa) had such populations, and by 2000, 171 cities on five continents (those above, plus Australia) had populations greater than 200,000 (Brinkhoff 2001). In 1900, only 10% of humans lived in cities; by 2000 nearly 50% did so and nearly 70% are expected to do so by 2050 (United Nations 1996, Brown et al. 1998). Thus by 2050 nearly as many humans are expected to live in cities (6.5 billion) as occupy all of Earth today (Brown et al. 1998).

In response to this population increase, cities often grow to cover increasingly large areas. In the United States, the increase in the areal extent of urban areas is increasing faster than urban population growth. For example, from 1950 - 1990 Chicago's population grew by 38%, but it spread over 124% more land (O'Meara 1999). Likewise, Cleveland's population increased by 21% while it spread over 112% more land (O'Meara 1999). Even when populations shrink, a city's area can expand. For example, from 1970 - 1990 Detroit's population declined 7%, but its urbanized area increased 28% (U. S. Census Bureau 2001). This disproportionately large increase in the spatial extent of urban and suburban settlement is termed "urban sprawl." In the United States from 1982 - 1992, urban sprawl resulted in the loss of >2.1 million ha of forest, 1.5 million ha of cropland, 0.94 million ha of pasture, and 0.77 million ha of rangeland (WRI 1996).

How does an increasingly urban human population affect biodiversity? Here I begin to address this question by reviewing and synthesizing the literature relating birds to urbanization. I use *urbanization* to mean the process by which human settlement increases in: 1) population density and 2) intensity of land use in an area. In this way I address the responses of birds to urbanization across a gradient of human settlement from low density exurban and rural settlement to what we more typically think of as *urban* (suburban and urban settlement; see Marzluff et al. 2001 for definitions).

First, I present a model of how urbanization occurs and alters habitats important to birds. Second, I summarize recent studies of birds conducted at various intensities of human settlement and describe avian community responses. Third, I suggest important ways we can increase our understanding of bird responses to urbanization. Last, I discuss important findings that managers, planners, and policy makers can use to more effectively conserve native birds in rapidly urbanizing areas.

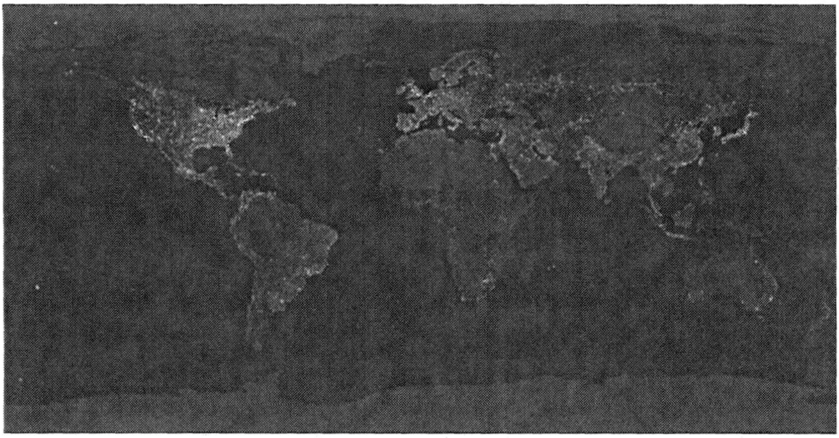

Figure 2.1. Radiation from human settlement as seen from space. The image is a composite of images taken from October 1994 to March 1995. Image by Craig Mayhew and Robert Simmon, NASA GSFC, based on data from the Defense Meteorological Satellite Program. Courtesy of Christopher Elvidge, NOAA National Geophysical Data Center.

2. THE HISTORICAL PATTERN OF URBANIZATION AND ITS EFFECTS ON BIRD HABITAT

A complex suite of social, political, and economic drivers determine when, where, how, and at what rate human settlement proceeds. Despite this complexity and the resulting variety of settlement patterns, colonization and settlement of large undeveloped tracts proceeds in a fairly regular way (Fig. 2.2; Berry 1990).

Initially, as frontiers are opened, small settlements are built. These are usually along coasts or large rivers to meet transportation needs. Resources are initially provided from previously settled areas until settlements can develop their own resource extraction and agricultural capabilities. Settlement spreads as roads are built into the resource producing area.

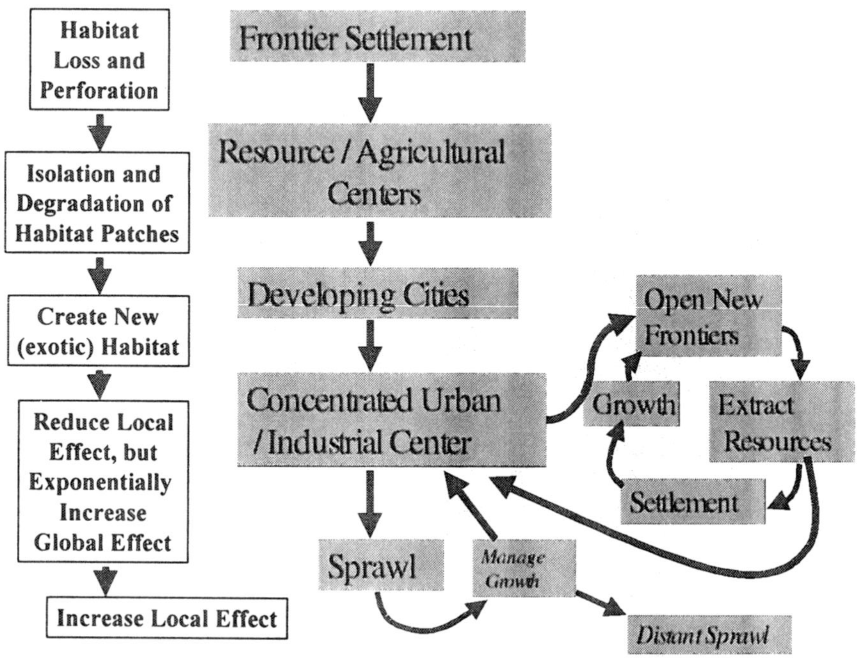

Figure 2.2. Model of the progression of human settlement (grey boxes) and resulting effects on bird habitat (white boxes). Habitat is lost and perforated, then isolated as frontier settlements grow into small cities. As cities grow and eventually sprawl, exotic habitats are created and remaining native habitat is further isolated and degraded at local and global scales.

During these initial phases of settlement, natural habitat (especially coastal and riparian habitat) is lost, perforated, and then isolated and degraded to varying degrees. The response of the biota likely depends on the natural dispersion of habitats; species adapted to natural-fragmented systems or systems that depend on periodic disturbance are likely to be better adapted to these changes than species adapted to large-scale, stable, contiguous systems (Marzluff and Ewing in press).

Growing needs for resources and their extraction, processing, and shipping often fuel substantial growth and eventually small settlements develop into small cities that often occupy strategic locations for transportation or defense. Increasing interactions among cities increase the likelihood that non-native plants, animals, and diseases spread. Developing cities themselves also provide new habitats for synanthropic animals. Paving, construction of impervious surfaces, and redirection of water flow associated with city development reduces habitat suitability for many species and disrupts nutrient cycling possibly undermining ecological functionality.

Established cities may grow or fade through time depending primarily on the market value of the services they provide. Often, they develop a centralized business or industrialized area surrounded by residential areas (Erz 1966). The resulting concentration of extreme changes in land cover may limit the spatial extent of habitat changes, but resources needed by growing human populations often fuel more distant development and resulting habitat modification. Thus, new frontiers are often opened, existing settlements augmented, or distant natural resources tapped in response to the growing needs of large urbanizing areas. As a result of this enlarging *ecological footprint* (Wackernagel and Rees 1996), cities have increasingly far-reaching, global effects on natural habitat.

Where surrounding land is available for development, growing cities often sprawl to fill it. In the United States, sprawl occurs where large residential suburbs occur around more concentrated city centers, industrial sites, and multifamily dwellings. Metropolitan Atlanta, Georgia, for example, grew to include an area over 110 miles long from north to south in 1998 (Bullard 2000). Sprawling cities often become polycentric (Alberti et al. 2001) as new businesses capitalize on economic advantages of siting and operating in the less expensive lands surrounding large cities. Sprawl accentuates the local effects of urbanization on land cover because expanses of native vegetation are converted to complex mosaics of exotic ground covers, impervious surface, and islands of native habitat. Pets, humans, and expanding non-native synanthropic species further degrade the quality of remaining native habitat in and around sprawling cities. The result is an increasingly large actual *city footprint* on natural habitat. This local effect likely is synergistic with a more global *ecological footprint* that reflects total resources used to sustain the lifestyle of city inhabitants. The result is that cities now degrade large amounts of habitat used by native birds in their immediate vicinity and globally.

Local and regional policies may be enacted to curtail or define the pattern of sprawl (Buzbee 2000). Some states have adopted coordinated growth management strategies (e.g., Florida, Washington, Oregon). This may limit the immediate extent of land conversion, but may also favor growth in nearby areas outside growth management jurisdiction (H. Blanco, U. Washington, Pers. Comm.). For example, Vancouver, Washington has grown in response to growth management in neighboring Portland, Oregon. Such policies are important because the pattern and intensity of development likely determines the magnitude of the local effect of sprawl on biodiversity. Sprawl that proceeds by perforating native vegetation with well-spaced housing and retains an intact matrix of native vegetation reserves and corridors may have less impact on biodiversity than sprawl that moves like a wave leveling native vegetation and replacing it with expanses of non-natives and anthropogenic structures. However, this remains largely untested.

Perhaps the most recent chapter in how we settle land is being written in response to sprawl in America. The relative ease of traveling, ability to telecommute, increased personal wealth, and the desire to avoid dense settlements is now fueling a "second conquest" (Gersh 1996) of many rural and wildland areas in the United States. Most notable is growth in resort areas, sprawling "bedroom communities" of urban areas, and subdivisions of formerly extensive ranches (10-50 ha "ranchettes"). Growth has been most rampant (20-65% from 1990-1995; US Census Bureau) in counties with significant wildlands (Rudzitis and Johansen 1989, Gersh 1996). Changes in these areas are dramatic and foreshadow changes to be expected throughout much of the world where development near wildlands is coveted for its amenity value and possible for an increasing number of people because of telecommuting. As an example, consider the East River Valley of Colorado. Theobald et al. (1996) calculated that 20% of this 35,000 ha peninsula of private land surrounded by public wild land is now divided into ranchettes 18 ha or smaller in size. During the last 30 years, 100 km of road have been added to the area primarily to serve the 171 new ranchettes and 1,371 new homes in high-density subdivisions. Thus, the current process of urbanization suggests that birds will be affected in two distinct settings: (1) large urban areas and their associated suburbs; and (2) exurban areas and their interface with rural areas and wildlands.

3. DOCUMENTED EFFECTS OF URBANIZATION ON BIRDS

Urbanization has complex direct and indirect effects on native flora and fauna. With respect to birds, Marzluff (1997) suggested that settlement can change: 1) ecosystem processes, 2) habitat, 3) food, 4) predators and competitors, and 5) disease. These effects lead to significant changes in the population biology of birds in urban areas with resulting effects on the structure and composition of bird communities. Species able to exploit urban environments may develop dense and stable populations because ameliorated climate, abundant food and water, reduced predators, and increased nest sites allow for lengthened breeding seasons, increased survival, and increased productivity (Gehlbach 1994, 1996). Many native species may decline with urbanization because of the scarcity of natural habitat, increases in predators, parasites, or competitors, or intolerance of human activity. As a result, urban bird communities are often described as supporting larger (greater biomass) and sometimes richer (more species), but always less evenly distributed avian communities dominated by a few, very abundant and often non-native, species (e.g., Emlen 1974, Beissinger and Osborne 1982, Mills et al. 1989).

I tested the generality of the above descriptions of bird populations and communities in urbanizing areas by searching the modern literature (1900-2000) to document observed effects of settlement on birds. I queried the electronic databases *Biosis, Wildlife Worldwide, and Current Contents* using *urban, birds, human effects on birds, fragmentation, settlement, suburbs, and development* as key words. I also manually reviewed the Literature Cited sections of each article for additional published works on the effects of urbanization on birds. I found a total of 101 articles that reported new empirical research on birds and urbanization (Appendix 1). The basic design of these studies was presented in Marzluff et al. (2001).

3.1 Effects on Community Composition

Three measures of community complexity are typically reported: the density of birds, the number of species (richness), and the equitability of individuals per species (evenness). Density and richness were reported in a large number of studies (n = 57 and 51, respectively), but evenness was rarely reported (n = 9 studies). Many studies (n = 29) reported community measures for only a single point along a gradient of urbanization (with references to changes through time at one place, n = 8; or with comparisons of a given degree of settlement to representative wildlands, n = 21). However, 29 studies of density, 16 studies of richness, and 3 studies of evenness compared measures from at least two points along the gradient, and often from three to four points (Fig. 2.3; these may be from several physical places, but they clearly were from two or more degrees of settlement; e.g., urban versus rural).

Avian community metrics responded inconsistently to urbanization. The majority of studies found density of birds increased, while richness and evenness decreased in response to urbanization (Fig. 2.3). However, these majorities were not overwhelming. Of 57 studies investigating density, 36 (63%) found density to increase with increasing settlement. Of the 51 studies investigating richness, 31 (61%) found richness to decrease with increasing settlement. While only 9 studies reported evenness, 5 (56%) reported it to decrease with increasing settlement. Increasing density and decreasing diversity were found in studies around the world and in a variety of environments (upland and riparian forests, coastal estuaries, shrublands, Sonoran desert, grasslands, and prairie woodlands). However, studies reporting decreasing density with increasing urbanization were primarily conducted in upland forests (n = 13) and only occasionally in riparian forests (n = 2) or coastal areas (n = 1). Study locations reporting increasing richness with urbanization were only slightly more diverse (n = 6 upland forests, 1 shrubland, 1 riparian desert area, and 1 grassland).

The range of settlement investigated does not appear to confound these results (Fig. 2.3). Studies that found no effect were usually only conducted at one point along the settlement gradient, but those finding positive, negative, and intermediate effects were conducted at sites varying widely in settlement. Likewise, duration of study does not appear to drive a study's finding. Studies reporting increases in community metrics with settlement ranged from 1 - 150 years in duration (density: 1 - 150 yr; richness: 1 - 140 yr, evenness: n = 1 study of 2 year duration). Those reporting decreases in community metrics with settlement ranged from 1 - 75 years in duration (density: 1 - 32 yr, richness: 2 - 75 yr, and evenness: 1 - 2 yr).

Studies reporting maximum density or richness at intermediate degrees of urbanization were rare (Fig. 2.3). Where a full range of settlement was investigated, density peaked at suburban sites (intermediate urbanization) in Australian shrub (Sewell and Catterall 1998), California forests (Blair 1996), and Massachusetts forests (DeGraaf and Wentworth 1986). Only one study reported highest diversity in suburban sites (Sewell and Catterall 1998).

3.2 Responses of Specific Groups of Birds

The most consistent effects of increasing settlement were increases in non-native species of birds, increases in birds that use buildings as nest sites (e.g., swallows and swifts), increases in nest predators and nest parasites

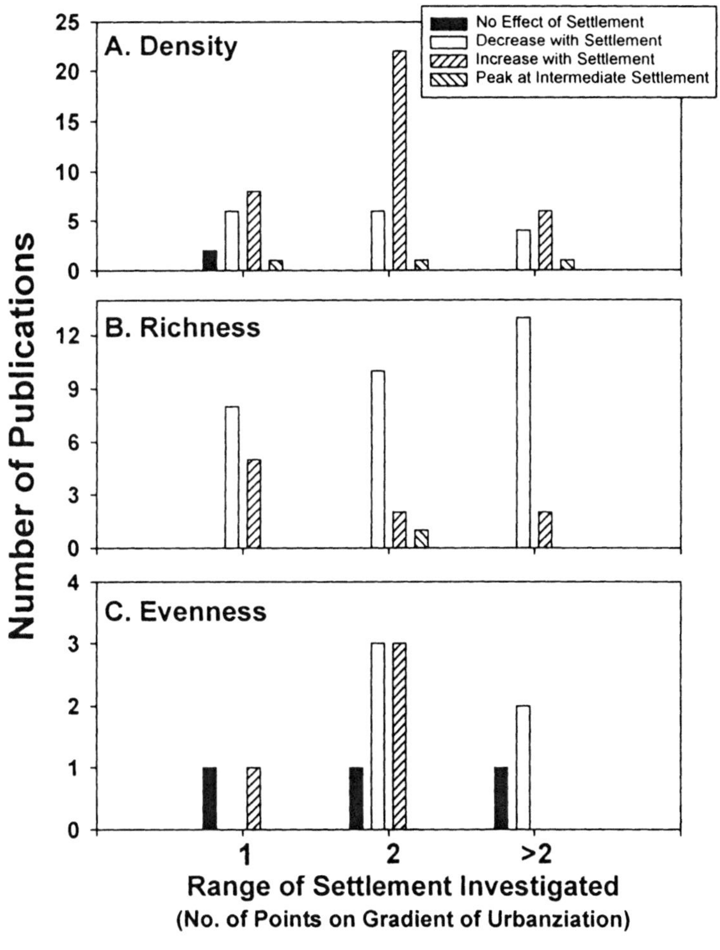

Figure 2.3. Documented effects of human settlement on measures of bird community density (A), richness (B), and evenness (C). Publications that documented no effect, positive effects, negative effects, and intermediate effects are plotted separated by the range of settlement intensity they investigated (a single point on the gradient of urbanization to greater than 2 points on the gradient). Note differences in scale of y-axis for each measure.

(Brown-headed Cowbirds, *Molothrus ater*), and decreases in interior- and ground-nesting species (Fig. 2.4). These results are relatively consistent around the world, in a variety of habitats, and appear independent of the method and duration of study. House Sparrows (*Passer domesticus*) and Rock Doves (*Columba livia*) are non-native in most cities, yet are consistently the most abundant species in urban settings worldwide. They represent an extreme example of homogenization, or convergence, of urban avifaunas discussed by Blair (2001).

A variety of reasons were postulated to explain these changes. Swallows and swifts that usually nest on cliffs, or in caves and hollowed trees appear to easily use eves, roofs, chimneys and other vertical substrates that are abundant in urban settings. Occasionally, increased competition with non-native species for these nest sites or reductions in insects because of pesticide use may reduce the suitability of urban areas for swallows and swifts (Walcott 1974). In contrast, loss of interior forest conditions in response to most urbanization is detrimental to species requiring interior

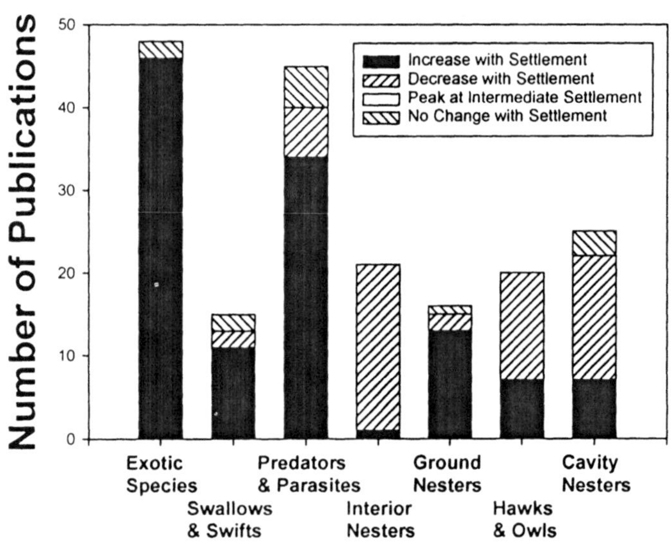

Figure 2.4. Documented effects of human settlement on specific groups of birds. Bar height represents the total number of publications found on each group of birds. Shading represents the number of publications on each group that reported species to increase, decrease, peak at intermediate levels, or not change with human settlement.

forest for nesting and likely contributes to the loss of ground nesting species. Reduction in the structural complexity of ground cover and shrub layers in urban settings may reduce the suitability of urban areas for some ground nesters. However, some ground nesters that can find suitable nest sites may also benefit from supplemental food available in urban settings (Horn 1985). Both studies that reported increases in ground nesters in response to urbanization were conducted in grassland settings where urbanization might actually result in an increase in ground cover diversity and structural complexity (Vale and Vale 1976, Bothelho and Arrowood 1996). Supplemental food (n = 9 studies) and habitat fragmentation (n = 3 studies) were identified as important to increasing the suitability of urban environments for corvid nest predators and cowbirds, but reduced persecution and possibly reduced predation may also be important.

Effects of urbanization on hawks, owls, and cavity nesters were less consistent (Fig. 2.4). Most studies of raptors (n = 13; 65%) and cavity nesters (n = 15; 65%) reported reductions in these species with increasing settlement. Variation in raptor responses to urbanization is linked to habitat. Raptors decreased most commonly when forests were urbanized (71% of 17 studies) and least when grasslands were urbanized (33% of 3 studies). Reductions occurred in forested settings primarily because nesting habitat and prey abundance were reduced (Tilghman 1987, Gehlbach 1996, Natuhara and Imai 1996). Raptors were also disturbed by humans in forests and prey was reduced (Bosakowski et al. 1993, Konstantinov 1996, Tella et al. 1996). True grassland raptors that nest on the ground and avoid woody areas (e.g., Ferruginous Hawk, *Buteo regalis*, and Rough-legged Hawk, *Buteo lagopus*) also decline with urbanization (Berry et al. 1998). In contrast, raptors typically found in forests or savannahs increased when grasslands were settled because trees used for nesting increased, insects and small mammals associated with agriculture increased, intraspecific disturbance decreased, and human persecution decreased (Parker 1996, Botelho and Arrowood 1996, Berry et al. 1998).

Cavity nesters responded to changes in nest sites and food. In forested settings, cavity nesters usually declined with urbanization because native vegetation (presumably snags) was reduced (DeGraaf and Wentworth 1986, Tilghman 1987, Natuhara and Imai 1996, Rottenborn 1999). These reductions occurred despite increases in food (Cramp and Tomlins 1966, Beissinger and Osborne 1982, Horn 1985). However, in five studies, urban forest cavity nesters increased in response to food supplementation. These may be short-term responses to feeders (breeding was not measured) or population responses by small species not limited by nest sites (e.g., Black-capped Chickadees, *Parus atricapillus*, Brittingham and Temple 1992).

Cavity nesters increased in grassland and desert settings (n = 2) in part because urbanization increased nest sites there (Vale and Vale 1976).

3.3 Reported Effects on Demography

Patterns of survival and reproduction of birds in response to urbanization were not consistent (Fig. 2.5); but few species showed no response. In two studies, one of Black-billed Magpies (*Pica pica*; Eden 1985) and one on Common Blackbirds (*Turdus merula*; Kentish et al. 1995), reproduction did not differ along an urban gradient. Only four species showed reduced reproduction with increasing urbanization (Herring Gull [*Larus argentatus*], Lesser Kestrel [*Falco naumani*], Great Tit [*Parus major*], and Cooper's Hawk [*Accipiter cooperii*]; Table 2.1) and only one showed reduced survival (one study of Common Blackbird; Batten 1973, Table 2.1). Reproduction and survival of most species increased with increasing settlement. Species

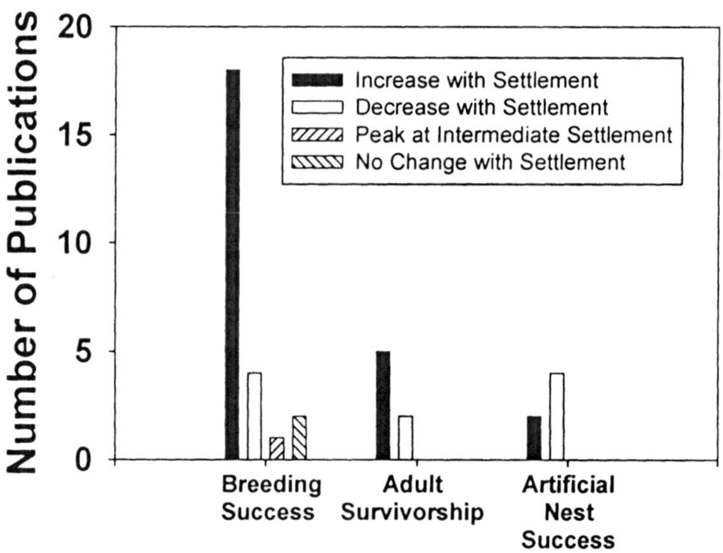

Figure 2.5. Documented effects of human settlement on three measures of avian demography (A: actual breeding success, B: adult survival, and C: breeding success indicated by artificial nest experiments).

that responded positively to urbanization included Black-billed magpies, Common Blackbirds, Herring Gulls, Great Tits, Cooper's Hawks, Screech Owls (*Otus asio*), Woodpigeon (*Columba palumbus*), Mississippi Kite (*Ictinia mississippiensis*), Burrowing Owl (*Speotyto cunicularia*), Red-shouldered Hawk (*Buteo lineatus*), House Sparrow, Rock Dove, European Kestrel (*Falco tinnunculus*), Wren (*Troglodytes troglodytes*), Carrion Crow (*Corvus corone*), and American Goldfinch (*Carduelis tristis*) (Table

Table 2.1. Documented increases (+) and decreases (-) in demographic parameters and density of bird species in response to increasing urbanization. Multiple signs per species indicate results of multiple published papers. Studies that showed no change in demography with respect to urbanization are given a "0". Those showing greatest reproduction at intermediate levels of urbanization are given a +/-.

Species	Reproduction	Survival	Density	References
Black-billed Magpie	0, +, +		+, +	Eden 1985, Plesnik 1990, Tatner 1982
Common Blackbird	0, +, +, +, +	-, +	+, +, +, +	Snow 1969, Landman 1991, Batten 1973, Konstantinov 1982, Luniak and Mulsow 1988
Herring Gull	-, +	+	-	Hunt 1972, Milone and Grotta 1983
Lesser Kestrel	-			Tella et al. 1996
Great Tit	-	+	+	Schmidt 1988, Horak and Lebreton 1998
Cooper's Hawk	-, +		+	Bosakowski et al. 1993, Rosenfield et al. 1996
Screech Owl	+	+	+	Gehlbach 1996
Woodpigeon	+	+	+	Tomialojc 1979
Mississippi Kite	+		+	Parker 1996
Burrowing Owl	+		+	Botelho and Arrowood 1996
Red-shouldered Hawk	+			Bloom and McCrary 1996
Cape Sparrow	+/-			Earle 1988
House Sparrow	+			Earle 1988
Rock Dove	+			Dabert 1987
European Kestrel	+			Plesnik 1990
Wren	+		+	Wesolowski 1983
Carrion Crow	+		+	Konstantinov 1996
American Goldfinch	+			Middleton 1988

2.1). Only the Cape Sparrow (*Passer melanurus*) showed an intermediate response to urbanization, having its highest rate of reproduction at intermediate levels of settlement (Earle 1988). The observation that most demographic studies of birds in urban settings show positive responses to settlement is not general. It simply reflects the fact that most demographic studies in urban settings have been done on species that are able to benefit from human activity, reduced persecution, and changes in land cover. In fact, studies of simulated songbird nests (n = 6) suggest that reproduction of many species is likely to be reduced in urban settings (n = 4, 67% of studies showed increased predation in urban settings).

Demographic studies in urban settings generally support the contention that urban bird community assemblages reflect the demographic performance of their constituent populations (Gehlbach 1988, Marzluff and Ewing in press). Magpies, Common Blackbirds, Screech Owls, Woodpigeons, Wrens, Cooper's Hawks, and Carrion Crows all showed consistent increases in either reproduction or adult survivorship that were paralleled by increases in density (Table 2.1). Density responses of Herring Gulls and Great Tits were less clear. In one study, Herring Gull density declined with urbanization apparently in response to reduced reproduction (Milone and Grotta 1983), despite potential increased survivorship (Hunt 1972). Conversely, Great Tit density increased apparently in response to increased adult survivorship (Horak and Lebreton 1998) despite potential reduced fecundity (Schmidt 1988).

3.4 Mechanisms Identified as Responsible for Changes in Bird Populations and Communities

Bird communities change in response to urbanization because some species are favored and others are reduced by settlement. The factors favoring species in urbanizing areas appear simpler than those reducing species (Fig. 2.6). Primary among factors identified to favor species is increasing availability of food (Table 2.2). Urban tits, pigeons and doves, sparrows, thrushes, gulls, waterfowl, and corvids have benefited from increased food (Hunt 1972, Tomialojc 1979, Tatner 1982, Earle 1988, Dabert 1987, Brittingham and Temple 1992, Jozkowica and Gorska-Klek 1996). Supplemental food is most often associated with increased fecundity, but occasionally with increased overwinter survival (Schmidt 1988, Brittingham and Temple 1992). Predator reduction, and to a lesser extent, reduced human persecution also have benefited species in urban areas (Table 2.2), most notably small raptors (kites, buteos, owls; Betelho and Arrowood 1996, Bloom and McCrary 1996, Parker 1996, Tella et al. 1996). Urban enhancements to the natural habitat also benefit some species in certain conditions (increasing woody vegetation in deserts and grasslands, increasing water in arid environments, and providing specialized ledge, chimney, and box nests; Table 2.2).

A complex web of factors, both direct and indirect, have been identified as reducing the suitability of urban areas to birds (Fig. 2.6). Important direct effects on fecundity are decreased habitat availability, reduced patch size, increased edge, increased non-native vegetation, and decreased vegetative complexity (Table 2.2). These factors work simultaneously to reduce the amount and quality of nesting sites to many native birds in urbanizing areas. Indirectly, nest predation increases in urban areas because many of the species favored by increased food and reduced predation (see above) are effective nest predators (and because of the reduction in the quality of nests sites). Competitors, brood parasites, and direct human disturbance also may reduce fecundity of native birds, but these factors are less often reported in the literature (Table 2.2). Survivorship of birds in urban areas also may be reduced because of increases in small raptors (see above) and non-native

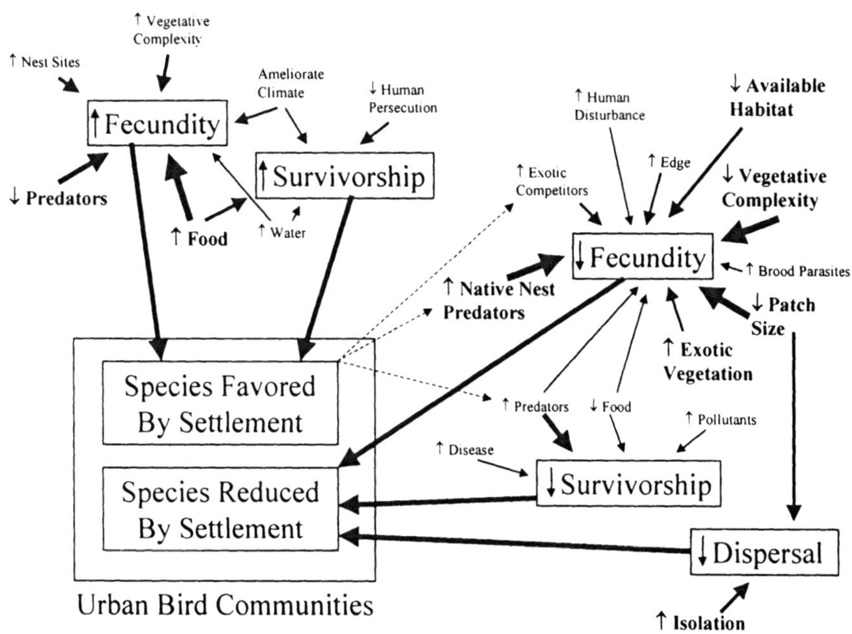

Figure 2.6. Direct (solid arrows) and indirect (dashed arrows) effects of human settlement on bird population parameters (isolated boxes for fecundity, survival, and dispersal) and eventual bird community assembly (joined boxes). The most frequently cited effects (Table 2.2) are in bold text. Relative width of arrows corresponds to the cited frequency of effects (Table 2.2).

predators (e.g. domestic cats; Batten 1973). Urbanization also may reduce arthropod populations that are important avian food resources (Ruszczyk et al. 1987) and increase exposure to diseases (Erskine 1992) and toxins (Walcott 1974, Schmidt 1988), all of which may reduce survival and/or fecundity (Table 2.2). Disruption of dispersal patterns through habitat fragmentation and changes in the intervening matrix types may influence colonization and extinction rates and thereby influence large-scale population regulation (Fig. 2.6).

Changes in the mechanisms and processes detailed in Fig. 2.6 are responsible for reported changes in bird communities (1) along gradients of urbanization, (2) through time as areas urbanize, and (3) through time as developments mature. As one samples different communities from wildland to urban settings the relationships in Fig. 2.6 strengthen so that many native species decline, a few capitalize on urban safety and food, and many non-natives colonize. This also unfolds through time at a single location as

Table 2.2. Mechanisms identified in the literature as important to increasing the density and decreasing the richness and evenness of bird communities in urbanizing areas.

Mechanism	No. times referenced
Food Supplementation	29
Fragmentation	
Decreased Patch Size	17
Increased Isolation	6
Increased Edge	4
Decreased Vegetative Complexity	16
Increased Nest Predation	14
Increased Exotic Vegetation	8
Decreased Predation	8
Decreased Habitat Availability	8
Increased Vegetative Complexity	5
Increased Exotic Competitors	5
Increased Exotic / Domestic Predators	5
Increased Human Disturbance	5
Amelioration of Climate (Heat Island)	4
Increased Nest Sites	4
Decreased Human Persecution	3
Increased Pollution / Toxins	2
Increased Nest Parasitism	2
Decreased Food	2
Increased Disease	1
Increased Water	1

settlement increases because the original native avifauna changes composition and is augmented by non-natives. Many neighborhoods also undergo a similar succession because initially extreme reduction in vegetative complexity and provision of novel nest sites favors few birds, but with increasing ornamental shrub and tree growth plus supplemental foods and regrowth of some native vegetation, bird density and richness increase (Vale and Vale 1976, DeGraaf and Wentworth 1986, Jones 1981, Murgenyembe et al. 1989).

3.5 Generalities

Humans have settled most of the lowland, especially the coastal, areas of the world that has a temperate climate (Fig. 2.1). In doing so, we transform land cover by perforating mostly contiguous forest and grassland with increasing frequency until we surround remnant forest and grass patches with sprawling cities and suburbs. Avian ecologists have documented the response of birds to this transformation by monitoring birds at a single place through time or by comparing birds along gradients of settlement from wildland to urban (Marzluff et al. 2001). In general, research characterizes bird communities as becoming increasingly dense, but impoverished as settlement increases. Most studies indicate that even where richness of urban bird communities is relatively high, a few species are disproportionately abundant.

Bird communities change in response to settlement because populations of some species thrive while others dwindle in response to changes in habitat, food, predators, climate, and disturbance (Fig. 2.6). Typically, as settlement increases, bird communities become dominated by synanthropic species (Johnston 2001). Most commonly these include small raptors, corvids, mynas, pigeons and doves, weavers, thrushes, blackbirds, swallows, swifts, and starlings. Despite the ability of a few especially successful species to dominate urban avifaunas (Rock Dove, European Starling, House Sparrow), the majority of species in urban areas are synanthropic native species.

The typical decline in richness and evenness associated with settlement applies best to forested settings. There, habitat is simplified, degraded, and isolated as settlement fragments and converts native forests to savanna-like settings. However, in grassland and desert areas, settlement often increases the structural complexity of mature vegetation by promoting diverse trees, shrubs, and ground covers. Important limiting factors, notably water and nest sites, are often provided by settlement. As a result, where some native habitat is maintained, it is often enhanced by settlement and many raptors, cavity nesters, ground nesters, and shrub nesters increase in diversity. This does not mean than urbanization benefits all grassland or desert species. In fact, many true prairie species are extremely susceptible to grassland loss and fragmentation (C. Bock, University of Colorado, Pers. Comm.). Similarly, many desert species require native vegetation (Mills et al. 1989). The overall impacts of urbanization in these two-dimensional settings may simply seem less because many species that select woody or riparian habitat find these conditions most abundant in urbanizing areas.

4. FUTURE DIRECTIONS

My review and synthesis of the literature, suggests several specific research needs that would help us better understand how birds respond to urbanization. We need more studies that compare bird population processes and resulting community composition along the full range of settlement intensity (wildland to urban). These studies could test the hypotheses that: 1) community diversity rarely peaks at intermediate levels of urbanization, 2) grassland and desert bird communities are less responsive to urbanization than forested communities, and 3) a complex suite of direct and indirect effects determine population- and community-level responses to urbanization. One component of such studies that has received little investigation is how the type of interface between anthropogenic and natural land cover affects bird populations. Resultant "edge effects" are typically strongest in settled areas (Marzluff and Restani 1999), but do such effects vary with the type of interface? Do edges affect more than nest predation and nest parasitism? Urban interfaces in grasslands, for example, might affect bird movements because birds that have evolved in two-dimensional habitats may be especially shy of three-dimensional, urban habitats (Carl Bock, University of Colorado, Pers. Comm.).

Demographic studies of bird populations are relatively common and long-term in urbanizing areas (Marzluff et al. 2001). However, they are usually done in isolation from community-level studies. Research that documents changes in avian communities in response to urbanization and explains these changes with studies of constituent population demographics are lacking. Such studies would increase our understanding of how urban bird communities are assembled and maintained. The mechanistic understanding they reveal would also increase our insights into how to manage for avian diversity in settled areas (Marzluff et al. 2000). Demographic studies of sensitive species are especially needed in urbanizing areas. Most existing studies document increased survivorship or reproduction by syanthropic species (Fig. 2.5). Few show how species fail to adapt to changes caused by settlement.

Few studies explicitly compare the far-reaching effects of settlement on birds. How far into the rural, exurban, and wildland landscapes do the effects of more dense settlement extend? Insights into these sorts of effects could be gained by using landscape ecological methods to relate the amount, type, and pattern of settlement in relatively large areas to the functioning of bird populations and structure of bird communities. Such studies would allow us to better anticipate settlement's effects and more effectively arrange conservation lands in a large planning landscape.

5. IMPLICATIONS FOR POLICY MAKERS, MANAGERS, AND PLANNERS

As the human population increases and becomes more urbanized, the world at night will shine even brighter. This much seems certain. The pattern of settlement, however, is far from certain. Urban planners and policy makers can profoundly affect how and where cities grow. They can direct the pattern of settlement within their jurisdictional boundaries thereby affecting the interspersion of vegetated and built land covers. Does this interspersion of "green space" with settlement matter to biodiversity in general and birds in specific? Unfortunately we have little information on which to base an answer. We know that bird communities change most with the intensity of settlement, but I found only three studies (DeGraaf and Wentworth 1986, Friesen et al. 1995, Nilon et al. 1995) that compared different patterns of settlement. The most explicit comparisons were in relatively unsettled areas (Friesen et al. 1995, Nilon et al. 1995). Clearly, we need to know how different patterns of urban, and especially suburban settlement affects bird populations and communities if we are to provide relevant science to managers, planners, and policy makers.

Despite a basic lack of information on how settlement pattern affects birds (and other components of biodiversity), planners and managers can find abundant advice on how to arrange settled and green space (e.g., Davis and Glick 1978, Knight 1990, Soulé 1991, Arendt 1996). These suggestions may be a good short-term solution to an information need, but as they are based on general application of island biogeographic theory they need to be viewed as hypotheses rather than as facts. The functioning of reserves and corridors needs more testing in urbanizing settings to be effectively applied. Planning based on mechanistic understanding of how bird populations respond to settlement pattern will more likely have its desired outcome.

Beyond general landscape-level advice for planners, there is some practical information available on how to manage existing urbanizing lands for birds (Marzluff and Ewing 2001). Key elements are: 1) maintaining native vegetation, 2) minimizing loss of structural diversity, and 3) actively reducing the impacts of non-native species (esp. cats and rodents) on bird productivity.

Managers, planners, and policy makers may find birds to be important signals of urban ecosystem health and general biodiversity. Using birds for this purpose is natural because birds are familiar, enjoyable, and important to millions of people. However, birds can easily deceive us into thinking that all is well because they have strong powers of dispersal. They are able to colonize, or at least inhabit, a wide variety of landscapes simply because they can fly over unsuitable habitat and search widely for suitable habitat. Careful study of the functioning of bird populations will be necessary before we conclude that the presence of birds is indicative of ecosystem health. Planners, managers, and policy makers should demand a demonstration of numerical stability and balanced survivorship and reproduction of bird populations before accepting the occurrence of birds as a pronouncement of overall health.

ACKNOWLEDGMENTS

The Urban Ecology group at the University of Washington continually stimulates my thoughts on the issues discussed in this paper. I thank Marina Alberti, Carl Bock, and Reed Bowman for constructive comments on an earlier draft of this paper. Roarke Donnelly commented on the paper and typeset it. My research was supported by The National Science Foundation (DEB-9875041), The University of Washington, and The Washington Department of Fish and Wildlife.

REFERENCES

Alberti, M., E. Botsford, and A. Cohen. 2001. Quantifying the urban gradient: linking urban planning and ecology, p. 89-115. *In* J. M. Marzluff, R. Bowman, and R. Donnelly [EDS.], Avian ecology and conservation in an urbanizing world. Kluwer Academic, Norwell, MA.

Aldrich, J. W., and R. W. Cofin. 1980. Breeding bird populations from forest to suburbia after thirty-seven years. American Birds 34:3-7.

Arendt, R. G. 1996. Conservation design for subdivisions. Island Press. Washington, D.C.

Askins, R. A., and M. J. Philbrick. 1987. Effect of changes in regional forest abundance on the decline and recovery of a forest bird community. Wilson Bull. 99:7-21.

Batten, L. A. 1972. Breeding bird species diversity in relation to increasing urbanisation. Bird Study 19:157-166.

Batten, L. A. 1973. Population dynamics of suburban blackbirds. Bird Study 20:251-258.

Beissinger, S. R., and D. R. Osborne. 1982. Effects of urbanization on avian community organization. Condor 84:75-83.
Berry, B. J. L. 1990. Urbanization, p. 103-119. *In* B. L. Turner II, W. C Clark, R. W. Kates, J. F. Richards, J. T. Mathews, and W. B. Meyers [EDS.], The earth as transformed by human action. Cambridge University Press, Cambridge, UK.
Berry, M. E., C. E. Bock, and S. G. Haire. 1998. Abundance of diurnal raptors on open space grasslands in an urbanized landscape. Condor 100:601-608.
Bezzel, E. 1985. Birdlife in intensively used rural and urban environments. Ornis Fenn. 62:90-95.
Bird, D. M., D. E. Varland, and J. J. Negro [EDS.]. 1996. Raptors in human landscapes. Academic Press, London, UK.
Blair, R. B. 1996. Land use and avian species diversity along an urban gradient. Ecol. Appl. 6:506-519.
Blair, R. B. 2001. Creating a homogeneous avifauna, p. 461-488. *In* J. M. Marzluff, R. Bowman, and R. Donnelly [EDS], Avian ecology and conservation in and Urbanizing World. Kluwer Academic Press, Norwell, MA.
Blanco, G., and T. Velasco. 1996. Bird-habitat relationships in an urban park during winter. Folia Zool. 451:35-42.
Bloom, P. H., and M. D. McCrary. 1996. The urban buteo: red-shouldered hawks in southern California. p. 31-39. *In* Bird, D. M., D. E. Varland, and J. J. Negro [EDS.]. Raptors in human landscapes. Academic Press, London, UK.
Bolger, D. T., T. A. Scott, and J. T. Rotenberry 1997. Breeding bird abundance in an urbanizing landscape in coastal southern California. Conserv. Biol. 11:406-421.
Bosakowski, T., R. Speiser, D. G. Smith, and L. J. Niles. 1993. Loss of Cooper's Hawk nesting habitat to suburban development - Inadequate protection for a state-endangered species. J. Raptor Res. 27:26-30.
Botelho, E. S., and P. C. Arrowood. 1996. Nesting success of western burrowing owls in natural and human-altered environments. p. 61-68. *In* Bird, D. M., D. E. Varland, and J. J. Negro [EDS.]. Raptors in human landscapes. Academic Press, London, UK.
Brinkhoff, T. 2001. The principle agglomerations of the world. http://www. Citypopulation.de/cities.html.
Brittingham, M. C., and S. A. Temple. 1992. Use of winter bird feeders by Black-capped Chickadees. J. Wildlife Manage. 56:103-110.
Brousseau, P., J. Lefebvre, and J. F. Giroux. 1996. Diet of Ring-billed Gull chicks in urban and non-urban colonies in Quebec. Colon. Waterbirds 19:22-30.
Brown, L. R., Gardner, G., and B. Halweil. 1998. Beyond Malthus: sixteen dimensions of the population problem. Worldwatch Paper 143. Worldwatch Institute, Washington, DC.
Bullard, R. D. 2000. Anatomy of sprawl. p. 1-19 *In* R. D. Bullard, G. S. Johnson, and A. O Torres. [EDS.]. Sprawl City. Island Press. Washington, D.C..
Buzbee, W. W. 2000. Urban sprawl and legal reform. p. 161-186. *In* R. D. Bullard, G. S. Johnson, and A. O Torres. [EDS.]. Sprawl City. Island Press. Washington, D.C..
Butcher, G. S., W. A. Niering, W. J. Berry, and R. H. Goodwin. 1981. Equilibrium biogeography and the size of nature preserves: an avian case study. Oecologia 49:29-37.
Caccamise, D. F., L. M. Reed, L. S. DeLay, K. A. Bennett, and J. J. Dosch. 1996. The avian communities of a suburban grassland reugium: population studies at an airport in Northeastern United States. Acta Ornith. 31:3-13.
Cicero, C. 1989. Avian community structure in a large urban park: controls of local richness and diversity. Land. Urban Plann 17:221-240.
Connell, J. H. Diversity in tropical rain forests and coral reefs. Science 199:1302-1010.

Cramp, S. 1980. Changes in the breeding birds of Inner London since 1900. Symposium on Urbanization.

Cramp, S., and A. D. Tomlins. 1966. The birds of Inner London 1951-65. British Birds 59:209-233.

Crooks, K. R. and M. E. Soulé. 1999. Mesopredator release and avifaunal extinctions in a fragmented system. Nature 400:563-566.

Dabert, J. 1987. Breeding ecology of the feral pigeon *Columbia livia f. domestica* in Poznan, Poland. Acta Ornith. 23:177-195.

Danielson, W. R., R. M. DeGraaf, and T. K. Fuller. 1997. Rural and suburban forest edges: effects on egg predators and nest predation rates. Landscape Urban Plann. 38:25-36.

Davis, A. M. and T. F. Glick. 1978. Urban ecosystems and island biogeography. Environ. Conserv 3:299-304.

DeGraaf, R. M. 1991. Winter foraging guild structure and habitat associations in suburban bird communities. Land. Urban Plann. 21:173-180.

DeGraaf, R. M., and J. M. Wentworth. 1986. Avian guild structure and habitat associations in suburban bird communities. Urban Ecology 9:399-412.

Dowd, C. 1992. Effect of development on bird species composition of two urban forested wetlands in Staten Island, New York. J. Field Ornith. 63:455-461.

Dulisz, B., and J. J. Nowakowski. 1996. The species diversity of the avifauna in built-up areas in the city of Olsztyn (NE Plonad). Acta Ornith. 31:33-38.

Earlé, R. A. 1988. Reproductive isolation between urban and rural populations of cape sparrows and house sparrows. Acta Congressus Internationalis 19:1778-1786.

Eden, S. F. 1985. The comparative breeding biology of magpies *Pica pica* in an urban and a rural habitat (Aves:Corvidae). J. Zool. 205:325-334.

Edgar, D. R., and G. P. Kershaw. 1994. The density and diversity of the bird populations in three residential communities in Edmonton, Alberta. Can. Field Nat. 108:156-161.

Elvidge, C.D., K. E. Baugh, E. A. Kihn, H. W. Kroehl, and E. R. Davis. 1997. Mapping city lights wit nighttime data from the DMSP Operational Linescan System. Photogramm. Eng. Rem. S. 63:727-734.

Emlen, J. 1974. An urban bird community in Tucson, Arizona: Derivation, structure, regulation. Condor 76:184-197.

Emlen, J. T. 1974. An urban bird community in Tucson, Arizona: derivation, structure, regulation. Condor 76:184-197.

Engels, T. M., and C. W. Sexton. 1994. Negative correlation of Blue Jays and Golden-cheeked Warblers near an urbanizing area. Conserv. Biol. 8:286-290.

England, A. S., J. A. Estep, and W. R. Holt. 1995. Nest-site selection and reproductive performance of urban-nesting Swainson's Hawks in the central valley of California. J. Raptor Res. 29:179-186.

Erskine, A. J. 1992. A ten-year urban winter bird count in Sackville, New Brunswick. The Canadian Field-Naturalist 106:499-506.

Erz, W. 1966. Ecological principles in the urbanization of birds. Ostrich Suppl. 6:357-363.

Fonaroff, L. S. 1974. Urbanization, birds, and ecological change in Northwestern Trinidad. Biol. Conserv. 6:258-262.

Fraissinet, M. 1989. Espansione della taccola, *Corvus monedula*, nei capoluoghi Italiani. Rivista Italiana Ornithologia 59:33-42.

Friesen, L. E., P. F. Eagles, and R. J. MacKay. 1995. Effects of residental development on forest-dwelling neotropical migrant songbirds. Conserv. Biol 9:1408-1414.

Gavareski, C. A. 1976. Relation of park size and vegetation to urban bird populations in Seattle, Washington. Condor 78:375-382.

Gehlbach, F. R. 1988. Population and environmental features that promote adaptation to urban ecosystems: the case of Eastern Screech-owls (*Otus asio*) in Texas. Proceedings Congressus Internationalis Ornithologica 19:1809-1813.

Gehlbach, F. R. 1994. The eastern screech owl: life history, ecology, and behavior in the suburbs and countryside. Texas A & M University Press, College Station, TX.

Gehlbach, F. R. 1996. Eastern screech owls in suburbia: a model of raptor urbanization. p. 69-74. *In* Bird, D. M., D. E. Varland, and J. J. Negro [EDS.]. Raptors in human landscapes. Academic Press, London, UK.

Germaine, S. S., S. S. Rosenstock, R. E. Schweinsburg, and W. S. Richardson. 1998. Relationships among breeding birds, habitat, and residential development in greater Tucson, Arizona. Ecol. Appl. 8:680-691.

Gersh, J. 1996. Subdivide and conquer: concrete, condos, and the second conquest of the American West. Amicus Journal 18:14-20.

Goszczynski, J., P. Jablonski, G. Lesinski, and J. Romanowski. 1993. Variation in diet of Tawny Owl *Strix aluco* L. along an urbanization gradient. Acta Ornith. 27:113-123.

Gotfryd, A., and R. I. C. Hansell. Prediction of bird-community metrics in urban woodlots. p. 321-326 *In* J. Verner, M. L. Morrison, and C. J. Ralph [EDS.] Wildlife 2000: Modeling Habitat Relationships of Terrestrial Vertebrates. University of Wisconsin Press, Madison, WI.

Green, R. J., C. P. Catterall, and D. N. Jones. 1989. Foraging and other behaviour of birds in subtropical and temperate suburban habitats. Emu 89:216-222.

Guthrie, D. A. 1974. Suburban bird populations in Southern California. Am. Midl. Nat. 92: 461-466.

Hogrefe, T. C., R. H. Yahner, and N. H. Piergallini. 1998. Depredation of artificial ground nests in a suburban versus a rural landscape. J. Penn. Acad. Science 72:3-6.

Hohtola, E. 1978. Diferential changes in bird community structure with urbanisation: A study in Central Finland. Ornis Scand. 9:94-100.

Hooper, R. G., E. F. Smith, H. S. Crawford, B. S. McGinnes, and V. J. Walker. 1975. Nesting bird populations in a new town. Wildlife Soc. Bull. 3:111-118.

Hõrak, P. and J.-D. Lebreton 1998. Survival of adult Great Tits *Parus major* in relation to sex and habitat: a comparison of urban and rural populations. Ibis 140:205-209.

Horn, D. J. 1985. Breeding birds of a central Ohio woodlot in response to succession and urbanization. Ohio J. Science 85:34-40.

Houghton, R. A. 1994. The worldwide extent of land-use change. BioScience 44:305-313.

Huhtalo, H., and O. Järvinen. 1977. Quantitative composition of the urban bird community in Tornio, Northern Finland. Bird Study 24:179-185.

Hunt, G. L., Jr. 1972. Influence of food distribution and human disturbance on the reproductive success of Herring Gulls. Ecology 53:1051-1061.

Johnston, R. F. 2001. Synanthropic birds of North America, p. 49-67. *In* J. M. Marzluff, R. Bowman, and R. Donnelly [EDS], Avian ecology and conservation in and Urbanizing World. Kluwer Academic Press, Norwell, MA.

Jokimäki, J., and J. Suhonen. 1993. Effects of urbanization on the breeding bird species richness in Finland - A biogeographical comparison. Ornis Fenn. 70:71-77.

Jokimäki, J., J. Suhonen, K. Inki, and S. Jokinen. 1996. Biogeographical comparison of winter bird assemblages in urban environments in Finland. J. Biogeogr. 23:379-386.

Jones, D. N. 1981. Temporal changes in the suburban avifauna of an inland city. Aust. Wildl. Res. 8:109-119.

Józkowicz, A. and Górska-Klek. 1996. Activity patterns of the mute swans *Cygnus olor* wintering in rural and urban areas: a comparison. Acta Ornithologica 31: 45-51.

Kentish, B. J., P. Dann, K. W. Lowe. 1995. Breeding biology of the Common Blackbird *Turdus merula* in Australia. Emu 95:233-244.

Kerpez, T. A., and N. S. Smith. 1990. Competition between European Starlings and native woodpeckers for nest cavities in saguaros. Auk 107:367-375.

Knight, R. L. 1990. Ecological principles applicable to the management of urban ecosystems, p. 24-34. *In* E. A. Webb, and S. Q. Foster [EDS], Perspectives in urban ecology. Denver Museum of Natural History and Thorne Ecological Institute, Denver, CO.

Konstantinov, V. M. 1996. Antropogenic transformations of bird communities in the forest zone of the Russian Plain. Acta Ornith. 31:53-58.

Konstantinov, V. M., V. G. Babenko,and I. K. Barysheva. 1982. Numbers and some ecological features of synanthropic populations of the corvidae under the conditions of intensive urbanization. Zoologichesky Zhurnal 61:1837-1845.

Konstantinov, V. M., W. Nowicki, and A. G. Pichurin. 1996. Birds in urban and suburban areas. Acta Ornithol. 31:59-66.

Lancaster, R. K., and W. E. Rees. 1979. Bird communities and the structure of urban habitats. Can. J. Zool. 57:2358-2368.

Landmann, A. 1991. Habitatpräferenzen, dynamik der raumnutzung und bestandsstruktur bei dorfamseln. J. Orn. 132:303-318.

Langen, T. A., D. T. Bolger, and T. J. Case. 1991. Predation on artificial bird nests in chaparral fragments. Oecologia 86:395-401.

Linehan, J. T. 1967. Breeding-bird populations in Delaware's urban wodlots. Audubon Field Notes 21:641-647.

Lo Valvo, M., T. La Mantia, and B. Massa. 1985. Bird population of Palermo's urban and suburban areas. Boll. Zool. 52:347-354.

Luniak, M. and R. Mulsow. 1988. Ecological parameters in urbanization of the European Blackbird. Proceedings Congressus Internationalis Ornithologica 19:1787-1793.

Major, R. E., G. Gowing, and C. E. Kendal. 1996. Nest predation in Australian urban environments and the role of the Pied Currawong, *Strepera graculina*. Aust. J. Ecol 21:399-409.

Marzluff, J. M. 1997. Effects of urbanization and recreation on songbirds, p. 89-102. *In* W. M. Block and D. M. Finch [EDS.], Songbird ecology in southwestern ponderosa pine forests: a literature review. USDA Forest Service General Technical Report RM-GTR-292. USDA Forest Service, Fort Collins, CO.

Marzluff, J. M., and K. Ewing. 2001. Restoration of fragmented landscapes for the conservation of birds: a general framework and specific recommendations for urbanizing landscapes. Restoration Ecology.

Marzluff, J. M., and N. Hamel. 2001. Land use issues, p. 659-673. *In* S. A. Levin [ED.], Encyclopedia of biodiversity. Academic Press, San Diego, CA.

Marzluff, J. M., K. J. McGowan, R. Donnelly, and R. L. Knight. 2001. Causes and consequences of expanding American Crow populations, p. 333-365. *In* J. M. Marzluff, R. Bowman, and R. Donnelly [EDS], Avian ecology and conservation in and Urbanizing World. Kluwer Academic Press, Norwell, MA.

Marzluff, J. M., and M. Restani. 1999. The effects of forest fragmentation on avian nest predation, p. 155-169. *In* J. A. Rochelle, L. A. Lehmann, and J. Wisniewski, [EDS.], Forest wildlife and fragmentation: management and implications. Brill, Leiden, Netherlands.

Marzluff, J. M., F. R. Gehlbach, and D. A. Manuwal. 1998. Urban environments: Influences on avifauna and challenges for the avian conservationist, p. 283-299. *In* J. M. Marzluff and R. Sallabanks [EDS.], Avian conservation: Research and management. Island Press, Washington, DC.

Marzluff, J. M., M. G. Raphael, and R. Sallabanks. 2000. Understanding the effects of forest management on avian species. Wild. Soc. Bull. 28:1132-1143.

Matthews, E. 1983. Global vegetation and land use: New high resolution data bases for climate studies. J. Climate Applications and Meterology 22:474-487.

Meyer, W. B., and B. L. Turner, II. 1992. Human population growth and global landuse/cover change. Annu. Rev. Ecol. Syst. 23:39-61.

Middleton, A. L. A. 1988. The urban environment and its potential as a refuge from brood parasitism by the brown-headed cowbird. Proceedings Congressus Internationalis Ornithologica 19:1802-1808.

Mills, G. S., J. B. J. Dunning, and J. M. Bates. 1989. Effects of urbanization on breeding bird community structure in southwestern desert habitats. Condor 91:416-428.

Milone, M., and M. Grotta. 1983. Year-to-year variation in numbers of Herring Gulls nesting in Campania. Boll. Zool. 50:25-27.

Mirabella, P., M. Fraissinet, and M. Milone. 1996. Breeding birds and territorial heterogenity in Naples city (Italy). Acta Ornith. 31:25-31.

Munyenyembe, F., J. Harris, and J. Hone. 1989. Determinants of bird populations in an urban area. Aust. J. Ecol. 14:549-557.

Natuhara, Y. and C. Imai. 1996. Spatial structure of avifauna along urban - rural gradients. Ecological Research 11:1-9.

Nilon, C. H., C. N. Long, and W. C. Zipperer. 1995. Effects of wildland development on forest bird communities. Landscape Urban Plann. 32:81-92.

Nowakowski, J. J. 1996. Changes in the breeding avifauna of Olsztyn (NE Poland) in the years 1968-1993. Acta Ornith. 31:39-44.

Nuorteva, P. 1971. The synanthropy of birds as an expression of the ecological cycle disorder caused by urbanization. Ann. Zool. Fennici 8:547-553.

O'Meara, M. 1999. Reinventing cities for people and the planet. Worldwatch Paper 147. Worldwatch Institute, Washington, D.C.

Parker, J. W. 1996. Urban ecology of the Mississippi Kite. p. 45-52. *In* Bird, D. M., D. E. Varland, and J. J. Negro [EDS.]. Raptors in human landscapes. Academic Press, London, UK.

Pitelka, F. A. 1942. High population of breeding birds within an artifical habitat. Condor 44:172-174.

Plesník, J. 1990. Long-term study of some urban and extra-urban populations of the Kestrel (*Falco tinnunculus* L.). Bird Census and Atlas Studies. Proceedings XIth International Conference on bird census and atlas work. K. Stastiny and V. Bejcek. Prague:453-458.

Richards, J. F. 1990. Land transformation, p. 163-178. *In* B. L. Turner II, W. C. Clark, W. R. Kates, J. F. Richards, J. T. Matthews, and W. B. Meyer [EDS.], The earth as transformed by human action. Cambridge University Press, Cambridge, UK.

Rolando, A., G. Maffei, C. Pulcher, and A. Giuso. 1997. Avian community structure along an urbanization gradient. Ital. J. Zool. 64:341-349.

Rosenberg, K. V., S. B. Terrill, and G. H. Rosenberg. 1987. Value of suburban habitats to desert riparian birds. Wilson Bull. 99:642-654.

Rosenfield, R. N., J. Bielefeldt, J. L. Affeldt, and D. J. Beckman. 1996. Urban nesting biology of Cooper's hawks in Wisconsin. p. 41-44. *In* Bird, D. M., D. E. Varland, and J. J. Negro [EDS.]. Raptors in human landscapes. Academic Press, London, UK.

Rottenborn, S. C. 1999. Predicting the impacts of urbanization on riparian bird communities. Biol. Conserv. 88:289-299.

Rudzitis, G. and H. Johansen. 1989. Migration into western wilderness counties: causes and consequences. Western Wildlands 3:19-23

Russo, C., and T. P. Young. 1997. Egg and seed removal at urban and suburban forest edges. Urban Ecosystems 1:171-178.

Ruszczyk, A., J. J. S. Rodrigues, T. M. T. Roberts, M. M. A. Bendati, R. S. del Pino, J. C. V Marques, and M. T. Q. Melo. 1987. Distribution patterns of eight bird species in the urbanization gradient of Porto Alegre, Brazil. Ciéncia e Cultura 39:14-19.

Schmidt, K. H. 1988. Site fidelity and isolation of great tits (*Parus major*) in urban habitats. Proceedings International Ornithological Congress 19:1794-1801.

Senra, A. and E. E. Alés. 1992. The decline of the white stork *Ciconia ciconia* population of western Andalusia between 1976 and 1988: causes and proposals for conservation. Biological conservation 61: 51-57.

Sewell, S. R., and C. P. Catterall. 1998. Bushland modification and styles of urban development: Their effects on birds in southeast. Wild. Res. 25:41-63.

Snow, D. W. 1969. An analysis of breeding success in the blackbird, *Turdus merula*. Ardea 57:163-171.

Sodhi, N. S. 1992. Comparison between urban and rural bird communities in prairie Saskatchewan - Urbanization and short-term population trends. Can. Field Nat. 106:210-215.

Soulé, M. E. 1991. Land use planning and wildlife maintenance. J. Amer. Plann. Assoc. 57:313-323.

Soulé, M. E., D. T. Bolger, A. C. Alberts, J. Wright, M. Sorice, and S. Hill. 1988. Reconstructed dynamics of rapid extinctions of chaparral-requiring birds in urban habitat islands. Conserv. Biol. 2:75-92.

Tatner, P. 1982. The breeding biology of magpies Pica pica in an urban environment. J. Zool. 197:559-581.

Tella, J. L., F. Hiraldo, J. A. Donázar-Sancho, and J. J. Negro. 1996. p. 53-60. *In* Bird, D. M., D. E. Varland, and J. J. Negro [EDS.]. Raptors in human landscapes. Academic Press, London, UK.

Theobald, D. M, H. Gosnell, and W. E. Riebsame. 1996. Land use and landscape change in the Colorado Mountains II: A case study of the East River Valley. Mountain Research and Development 7:101-107.

Tilghman, N. G. 1987a. Characteristics of urban woodlands affecting breeding bird diversity and abundance. Landscape Urban Plann. 14:481-495.

Tilghman, N. G. 1987b. Characteristics of urban woodlands affecting winter bird diversity and abundance. Forest Ecol. Manag. 21:163-175.

Tomialojc, L. 1979. The impact of predation on urban and rural Woodpigeon (*Columba palumbus* L.) populations. Polish Ecological Studies 5:141-220.

U. S. Census Bureau. 2001. Urbanized areas in the United States. http://www/sprawlcity.org/studyUSA/USAexecutivesummary.html

United Nations. 1996. World urbanization prospects: the 1996 revision. United Nations, New York, NY.

Vale, T. R. and G. R. Vale. 1976. Suburban bird populations in west-central California. J. Biogeogr. 3:157-165.

Vitousek, P. M., H. A. Mooney, J. Lubchenco, and J. M. Melillo. 1997. Human domination of the earth's ecosystems. Science 277:494-499.

Vitousek, P. M., P .R. Ehrlich, A. H. Ehrlich, and P. A. Matson. 1986. Human appropriation of the products of photosynthesis. Bioscience 36:368-373.

Wackernagel, M. and W. E. Rees. 1996. Our ecological footprint: reducing human impact on the earth. New Society Publishers, Philadelphia, PA.

Walcott, C. F. 1974. Changes in bird life in Cambridge, Massachussetts from 1860 to 1964. Auk 91:151-160.

Wesolowski, T. 1983. The breeding ecology and behaviour of Wrens *Troglodytes troglodytes* under primeval and secondary conditions. Ibis 125:499-515.

Whitcomb, R. F., J. F. Lynch, M. K. Klimkiewicz, C. S. Robbins, B. L. Whitcomb, and D. Bystak. 1981. Effects of forest fragmentation on avifauna of eastern deciduous forest, p. 125-205. *In* R. L. Burgeos and D. M. Sharpe [EDS.], Forest island dynamics in man-dominated landscapes. Springer-Verlag, New York, NY.

Wilcove, D. S. 1985. Nest predation in forest tracts and the decline of migratory songbirds. Ecology 66:1211-1214.

Woolfenden, G. E. and S. A. Rohwer 1969. Breeding birds in a Florida suburb. Bull. Florida State Museum 13:1-83.

WRI (World Resources Institute). 1996. World resources: a guide to the global environment: the urban environment. Oxford University Press. Oxford, UK.

Yaukey, P. H. 1996. Patterns of avian population density, habitat use, and flocking behavior in urban and rural habitats during winter. Prof. Geogr. 48:70-81.

Zalewski, A. 1994. Diet of urban and suburban tawny owls (*Strix aluco*) in the breeding season. J. Raptor Research 28:246-252.

Table 1. append

Author	Date	Location	Habitat	Duration (yrs)
Aldrich and Coffin	1980	Virginia	forest	37
Askins and Philbrick	1987	Connecticut	forest	32
Batten	1972	United Kingdom	forest	140
Batten	1973	United Kingdom	forest	9
Beissinger and Osborne	1982	Ohio	forest	2
Berry et al.	1998	Colorado	grassland	3
Bezzel	1985	Central Europe	forest	130
Blair	1996	California	forest	2
Blanco and Velasco	1996	Spain	forest	1
Bloom and McCrary	1996	California	forest	25
Bolger et al.	1997	California	shrubland	1
Bosakowski et al.	1993	New Jersey	forest	1
Botelho and Arrowood	1996	New Mexico	grassland	1
Brittingham and Temple	1992	Wisconsin	forest	3
Brousseau et al.	1996	Quebec	islands	2
Butcher et al.	1981	Connecticut	forest	23
Caccamise et al.	1996	New Jersey	grassland	4
Cicero	1989	California	aquatic	1.5
Cramp	1980	United Kingdom	forest	75
Cramp and Tomlins	1966	United Kingdom	forest	15
Dabert	1987	Poland	urban	3
Danielson et al.	1997	Massachuchetts	forest	2
DeGraaf	1991	Massachuchetts	forest	5
DeGraaf and Wentworth	1986	Massachuchetts	forest	5
Dowd	1992	New York	riparian forest	2
Dulisz and Nowakowski	1996	Poland	forest	3
Earle	1988	South Africa	shrubland	3
Eden	1985	UK	forest	1

Author	Date	Location	Habitat	Duration (yrs)
Edgar and Kershaw	1994	Canada	forest	1
Emlen	1974	Arizona	sonoran desert	1
Engels and Sexton	1994	Texas	forest	2
England et al.	1995	California	forest	5
Erskine	1992	Canada	forest	10
Fonaroff	1974	Trinidad	forest	8
Fraissinet	1989	Italy	forest	30
Friesen et al.	1995	Canada	forest	2
Gavareski	1976	Washington	forest	1
Gehlbach	1996	Texas	riparian forest	16
Germaine et al.	1998	Arizona	desert	1
Goszczynski et al.	1993	Poland	forest	9
Gotfryd and Hansell	1986	Canada	forest	4
Green et al.	1989	Australia	forest	3
Guthrie	1974	California	shrubland	1
Hogrefe et al.	1998	Pennsylvania	forest	1
Hooper et al.	1975	Virginia	forest	1
Horak and Lebreton	1998	Estonia	forest	7
Horn	1985	Ohio	forest	11
Hotola	1978	Finland	forest	3
Huhtalo and Jarvinen	1977	Finland	forest	1
Hunt	1972	Maine	coast/island	3
Jokimaki and Suhonen	1993	Finland	forest	20
Jokimaki et al.	1996	Finland	forest	1
Jones	1981	Australia	forest	1
Jozkowica and Gorska-Klek	1996	Poland	aquatic	3
Kentish et al.	1995	Australia	forest	24
Konstantinov et al.	1982	Russia	forest	18
Konstantinov	1996	Russia	forest	30
Konstantinov et al.	1996	Russia and Poland	forest	10
Lancaster and Rees	1979	Canada	forest	1
Landman	1991	Switzerland	forest	7
Langen et al.	1991	California	shrubland	1
Linehan et al.	1967	Delaware	forest	2
Luniak and Mulsow	1988	Europe	forest	150
Major et al.	1996	Australia	shrubland	1
Middleton	1988	Canada	forest	5
Mills et al.	1989	Arizona	sonoran desert	1
Milone and Grotta	1983	Italy	beach/coast	18
Mirabella et al.	1996	Italy	forest	4
Munyenyembe et al.	1989	Australia	forest	2
Natuhara and Imai	1996	Japan	forest	1
Nilon et al.	1995	Missouri	forest	2
Nowakowski	1996	Poland	forest	25
Nuorteva	1971	Finland	forest	1
Parker	1996	Kansas	prairie woodlots	17

Author	Date	Location	Habitat	Duration (yrs)
Pitelka	1942	California	beach/estuary	1
Plesnik	1990	Poland	forest	14
Rolando et al.	1997	Italy	forest	5
Rosenberg et al.	1987	Arizona	riparian desert wash	2
Rosenfield et al.	1996	Wisconsin	forest	14
Rottenborn	1999	California	riparian forest	1
Russo and Young	1997	New York	forest	2
Ruszczyk	1987	Brazil	forest	1
Schmidt	1988	Germany	forest	16
Senra and Ales	1992	Spain	riparian forest	13
Sewell and Catterall	1998	Australia	shrubland	1
Snow	1969	United Kingdom	forest	15
Sodhi	1992	Canada	prairie	3
Soule et al.	1988	California	shrubland	1.3
Tatner	1982	United Kingdom	forest	3
Tella et al.	1996	Spain	grassland	6
Tilghman	1987a	Massachuchetts	forest	2
Tilghman	1987b	Massachuchetts	forest	2
Tomialojc	1979	Poland	forest	13
Vale and Vale	1976	California	shrubland, grassland, ag fields	2
Walcott	1974	Massachuchetts	forest	104
Wesolowski	1983	Poland	forest	6
Whitcomb et al.	1981	Maryland	forest	2
Wilcove	1985	Maryland	forest	1
Woolfenden and Rohwer	1969	Florida	shrubland	2
Yaukey	1996	Georgia	forest	2
Zalewski	1994	Poland	forest	3

Chapter 3

Synanthropic birds of North America

Richard F. Johnston
Natural History Museum and Department of Ecology and Evolution, University of Kansas, Lawrence, KS 66045 USA, e-mail: rfj@falcon.cc.ukans.edu

Key words: Adaptation, evolutionary rate, fecundity, habitat, longevity, succession, symbiosis, urbanization

Abstract: Biological synanthropy refers to human-mediated symbioses. Synanthropic birds are thus avian symbionts of humans. Such birds have a wide degree of relationship to humans, ranging from the almost obligate to the tangential. Avian synanthropy, which includes urbanized and urbanizing birds, is a common ecological relationship, to which at least 25% of North American birds can be referred. Species known to be specialists on early stages of ecological succession, as well as habitat generalists, may be predisposed to be synanthropic. This paper presents certain details on the scope of the behavior and discusses the taxonomic range of birds that can be judged to be synanthropic. Synanthropy may have demographic consequences, such as decreased longevity and increased fecundity.

1. INTRODUCTION

Humans have been changing the biology of the world for millennia. This is true even in North America, although intensive ecological change has been restricted to the past few hundred years. Such pressure on natural landscapes has been a nagging concern to humans of many cultures. Some, or even most, consequences of human ecological modification are viewed as unnatural by people representative of Western Culture, and the view is not wholly absent from people otherwise. One perhaps unexpected consequence of human ecological change is that some other organisms treat the changes as ecological opportunities and may even become specialists in exploiting them.

Organisms that cohabit with humans may be called "synanthropic" (Tomialojc 1970). The term *synanthropy* means "with humans," referring to the subset of a biotic community that can cohabit with humans; a synanthropy is therefore a human-mediated symbiosis, usually of mutualism or commensalism. North American studies have used the latter terms, but owing to their distribution over many fields of study the widely-scattered literature has not yet been summarized under any term.

Synanthropy is a relatively recent phenomenon, and North America had no avian or other synanthropic interactions prior to perhaps 30,000 years ago. Following the arrival of humans somewhere between 50,000 and 15,000 years ago (e.g., Mead and Meltzer 1985, Nemecek 2000) most of the early relationships between birds and humans were ones of predation. But humans eventually began to affect the original biota in a number of ways: the institution of agriculture, overwinter storing of foodstuffs, modification of surface and subsurface water, use of fire, fragmentation of habitats, the introduction of alien organisms. All of these things could eventually generate synanthropic responses.

For birds, synanthropy includes a wide degree of relationship to humans, ranging from obligate commensalism, such as in House Sparrows, *Passer domesticus* (Summers-Smith 1988), to species only very marginally commensal, such as the Ash-throated Flycatcher, *Myiarchus cinerascens*. The *scope* of this interaction is profound, and humans affect habitats so persistently and in such a variety of ways that a truly large number of opportunities for symbioses is created. The number of synanthropic bird species is probably in excess of what is discussed here, and it has been suggested, only halfway in jest, that any species still extant is symbiotic with humans (Bazzaz et al. 1998).

Although a large set of bird species is affected negatively by human activities, and a multitude of extinctions have been recorded (Hughes et al. 1997), extinction is not here a matter of concern. However, I should note that avian synanthropy approaches the importance of extinction, because mutualistic symbiotic relationships may affect rates of evolutionary change in symbiotic birds (Johnston and Selander 1964, Martin and Clobert 1996).

2. METHODS

This paper summarizes a review of the literature concerning the avian species, chiefly of western North America, that show some positive response to human intervention in ecological steady-states. The review has grown out of a personal long-term interest in House Sparrows and Rock Doves, *Columba livia,* two classic avian symbionts. The review also was stimulated by studies of European ecologists, whose work generally has been sheltered from North American scrutiny by use of the term *synanthropy*. Europeans usually consider only urban habitats in their studies of synanthropy, but I have included rural, rangeland, and other habitats remote from cities but impacted by humans. The urbanizing gradient, an important concept that is addressed by other contributions to this volume, extends well beyond conventional signs of human settlement, as in grazing and clear-cutting of forests.

The geographic concern of this review is North America north of Mexico. All species of birds in North America (American Ornithologists' Union 1998), including those of the marine littoral, have been surveyed in what I considered to be the pertinent literature. A significant fraction of this literature concerns research funded by the Federal Government; this has resulted in heavy emphasis on studies done on public land, most of which is in the western USA. The result, which includes 213 species, may therefore be more complete concerning birds living in grazing country and where forestry is practiced, than for birds generally.

Criteria for being deemed synanthropic are focussed on whether a bird species gains some benefit from use of habitats modified by humans. Degrees of synanthropy are recognized as follows: a *full synanthrope* is a species in which most populations have major dependence on variables influenced by humans; this usually involves food, structural habitat, or dispersal. A *casual synanthrope* is a species in which populations exploit human ecology without being dependent. A *tangential synanthrope* is a species in which individuals occassionally exploit human ecology. However, allocating species to these classifications can be messy. For example, Dark-eyed Juncos (*Junco hyemalis*) exhibit different degrees of synanthropy depending on whether they are visiting feeders in urban Alberta or nesting in backyards in Mendocino County, California. Nonetheless, most classifications can be justified. Some positive synanthropic relationships initially may have been coincidental, simply as part of a general phenotypic capability; some probably include learning; and some appear to be so fundamental an aspect of a species that they are probably genetically embedded. Additionally, any species introduction or reintroduction (human-mediated dispersal) also represents synanthropy.

It is important to note that some habitat modifications, such as those resulting from cattle grazing, may have either positive or negative effects on a bird's life history. For any such instance, all observed consequences are listed here. The fact that a species has shown positive, negative, or no apparent synanthropic response may mean that the fieldwork supporting the studies has varied from one set of researchers to another, or that a species has different synanthropic capabilities at different times or in different parts of its geographic distribution.

3. RESULTS

The results of the review are presented in Appendix 3.1. Each entry notes the degree of synanthropic symbiosis shown by a species; the serial listing of all species shows supraspecific distributions of the behavior. The degree of symbiosis is a personal judgement, because no conventional usage has been developed for the behavior. Taxonomic distribution is simply adjudicated, and is provided by the within-taxon range of occurrence of the symbiosis; this is independent of the degree found in any species.

On the order of a quarter of the birds of North America appear to be human symbionts to a greater or lesser degree; certain relationships are worth comment.

Gulls, *Larus*, are inordinately well-represented, largely exploiting human garbage dumps (Pierotti and Annett 2001); some species exploit agriculture, mostly associated with plowing of fields.

The Rock Dove has perhaps the longest-standing symbiotic relationships with humans of any bird. The species is nevertheless notable in the degree to which even feral populations (escapes from human confinement that still show some sign of the physiology and behavior of captives) are capable of living uncomplicated wild lives independent of humans. Five of the seven species of columbids listed in Appendix 3.1 were introduced to North America by humans.

The Chimney Swift, *Chaetura pelagica*, is one of the few nearly obligate human symbionts, using chimneys for both nesting sites and nocturnal roosting of flocks in autumn.

A number of woodpeckers respond positively to the more open forest and woodland habitats created by human felling and thinning of trees, and this also is true of several flycatchers. Some of the flycatchers favor the more open habitats created by grazing.

Corvids seem mostly to exploit food made available by humans (Marzluff et al. 2001, McGowan 2001, Bowman and Woolfenden 2001), but ours are not as adept at exploiting full urban habitats as are some of their European relatives (Jerzak 2001).

Swallows use a variety of nestsites created by human activities. Eastern North American populations of the Purple Martin, *Progne subis,* rely almost exclusively on nestboxes provided by humans, but western populations are nearly as exclusively restricted to nesting in natural cavities.

Some of the warblers use woodlands in which the understory vegetation has been thinned or removed, and this is true also of emberizine finches and blackbirds. Fringillids conspicuously use feeding stations operated by humans.

A dominant thread in these symbioses that transcends taxonomic relationships is a positive response to thinning of vegetation, whether the thinning is of woodlands or grasslands. This response is owing to the tendency for many avian species to be capable of adjusting to large-scale modifications in natural habitats, a tendency that will be further examined below.

4. DISCUSSION

4.1 Causes of Synanthropy

Organisms need to be adjusted to ecological disturbance, because disturbances are, and have been, naturally-recurring events throughout the world, quite independent of humans. Change is constant, whether that of the earth or its vegetational cover.

The geological record shows that life has persisted and evolved for more than three billion years, during which time change has occurred consistently. Living creatures are effective in living within temporal or spatial or ecological change; such change is the cause of organismic variation, why the fossil record shows a succession of organisms, and why persistent Darwinian evolution is necessary for the continuation of life.

Some bird species actually live in environments that are in a continual state of change, such as the Bell's Vireos (*Vireo bellii)* using successional old fields (Barlow 1962), and some require successional habitats periodically in their annual cycles, such as American Golden Plovers (*Pluvialis dominica)* on golf courses. Other causes of environmental variation include earthquakes, vulcanism, shifts in the axis of the earth's rotation, windstorms, montane avalanches, wildfires, or asteroidal impacts. The successional results of any of these are different from anthropogenic changes only in degree. These natural changes ultimately are responsible for the genetic variation permitting the initial selective responses by organisms to human-generated ecological change.

Most of the species in Appendix 3.1 are casual synanthropes. They are capable of using human-generated habitat variation when they chance upon it. These species seem to respond to such habitat variation as if it were "natural," and treat habitats resulting from natural or human means as more-or-less equivalent.

4.2 Consequences of Synanthropy

4.2.1 Life History Variables

One important consequence of birds becoming symbiotic with humans is life history evolution of a fundamental sort. What may we expect of some of the species of North American birds currently showing casual or tangential symbiotic response to human-generated habitats? The answer is life history evolution, specifically a good deal of fine-tuning of the relationship. This is even probable, if we may judge by information on European birds. Martin and Clobert (1996) have shown that a number of European birds, which began living in human ecosystems at least a thousand years earlier than birds in North America, today show profound adaptation of life history characteristics in response to living with humans. Using North American relatives as phyletic controls, Martin and Clobert showed that relatively rapid evolution of avian demographic characters (fecundity and longevity) has occurred in some of the European species. Given an equal amount of time, similar characteristics of fecundity and longevity might evolve in phyletically close relatives in North America.

What is involved is the general inverse relationship between longevity and reproduction in animals. Certainly in birds, species with long life expectancies tend to have modest reproductive output at any one point in their lives; those with short life expectancies tend to have large reproductive efforts, such as large or multiple clutches per season. These demographic relationships in birds have been discussed in detail over the past decade and are summarized by Martin (1998).

The shift to short life/high fecundity in European birds is probably a result of increased predation due to human and associated predators. European birds are now rewarded for reproducing early in their lives, and/or often, and/or with large broods. Birds not so doing have left fewer offspring and they are now rare in populations. In North America, where human densities are not as high as in Europe, bird longevity is greater, and measures of fecundity more modest, compared to Old World relatives.

4.2.2 Rapid Evolution

Symbioses have been identified as potential causes for increased rates of evolutionary change (Pirozynski and Malloch 1975, Janzen 1985, Margulis and Fester 1991, Lutzoni and Pagel 1997), and rapid evolution seems to be a likely consequence for birds that become symbiotic with humans (Johnston and Selander 1964). The evolutionary model is classical. It involves metapopulations being formed in human generated environmental patches.

Opportunities for testing the viability of novel combinations of characters would be increased in such population mosaics, in accord with the shifting balance theory of Sewall Wright (1978, Wade and Goodnight 1998). The generation of continentally-distributed habitat patches of different sizes, of varying distances from one another, persisting for varying lengths of time would support genetic differentiation even in the absence of selection; the patches may be considered to be crude equivalents of geneticists' population cages.

Because the United States is now the third most populous country in the world, we may conclude that in the near future our birds should experience ever increasing anthropogenic mortality and could be more and more rewarded for high fecundity. For some species in some localities, the process might already be underway (Marzluff et al. 2001).

The conditional phrasing above is necessary, because such scenarios apparently have not occurred within all affected European species. Conditions of population structure, degree of migratory behavior, degree of selective predation, etc., need to be in some contingent balance before microevolutionary changes could occur. And, it is clear that the degree of isolation of populations living under human-generated selective pressure needs to be reasonably pronounced for reassorted genetically-controlled variables to become fixed. However, fortunately we already know that adaptive selective rearrangement of fundamental biological variables has occurred in some populations of some European species.

5. POLICY AND MANAGEMENT IMPLICATIONS

As shown above, only a small number of North American birds can be considered fully symbiotic with humans, and many of these species were intentionally brought to the Western Hemisphere by humans. Their populations then (if not now) were under the care of game managers or an interested public. Each such species has, so to speak, made its separate peace with humans. It is the remainder, an appreciable fraction of the autochthonous avifauna, with which we are now concerned.

If this fraction of the birds of North America eventually responds to human-generated ecological variation in a fashion similar to that of some of their European relatives (short lives with high fecundities) North Americans would not need to decide whether to modify their environmental impacts as far as those birds are concerned. This would be a "do nothing" decision, which at least has the virtue of considering humans to be natural components of the world's biota. The dangers of doing nothing stem from not really knowing how or which North American birds will respond to the ecological complexities yet to be wrought by humans in the 21st century. For one thing, the rate of human-generated ecological change today is substantially higher than in pre-industrial Europe, and, instead of the evolutionary elaboration of short lives at high fecundities, multiple instances of extinction could occur. Thus, tempo of change and mode of response are tightly connected.

Possibly many of the casual and tangential synanthropes identified here will be able successfully to pursue coexistence with humans. These species have responded positively to at least one aspect of human-generated ecological change. For such species, an attempt to control that change could be a management tool. If changes are imposed on a population gradually, the potential for selective adjustment leading to adaptive behavior and physiology would not be as likely to fail as otherwise might be the case.

Again, the model for such a protocol is European; the first thousand years of concerted increases in human population numbers in Europe occurred in societies in which superstition and magic were held in high esteem, that were without modern medicine, and had inefficient agriculture and animal husbandry. Human populations therefore increased only irregularly in that millennium, and reached truly heavy densities perhaps just in the past 200 years.

In North America, human densities have reached levels that are critical to relatively few bird species, but the overall history of bird extinction is grim (Bibby 1995), and it is not without significance that we note repeated prospects of local extinction (e.g., Atwood 1993) and little approximation by birds of elaborating the short-life/high fecundity solution to human pressures. For the latter to occur, something other than an explosive rate of human population growth and environmental impact probably is necessary.

At least for that reason, we cannot profitably think of synanthropism as a possible solution to the problem of maintaining avian biological diversity. The encouraging fact that an appreciable fraction of North American birds have positive responses to human environmental modifications is simply that—encouraging—because another, larger, fraction of our avifauna is not known to show a positive response to what we do. This should prudently be interpreted to mean that responses of the larger fraction, when known, might be negative.

ACKNOWLEDGEMENTS

Discussion with Marian Janiga was helpful in organizing my thinking about synanthropic birds. Advice from Bob Holt, John Marzluff, Michael Moulton and Michael North corrected some of my misperceptions and focused attention on some of the more important aspects of the topic. I am pleased to have had their help.

REFERENCES

American Ornithologists' Union. 1998. Check-list of North American Birds. 7th Edition. American Ornithologists' Union, Washington, DC. (1a)

Atwood, J. L. 1993. California Gnatcatchers and coastal sage scrub: the biological basis for endangered species listing, p.149-169. *In* J. E. Keeley [ED.], Interface between ecology and land development in California. Southern Calif. Acad. Sci., Los Angeles.

Baker, D. L., and F. S. Guthery. 1990. Effects of continuous grazing on habitat and density of ground-foraging birds in south Texas. J. Range Manag. 43:2-5. (1b)

Barlow, J. C. 1962. Natural history of the Bell Vireo Vireo belli Audubon. University of Kansas Publishing, Museum of Natural History 12: 241-296.

Bazzaz, F., G. Ceballos, M. Davis, R. Dirso, P. Ehrlich, T. Eisner, S. Levin, J. Lawton, J. Lubchencko, P. Matslon, P. Raven, J. Roughgarden, J. Sarukhan, G. Tilman, P. Vitousek, J. Walker, D. Wall, E. Wilson, and G. Woodwell. 1998. Ecological science and the human predicament. Science 282:879.

Bibby, C. J. 1995. Recent past and future extinctions in birds, p.98-110. *In* J. H. Lawton and R. M. May [EDS.], Extinction rates. Oxford University Press, United Kingdom.

Blockstein, D. 1998. Cormorant/aquaculture workshop, p.5. *In* 1998 Annual Report. Ornithological Council, Washington, DC. (2a)

Bock, C. E., and J. H. Bock. 1988. Grassland birds in southeastern Arizona: impacts of fire, grazing, and alien vegetation, p.43-48. *In* P. Goriup [ED.], Ecology and conservation of grassland birds. Internat. Council for Bird Preserv., Publ. no. 7, Cambridge, United Kingdom. (2b)

Bock, C. E., J. H. Bock, W. R. Kenny, and V. M. Hawthorne. 1984. Responses of birds, rodents, and vegetation to lifestock exclosure in a semidesert grassland site. J. Range Manag. 37:239-242. (3)

Bowman, R. and G. Woolfenden. 2001. Nest success and the timing of nest failure of Florida Scrub-jays in suburban and wildlands, p. 385-404. *In* J. M. Marzluff, R. Bowman, and R. Donnelly [EDS.], Avian ecology and conservation in an urbanizing world. Kluwer Academic, Norwell, MA.

Canterbury, G. E., and D. E. Blockstein. 1997. Local changes in a breeding bird community following forest disturbance. J. Field Ornith. 68:537-546. (4a)

Chambers, C. L., and W. C. McComb. 1997. Effects of silvicultural treatments on wintering bird communities in the Oregon Coast Range. Northwest Sci. 71:298-304. (4b)

Crouch, G. L. 1982. Wildlife on ungrazed and grazed bottomlands on the South Platte River, northeastern Colorado, p.188-197. *In* Wildlife-livestock relationships symposium. Forest Wildlife and Range Exp. Sta., Moscow, ID. (5a)

Davis, J. N. 1999. Lawrence's Goldfinch, *Carduelis lawrencei*, p.1-20, *In* A. Poole and F. Gill [EDS.], The Birds of North America, No. 480. The Birds of North America, Inc., Philadelphia, PA. (5b)

Dawson, K. J. 1996. Interior and exterior land use as factors in landscape biodiversity from case studies, p.513-530. *In* R. C. Szaro and D. W. Johnston [EDS.], Biodiversity in managed landscapes. Oxford University Press, New York, NY. (6a)

Dellasala, D. A., J. C. Hagar, K. A. Engel, W. C. McComb, R. L. Fairbanks, and E. G. Campbell. 1996. Effects of silvicultural modifications of temperate rainforest on breeding and wintering bird communities, Prince of Wales Island, southeast Alaska. Condor 98:706-721. (6b)

Duebbert, H. F., and J. T. Lokemoen. 1977. Upland nesting of American Bitterns, Marsh Hawks, and Short-eared Owls. Prairie Natur. 9:33-40. (7)

Ehrlich, P. R., D. S., Dobkin, and D. Wheye. 1988. The Birder's Handbook. Simon and Schuster, New York, NY. (8)

Feare, C.J. 1984. The starling. Oxford Univ. Press, Oxford, United Kingdom. (9a)

Frank, S. City Peregrines. Hancock House, Blaine, WA. (9b)

Gleason, R. S. 1978. Aspects of the breeding biology of Burrowing Owls in southern Idaho. MS thesis, Univ. Idaho, Moscow. (10)

Graul, W. D. 1975. Breeding biology of the Mountain Plover. Wilson Bull. 87:6-31. (11a)

Haywood, G. D., and P. H. Haywood. 1993. Boreal Owl (*Aegolius funereus*). *In* A. Poole and F. Gill [EDS.], The Birds of North America. Academy of Natural Sciences, Philadelphia, PA. (11b)

Hejl, S. J., R. L. Hutto, C. R. Preston, and D. M. Finch. 1955. Effects of silvicultural treatments in the Rocky Mountains, p.220-244. *In* T. E. Martin and D. M. Finch [EDS.], Ecology and management of Neotropical migratory birds. Oxford University Press, New York, NY. (12a)

Hochachka, W. M., and K. V. Rosenberg. 1998. Project FeederWatch Annual Report 1997-98. Birdscope 12:1-7. (12b)

Howlin, S. and J. C. Hager. 1999. Changes in bird densities associated with thinning of Douglas-fir stands in the Oregon Cascades, p.34. *In* Abstracts, 1999 Annual Meeting, Cooper Ornithological Society, Portland, OR. (12c)

Hughes, J. B., G. C. Daily, and P. R. Ehrlich. 1997. Population diversity: its extent and extinction. Science 278:689-692.

Janzen, D. H. 1985. The natural history of mutualisms, p.40-99. *In* D. H. Boucher [ED.], The biology of mutualism: ecology and evolution. Oxford University Press, New York, NY.

Jerzak, L. 2001. Synurbanization of the Magpie in the Palearctic, p. 405-427. *In* J. M. Marzluff, R. Bowman, and R. Donnelly [EDS.], Avian ecology and conservation in an urbanizing world. Kluwer Academic, Norwell, MA.

Johnston, R. F. Personal observations. (13)

Johnston, R. F., and M. Janiga. 1995. Feral pigeons. Oxford University Press, New York, NY. (14)

Johnston, R. F., and R. K. Selander. 1964. House sparrows: Rapid evolution of races in North America. Science 141:548-550.

Kantrud, H. A. 1981. Grazing intensity effects on the breeding avifauna of North Dakota native grasslands. Canad. Field Natur. 95:403-417. (15)

Kantrud, H. A., and K. F. Higgins. 1992. Nest and nest site characteristics of some ground-nesting, non-passerine birds of northern grasslands. Prairie Natur. 24:67-84. (16)

Kantrud, H. A., and R. L. Kologiski. 1982. Effects of soil and grazing on breeding birds of uncultivated upland grasslands of the northern Great Plains. USDI Fish Wldlf. Serv. Res. Rep. no. 15, Washington, DC. (17)

Knopf, F. J., J. A. Sedgwick, and R. W. Cannon. 1988. Guild structure of a riparian avifauna relative to seasonal cattle grazing. J. Wildl. Manage. 52:280-290. (18)

Kochert, M. N. 1989. Responses of raptors to livestock grazing in the western United States. Proc. Western Raptor Manag. Symp. Workshop, p.193-203. National Wldlf. Fed., Washington, DC. (19)

Littlefield, C., S. Thompson, and B. Ehlers. 1984. History and present status of Swainson's Hawk in Southeastern Oregon. J. Raptor Res. 18:1-5. (20)

Lokemoen, J. T., and H. F. Duebbert. 1976. Ferruginous Hawk nesting ecology and raptor populations in South Dakota. Condor 78:463-470. (21)

Lutzoni, F., and M. Pagel. 1997. Accelerated evolution as a consequence of transitions to mutualism. Proc. Nat. Acad. Sci. 94:11422-11427.

Maher, W. J. 1979. Nesting diets of prairie passerine birds at Matador, Saskatchewan, Canada. Ibis 121:437-452. (22)

Margulis, L., and R. Fester. 1991. Symbiosis as a source of evolutionary innovation: speciation and morphogenesis. MIT Press, Cambridge, MA.

Martin, T. 1995. Avian life history evolution in relation to nest sites, nest predation and food. Ecol. Monogr. 65:101-127.

Martin, T. 1998. Environmental constraints and mechanistic pathways underlying the evolution of life histories, p.161. *In* Abstracts, North American Ornithological Conference, St. Louis, MO.

Martin, T., and J. Clobert. 1996. Nest predation and avian life-history evolution in Europe versus North America: a possible role of humans. Amer. Natur. 147:1028-1046.

Martin, T. E., and D. M. Finch. 1995. Ecology and management of Neotropical migratory birds. Oxford University Press, Oxford, United Kingdom. (23)

Martin, J. W. 1987. Behavior and habitat use of breeding Northern Harriers in southwestern Idaho. J. Raptor Res. 21:57-66. (24a)

Marzluff, J. M., K. McGowan, C. Caffrey, R. Donnelly, and R. Knight. 2001. Causes and consequences of expanding American Crow populations, p. 333-365. *In* J. M. Marzluff, R. Bowman, and R. Donnelly [EDS.], Avian ecology and conservation in an urbanizing world. Kluwer Academic, Norwell, MA.

Mayfield, H. F. 1992. Kirtland's Warbler, *Dendroica kirtlandii*, p.1-16. *In* A. Poole, P. Stettenheim, and F. Gill [EDS.], The Birds of North America. Academy of Natural Sciences, Washington, DC. (24b)

McGowan, K. 2001. Demographic and behavioral comparisons of suburban and rural American Crows, p. 367-383. *In* J. M. Marzluff, R. Bowman, and R. Donnelly [EDS.], Avian ecology and conservation in an urbanizing world. Kluwer Academic, Norwell, MA.

McIntyre, J. W., and J. F. Barr. 1997. Common Loon, *Gavia immer*, p.1-32. *In* A. Poole and F. Gill [EDS.], The Birds of North America. The Birds of North America, Inc., Phladelphia, PA. (24c)

Mead, J. I., and D. J. Meltzer. 1985. Environments and extinctions: man in late glacial North America. University of Maine Center Study of Early Man, Orono, ME.

Medin, D. E. 1986. Grazing and passerine breeding birds in a Great Basin low-shrub desert. Great Basin Natur. 46:567-572. (25)

Medin, D. E., and W. P. Clary. 1990. Bird and small mammal populations in a grazed and ungrazed riparian habitat in Idaho. USDA Forest Serv. Intermt. Res. Sta. Pap., INT-441. (26)

Mosconi, S. L., and R. L. Hutto. 1982. The effects of grazing on land birds of a western Montana riparian habitat, p.221-223. *In* J. M. Peek and P. D. Dalke [EDS.], Wildlife-livestock relationships symposium. Univ. Idaho Forest Wldlf. Range Exp. Sta., Moscow, ID. (27)

Nemecek, S. 2000. Who were the first Americans? Sci. Amer. 283:80-87.

Nydegger, N. C., and G. W. Smith. 1986. Prey populations in relation to Artemisia vegetation types in southwestern Idaho, p.152-156. *In* E. D. McArthur and B. L. Welch [EDS.], Symposium on the biology of Artemisia and Chrysothamnus. USDA Forest Serv. Gen. Tech. Rep., INT-200. Provo, UT. (28)

Osborn, S. A. H. 1999. Effects of streamside development on the American Dipper in western Montana, p.49. *In* Abstracts, 1999 Annual Meeting, Cooper Ornithological Society, Portland, OR. (29)

Owens, R. A., and M. T. Myres. 1973. Effects of agriculture on populations of native passerine birds of an Alberta fescue grassland. Can. J. Zool. 51:697-713. (30)

Page, J. L., N. Dodd, T. O. Osborne, and J. A. Carson. 1978. The influence of livestock grazing on non-game wildlife. Cal. Nov. Wldlf. 1978:159-173. (31)

Pierotti, R. and C. Annett. 2001. The ecology of Western Gulls in habitats varying in degree of urban influence, p. 309-331. *In* J. M. Marzluff, R. Bowman, and R. Donnelly [EDS.]. Avian ecology and conservation in an urbanizing world. Kluwer Academic, Norwell, MA.

Pirozynski, K. A., and D. W. Malloch. 1975. The origin of land plants: A matter of mycotrophism. Biosystems 6:153-154.

Reynolds, T. D. 1980. Effects of some different land management practices on small mammal populations. J. Mammal. 61:558-561. (32)

Reynolds, T. D., and T. D. Rich. 1978. Reproductive ecology of the Sage Thrasher (*Oreoscoptes montanus*) on the Snake River plain in south-central Idaho. Auk 95:580-582. (33)

Reynolds, T. D., and C. H. Trost. 1981. Grazing, crested wheatgrass, and bird populations in southeastern Idaho. Northwest. Sci. 55:225-234. (34)

Rich, T. D. 1986. Habitat and nest-site selection by burrowing owls in the sagebrush steppe of Idaho. J. Wildl. Manage. 50:548-555. (35)

Risser, P. G., E. C. Birney, H. D. Walker, et al. 1981. The true prairie ecosystem. Hutchinson Ross Publ. Co., Stroudsburg, PA. (37)

Ryder, J.P., and R.T. Alsauskas. 1955. Ross' Goose (*Chen rossi*). *In* A. Poole and F. Gill [EDS.], The Birds of North America. Academy of Natural Sciences, Philadelphia, PA. (39)

Ryder, R. A. 1980. Effects of grazing on bird habitats, p.51-56. *In* R. Degraff and N. Tilghman [EDS.], Management of western forests and grasslands for nongame birds. USDA Forest Service Gen. Tech. Rep. INT-86. Intermountain Forest and Range Exp. Sta., Ogden, UT. (40)

Schulz, T. T., and W. C. Leininger. 1991. Nongame wildlife communities in grazed and ungrazed montane riparian sites. Great Basin Natur. 51:286-292. (42)

Sedgwick, J. A. and F. L. Knopf. 1987. Breeding bird response to cattle grazing of a cottonwood bottomland. J. Wildl. Manage. 51:230-237. (43)

Skinner, R. M. 1975. Grassland use patterns and prairie bird populations of Missouri, p.171-180. *In* M. Wati [ED.], Prairie: a multiple view. University of North Dakota Press, Grand Forks, ND. (44)

Snyder, N. F., and H. A. Snyder. 1975. Raptors in range habitat, p.190-209. *In* Symposium on management of forest and range habitats for nongame birds. USDA Forest Service, Gen. Tech. Rep. WO-1. Washington, DC. (45a)

Stout, B. E., and G. L. Neuchterlein. 1999. Red-necked Grebe, *Podiceps grisegena*, p.1-32. *In* A. Poole and F. Gill [EDS.], The Birds of North America. Academy of Natural sciences, Philadelphia, PA. (45b)

Summers-Smith, J. D. 1988. The sparrows. T. and A. D. Poyser, Staffordshire, United Kingdom. (46)

Taylor, D. M. 1986. Effects of cattle grazing on passerine bird nesting in riparian habitat. J. Range Manag. 39:253-258. (47)

Tester, J. R., and W. H. Marshall. 1961. A study of certain plant and animal interrelations on a native prairie in northwestern Minnesota. University of Minnesota Museum of Natural History, Occas. Pap. no. 8. (48)

Tomialojc, L. 1970. Quantitative studies on the synanthropic avifauna of Legnica and its environs. Acta Ornitologica 2:209-290.

Wade, M. J., and C. J. Goodnight. 1998. Perspective: the theories of Fisher and Wright in the context of metapopulations: when nature does many small experiments. Evolution 52:1537-1553.

Webb, E. A., and C. E. Bock. 1990. Relationship of the Botteri's Sparrow to sacaton grassland in southeastern Arizona, p.199-209. *In* P. Krausman and N. M. Smith [EDS.], Managing wildlife in the southwest. Arizona Chapter, Wildlife Society, Phoenix, AZ. (49)

Wiens, J. A. 1973. Pattern and process in grassland bird communities. Ecol. Monogr., 43:237-270. (50)

Wright, S. 1978. Evolution and the genetics of populations. Vol. 4, Variability within and among natural populations. University of Chicago Press, IL.

Appendix 3.1. Responses of some North American birds to modification of ecological variables by humans.

Species[ab]	Habitat[c]	Variable	Resp.[d]	Reference[e]
Gavia immer T	L	Nestsite	+	24c
Podilymbus podiceps T	L,R	Artific. pond	+	8
Podiceps grisegena T	L	Nestsite, pond	+	45b
Fulmarus glacialis T	MAR	Fishing waste	+	8
Pelecanus occidentalis T	MI	Fishing waste	+	8
Phalacrocorax auritus C	MI,L	Aquaculture	+	2a,8
Ardea herodias T	MI,L	Artif. feeding	+	13
Bulbulcus ibis C	GM,M,Ru	Forage	+	13
Coragyps atratus T	Ru	Roadkill	+	13
Cathartes aura T	Ru	Roadkill	+	13
Chen caerulescens C	L,Ru	Agric.,Managed	+	8
Chen rossii C	L,Ru	Agric.,Managed	+	39
Branta canadensis C	L,G,U	Habitat mod.	+	13
Cygnus olor F	L,R,U	Human dispersal	+	13
Aix sponsa C	R	Nesting boxes	+	13
Anas platyrhynchos T	L,U	Habitat mod.	+	13
Anas discors T	L	Habitat mod.	+	13
Pandion haliaetus T	L,R	Habitat mod.	+	8
Elanus leucurus T	G,W	Agriculture	+	13
Circus cyaneus T	S	Grazing	+	24a,34
Accipiter gentilis T	F	Clearcut	+,-	12a
Buteo regalis T	GM	Grazing	+	21
Buteo jamaicensis T	S	Grazing	+,-	19
	F	Thinned, clrcut	+,-	4b
	U	Human attitude	+	9b
Buteo swainsoni T	S	Grazing	+	20,34
Aquila chrysaetos T	S	Grazing	+,-	28,19
Falco sparverius T	F	Clearcut	+	12a
	S	Grazing	+,-	19,31,34
	R	Grazing	+,0,-	5a,26,27

Species[ab]	Habitat[c]	Variable	Resp.[d]	Reference[e]
Falco peregrinus C	U	Nestsites	+	9b
Tetrao himalayensis F	GA	Dispersal	+	1a
Perdix perdix F	A,G	Human dispersal	+	8
Alectoris chukar F	GD	Human dispersal	+	8
Phasianus colchicus F	GM	Human dispersal	+	8
Bonasa umbellus C	F	Managed	+	13
Tympanuchus cupido C	A,GM	Grazing, managed	+	13
Meleagris gallopavo C	W	Managed	+	13
Callipepla californica C	W,G	Managed	+	13
Colinus virginianus C	W,G	Managed	+	13
Tryngites subruficollis T	A	Mowed alfalfa	+	13
Pluvialis dominica T	GM	Mowing, fires	+	13
Charadrius vociferus T	GS,GM	Grazing	+	17,40
	R	Grazing	+	26,42
Charadrius montanus T	GS,GM	Grazing	+	11a,17,40
Bartramia longicauda C	GM	Grazing	+,0,-	15,16,17
Numenius americanus T	GM,GS,S	Grazing	+,0,-	17,34,40
	R	Grazing	+	26
Larus atricilla C	A	Plowing	+	1a
Larus pipixcan C	A	Plowing	+?	7
Larus ridibundus C	MI	Garbage	+?	8
Larus canus C	MI,U	Garbage	+?	8
Larus delawarensis C	A,U	Plowing, trash	+,+?	8
Larus californicus C	U	Garbage	+?	8
Larus argentatus C	MI,U	Garbage	+?	8
Larus thayeri C	MI,U	Garbage	+?	1a
Larus fuscus C	MI,U	Garbage	+?	1a
Larus occidentalis C	MI,U	Garbage	+?	1a
Larus glaucescens C	MI,U	Garbage	+?	8
Larus hyperboreus C	MI,U	Garbage	+?	1a
Larus marinus C	MI,U	Garbage	+?	1a
Rissa brevirostris C	P	Fish waste	+	8
Sterna nilotica C	A	Plowing	+	8
Sterna antillarum T	R,A	Nestsites	+	1a
Rynchops niger T	A	Nestsites	+	1a
Columba livia F	U,Ru,A	Overall ecology	+	14
Columba fasciata C	F,Ru,A	Food supply	+	13
Streptopelia risoria C	Ru	Human dispersal	+	1a
Streptopelia decaocto C	Ru,A	Human dispersal	+	1a
Streptopelia chinensis C	Ru,A	Human dispersal	+	1a
Zenaida macroura T	GM,GT	Grazing	+,0	1b,15
	GD,S	Grazing	+,-	3,31
	R	Grazing	+,0,-	5a,27,31
Columbina inca T	GD,U,Ru	Food, water	+	13
Myiopsitta monachus C	U	Human dispersal	+	1a
Amazona viridigenalis C	F	Human dispersal	+	1a
Coccyzus americanus T	R	Grazing	+	5a
Tyto alba C	Ru	Nest,roost sites	+	13

Species[ab]	Habitat[c]	Variable	Resp.[d]	Reference[e]
Otus asio T	U,Ru	Nestsites	+	13
Bubo virginianus T	W,Ru	Nests,roosts	+	13
Athene cunicularia T	GM	Grazing	+,0	17
	S	Grazing	+,0,-	10,19,35,45
Aegolius funereus T	F	Nestsites	+	11b
Chordeiles minor T	GS,GM,S	Grazing	+,0,-	17,34,40
	R	Grazing	+	5a
Phalaenoptilus nuttallii T	S	Grazing	+	34
Chaetura pelagica F	U,Ru	Nestsites	+	13
Archilochus colubris T	U,Ru	Food	+	13
Calypte anna T	U,Ru	Food	+	12b
Stellula calliope T	F	Tree cutting	+	31
	R	Grazing	+	31
Selasphorus platycercus T	F	Clearcut	+	12a
	R	Grazing	+,-	26,42
Selasphorus rufus T	F	Var. cutting	+,0	6b,12a
	R	Grazing	+	31
Colaptes auratus T	F	Var. cutting	,-	4b,12a
	R,S	Grazing	+,0,-	26,27,31,42
Melanerpes lewis T	R	Grazing	+	26,27
Melanerpes carolinus T	U,Ru	Food	+	12b
Sphyrapicus thyroideus T	F	Cutting	0	12a
Sphyrapicus nuchalis T	F	Var. cutting	+,-	12a
	R	Grazing	+,-	27,31,42
Sphyrapicus ruber T	F	Thinned	+	4b
Picoides nuttallii T	R	Managed woods	+	6a
Picoides pubescens T	R	Managed woods	+	6a
Picoides villosus T	U,Ru	Food	+	12b
Contopus cooperi T	F	Clearcut	+	12a
Contopus sordidulus T	R	Grazing	+,-	27,31
Empidonax trailii T	R	Managed woods	+	6a
Empidonax minimus T	R	Grazing	+	27
Empidonax hammondii T	F	Thinning	+	12c
Empidonax wrightii T	S	Grazing	+,-	31,34
Empidonax oberholseri T	F	Clearcut	+	12a
Empidonax difficilis T	R	Managed woods	+	6a
Empidonax occidentalis T	F	Partial cut	0	12a
Sayornia nigricans T	Ru	Nestsites	+	1a
Sayornis phoebe T	Ru	Nestsites	+	1a
Sayornis saya T	S,R	Grazing	+	5a,34
Myiarchus cinerascens T	R	Managed woods	+	6a
Lanius ludovicianus T	S	Grazing	+,-	26,31
Vireo bellii T	G-W	2nd succession	+	13
Vireo huttoni T	F	Hvy. cut	+,-	4b
Vireo gilvus T	R	Grazing	+,-	26,27,31,42
Vireo olivaceus T	R	Grazing	+	27
	F	Thinned	+	4a
Perisoreus canadensis T	U,Ru	Food	+	12b
Cyanocitta cristata T	U,Ru	Food	+	12b

Species[ab]	Habitat[c]	Variable	Resp.[d]	Reference[e]
Cyanocitta stelleri T	F	Thinned	+	4b,6b
	U,Ru	Food	+	12b
Aphelocoma californica T	U,Ru	Food	+	12b
Nucifraga columbiana T	F	Partial cut	+	12a
Pica pica T	Ru	Food	+	12b
Corvus corax T	Ru	Food	+	12b
Corvus brachyrhynchos T	U,Ru	Food	+	13
Corvus ossifragus T	U	Food	+	1a
Eremophila alpestris C	Ru	Nestsites	+	13
	G	Grazing	+	3,17,37,40
	S	Grazing	+,-	31,32
Progne subis F	U	Nestsites	+	13
Tachycineta bicolor T	F	Clearcut	+	12a
	R	Grazing	+,-	26,31,42
	Ru	Nestsites	+	13
Stelgidopteryx serripennis T	Ru,R	Nestsites	+	1a
Riparia riparia T	Ru	Nestsites	+	13
Petrochelidon pyrrhonota T	Ru,U	Nestsites	+	1a
Petrochelidon fulva T	Ru	Nestsites	+	1a
Hirundo rustica T	Ru,U	Nestsites	+	13
Poecile carolinensis T	U,Ru	Food	+	1a
Poecile atricapillus T	U,Ru	Food	+	12b
Poecile hudsonicus T	Ru	Food	+	12b
Baeolophus bicolor T	U,Ru	Food	+	12b
Sitta canadensis T	U,Ru	Food	+	12b
	F	Thinned woods	+	4b
Sitta carolinensis T	U,Ru	Food	+	12b
	F	Managed park	+	6a
Certhia americana T	U,Ru	Food	+	13
Troglodytes aedon T	U,Ru	Nestsites	+	13
	F	Partial cut	+	4b,12a
	R	Grazing	+,0,-	5a,26,27,42, 43,47
Troglodytes troglodytes T	F	Cut, thinned	+	4b
Cinclus mexicanus T	R	Nestsites	+	29
Pycnonotus jocosus F	U	Dispersal	+	1a
Regulus calendula T	F	Succession	+	6b
Sialia sialis T	W,G	Nestsites	+	13
Sialia mexicana T	W	Partial cut	+	12a
Sialia currucoides T	F	Clearcut	+	12a
	S	Grazing	+	31
	R	Grazing	+,0	5a,42
Myadestes townsendi T	F	Clearcut	+	12a
Catharus ustulatus T	F	Succession	+	6b
Turdus migratorius C	F	Clearcut	+,-	4b,12a
	F	Partial cut	+	12a
	F	Second growth	+	6b
	R	Grazing	+,-	26,27,42,43
	R	Grazing	+,0	5a,26,47

Species[ab]	Habitat[c]	Variable	Resp.[d]	Reference[e]
	U,Ru	Food, nesting	+	12b
Ixoreus naevius T	U,Ru	Food	+	12b
Dumetella carolinensis T	R	Grazing	+,-	5a,27
Mimus polyglottos T	GD	Grazing	+	3
	U,Ru	Food	+	12b
Oreoscoptes montanus T	S	Grazing	+,-	31,33
Sturnus vulgaris F	U,Ru	Nests, food	+	9a,12b
	F	Clearcut	+	4b
Acridotheres cristatellus F	U	Urbanization	+	13
Anthus spragueii T	GM	Grazing	+,-	15,17,22,30
Vermivora celata T	F	Second growth	+	6b
Dendroica petechia T	R	Grazing	+,0,-	26,27,31,47
Dendroica pensylvanica T	F	Thinning	+	4a
Dendroica coronata T	R	Grazing	+,-	5a,27,31,42
Dendroica kirtlandii C	W	Cowbird trapping	+	24b
Setophaga ruticilla T	R	Grazing	+	5a,27
Oporornis philadelphia T	F	Thinning	+	4a
Oporornis tolmei T	F	Clearcut	+,-	34
	F	Partial cut	+	12c,34
	R	Grazing	+,-	26,27,42
Geothlipis trichas T	F	Thinning	+	4a
Wilsonia pusilla T	F	Clearcut	+,-	12a
	F	Second growth	+	6b
	R	Grazing	+,-	5a,31,42
Piranga ludoviciana T	R	Grazing	+,-	27,42
Pipilo chlorurus T	R	Grazing	+	31
Pipilo maculatus T	U,Ru	Food	+	12b
	F	Cut	+	4b
	R	Grazing	+,-	26,31,43
Pipilo crissalis T	U,Ru	Food	+	12b
Aimophila botterii T	GD	Grazing	+,-	49
Spizella passerina T	F	Clearcut	+	12a
	R	Grazing	-	5a
Spizella pallida T	GM	Grazing	0,-	15,17,30
Spizella breweri T	GD	Grazing	+,0	31
	GM	Grazing	+,0,-	3,17
	S	Grazing	+,-	31
Pooecetes gramineus T	GD	Grazing	0	3
	GM	Grazing	+,0,-	15,17,32
	S	Grazing	+,-	31
	R	Grazing	+	42
Chondestes grammacus T	GM	Grazing	0,-	17,40
Amphispiza bilineata T	GD	Grazing	+	3
Amphispiza belli T	S	Grazing	+,-	31,32
Calamospiza melanocorys T	GM	Grazing	+,0	15,17
Passerella iliaca T	F	Clearcut	+,-	12a
	F	2nd growth	+	6b
Melospiza melodia T	F	Clearcut	+	4b,12a
	F	Thinned	+	4a,4b,6b

Species[ab]	Habitat[c]	Variable	Resp.[d]	Reference[e]
	R	Grazing	+,-	18,26,27
Melospiza lincolnii T	F	Clearcut	+,-	12a
Zonotrichia albicollis T	U,Ru	Food	+	12b
Zonotrichia leucophrys T	F	Clearcut	+	12a
	R	Grazing	+,-	18,26,31,42
	S	Grazing	+,-	31,34
Junco hyemalis T	F	Clearcut	+	4b,12a
	F	2nd growth	+	6b
	R	Grazing	-	5a,31,42
	U,Ru	Food	+	12b
Calcarius mccownii T	G	Grazing	+	17,22,23,40
Calcarius ornatus T	GD	Grazing	+	2b
	GM	Grazing	+,0	15,17,30,40
Cardinalis cardinalis T	U,Ru	Food	+	12b
Pheucticus melanocephalusT	R	Grazing	+,-	27,47
Guiraca caerulea T	R	Managed area	+	6a
Passerina amoena T	R	Managed area	+	6a
Spiza americana T	GT	Grazing	+,-	37,44
Dolichonyx oryzivorus T	GM	Grazing	-	15,17
	GT	Grazing	+,0,-	15,44,48
Agelaius phoeniceus T	GM	Grazing	-	15
	GT	Grazing	+	44
	S	Grazing	-	31
	R	Grazing	-	5a,18,47
Sturnella magna T	GT	Grazing	+,-	1b,37,44
	GD	Grazing	0	3
Sturnella neglecta T	GS	Grazing	-	40,50
	GM	Grazing	0,-	15,17,50
	S	Grazing	+,-	31,32
	R	Grazing	+,-	5a,26
Xanthocephalus xanthocephalus T	R	Grazing	+	47
Euphagus cyanocephalus T	S	Grazing	+,-	25,31
	R	Grazing	+,0,-	5a,31,47
Quiscalus quiscula T	U,Ru	Food	+	12b
Molothrus ater T	GM	Grazing	0	15,17
Icterus cucullatus T	R	Grazing	+	31
Icterus bullockii T	R	Grazing	+,0,-	5a,26,27,31,47
Leucosticte tephrocotis T	Ru	Food	+	1a
Leucosticte atrata T	Ru	Food	+	1a
Leucosticte australis T	Ru	Food	+	1a
Pinicola enucleator T	F	2nd growth	+	6b
	U,Ru	Food	+	12b
Carpodacus purpureus T	U,Ru	Food	+	1a
Carpodacus cassinii T	F	Clearcut	+,-	31
	F	Partial cut	+	31
	R	Grazing	+,0,-	26,27,42,47
	U,Ru	Food	+	12b
Carpodacus mexicanus C	U,Ru	Food, nesting	+	12b

Species[ab]	Habitat[c]	Variable	Resp.[d]	Reference[e]
Loxia curvirostra T	F	2nd growth	+	6b
Carduelis flammea T	U,Ru	Food	+	12b
Carduelis pinus T	R	Grazing	+	27,42
	F	Clearcut	+	4b
	U,Ru	Food	+	12b
Carduelis psaltria T	U,Ru	Food	+	13
Carduelis lawrencei T	W,A	Grazing	+,-	5b
Carduelis tristis T	U,Ru	Food	+	12b
Coccothraustes vespertinus T	F	Thinning	+	4b
	U,Ru	Food	+	12b
Passer domesticus F	U,Ru	Food, nesting	+	12b,46
Passer montanus F	U,Ru	Food, nesting	+	13

[a]Names are those of the American Ornithologists' Union (1998).
[b]Degree of synanthropy: *F*, full synanthrope with humans; *C*, casual synanthrope; *T*, tangential synanthrope.
[c]A, agricultural; F, forest; G, grassland; GD, desert grassland; GM, mixed tall and short grassland; GS, shortgrass prairie; GT, tallgrass prairie; L, lacustrine; M, marsh; MI, marine littoral; R, riparian; Ru, rural; P, pelagic; S, shrubsteppe; U, urban; W, woodland.
[d]Response: +, increase in numbers; 0, no numerical response noted.; -, decrease in numbers.
[e]Refers to numbered entries in the references section.

Chapter 4

Human perception and appreciation of birds: A motivation for wildlife conservation in urban environments of France

Phillipe Clergeau[1], Gwanaelle Mennechez[1], Andre Sauvage[2], and Agnes Lemoine[2]
[1]*I.N.R.A. Faune sauvage & U.M.R. EcoBio, Campus de Beaulieu, 35 042 Rennes cedex, France, e-mail: clergeau@univ-rennes1.fr*

[2]*Universite Rennes2, Laboratoire de recherche en science sociales, 4 palce Saint Melaine, 35 000 Rennes, France*

Key words: Avian diversity, France, public appreciation, sociology, urban ecology, wildlife management

Abstract: Conservation of biodiversity in urban areas needs both scientific justification and public interest. However, data about the public's perception of urban wildlife is scarce in Western Europe. To identify human perception and appreciation of urban wildlife, we studied avian communities and the public's perceptions about birds in general, in Rennes, France, at four sectors along a gradient of decreasing urbanization: urban, and adjoining suburban, distant suburban and rural sectors. The diversity and density of birds were analyzed in Rennes during winter and spring (6 censuses). Avian diversity was highest in the distant suburban sector and lowest in urban areas. In the two intermediate sectors (adjoining and distant suburban), diversity was similar and varied little between seasons (Rennes has a relatively temperate climate). The density of birds was highest in urban areas because of the abundance of some species, including Rock Doves (*Columba livia*) and Swifts (*Apus apus*) and lowest in the rural sector. Sociological analyses of surveys conducted in Rennes (200 questionnaires) showed that human perception of birds varied with the sector of respondents. Urban dwellers perceived the different bird species (more diversity than density) and appreciated their presence, except for some species, such as Herring Gulls *Larus argentatus* or European Starlings *Sturnus vulgaris*. According to our results, efforts to increase or maintain avian biodiversity in cities appears to be justified in terms the perceived benefit to humans, both in terms of the perceived environmental and personal benefits to humans. However, when a species must be managed (e.g. roosting European Starlings), different attitudes by the public in different parts of the city may require different styles of communicating management objectives.

1. INTRODUCTION

Throughout history, municipalities of Western Europe have tried to address the social needs of providing nature in urban environments. Today, one of the motivations for greenspace in towns is also to increase species diversity in flora and fauna (Congress of Starsbourg 1994, Congress of Leicester 1995) and to achieve sustainable development in the city (growth according to environmental, economic and socio-cultural needs) (Middleton 1994, Sénécal 1996). In this way, urban planning can be directed, for example planning for urban greenways to preserve the presence of numerous wildlife species even within the city center (Davis and Glick 1978, Saunders and Hobbs 1991, Fabos and Ahern 1996).

However, relatively little is known about the public's perception of biodiversity in urban areas. Are the species associated with green spaces (parks, gardens, greenways, etc) perceived and appreciated by city dwellers? Does this perception vary with location along an urban gradient or do these perceptions change as biodiversity increases (Terrasson 1991, Clergeau 1996)? Appreciation of birds and even individual species may vary from peaceful coexistence with animals (including feeding them) to a perception of birds as pests (Clergeau et al. 1996). Avian diversity in urban areas may be maintained by identifying those species that can survive in the city, and those species most appreciated by humans. Most research on human-wildlife relations in urban environment has been conducted in North America (Dagg 1970, Brown et al. 1979, Shaar 1979, Witter et al. 1981, Kellert 1984, Adams 1994, Vandruff et al. 1995, Conover 1997). Little has been done elsewhere (Gilbert 1989, Ilyichev et al. 1990). Studies of human perceptions of wildlife in urban areas are rare in Latin countries of Western Europe, including France. Understanding social attitudes is important for the effective management of urban green space and urban fauna. This is even more relevant today as urban areas expand, both geographically and politically (Brown et al. 1979).

We chose birds as our biological model, because they are good ecological indicators and easy to observe (Furness and Greenwood 1993), and because they are potential targets of conflicts with humans. Humans may feed birds in some areas, while in other areas they may demand their eradication, for example, French dairy farmers against European Starlings (*Sturnus vulgaris*) in Caen (1974) or fish farmers against Cormorants (*Phalacrocorax carbo*) in Strasbourg (1996).

Previous studies have shown the effects of the density of buildings and human populations on the behavior of humans (Sauvage 1992) or of birds (Clergeau 1995). Thus, we have chosen these two urban features to simultaneously analyze avian community structure and the perception of the bird community by humans. We account for variations in the density of these urban features by using a gradient of urbanization from city center to rural area (McDonnell and Pickett 1990; Alberti et al. 2001).

The objective of this study was to answer (a) how does bird diversity and abundance differ between urban sectors in a middle-size city such as Rennes? (b) How familiar are city dwellers with the identity of species of birds in the city landscape? (c) Do city dwellers in different urban sectors differ in their levels of appreciation for birds?

2. METHODS

Rennes (pop. 200.000) is a middle-sized city located in Brittany, western France (Fig. 4.1). It has a temperate climate. In older European cities, a gradient of building density and decreasing building age can be represented by three circular sectors in concentric rings of decreasing urbanization around the city center (Erz 1966, Marchand and Riquet 1996): pericentral sector, peripheral sector and periruban sector. In Rennes, these sectors were previously established by Guéguan-Roué (1994) (Fig. 4.2). In this paper, we refer to these areas surrounding the urban city center as adjoining suburban, distant suburban, and rural for consistency with others (Marzluff et al. 2001; Table 1.1).

In each of these four sectors, including the city center, we selected one or two sites of approximately 10 ha, sufficiently large to permit both ecological and sociological analyses. We specifically avoided areas with large parks or ponds so that the study sites reflected the urban environment rather than natural areas within the urban environment. To obtain an urban morphology (building age, building height, percent green space) and to keep historical and sociological features within each site relatively homogenous, we analyzed aerial photographs (National Geographic Institute source, Mennechez 1999) (Figure 4.1, Table 4.1) and previous studies on Rennes (Allain and Baudelle 1991, Allain 1992, Guéguan-Roué 1994).

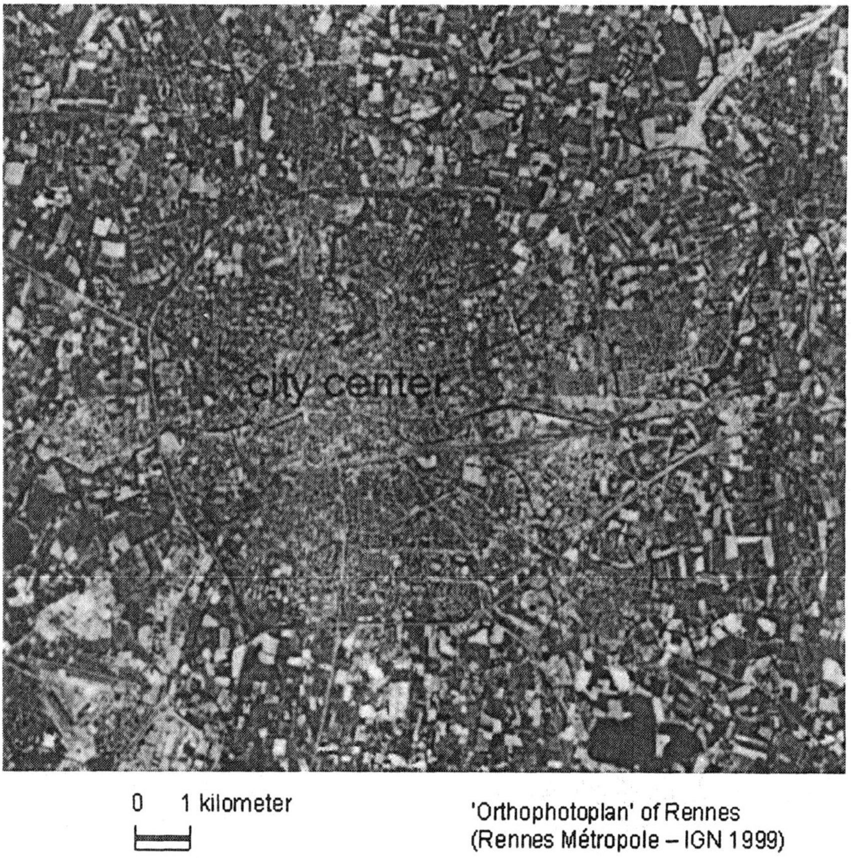

Figure 4.1. Aerial image of the urban city center and some of the surrounding sectors of Rennes, France.

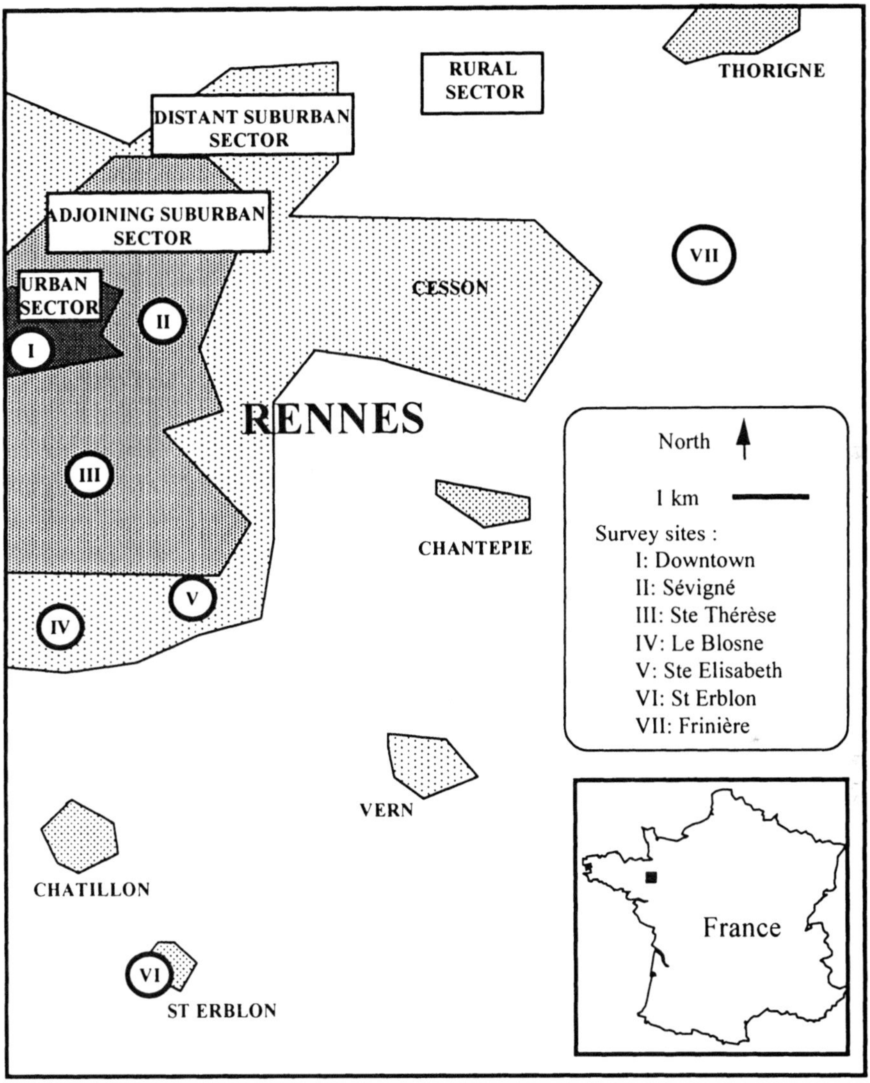

Figure 4.2. Simplified map of the study area showing the relative positions of the different urban sectors and each of the seven study sites in Rennes, France, 1999.

In the urban sector, we selected one site, Downtown (I), that included the historic center of the town with more or less continuous buildings. It was characterized by high building density and a very low proportion of green spaces, with hardly more than one percent represented by small grasslands and a few trees. Prosperous shopkeepers and managers lived in these old buildings.

In the adjoining suburban sector (pericentral), we selected two sites: Sévigné (II), that consisted of town houses built in the 19th and the 20th centuries with ornamental gardens hidden from the street occupied by relatively wealthy inhabitant and Sainte Thérèse (III), built during the same period but occupied by less wealthy human inhabitants and characterized by many small kitchen, rather than ornamental, gardens. In both sites, the human density was relatively similar.

A bit further from the city center, in the distant suburban sector (peripheral), we selected two sites: Le Blosne (IV), that consisted of large residential blocks completed in 1980 (large lawns were a very important element of this site), and Sainte Elisabeth (V), also a very recently developed site but comprised of detached houses built in the middle of small yards. In these two neighborhoods, the population was made up primarily of middle class workers.

Table 4.1. Characteristics of seven study areas in four sectors along a gradient of decreasing urbanization from the city center of Rennes, France. Percentages of landscape elements determined from digitized aerial images, 1999.

	% cover of buildings	% cover green space	% cover paved areas	# stories of buildings
Urban (city center)				
Downtown	51.0	1.1	47.9	3-5
Adjoining suburban (pericentral)				
Sévigné	24.0	35.2	40.8	1-2
Ste Thérèse	24.5	36.6	38.9	1-2
Distant suburban (peripheral)				
Ste Elisabeth	25.0	39.0	36.0	1-2
Le Blosne	15.9	53.1	44.7	4-17
Rural (periurban)				
St Erblon	15.0	47.0	38.0	1-2
Frinière	4.5	76.6	18.9	1-2

In the rural sector (periurban) surrounding the town up to several kilometers, we selected two sites: Saint Erblon (VI), a small town 10 km away from the city center in Rennes similar to Sainte Elisabeth but surrounded by agricultural land, and Frinière (VII), a few kilometers away from Rennes, comprised of traditional farms and houses, both renovated and recent. Site VI was characteristic of the building density in the southern part of the study area (few houses within an agricultural landscape), and site VII characterized the scattered buildings, farms and houses standing alone or in hamlets in the rural area.

We conducted avian censuses at each site twice during the winter period (February and March 1995) and four times during the spring period (late April, early May, early June, late June 1995). During each of these six counts we mapped the location of all birds seen or heard during a period from one and a half hours to two hours in duration (for mapping method see Svensson 1974). During each census, we walked different paths within a site and attempted to cover the entire site; walks took place in the morning when birds were more active. A least one and a half hours was spent in each site to insure minimal coverage, and up to 2 hr was spent in sites with high numbers of birds to ensure complete coverage of the site. Care was taken to avoid counting the same birds twice.

Our sociological approach consisted of two successive analyses performed in each of the four sectors, but in only one site per sector: sites I, III, IV and VI. First, we conducted 40 oral interviews (10 per site in April 1995), each lasting one and a half hours, and covering four subjects: the neighborhood, nature in general, nature in the city, and birds. These 'open' interviews permitted us to construct a written questionnaire. Second, a quantitative study (a sample survey) was performed to test general hypotheses generated during the interviews. Two hundred questionnaires (50 per site in May 1995, duration of half an hour) consisted of 51 questions on the perception (sense of 'I have seen or not'), observation (sense of 'I have watched or not') and appreciation (sense of 'I like or not') of birds, on the human behavior of feeding birds and on the place of birds in the city. Subjects were interviewed in their home. Subjects were selected randomly from homes in the center of each site, but we stratified our sample design to ensure interviewees represented the overall age and sex structure of the population. Results presented in this study derive only from the questionnaires.

We used the Shannon-Wiener index to measure bird diversity and included estimates of species evenness and richness (Magurran 1988) in our community comparisons. We compared answers of respondents from different sectors using Chi-square tests.

3. RESULTS

3.1 Habitat Features

The seven sites could be ranked along an urban gradient by the relative proportion that green spaces and human-made structures (buildings and roads) occupied. This gradient was not linear (Table 4.1). In the city center, only 1% of the areas was comprised of green spaces, but in the other sites, green areas comprised at least 35% (suburban Sévigné) and as much as 77% (rural Frinière).

3.2 Avian Surveys

Avian diversity increased from most to least urbanized sectors in both spring and winter (Table 4.2), although the trend was not linear. A significant correlation existed between the amount of green space and each bird diversity index (Spearman r always > 0.7; P always < 0.04) during spring censuses but not in winter. In most sites, diversity was similar between spring and winter, although species richness tended to be slightly lower during the winter. Evenness was also similar between seasons, except in the urban sector, where species evenness was lower in the winter than in the spring, suggesting the winter bird community was dominated by just a few species.

In spring and winter, the urban sector had the highest density of birds, whereas less urbanized sectors supported fewer birds, but again the trend was not linear. At all sites, bird densities were lower in winter than in spring (Table 4.2).

In the urban sector, we observed 21 species (Appendix 4.1), but of these only 12-14 were resident throughout the year. Rock Doves (*Columba livia*) and Herring Gulls (*Larus argentatus*) dominated the community in winter (78.3% of observed individuals). House Sparrows (*Passer domesticus*) and European Starlings also were common (16.3%) in winter, whereas other species were rarely observed. In spring, Swifts (*Apus apus*) that returned from their wintering area became common in the city center (27.2%). Most of the dominant species in this sector were cavity-nesting birds (pigeons, starlings, sparrows, swifts).

In the suburban sector, bird diversity was similar between sites and between winter and spring at both sites. In winter, European Starlings (Sévigné) or House Sparrows (Sainte Thérèse) were the most common species, but numerous garden species including Blackbirds (*Turdus merula*) and Robins (*Erithacus rubecula*) were observed. Rock Doves and Wood Pigeons (*Columba palumbus*) were always plentiful (more than 10%). In spring, Swifts became very common (more than 50% in Sainte Thérèse), but about 10 garden species were distributed over the two sites. Birds associated with shrubs (e.g. Wrens *Troglodytes troglodytes*) were abundant at both sites. In Sainte Elisabeth, House Sparrows were dominant (31.6% of observed individuals), but European Starlings, Blackbirds and Dunnocks (*Prunella modularis*) also were numerous. In Blosne, no species dominated in winter, but pigeons and several garden birds, such as Blackbirds, Greenfinchs (*Carduelis chloris*), Great and Blue Tits (*Parus major, P. caeruius*) were relatively common.

In the rural sector, bird diversity was similar between sites, although Frinière had slightly higher diversity in winter than did St. Erblon. Diversity in both sites was higher than in other sectors during both winter and spring. Species evenness also was relatively high suggesting that the avian

Table 4.2. Avian community characteristics in spring and winter at seven study sites in four sectors along a gradient of decreasing urbanization from the city center of Rennes, France.

	SPRING				WINTER			
	Shannon	Evenness	Richness	Bird density/ 10 ha	Shannon	Evenness	Richness	Bird density/ 10 ha
Urban								
Downtown	1.75	0.62	17	281.4	1.30	0.49	14	249.0
Adjoining suburban								
Sévigné	2.42	0.77	23	155.0	2.50	0.83	19	100.7
Sainte Thér.	2.20	0.72	21	213.5	2.45	0.80	20	113.2
Distant suburban								
Ste Elisabeth	2.42	0.77	23	182.7	2.30	0.74	21	163.3
Le Blosne	2.30	0.72	24	137.6	2.38	0.81	19	115.5
Rural								
St Erblon	2.62	0.80	26	124.6	2.46	0.83	18	105.3
Frinière	2.69	0.79	30	117.3	2.52	0.79	24	109.6

community was less dominated by the synanthropic species (see Johnson 2001) associated with the urban sector (pigeons, gulls, sparrows).

3.3 Human Perceptions of Birds

In the urban sector, 12% of people interviewed reported never seeing birds (Table 4.3). Many inhabitants of this sector did not intentionally observe birds (≅ 45%), thus their interactions with birds occurred by chance. Two-thirds of the interviewees did not perceive a rhythm or a periodicity to their encounters with birds; they noticed only the noise made by the birds. Most interviewees (85%) reported that they never attempted to read books about birds either for identification or to add to their own knowledge.

In the urban sector, most interviewees reported that they did not like gulls, but only 21% reported not liking European Starlings, which were the least appreciated species in other sectors (Table 4.4). Most urban dwellers considered birds in their sector to be "wild " (Table 4.3). Few respondents reported regularly feeding birds (9.1%), the lowest among all sectors. Complaints about birds were relatively common among residents of this sector, with 47% complaining about the dirt produced by birds and 25% about noise. However, it appeared that urban residents had ambivalent feelings about birds; 67% reported that birds were a source of personal pleasure, similar to the rate of this response in other sectors. Few inhabitants of the urban sector understood the rationales for the control of bird populations; 64% reported that killing birds was wrong, a much higher percentage than in other sectors (Table 4.3).

In the suburbs adjoining the urban area, few respondents reported never having seen birds and about 50% reported having seen many birds (Table 4.3). A large proportion (43%) intentionally observed birds and noticed some periodicity (85%) to their observations. About 50% consulted books to learn more about birds. Fewer residents referred to birds as "wild" and more considered birds to be familiar. Starlings were the least appreciated species, followed by gulls and pigeons. Complaints about birds were relatively common, especially dirt and damage associated with birds, but respondents reported regularly feeding birds (36%) and, as in the urban sector, most respondents perceived birds as a source of personal pleasure (> 70%), indicating some ambivalence in attitudes towards birds.

In the more distant suburbs, many respondents (98%) reported having seen birds, even if the inhabitants observed them only occasionally (Table 4.3). Over 80% of respondents reported occasionally or regularly observing

Table 4.3. Distribution of responses to questions on human perceptions of birds relative to the urban sector of the respondents.

Sectors	Urban	Adjoining suburbs	Distant suburbs	Rural	χ^2 test
Have you seen birds in your neighborhood?					
Many	51.5%	49.5	59.1	53.6	
Few	36.4%	49.0%	38.7%	46.4%	$\chi^2 = 12.8$
No	12.1%	1.5%	2.2%	0%	$P = 0.047$
Do you observe these birds?					
Regularly	15.1%	42.6%	36.1%	50.4%	
Occasionally	40.4%	38.7%	56.8%	32.8%	$\chi^2 = 29.0$
Never	45.1%	18.5%	7.1%	16.8	$P < 0.001$
Can you recognize individual species present in your environment?					
Yes	13.8%	47.2%	64.2%	46.2%	$\chi^2 = 53.4$
No	86.2%	52.8%	35.8%	53.6%	$P < 0.001$
Do you try to identify these birds, for example with books?					
Yes	15.1%	50.2%	52.5%	40.0%	$\chi^2 = 36.4$
Never	84.9%	49.8%	47.5%	60.0%	$P < 0.001$
Have you ever seen birds at particular times of the day or year?					
Yes	31.1%	84.6%	81.2%	73.7%	$\chi^2 = 40.9$
No	68.9%	15.4%	18.8%	26.3%	$P < 0.001$
In general, would you say that most birds in your environment are					
Domestic	6.1%	1.6%	5.2%	0%	
Tame	60.6%	41.0%	33.7%	64.0%	
Wild	27.2%	52.9%	53.5%	33.6%	$\chi^2 = 19.5$
Don't know	6.1%	4.5%	7.6%	2.4%	$P = 0.021$
Do your feed birds?					
Regularly	9.1%	35.8%	43.7%	44.7%	
Occassionally	48.5%	43.7%	26.6%	32.8%	$\chi^2 = 20.6$
Never	42.4%	20.5%	29.7%	22.5%	$P = 0.002$
For yourself, birds are					
A source of pleasure	66.7%	71.6%	60.9%	75.3%	
A source of disturb.	0%	0%	0%	0%	w/o pt. 2
Sometimes 1 or 2	21.2%	22.1%	30.2%	16.7%	$\chi^2 = 4.5$
Neither 1 or 2	12.1%	6.3%	8.9%	8.0%	$P > 0.05$
What kind of disturbance do birds create?					
Dirt	46.9%	34.1%	35.5%	29.2%	
Noise	25.0%	11.8%	10.9%	11.6%	
Damage	18.8%	18.5%	22.6%	27.7%	
Health	3.1%	7.8%	8.9%	7.6%	$\chi^2 = 30.9$
Others	6.2%	18.7%	22.1%	23.8%	$P = 0.020$
What do you think about lethal control of bird populations?					
It is normal	0%	1.5%	0%	0%	
I can understand	24.2%	71.8%	58.5%	78.4%	w/o pt. 1
I disagree	63.6%	23.6%	37.9%	19.2%	$\chi^2 = 76.1$
I don't know	12.2%	3.1%	3.6%	2.4%	$P < 0.001$

birds at particular times of the day. Many respondents (53%) searched for additional information and consulted books about birds. Similar to the adjoining suburbs, 54% of respondents referred to birds as familiar and 44% regularly fed birds. Starlings, gulls, and pigeons were the least appreciated birds. Complaints about birds were as common in this sector as in the more urban sectors and the complaints were similar, with those regarding dirt and damage the most frequent. Again, some ambivalence towards birds was expressed by respondents; although 61% stated that birds were a source of personal pleasure, 30% said that sometimes birds could also be a source of disturbance.

In the rural sector, virtually all respondents (100%) reported having seen birds, although the percentage reporting seeing many birds was similar to that in the suburbs, but only 74% noticed some periodicity to their observations. About 40% consulted books to learn more about birds and 45% reported regularly feeding birds, but only 34% of respondents referred to birds as familiar. By a relatively large margin, European Starlings were the least appreciated species. Complaints about birds were less frequent in this sector than in others and a large percentage (75%) reported birds as a source of personal pleasure. Many respondents believed they could recognize birds individually (47%) and they could name over twenty individual species living in their neighborhood. They based their observations on numerous characteristics (differences in morphology, rhythms, vocalizations, etc) and considered bird behaviors to distinguish species. Some respondents (21%) thought that birds were too numerous, yet, 75% of respondents understood the need for population control.

Table 4.4. Variation in the appreciation of the four most common species in Rennes according to sectors. Only the rates of negative appreciation are reported here.

Species	Urban	Adjoining suburbs	Distant suburbs	Rural
Negative appreciation				
European Starling	21.2%	71.7%	47.0%	52.8%
Herring Gull	48.5%	29.2%	44.4%	28.1%
House Sparrow	9.1%	12.6%	11.6%	12.8%
Rock Dove	9.1%	28.2%	38.4%	15.3%

3.4 Human Perception of Individual Species

The identification of the species most common to each sector was relatively accurate for the three most dominant species in each sector. Regardless of their abundance, House Sparrows, Blackbirds and Rock Doves also were widely recognized by respondents (> 80%). Birds that were reported as frequently being observed, were also most likely to be correctly identified (House Sparrows, Blackbirds, Rock Doves, Robins and Great Tits). In addition, large birds, whether common or not, such as crows and gulls, were likely to be correctly identified.

Respondents could identify 11 bird species in the urban, 19 in the adjoining suburbs, 11 in the distant suburbs and 23 in the rural sector (obtained from our first 40 oral interviews). We observed that the number of different species recognized by inhabitants was closest to the actual bird diversity when the sector was characterized by numerous private gardens: Ste-Thérèse (79% of identified species) and St-Erblon (85%) versus city center (52%) and Blosne (46%).

4. DISCUSSION

Our results support previously described patterns in avian diversity along an urban gradient (Savard 1978, Luniak 1980, Gilbert 1989, Blair 1996); the more an urban site is composed of heterogeneous greenspaces, the more diverse the bird community. In the suburbs of Rennes, the amount of greenspace did not vary greatly (35%-55%), whereas in other towns with a similar degree of urbanization, the amount of greenspace varies from 20 to 70% (Clergeau et al. 1998). Consequently, the bird diversity in these sectors of Rennes did not vary greatly. Still species richness appeared to decline with increasing urbanization. In part, this might be explained by the relative closeness of potential source areas of native birds; the distance between the city center and rural areas was less than five km. However, previous comparative work in Québec City (Canada) showed that the size and quality of urban habitats were more important in determining bird species diversity that the distance to source areas (Clergeau et al. 1998). In addition, the relatively stable, temperate climate in Rennes explained the small differences in bird communities we observed between winter and spring in these sectors.

The perception of birds by inhabitants appeared to be less related to the abundance of birds than to their diversity. In the urban sector, where bird abundance is high but diversity is low, residents were the most likely to never perceive birds. In the suburbs, with fewer total birds, but much great avian diversity, virtually all residents perceive birds. These patterns, like many of the other patterns we described, were not linear. Avian diversity declined dramatically in the urban sector, as did the perception of birds.

The sociological differences between sectors exist not only in the perception of birds, but also in their appreciation. In the urban sector, responses to wildlife differed greatly from the other sectors. Fewer residents were tuned into the annual cycles of birds, fewer fed birds, and fewer ever read about birds. In the suburbs, which shared similar bird communities and urban morphologies, respondents showed a similar appreciation of birds. For instance, residents classified species (tame, wild...) similarly, fed birds similarly, and a similar proportion of residents used books as a source of information about birds. However, in other ways, human attitudes in the suburbs were more similar to the rural areas. For example, a similar percentage of people viewed birds as a source of personal pleasure and respondents ranked individual species similarly in their dislike for them. In both of these sectors, many residents have kitchen or ornamental gardens, more so than in areas closer to urban centers.

Birds are sometimes perceived by urban residents as useful, as epitomized by the remark "instead of throwing crumbs away, we give them to the birds". However, the boundary between tolerance and intolerance of birds is a threshold that might be easily overstepped. This is evident in the ambivalence of some responses, such as the large percentage of complaints about birds, but also the large percentage of reports of birds as a source of pleasure. In addition, the appreciation of different species is likely context specific. In many of the urban surveys, gulls were little appreciated by respondents. However, in coastal towns, where gulls are part of the expected scenery, the presence of gulls might be better appreciated. Previous sample surveys conducted in French coastal towns, Le Havre and Cherbourg, showed that breeding gulls are a nuisance for half the population and a pleasure for a third (Ville du Havre 1993).

Although some species (e.g.; starlings or gulls) are seldom appreciated and some practices (e.g.; feeding birds) not common in certain urban sectors, in general, people appear to appreciate the urban avifauna in Rennes: 74% of the interviewees thought that the presence of birds in town is a good thing, 69% found pleasure in this, and 72% have fed birds before.

4.1 Policy and Management Implications

Interactions with birds can affect human values and the quality of life of individuals (Vandruff et al.1995), and human behavior and actions certainly have an effect on avian community structure (Marzluff 2001). Feeding birds like pigeons or gulls is very likely one of the major causes for their proliferation in cities. Municipalities may continue to limit bird populations by direct means, such as trapping, but some work suggests that directing efforts at altering human behaviors also could be effective (Haag 1993, Clergeau 1997). Our results suggest that the relationships between humans and birds vary even within a city and that successful and accepted management of urban avifauna, such as ridding cities of roosting starlings, might require an understanding of the range of perceptions among the human residents. Communication of management actions might require a different approach depending on the urban sector of the public, especially for those residents in the city center.

Our study suggests that the quality of life of human urban residents is one among many justifications for the protection and restoration of urban avifauna in France, because people perceive and appreciate birds as part of their urban environment. Indirect benefits also may accrue to humans, such as increased land values or decreased property taxes (Haskell et al. 2001). These results corroborate previous work in the United States, that showed that people who live in cities appreciate the chance to observe birds (Brown et al. 1979, Witter et al. 1981, Kellert 1984, Conover 1997). In various cultural contexts, there seems to be a clear and preponderant favor of birds in cities. Middleton (1994) suggests than exposure to biodiversity in town can influence decisions relating to biodiversity at national level. Whether this is true or not, education is a key to maintaining diversity in urban wildlife and to sustain urban development.

ACKNOWLEDGEMENTS

We thank F. Dubs, S. Chevrier and J-P Marchand for their collaboration during the study, C. Williamson who improved the English, and R. Bowman who carefully edited the manuscript. This analysis was supported by grant from Centre National de la Recherche Scientifique (P.I.R.Villes Paris).

REFERENCES

Adams, L.W. 1994. Urban wildlife habitats, a landscape perspective. Univ. of Minnesota Press, MN.

Alberti, M., E. Botsford, and A. Cohen. 2001. Quantifying the urban gradient: linking urban planning and ecology. *In* J. M. Marzluff, R. Bowman, and R. Donnelly [EDS.]. Avian ecology and conservation in an urbanizing world. Kluwer Academic, Norwell, MA.

Allain, R. 1992. La maison et la ville en Bretagne. Ph.D. Thesis of Univ. Rennes2, Rennes, France.

Allain, R., and G. Baudelle. 1991. La structure résidentielle de Rennes. Géographie sociale 11:431-444.

Blair, R. B. 1996. Land use and avian species diversity along an urban gradient. Ecol. Appl. 6: 506-519.

Brown, T., D. Dawson, and R. Miller. 1979. Interests and attitudes of metropolitan New York residents about wildlife. Trans. North Am. Wildl. Nat. Resour. Conf. 44:289-297.

Clergeau, P. 1995. De nouveaux oiseaux dans la ville? Colonisation des zones urbaines par les étourneaux. Report Ministère de l'Environnement/Ministère du Logement, Paris, France.

Clergeau, P. 1996. Urban biodiversity: is there such a thing? Le courrier du CNRS, special issue Cities- Habitat II Istanbul, CNRS ed., Paris.

Clergeau, P. 1997. Oiseaux à risques en ville et en campagne; vers une gestion intégrée des populations. INRA ed., Paris, France.

Clergeau, P., D. Esterlingot, J. Chaperon, and C. Lerat. 1996. Difficultés de cohabitation entre l'homme et l'animal: le cas des concentrations d'oiseaux en site urbain. Natures, Sciences, Sociétés 4:102-115.

Clergeau, P., J. P. L. Savard, G. Mennechez, and G. Fallardeau. 1998. Bird abundance and diversity along an urban-rural gradient: a comparative study between two cities on different continents. Condor 100:413-425.

Congress of Leicester. 1995. Recent advances in urban and post-industrial wildlife conservation and habitat creation. Land Contamination and Reclamation 3:59-153.

Congress of Starsbourg. 1994. Vers une gestion raisonnée des espaces verts. Le Centre National de la Fonction Publique ed., Paris.

Conover, M. R. 1997. Wildlife management by metropolitan residents in the United States: practices, perceptions, costs and values. Wildl Soc. Bull. 25:306-311.

Davis, A. M., and T. F. Glick. 1978 Urban ecosystems and island biogeography. Environ. Conserv. 5:299-304.

Dagg, A. I. 1970. Wildlife in an urban area. Nature Can. 97: 201-212.

Erz, W. 1966. Ecological principles in the urbanization of birds. Ostrich Suppl. 6: 357-363.

Fabos, J. G., and J. Ahern. 1996. Greenways, the beginning of an international movement. Elsevier, Amsterdam.

Furness, R. W., and J. J. D. Greenwood. 1993. Birds as monitors of environmental change. Chapman and Hall, London.

Gilbert, O. 1989. The ecology of urban habitats. Chapman and Hall, London.

Guégan-Roué, A. 1994. La dynamique spatiale de l'agglomération rennaise: la télédétection comme outil d'analyse et de gestion de l'espace. Ph.D. Thesis of Univ. Rennes2, Rennes.

Haag, D. 1993. Street pigeons in Basel. Nature 361:200.

Haskell, D. G., A. M. Knupp, and M. C. Schneider. 2001. Nest predator abundance and urbanization, p. 245-260. *In* J. M. Marzluff, R. Bowman, and R. Donnelly [EDS], Avian ecology and conservation in and Urbanizing World. Kluwer Academic Press, Norwell, MA.

Hunt, J. 1994. The social construction of precaution, p. 117-125. *In* O'Riordan, T., and J. Cameron [EDS.], Interpreting the precaution principle. Earthscan, London.

Ilyichev V. D., V. M Konstantinov, and B. M. Zvonov. 1990. The urbanized landscape as an arena for mutual relations between man and birds, p. 122-130. *In* M. Luniak [ED.], Proceedings of international symposium on urban ecological studies in central and eastern Europe. Pol. Acad. Sci., Warszawa.

Johnston, R. F. 2001. Synanthropic birds of North America, p. 49-67. *In* J. M. Marzluff, R. Bowman, and R. Donnelly [EDS], Avian ecology and conservation in and Urbanizing World. Kluwer Academic Press, Norwell, MA.

Kellert S. R. 1984. Urban American perception of animals and the natural environment. Urban Ecol. 8:209-228.

Luniak, M. 1980. Birds of allotment gardens in Warsaw. Acta Ornithologica 17: 297-320.

McDonnell, M. J., and S. T. A. Pickett. 1990. Ecosystem structure and function along urban-rural gradients: an unexploited opportunity for ecology. Ecology 71:1232-1237.

Magurran, A. E. 1988. Ecological diversity and its measurement. Princeton Univ. Press, Princeton, NJ.

Marchand, J-P., and P. Riquet. 1996. North Europe, Median Europe. Tome IX, Geographie Universelle. Belin-Reclus, Montpellier.

Marzluff, J. M. 2001. Worldwide increase in urbanization and its effects on birds, p. 19-47. *In* J. M. Marzluff, R. Bowman, and R. Donnelly [EDS], Avian ecology and conservation in and Urbanizing World. Kluwer Academic Press, Norwell, MA.

Marzluff, J. M., K. J. McGowan, R. Donnelly, and R. L. Knight. 2001. Causes and consequences of expanding American Crow populations, p. 333-365. *In* J. M. Marzluff, R. Bowman, and R. Donnelly [EDS], Avian ecology and conservation in and Urbanizing World. Kluwer Academic Press, Norwell, MA.

Mennechez, G. 1999. Urbanisation et espèces généralistes : Biologie comparée de l'étourneau *Sturnus vulgaris* lors de la reproduction sur un gradient rural-urbain. Ph.D. Thesis of Univ. Rennes1, Rennes.

Middleton, J. 1994. Effects of urbanization on biodiversity in Canada, p. 127-132. *In* Biodiversity in Canada: a science assessment for Environment Canada, Environment Canada, Ottawa.

Saunders, D. A., and R. J. Hobbs. 1991. Nature conservation: the role of corridors. Surrey Beatty and sons, Australia.

Sauvage, A. 1992. Les habitants, de nouveaux acteurs sociaux. Editions l'Harmattan, Paris.

Savard, J.-P. L. 1978. Birds in metropolitan Toronto: distribution, relationships with habitat features, and nesting sites. Ph.D. Thesis of Univ. of Toronto, Toronto.

Sénécal, G. 1996. Champs urbains et développement durable: les approches canadiennes de la ville écologique. Natures, Sciences, Sociétés 4:61-74.

Shaar, M. 1979. Attitudes and behaviors of residents in the Québec City region towards urban wildlife. Rep. of Can. Wildl. Serv., Québec.

Svensson, S. E. 1974. Interpersonal variation in species map evaluation in bird census work with the mapping method. Acta Ornithol. 14:322-338.

Terrasson, F. 1991. La peur de la Nature. Le Sang de la Terre, Paris.

Vandruff, L. W., D. L. Leedy, and F. W. Stearns. 1995. Urban wildlife and human well-being, p. 203-211. *In* H. Sukkopp, M. Numata and A. Huber [EDS.], Urban ecology as the basis of urban planning. SPB Academic Publ., The Hague.

Ville du Havre. 1993. Colonies de Goélands en zones urbaines. Report of scientific and technical workshop, Le Havre.

Witter, D. J., D. L. Tylka, and J. E. Wermer. 1981. Values of urban wildlife in Missouri. Trans. North Am. Wildl. Nat. Resour. Conf. 46 424-431.

Appendix 4.1. Density of bird species observed in winter and spring in Rennes. Means are calculated on individual numbers per 10 ha (eight counts).

		URBAN	ADJOINING SUBURBS		DISTANT SUBURBS		RURAL	
		Downtown	Sévigné	StThér.	StElis.	Blosne	StErblon	Frinière
Rock Dove	*Columba livia*	123.46	9.16	13.0	6.97	18.33	2.28	3.10
Herring Gull	*Larus argentatus*	38.46	2.51	5.36	0.83	1.86	0.44	0.16
Swift	*Apus apus*	38.11	22.46	40.58	2.50	1.04	4.30	2.04
House Sparrow	*Passer domesticus*	24.69	24.37	32.04	49.40	30.55	20.20	27.94
Eur. Starling	*Sturnus vulgaris*	13.66	9.35	8.43	20.80	8.39	6.94	8.58
Blackbird	*Turdus merula*	4.99	10.70	14.38	21.42	21.96	11.68	9.39
Carrion Crow	*Corvus corone*	2.96	3.57	3.47	1.25	1.55	1.14	1.47
Collared Dove	*Streptopelia dec.*	2.96	2.79	4.27	0.94		1.05	0.74
Black Redstart	*Phoenicurus och.*	2.61	0.10					
Greenfinch	*Carduelis chloris*	1.31	1.64	6.85	15.91	9.32	4.04	4.90
House Martin	*Delichon urbica*	1.31						
Great Tit	*Parus major*	0.95	5.49	6.45	8.53	8.70	8.70	7.76
Wren	*Troglodytes trog.*	0.71	11.37	8.63	6.24	1.45	9.31	7.27
Goldfinch	*Carduelis card.*	0.59	0.29	1.19	5.0	2.59	6.85	1.23
Robin	*Erithacus rubec.*	0.47	5.11	1.39	4.16	3.42	10.45	10.78
White Wagtail	*Motacilla alba*	0.47			0.42	0.62		0.16
Magpie	*Pica pica*	0.24	0.58	2.18	3.54	2.80	0.44	1.06
Dunnock	*Prunella modularis*	0.24	2.89	2.28	12.58	2.17	3.16	4.58
Serin	*Serinus serinus*	0.24	1.54	0.20	0.83	0.93	1.58	
Grey Heron	*Ardea cinerea*	0.24		0.10	0.31	0.52	0.09	0.25
Blue Tit	*Parus caerulus*	0.12	4.24	2.28	3.12	3.11	4.30	4.00

		URBAN	ADJOINING SUBURBS		DISTANT SUBURBS		RURAL	
		Downtown	Sévigné	StThér.	StElis.	Blosne	StErblon	Frinière
Chaffinch	*Fringilla coelebs*		2.79	6.05	5.41	8.08	10.89	12.42
Song Thrush	*Turdus phillomelos*		1.93	1.19	0.52		0.09	0.33
Chiffchaff	*Phylloscopus coll.*		1.73	1.49	1.04	0.93	9.72	2.37
Blackcap	*Sylvia atricapilla*		0.96	0.99		0.52	0.09	0.57
S-T Tree Creeper	*Certhia brachidactila*		0.58				0.09	0.98
Nuthatch	*Sitta europaea*		0.48					0.08
Goldcrest	*Regulus regulus*		0.19	0.20	0.21		0.18	
Kestrel	*Falco tinnunculus*			0.20			0.35	0.16
Linnet	*Carduelis cannabina*				0.42			0.33
Green Woodpeck.	*Picus viridis*				0.21			
Garden Warbler	*Sylvia borin*				0.10		0.09	
Mistle Trush	*Turdus viscivorus*				0.10			0.08
S. Flycatcher	*Muscicapa striata*					0.24		
Marsh Tit	*Parus palustris*					0.21		
Jay	*Garrulus glandarius*					0.09		0.65
Mallard	*Anas platyrhynchos*							0.57
G.S.Woodpeck.	*Dendrocopus major*							0.16
Swallow	*Hirundo rustica*						0.18	2.78
Buzzard	*Buteo buteo*							0.25
Yellowhammer	*Embreriza citrinella*							0.08
Melod. Warbler	*Hippolais polyglotta*							0.08
Total density		258.31	127.31	162.50	171.69	129.56	111.65	113.89
Species number		21	25	24	27	24	27	34

Chapter 5

Quantifying the urban gradient: Linking urban planning and ecology

Marina Alberti, Erik Botsford, and Alex Cohen
Department of Urban Design and Planning, University of Washington, Box 355740, Seattle, WA 98195 USA, e-mail: malberti@u.washington.edu

Key words: Contagion, gradient analysis, heterogeneity, landscape ecology, pattern metrics, urban ecology, urban patterns

Abstract: Ecologists have suggested that ecological conditions in urbanizing landscapes can be described by a complex urban-to-rural gradient (McDonnell and Pickett 1990). The gradient paradigm offers a useful framework to test hypotheses on the impacts of urban development on ecological processes. These studies, however, tend to simplify the actual urban structure into monocentric agglomerations characterized by concentric rings of development surrounding a dense core. The assumption of gradient analysis is that the overall urban exposure changes predictably with distance from the urban core. Due to such simplification, current gradient studies fail to capture the effects of alternative urban development patterns on ecological processes. In this paper we argue that urban-to-rural gradients cannot be represented by the distance from the urban core. Rather they can best be described using a series of pattern metrics that link urban development to ecological conditions. Based on an analysis of land-use and land-cover patterns in the Seattle metropolitan area we propose a strategy to quantify urban patterns. We examine the behavior of various pattern metrics and propose a set of metrics useful to test formal hypotheses on the relationships between urban patterns and ecological disturbances. Finally, we discuss the implications of this empirical study for gradient analysis of metropolitan areas and for future urban ecological research.

1. INTRODUCTION

Urban development poses major challenges to the earth's ecology. It alters natural habitats and species composition, disrupts hydrological systems, and modifies energy flow and nutrient cycles. Environmental changes associated with urbanization have been significant during the last half century and are expected to continue through the next decades. In the United States alone the urban population nearly doubled between 1950 and 1990, increasing from 96 million to 187 million (US Census 1995). Of this increase more than 70% occurred in the suburbs, increasing the pressure on the urban fringe (Platt 1996). The National Resources Conservation Service (1999) has documented that in the five years between 1992 and 1997 more land was developed (nearly 16 million acres) than between 1982 and 1992 (more than 13 million acres).

Parallel to this trend, ecologists are paying increasing attention to urban areas (McDonnell and Pickett 1993). The study of urban impacts on ecosystems is seen as an opportunity to expand ecological knowledge and address critical contemporary environmental problems (McDonnell and Pickett 1990). To study ecological systems in urbanizing landscapes McDonnell et al. (1993) have proposed that ecological conditions can be described by a complex urban-to-rural gradient. Using a gradient approach, they compare the spatial heterogeneity of the urban landscape (with sharp boundaries) to that of more gradual ecotones in environments with minimal human activities. Recent empirical studies describe urban-to-rural gradients in forest structure (Medley et al. 1995, McDonnell et al. 1997), bird diversity (Blair 1999, Bolger et al. 1996, Clergeau 1998), and watershed integrity (Harding et al. 1998, Lammert et al. 1999, Stevens and Cummins 1999).

Gradient analysis offers a useful approach to test hypotheses on the impact of urban development on ecological processes (McDonnell and Pickett 1990). Urban landscapes typically exhibit a gradient of land-uses which results in extraordinary amounts of environmental heterogeneity produced by land conversion and pollution loads. Current studies, however, tend to simplify the urban structure by using aggregated measures of urbanization and assuming a monocentric model which rarely occurs in present urban landscapes. The assumption of gradient analysis is that exposure to urban disturbance changes predictably with distance from the urban core. This apparent oversimplification causes gradient analysis to fail to discriminate among the impacts that alternative patterns of urban development have on ecosystems.

Urban ecological studies need to capture the disturbances associated with alternative patterns of development if they are to be relevant to the future of cities and help answer critical questions about strategies for achieving more sustainable urban forms. One of the tenets in support of compact urban development rather than sprawl is that compact forms (concentrated activity centers where a mixture of complementary land-uses exist in close proximity at relatively high densities) are more sustainable than others (Duany and

Plater-Zyberk 1991, Calthorpe 1993). But the empirical evidence on the relationships between urban form and ecological conditions is still too limited to support such a claim. While ecologists are paying increasing attention to urban areas, we still lack solid empirical studies that help discriminate alternative urban development patterns on an ecological basis.

Here, we argue that to understand urban ecological gradients we need to explicitly represent urban patterns in ecological studies. Urban-rural gradients cannot be described using a simple geographical transect from the urban core to exurban areas. Rather, we suggest that they can best be described using a series of urban pattern metrics: some describe spatially-aggregated variables (e.g., percent land by land-cover or land-use type), while others describe spatial distributions of land-use or land-cover types (e.g., contagion and fragmentation). We analyze land-use and land-cover patterns in the Seattle, WA metropolitan area to show that distance from the urban core alone is not a good predictor of ecological disturbance. We then propose pattern metrics that can be used to test formal hypotheses of the relationships between urban development and ecological conditions. Finally, we discuss the implications of our empirical study for gradient analysis of metropolitan areas and for future urban ecological research.

1.1 The Gradient Paradigm

The gradient paradigm provides a useful framework for developing and testing hypotheses on the interactions between urban development and ecological processes. Not only can gradient analysis explain ecological conditions associated with human disturbances in urbanizing environments, it also can reveal thresholds in a biotic and ecological responses to human-induced stresses. The interactions among human factors and between them and the biophysical factors make urban-to-rural gradients potentially quite complex and produce a new array of virtually unexplored conditions. McDonnell et al. (1997) use indirect multivariate gradient analysis to characterize the complex urban-rural gradient that exists in the New York metropolitan area. They show that urban forests, for example, exhibit unique ecosystem structures and functions that can be associated with stresses of urbanizing regions such as air pollution, elevated levels of heavy metals in the soil, the positive effects of the heat island phenomenon, and the presence of earthworms.

Ecological research, however, has tended to simplify the actual spatial pattern of disturbance in urbanizing landscapes to a few aggregated variables such as population density or built-up density which are expected to change predictably with distance from the urban core. Cities are described as monocentric agglomerations with concentric rings of development surrounding a dense central business district (CBD). The assumption for current applications of gradient analysis is that the overall urban-induced disturbances decline with distance from the urban core. However most US metropolitan areas over the last few decades have changed from a monocentered to a polycentric structure (Gordon et al. 1986, Giuliano and Small 1991, Song 1992, Cervero and Wu 1995). Not only is exposure to urban distance not necessarily a function of distance from the core; in addition the interactions between human and biophysical processes also vary with patterns of urbanization.

This simplification makes it difficult to disentangle the factors associated with urbanization that generate ecological heterogeneity. Medley et al. (1995) describe quadratic relationships between population density, land-use heterogeneity, and highway connectivity, and the urban-rural distance. They identify distinct characteristics of urban and suburban environments based on the landscape structure of native forest vegetation. They conclude that while urban exposure parallels the transect, the disturbances associated with urbanization show a complex spatial pattern not clearly related to distance from the urban core alone (Medley et al. 1995).

1.2 Complex Landscape Patterns

Urbanization results in a complex pattern of intermixed high- and low-density built-up areas. Alternative urban development patterns affect ecological processes both directly—by replacing native habitat with simplified, human-dominated systems—and indirectly—by rearranging the biophysical attributes that cause a variety of interrelated local and global effects (Godron and Forman 1983). Recent ecological studies have demonstrated that the ecological conditions of any patch are related to patch structure (Turner 1989), the size, composition, persistence, and interconnectivity of the patch, which in turn is important to species survival. Since ecological conditions are linked to patch structure and function, landscape changes produced by land conversion and pollution loads associated with patterns of urbanization have differential impacts on ecological conditions.

If gradient analyses are to provide a useful organizing framework to investigate variations in ecological conditions related to variations in exposure to urban disturbance, we will need to explicitly represent the complex urban landscape patterns. Current applications of gradient analysis disregard the elements and spatial configuration of disturbances associated with urbanization that drive ecological conditions. We wish to present a challenge to the current gradient paradigm. Is it the population density or the intensity of their activities and their spatial distribution that drive ecological conditions?

2. MEASURING URBAN ECOLOGICAL GRADIENTS

To understand relationships between urban patterns and ecological conditions, we propose a set of pattern metrics that we believe are linked to ecological processes in urbanizing landscapes. Urban development affects the composition and configuration of habitat patches (Marzluff 2001). Several ecological studies have demonstrated that the ecological conditions of any patch are related to ecological patterns at the landscape scale (Turner 1989, McDonnell et al. 1997). Researchers in landscape ecology have developed a large number of possible metrics for quantifying such patterns and their effects on disturbance regimes (O'Neill et al. 1988, Turner 1990, Gustafson and Parker 1992, Li and Reynolds 1993, McGarigal and Marks 1995, Gustafson 1999; Table 5.1). Each metric addresses some aspect of the structure, function, or change which can be observed in a landscape, and is useful for understanding ecosystem processes when a functional link can be demonstrated or hypothesized (Turner 1991).

From an ecological perspective, urban development produces a variety of unprecedented and intense disturbances through physical changes in the landscape. Landscapes are mosaics of patches. Patches are relatively discrete areas of relatively homogeneous environmental conditions that are perceived by or relevant to the organism or ecological phenomenon under consideration (McGarigal and Marks 1995). The definition of a patch should consider the scales at which variation is perceived and at which heterogeneity is affected by process. Metrics of landscape patterns aim to measure two major characteristics of the landscape: its *composition* and its *spatial configuration* (Turner 1989). Landscape composition refers to the presence and amount of different patch types within the landscape, without explicitly describing its spatial features (i.e., % land of a certain cover). Landscape configuration refers to the spatial distribution of patches within the landscape (i.e., degree of contagion).

In landscape ecology these metrics are good predictors of the ecosystem's ability to support important ecosystem functions (Turner and Gardner 1991). Ecological studies have shown, for example, that patch size is positively correlated to species and habitat diversity. Edge-to-interior ratio and nearest-neighbor probabilities reflect the degree of landscape fragmentation. Fractal dimension reflects the extent of human impacts. Contagion is an important measure of contiguous habitat types (O'Neill et al. 1988, Turner and Gardner

Table 5.1 Landscape metrics.

Metric	Equation
Pland Sum of the area of all patches of the corresponding patch type divided by total landscape area.	$pLand = \frac{\sum_{j=1}^{n} aij}{A}$
Patch density Number of patches in the landscape divided by total landscape area.	$PD = \frac{N}{A}$
Mean patch size Sum of the areas of all patches divided by the number of patches.	$MPS = \frac{\sum_{j=1}^{n} aij}{ni}$
Shannon diversity Minus the sum, across all patch types, of the proportional abundance of each patch type multiplied by that proportion	$SHDI = -\sum_{i=1}^{m}(P_i - \ln P_i)$
Dominance Deviation from the maximum possible landscape diversity where m is the number of land cover types, P_l is the proportion of the landscape in land cover i and $H_{max} = \ln(m)$ is the maximum diversity when all land cover classes are present in equal proportion.	$D = \frac{H_{\max} + \sum_{i=1}^{m}(P_i)\ln(P_i)}{H_{\max}}$
Contagion Probability of two cells of type i and j to be adjacent where m is the number of land cover types, P_{ij} is the proportion of cells in land cover i adjacent to cells of type j and 2 ln(m) is the maximum when all possible adjacencies of class i and j occur with equal probability.	$C = \frac{2\ln(m) + \sum_{i=1}^{m}\sum_{j=1}^{m} P_{ij} \ln P_{ij}}{2\ln(m)}$
Interspersion Minus the sum of the length (m) of each unique edge type of corresponding patch type divided by the total length (m) of edge (m) of the same type, multiplied by the logarithm of the same quantity, over each unique edge type; divided by the logarithm of the number of patch types minus 1.	$IJI = \frac{-\sum_{i=1}^{m}\sum_{k-i+1}^{m}\left[\left(\frac{e_{ik}}{E}\right) - \ln\left(\frac{e_{ik}}{E}\right)\right]}{\ln(\frac{1}{2}[m(m-1)])} \cdot 100$
Fractal dimension Double logarithm of patch perimeter divided by the logarithm of patch area	$FRACT = \frac{2\ln p_{ij}}{\ln a_{ij}}$

Sources: O'Neill et al. 1988, Turner 1990, Gustafson and Parker 1992, Li and Reynolds 1993, McGarigal and Marks 1995, Gustafson 1999.

1991). These metrics allow us to measure the pattern of ecological disturbances in the natural and built environments.

2.1 Composition

Percent Land of a certain class can be used to measure both land-use and land-cover composition. The percent of land-cover occupied by each patch type (i.e., paved land, forest, or grass) is considered an important indicator of ecological conditions since some ecological properties of a patch can be influenced by the composition of the patches and abundance of similar patches within the landscape. The percent of land-use occupied by each land-use type (i.e., single family residential, commercial, or industrial) is an important indicator of the land-use structure and can indicate the dominance of a certain use. Two other important metrics of landscape composition are the *Shannon Diversity Index and Dominance*. The Shannon diversity index measures the degree of diversity in a given landscape. Dominance measures the deviation from the maximum possible landscape diversity.

2.2 Heterogeneity

Patch Density and *Mean Patch Size* are useful measures of landscape configuration. The number of patches of a specific land-use and land-cover type is a good index of the urban landscape's spatial heterogeneity. From an ecological perspective it can affect population dynamics and disturbance regimes. Patch density is a fundamental aspect of landscape structure; it expresses the number of patches on a per-unit-area basis, which facilitates comparisons among landscapes of varying size. Patch density for a particular patch type could serve as a good fragmentation index. Similarly, the density of patches in the entire landscape mosaic could serve as a good heterogeneity index. Mean patch size can serve as a habitat fragmentation index. For example, a landscape with a smaller mean patch size for the target patch type might be considered more fragmented than another landscape with a larger size. (McGarigal and Marks 1995).

2.3 Contagion

The dispersion of the urban structure can be effectively measured by the *Contagion* index (Turner 1989, Li and Reynolds 1994). Contagion is the probability that two randomly chosen adjacent cells belong to the same class. This is calculated by the product of two probabilities: the probability that a randomly chosen cell belongs to category type i, and the conditional probability that given a cell is of category type i, one of its neighboring cells will belong to category type j. The common form of a contagion index takes the form (Li and Reynolds 1994). Contagion is based on cell adjacencies as opposed to patch adjacencies. It measures both patch dispersion and interspersion. Landscapes consisting of large patches of similar land-cover or land-use category have a greater number of adjacent cells. Where contagion is low, urban areas can be said to be comprised of many small and dispersed patches of various land-cover or land-use categories. However, contagion indices do not offer any indication of the degree of connectivity between patches or land-use and in this respect they may fall short as indicator of the urban structure. Furthermore since contagion is related to pixel aggregation both patch size and shape influence this measure. Simpler patch configurations and larger patch sizes result in higher contagion values for landscapes of the same composition.

2.4 Interspersion

The *Interspersion and Juxtaposition Index* (IJI) is a measure of adjacencies of each patch type with other patch types over the maximum possible interspersion for the given number of patch types. It can be applied to the landscape and class level. When applied to land-use it measures the degree of interspersion of land-uses. Contrary to Contagion, this index measures patch adjacencies, not cell adjacencies, and can be more appropriate in representing the urban structure. However the interspersion index only measures interspersion and it is not affected by the size, contiguity, or dispersion of patches; thus it captures only one aspect of sprawl.

2.5 Complexity

In landscape ecology *Fractal Dimension* is used to measure landscape complexity. Applied to land-cover it gives a measure of human impact on the landscape. The assumption is that natural boundaries have more complex shapes and that as human disturbance increases fractal dimension decreases. Since edges of one patch are also edges of adjacent patches, edge/area ratios also yield information about the overall shape complexity of a landscape. Fractal dimension represents shape complexity (a departure from Euclidean geometry) that has been effectively used to represent human induced disturbance.

One of the greatest challenges in measuring urban patterns with landscape metrics is differentiating the spatial patterns of patch dispersion. Using simulated landscape patterns, it has been shown that even if pattern metrics provide useful information on the size, shape and distance relationship on a landscape, none of a selected list of pattern metrics--including contagion, mean proximity index and fractal dimension--can distinguish the overall landscape pattern caused by a unique spatial distribution of patches. Each of the measures presented here quantifies an important component of the landscape that can be used to investigate formal hypotheses of the relationships between urban patterns and ecological conditions.

3. GRADIENT ANALYSIS OF METROPOLITAN SEATTLE

To illustrate the limitations of the current gradient approach and the need to explicitly represent urban spatial patterns, we analyze the urban gradient in the Seattle metropolitan area in Washington State. We start with the hypothesis that recent urban change in the metropolitan area has resulted in a complex urban structure where gradients, when they exist, may not be highly correlated with distance from the CBD. To test this hypothesis we analyze the behavior of several landscape metrics in the Seattle metropolitan region and its associated suburbs; this area crosses a variety of land-uses and land-covers from the CBD to the urban fringe. The region which currently houses a population of 3,215,300 has grown dramatically in the 1990s (17% increase), and is expected to continue to grow in the next decades (PSRC 2000). The population is distributed unevenly within the region, with most of the concentration occurring in Seattle, Tacoma and along the east side of Lake Washington.

3.1 Study Area Selection

We focused on a fan-shaped swath extending southeast approximately 50 km from downtown Seattle towards Mt Rainier National Park in the central Puget Sound region (Fig. 5.1). We delineated three transects encompassing a cumulative area of 600 square km. We selected an area that encompasses a range of land-covers and uses ranging from very high-density office development in Downtown Seattle, mature inner-ring residential neighborhoods, industrial warehouse developments in Renton, growing suburban residential areas in Kent and rural and forested areas east of the metropolitan area. A detailed description of each transect is provided in Appendix 1. The constrained nature of central Seattle caused us to use transects that pivoted about a single origin, rather than parallel transects. The location of these transects allowed for the analysis of a clear urban-rural gradient without a high degree of influence from topographical features such as lakes or abrupt areas of high elevation. We conducted a subsequent exploration of three transects placed Northeast, East, and Southeast of the Seattle CBD that confirmed our results.

3.2 Methodology

We selected a set of landscape and patch metrics to quantify various aspects of land-use and cover patterns that we hypothesize are relevant to ecological conditions. In this preliminary study we used the percent land, patch density, mean patch size, the Shannon diversity index, and the contagion index to quantify both land-use and land-cover patterns. Land-use and land-cover serve as proxies for the functional and structural characteristics of the urban pattern. These metrics are used here to explore the spatial distribution of land-use and land-cover composition from the

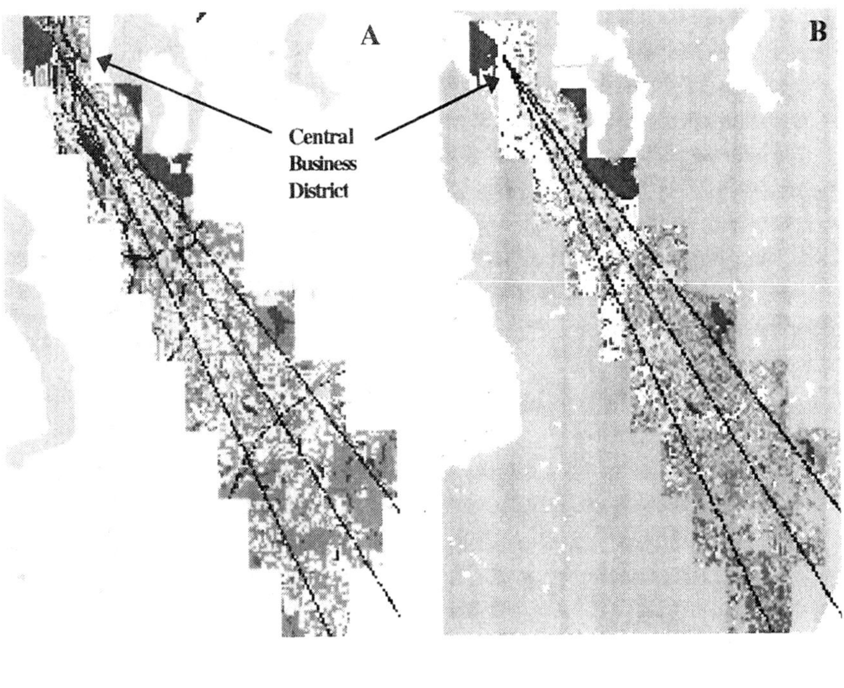

Figure 5.1. Land-use (A) and land-cover (B) study areas along gradients 1, 2 and 3 originating in the Seattle CBD and extending approximately 60 km to the SE.

Seattle urban core to the exurban areas. We calculated and analyzed a suite of class and landscape metrics using a 5x5 km window at regular intervals along each of three transects. We also systematically varied resolution to understand the role that scale plays in correlations between land-use and land-cover patterns and between them and distance from the urban core. We analyzed a total of 24 5x5 km windows for the coarse analysis and 389 1x1 km windows for the finer resolution analysis.

Along each transect, beginning in the Seattle CBD, we delineated 5x5 and 1x1 km boxes for measurement of landscape patterns. The initial box on each transect was centered at the origin of the transect line in downtown Seattle. All subsequent boxes were placed along each line so that their centroids fell on the transect and so that each box was immediately adjacent to the previous one. The extent of available land-cover data was limited and irregular, resulting in transects of different lengths and a variable number of 5x5 km study areas along each transect (Transect 1, 45.1 km, 8 boxes; Transect 2, 47.2 km, 9 boxes; Transect 3, 42.1 km, 7 boxes).

3.3 Data Sources and Preparation

The primary data inputs for this study were digital land-cover and land-use coverages. Land-cover data were obtained from a University of Washington effort PRISM to classify land-cover for the Puget Sound Lowlands of Washington State. Land-cover data were interpreted from Landsat Thematic Mapper (TM) imagery for the Puget Sound region for 1998. Landsat TM data were preprocessed to mosaic the two image swaths (path 46: row 26-27, and path 47: row 26-27) and were corrected for atmospheric and terrain distortion. The georeferencing procedure and terrain correction used the 30 meter USGS hydrological corrected digital elevation model (DEM). After the initial preprocessing, data were interpreted using a classification technique developed by Hill et al. (unpublished) to represent broad heterogeneous urban land-cover classes as well as more traditional homogeneous non-urban classes. The procedure creates an 8-class land-cover system which discriminates between 3 classes of urban land-cover characterized by varying levels of impervious surface and vegetation coverage. These are: *paved urban* (approximately 100% paved cover), *grass/shrub urban* (characteristic of newer suburban areas with limited tree canopy and relatively large lawn coverage) and *forested urban* (characteristic of mature residential neighborhoods with a high degree of canopy cover). In addition the procedure discriminates among 3 types of non-urban vegetation (*grass/shrubs/crops*, *deciduous forest*, and *coniferous forest*) and *water*.

Land-use data were derived from a digital file of King County, Washington tax parcels and a database of associated parcel-level land-uses, produced by the King County GIS Center, Department of Information and Administrative Services. The PRISM project at the University of Washington provided the data set after significant preprocessing. The King County parcel data contain detailed land-use information in the form of 3-digit land-use codes, which are geocoded to each parcel in the county. The preprocessing was necessary for several reasons. First, the parcel polygons had to be joined with the separate database file containing the full set of land-use codes. Further processing was necessary to assign land-use classifications to polygons with no associated land-uses. The data were then converted to a grid of 30-m cell size. We also aggregated the highly detailed King County data (of nearly 250 unique land-use codes) into 16 broader, intuitive classes based on similarity of functional characteristics (Table 5.2). Land-cover data were imported into ESRI ARC/INFO as a raster file and were clipped using the LATTICECLIP function to our study area boundary. The land-use dataset was imported into ARC/VIEW and was converted to a raster format. The dataset was then imported into ARC/INFO as a raster file and clipped using the LATTICECLIP function to the study area boundary. The land-use and land-cover raster, or grid, files were then further processed in ARC/INFO to convert them into a format which could be analyzed by the GRASS image analysis application. A Value Attribute Table (.VAT) file

Table 5.2. Land-use classification.

LU ID	LU NAME	LU DESCRIPTION
1	SFR	Single family residential
2	MFR	Multi-family residential
3	MXU	Mixed use
4	COM	Commercial
5	COH	Heavy use commercial
6	AGR	Agricultural
7	IND	Industrial
8	INH	Heavy use industrial
9	OFF	Office
10	OFH	Heavy use office
11	INS	Institutional
12	PKG	Parking
13	VCB	Vacant buildings
14	OSP	Open space
15	TRA	Transportation
16	TRH	Heavy use transportation

was built for each raster coverage using the BUILDVAT command. The .VAT file consists of a list of attribute values for each cell in a grid coverage. Each raster file was then converted into an ASCII file using the GRIDASCII function in ARC/INFO.

3.4 Metric Calculation

We used the Geographic Resource Analysis Support System (GRASS) and the r.le add-on programs for GRASS. The GRASS system is a GIS system that analyzes raster and vector data. The r.le module extends GRASS' capabilities and allows for sophisticated image processing techniques useful in landscape analysis. The r.le programs can sample raster images at several simultaneous scales using a variety of sampling methods (Baker 1997).

Once the GRASS analysis environment and directory structure were properly configured, we imported both the land-cover and the land-use datasets into GRASS using the r.in.ascii command. To obtain reliable and useful metrics values, we had to exclude certain land-use and land-cover classes from the analysis. We used the r.mask function to exclude the WATER class from the land-cover dataset and to exclude the WATER, MDV (miscellaneous developed land), MRU (miscellaneous rural land) and VCB (vacant building) classes from the land-use dataset. Inclusion of areas such as large water bodies in landscape pattern analysis could result in metrics values that do not represent the terrestrial landscape patterns of interest. Similarly, including vaguely defined or ambiguous classes, such as MDV or MRU, may result in measurement of patterns that do not reflect actual landscape conditions.

We then used the r.le.setup command to set the parameters for the raster image analyses. We used a systematic contiguous image analysis technique, which overlays a fixed grid frame over the raster image and, beginning in a designated location (usually the upper left cell of the frame), calculates the chosen metrics within each grid frame cell. The analysis moves systematically through each grid frame cell in the designated study region. We established 2 scales of analysis, one utilizing 1x1km fixed windows and the other using 5x5km fixed windows.

For each dataset at each of the two scales, we calculated a series of landscape metrics using the r.le.pixel and r.le.patch functions. The metrics calculated are listed in Table 5.3.

We calculated the contagion index metric (t1) in GRASS/r.le; however we chose to recalculate this metric based upon a more conventionally-accepted definition of contagion. Li and Reynolds (1993) devised a contagion index that increases in value as a landscape is dominated by a few large patches, and decreases in value with increasing interspersion of patch types. This revised contagion metric is based on the natural log of the sampling area richness (metric d1).

The GRASS/r.le program calculated the values for each metric as separate files together with a vector coverage of the sampling frame for each scale used in the analysis (5x5 km and 1x1 km). Metrics files were imported into SPSS format and combined into a single file for each sampling scale. The grid frame vector files were exported from GRASS/r.le in ASCII format and imported into ARC/INFO. Using the ASCIIGRID and GRIDPOLY functions, these frames were converted into grid and then vector coverages which we could then use to visually interpret the results of the analysis.

We generated a grid coverage of distances from downtown Seattle by creating a point coverage in ARC/VIEW with one data point representing downtown Seattle (located at the approximate center of the CBD). Using the *find distance* command in ARC/VIEW, we created a grid coverage of distances from this point at a resolution of 30m. We then resampled this coverage to match the resolution of the metrics output (approximately 1x1 km window). Using ARC/INFO, we created a polygonal coverage of our study area divided into a 1 km window. We then joined the distance database file with the polygonal grid frame, resulting in a polygonal coverage with distances assigned to each 1 km window. The distances associated with each cell were then joined, using the tables module ARC/VIEW, to each of the metrics output files. Once this was complete, each metric value was

Table 5.3 Metrics calculated using the GRASS/r.le programs (Baker 1997).

Metric ID	Metric name	Metric description
r.le.patch		
a5	Cover by group	Decimal fraction of sampling area occupied by each class
a7	Total density	Total density of patches within each sampling area
s1	Mean patch size	Mean area, in pixels, of all patches in each sampling area
r.le.pixel		
d2	Shannon index	Shannon Index of Diversity, a combination of richness and evenness
T1	Contagion	Measure of the probability of adjacency of classes
CONT	Revised contagion	Modified contagion index, using the CONT=2(ln(m)) formula, where m represents richness

assigned a corresponding distance from the Seattle CBD and could be analyzed for distance-related correlations. We then exported the data to SPSS and calculated both descriptive and correlational statistics for each set of metrics.

The data analysis involved both correlation and multiregression analysis between the land-cover and land-use metrics and distance to the urban core. We hypothesized that land-use metrics are better predictors of land-cover patterns than distance from the urban core.

4. RESULTS AND DISCUSSION

Land-use and land-cover composition, as measured by the class metric Percent Land show different trends for a number of classes and resolutions (Tables 5.4 and 5.5). Land-cover measured as percent paved area within the 5x5 km windows decreases with distance from the Seattle CBD on the three transects (r = -0.68). Not surprisingly high positive correlations are also observed between distance and deciduous and coniferous forests (r = 0.77). However, land-cover patterns are not as strongly correlated to distance from the Seattle core as other urban gradient studies have indicated. The Seattle metropolitan area shows a polycentric structure. The Southeast gradient from the Seattle CBD is interrupted by the city of Renton. Furthermore the three urban land-cover classes (paved urban, forest urban, and grassy urban) show mixed patterns. A statistically significant relationship is observed at the 5x5 km scale for the land-cover class Grassy urban but the correlation coefficient is lower (r = -58). The relationship between the Forest urban cover class and distance from CBD is not statistically significant indicating a much more complex patch structure.

Complex patterns of land-cover on the urban to rural gradient are revealed by the correlations at finer resolution (Table 5.4). Except for the land-cover classes Paved urban and Percent forest, which show respectively a negative and positive relationship with distance from CBD (r = -0.64, and r = 0.56), all the other land-cover classes have weak correlation with distance from CBD. A very low positive correlation exist between percent of Forest urban and distance from Seattle CBD. The correlation is not statistically significant for the land-cover class Bare soil.

Such complexity is even more evident in the correlation coefficients between land-use and distance from Seattle CBD (Tables 5.4 and 5.5). Intense urban activity declined with distance from the CBD. However while multi-family residential and mixed uses are negatively correlated with distance (r = -0.57 and r = -0.58) at the 5x5km scale, the positive correlation between single family residential uses and distance is lower (r = 0.40) and decreases at the higher resolution (r = 0.25). Higher correlations are estimated for open space and transportation with distance from CBD. Open space shows a high positive correlation with distance from CBD at both 5x5km (r = 0.90) and 1x1km (r = 0.65). Transportation is negatively correlated with distance from CBD at both 5x5km (r = -0.87) and 1x1km (r = -0.60).

Land-cover patterns measured by the number of patches, mean patch size, patch density, and contagion are also not strongly correlated with distance from the urban core even at the coarser scale (Table 5.6). The relationship between Mean patch size and distance is not statistically significant. The only patch configuration metric to show a high significant correlation is

Table 5.4. Correlation of pattern metrics with land-cover and land-use using a 1x1 km window.

	Distance from central business district	Total patch density	Shannon's diversity	Mean patch size	Contagion
% Land-cover					
Forested urban	0.37**	0.435**	0.690**	-0.430**	-0.433**
Grass urban	-0.44**	0.646**	0.518**	-0.614**	-0.352**
Paved urban	-0.64**	0.072	-0.284**	-0.208**	-0.056
Forest	0.56**	-0.144	0.308**	0.246**	-0.133*
Grass shrub crops	0.47**	0.240**	0.448**	-0.164**	-0.362**
Bare soil	-0.32	0.094	-0.017	-0.131**	-0.198**
% Land-use					
Single family residential	0.25**	-0.041	-0.196**	0.077	-0.186**
Multi-family residential	-0.26**	0.181*	0.316**	-0.173	-0.379**
Mixed use	-0.28**				
Commercial and heavy commercial	0.16**	0.079	0.346**	-0.097	-0.354**
Industrial and heavy industrial	-0.37**	-0.304**	0.174*	0.221**	0.045
Office and heavy office	-0.27**	0.044	0.358**	0.106	-0.396**
Open space	0.65**	-0.529**	0.444**	0.523**	0.032
Transportation	-0.60**	0.589**	0.595**	-0.757**	-0.415**

* Correlation is significant at the 0.05 level (2-tailed).

** Correlation is significant at the 0.01 level (2-tailed).

mean patch size at the 5x5 km scale (r = 0.63). On the contrary, land-use patterns measured by the number of patches, mean patch size, patch density, and the Shannon index at the coarser scale are strongly correlated with distance from the urban core (Table 5.6). Patch density and Shannon diversity decrease with distance while mean patch size and contagion increase. The correlations are much lower at the higher resolution due to the sensitivity of these metrics to scale (Table 5.6).

It is clear that while distance plays a role in defining land-cover composition and intensity of land-use, it is not one of the most important factors in explaining the land-use and land-cover structure and the potential ecological exposure to urban disturbance. The data show higher correlations

Table 5.5. Correlation of pattern metrics with land-cover and land-use using a 5x5 km window.

	Distance from central business district	Total patch density	Shannon's diversity	Mean patch size	Contagion
% Land-cover					
Forested urban	0.23	-0.248	-0.048	-0.032	-0.068
Grass urban	-0.58**	0.545**	-0.476**	-0.796**	-0.525**
Paved urban	-0.68**	0.101	0.018	-0.301	0.034
Forest	0.77**	-0.365	-0.448**	-0.070**	0.441*
Grass shrub crops	0.59**	-0.088	0.100	0.470*	0.000
Bare soil	-0.49**	0.581**	0.463*	-0.485**	-0.417*
% Land-use					
Single family residential	0.40**	-0.435**	-0.675**	0.129	0.551**
Multi-family residential	-0.57**	0.781**	0.768**	-0.630**	-0.744**
Mixed use	-0.58**				
Commercial and heavy commercial	-0.56**	0.625**	0.797**	-0.512**	-0.712**
Industrial and heavy industrial	-0.47*	0.375*	0.659**	-0.198	-0.575**
Office and heavy office	-0.48**	0.651**	0.731**	-0.461	-0.742**
Open space	0.90**	-0.854**	-0.838**	0.661**	0.852**
Transportation	-0.87**	0.905**	0.725**	-0.485**	-0.810**

* Correlation is significant at the 0.05 level (2-tailed).

** Correlation is significant at the 0.01 level (2-tailed).

among pattern metrics derived from land-use and land-cover (Tables 5.4, 5.5, and 5.7). Intense urban activities are highly positively correlated with patch density. Transportation infrastructure is positively correlated with patch density and Shannon diversity, while open space is negatively correlated with both these metrics. Furthermore, Patch density and Shannon diversity of land-use are both highly correlated with contagion. Not surprisingly Patch density and mean patch size are respectively highly positively and negatively correlated with Shannon diversity in both land-cover and land-use.

Furthermore, the percent of paved land is negatively correlated with single family residential use at both the 5x5km resolution (r = -0.73) and

Table 5.6. Correlation of distance from CBD with pattern metrics derived from land-cover and land-use.

Pattern metric	Correlation coefficient for 1x1 window	Correlation coefficient for 5x5 window
% Land-cover		
Contagion	-0.24**	0.13
Shannon's diversity	0.33**	0.07
Mean patch size	0.07	0.63**
Total patch density	-0.16**	-0.19
% Land-use		
Contagion	0.11**	0.80**
Shannon's diversity	-0.59**	-0.81**
Mean patch size	0.38**	0.66**
Total patch density	-0.60**	-0.86**

* Correlation is significant at the 0.05 level (2-tailed).
** Correlation is significant at the 0.01 level (2-tailed).

Table 5.7. Correlation among pattern metrics calculated using 1x1 and 5x5 windows.

	Shannon diversity		Mean patch size		Contagion	
% Land-cover	1x1	5x5	1x1	5x5	1x1	5x5
Total patch density	0.784**	0.684**	-0.949**	-0.638**	-0.703**	-0.725**
Shannon diversity			-0.539**	-0.677**	-0.736**	-0.974**
Mean patch size					0.315**	0.777**
% Land-use						
Total patch density	0.745**	0.848**	-0.726**	-0.678**	-0.247**	-0.878**
Shannon diversity			-0.606**	-0.706**	-0.334**	-0.897**
Mean patch size					0.376**	0.567**

* Correlation is significant at the 0.05 level (2-tailed).
** Correlation is significant at the 0.01 level (2-tailed).

1x1km (r = -0.63) and positively correlated with mixed use (r = 0.79, r = 0.37) (Table 5.8). The variation across the transect lines however suggest that the explanatory power of the linear model is not great. The observations at the two resolutions indicate strong positive relationships between industrial use and percent paved land (r = 0.80, r = 0.76). It also shows a weaker positive relationship between transportation and percent paved land (r = 0.65 and r = 0.51).

The explanatory power of the statistical relationships between some urban land-uses and land-cover patterns is greater than the distance from the Seattle CBD. Percent land in industrial and heavy industrial uses explains 57% of the variation in percent paved land ($R^2 = 0.57$, $P < 0.01$) in contrast to 40% explained by distance from CBD ($R^2 = 0.40$ $P < 0.01$) at the 1x1km resolution. After controlling for industrial land-use, the partial correlation between percent paved land and distance from CBD is 0.49 ($P < 0.01$).

These patterns are even more evident at the 5x5 km resolution. Percent land-use in industrial use and single family residential use explain respectively 62% and 60%of the variation in percent paved land in contrast to 50% explained by distance from CBD. Percent industrial and residential land-uses (both multi-family and single family) are thus good predictors of percent paved land and are ecologically important measures of urban exposure.

This can be explained by the large variability of land-use patches across the transect in terms of their land-cover composition and configuration. By examining the composition of land-cover patterns across and within different land-use categories we also found substantial variation with values for the land-cover category paved urban, ranging from 11% to 86% for single-family residential and 35% to 93% for multi-family residential and commercial. Our results confirm that urbanization shows a complex spatial

Table 5.8. Correlation between % paved urban and land-use.

% Land-use type	Correlation coefficients for 1x1 window	Correlation coefficients for 5x5 window
Single family residential	-0.63**	-0.73**
Multi-family residential	0.09	0.54**
Mixed use	0.37**	0.79**
Commercial and heavy commercial	0.30**	0.70**
Industrial and heavy industrial	0.76**	0.80**
Office and heavy office	0.42**	0.72**
Open space	-0.51**	-0.75**
Transportation	0.35**	0.65**

pattern not clearly related to distance from the urban core alone. Instead they suggest possible directions to investigate the extent to which land-use and land-cover pattern metrics perhaps are better predictor of ecological heterogeneity.

5. IMPLICATIONS FOR FUTURE RESEARCH

In this paper we have suggested that the quantification of urban patterns is critical to advance urban ecological research. To understand how species populations and community characteristics change in response to gradients of urban development we need to expand our knowledge about ecosystem structure and functions at the urban end of the gradient. Mechanistic explanations of ecological phenomena are traditionally based on knowledge of physical environmental constraints operating on the ecosystem, the biota operating within these constraints, and the interactions between the biota and the environment. The question of how the biotic structure and process rates are modified in a highly dominated urban environment requires us to consider how the structure and function of an ecosystem are modified by alternative patterns of urban development. Four research areas need to be addressed in urban ecological gradient analysis.

First, we need to identify specific ecological processes that we hypothesize are affected by urban patterns and select appropriate metrics that can be used for empirical tests. Landscape ecology studies provide a vast set of metrics that can be employed to quantify the urbanizing landscape. Building on landscape ecology, we can explore how spatial patterns of urban development affect disturbance regimes and influence the movement of organisms, distribution of materials and nutrients, and landscape processes.

A second implication concerns the classification of land. We need to characterize patterns to include both biophysical and socio-economic factors of urban environments and their structural, functional, and management characteristics. Land-cover data need to be enhanced with ancillary data to make them more useful and/or accurate. Additional information may be obtained by combining two or more datasets into a hybrid land-use/land-cover classification. A supervised classification combined with GIS map algebra techniques on a broad set of input data layers can better represent complex urban land characteristics. Such a synthetic classification system would convey information on both the human use of the land and the actual cover types on the ground.

A third implication concerns scale. Most metrics are scale dependent (e.g., relative to the measurement or map scale used), or are relevant to processes operating only at specific spatial scales (Krummel 1987, O'Neill

1988, Bartel and Brenkert 1991). The relationship between urban patterns and ecosystem dynamics depends on scale. Urban patterns are relevant to environmental processes operating at multiple scales. Scale considerations include both the resolution of a given urban pattern measurement and the geographic extent or boundary of the area under consideration. Spatial questions may have different implications in urbanizing areas than in non-urban landscapes, and our position is that they must be explicitly considered, to explain links between urban patterns and ecological processes. Because the study of urban patterns and ecological processes requires crossing spatial and hierarchical scales, a nested approach is needed to explore the scale at which these interactions are controlled.

Finally we might investigate whether there are relevant thresholds in relationships between spatial patterns and specific ecological processes. We can also search for a threshold in the relationship between the spatial extent or resolution at which a pattern is measured and its correlation with an ecological process. Knowing this threshold would allow an investigator to optimize the geographic area included in pattern measurements. It might suggest new hypotheses about relationships between urban pattern and process.

The study of the dynamic interactions between urban and ecological processes provides a great opportunity to expand the frontiers of ecological research. It challenges urban and ecological scientists to explore new territories and research questions at the border of their disciplines. The quantification of urban patterns is a first step towards a new collaboration between urban and ecological scientists in linking urban planning to ecology.

ACKNOWLEDGMENTS

The research was supported by funding provided by the National Science Foundation (DEB-987504) and the University of Washington Puget Sound Integrated Synthesis Model. We thank John Marzluff and two anonymous referees for their comments on an earlier version of this paper. We also thank Stefan Coe for his input and help with the data processing.

REFERENCES

Alperovich, G., and J. Deutsch. 1992. Population density gradients and urbanisation measurement. Urban Studies 29:1323-1328.

Baker, W. L. 1997. The r.le programs. GRASS/r.le 2.2. http://www.baylor.edu/~grass/gdp/terrain/r_le_22.html.

Batty, M. 1991. Generating urban forms from diffusive growth. Environment and Planning A 23.

Batty, M., and P. A. Longley. 1987. Urban shapes as fractals. Area 19:215-221.

Batty, M., and P. A. Longley. 1998. The morphology of urban land-use. Environ. Plann. B 15:461-488.

Breheny, M. J. [ED.]. 1992. Sustainable development and urban form. Pion Limited, London, UK.

Calthorpe, P. 1993. The next American metropolis: Ecology, community and the American dream. Princeton Architectural Press, Princeton, NJ.

Cervero, R. 1988. Land-use mixing and suburban mobility. Transportation Q. 42.

Cervero, R. 1989. America's suburban centers: The land-use-transportation link. Unwin-Hyman, Boston, MA.

Clergeau, P., J. P. L. Savard, G. Mennechez, and G. Falardeau. 1998. Bird abundance and diversity along an urban-rural gradient: A comparative study between two cities on different continents. Condor 100:413-425.

Cummins, K. W. 1999. Effects of long-term disturbance on riparian vegetation and in-stream characteristics. J. Freshwat. Ecol. 14:1-17.

Dooley, J. L. J., and M. A. Bowers. 1998. Demographic responses to habitat fragmentation: Experimental tests at the landscape and patch scale. Ecology 79:969-980.

Duany, A., and E. Plater-Zyberk. 1991. Towns and town-making principles. Rizzoli, New York, NY.

Ewing, R. 1995. Beyond density, mode choice, and single-purpose trips. Transportation Q. 49:15-24.

Flores, A., S. T. A. Pickett, W. C. Zipperer, R. V. Pouyat, and R. Pirani. 1998. Adopting a modern ecological view of the metropolitan landscape: The case of a greenspace system for the New York City Region. Landscape Urban Plann. 39:295-308.

Forman, R. 1995. Land mosaics: The ecology of landscapes and regions. Cambridge University Press, Cambridge, UK.

Forman, R., and M. Godron. 1981. Patches and structural components for a landscape ecology. Bioscience 31:733-740.

Forman, R., M. and Godron. 1986. Landscape ecology. John Wiley and Sons, New York, NY.

Frohn, R. C. 1998. Remote sensing for landscape ecology. Lewis Publishers, New York, NY.

Geoghegan, J., L. A. Wainger, and N. E. Bockstael. 1997. Spatial landscape inices in hedonic framework: An ecological economics analysis using GIS. Ecol. Econ. 23:251-264.

Godron, M., and R. T. T. Forman. 1983. Landscape modification and changing ecological characteristics, p. 12-28. *In* H. A. Mooney and M. Godron [EDS.], Disturbance and ecosystems: Components of response. Springer-Verlag, New York, NY.

Gustafson, E. J. 1998. Quantifying landscape spatial pattern: What is the state of the art? Ecosystems 1:143-156.

Haines-Young, R., and M. Chopping. 1996. Quantifying landscape structure: A review of landscape indices and their application to forested landscapes. Prog. Phys. Geog. 20:418-445.

Hargis, C. D., J. A. Bissonette, and J. L. David. 1998. The behavior of landscape metrics commonly used in the study of habitat fragmentation. Landscape Ecol. 13:167-186.

Heikkila, E., P. Gordon, J. I. Kim, R. B. Peiser, and H. W. Richardson. 1989. What happened to the CBD-distance gradient?: Land values in a polycentric city. Environ. Plann. A 21:221-232.

Hill, K. Booth, D. and E. Botsford. Unpublished. A rapid land-cover classification method for use in urban watershed analysis.

Ives, A. R., M. G. Turner, and S. M. Pearson. 1998. Local explanations of landscape patterns: Can analytical approaches approximate simulation models of spatial processes? Ecosystems 1:35-51.

Jenks, M., E. Burton, and K. Williams [EDS..]. 1996. The compact city: A sustainable urban form? Chapman and Hall, London, UK.

Lammert, M., and J. D. Allan. 1999. Assessing biotic integrity of streams: Effects of scale in measuring the influence of land-use/cover and habitat structure on fish and macroinvertebrates. Environ. Manage. 23:257-270.

Li, H., and J. F. Reynolds. 1994. A simulation experiment to quantify spatial heterogeneity in categorical maps. Ecology 75:2446-2455.

Marzluff, J. M. 2001. Worldwide increase in urbanization and its effects on birds, p. 19-47. *In* J. M. Marzluff, R. Bowman, and R. Donnelly [EDS.], Avian ecology and conservation in an urbanizing world. Kluwer Academic, Norwell, MA.

McDonald, J. F. 1987. The identification of urban employment subcenters. J. Urban Economics 21:242-258.

McDonnell, M., S. Pickett, P. Groffman, P. Bohlen, R. Pouyat, W. Zipperer, R. Parmelee, M. Carreiro, and K. Medley. 1997. Ecosystem processes along an urban-to-rural gradient. Urban Ecosystems 1:21-36.

McDonnell, M. J., S. T. A. Pickett. 1990. Ecosystem structure and function along urban-rural gradients: An unexploited opportunity for ecology. Ecology 71:1232-1237.

McDonnell, M. J., S. T. A. Pickett, and R. V. Pouyat. 1993. The application of the ecological gradient paradigm to the study of urban effects, p. 175-189. *In* M. J. McDonnell and S. T. A. Pickett [EDS.], Human as a component of ecosystems. Springer Verlag, New York, NY.

Medley, K. E., M. J. McDonnell, and S. T. A Pickett. 1995. Forest-landscape structure along an urban-to-rural gradient. Prof. Geogr. 47:159-168.

Mesev, T. V., and P. A. Longley. 1995. Morphology from imagery: Detecting and measuring the density of urban land-use. Environ. and Plann. A 27:759-780.

Messenger, T., and R. Ewing. 1996. Paper presented at the 75th Annual Meeting of the Transportation Research Board.

Moilanen, A. 1999. Patch occupancy models of metapopulation dynamics: Efficient parameter estimation using implicit statistical inference. Ecology 80:1031.

Nikora, V. I., C. P. Pearson, and U. Shankar. 1999. Scaling properties in landscape patterns: New Zealand experience. Landscape Ecol. 14:17-33.

O'Neill, R., D. DeAngelis, D. Waide, and T. Allen. 1986. A hierarchical concept of ecosystems. Princeton University Press, Princeton, NJ.

O'Neill, R. V., C. T. Hunsaker, K. B. Jones, K. H. Riitters, J. D. Wickham, P. M. Schwartz, I. A. Goodman, B. L. Jackson. and W. S. Baillargeon. 1997. Monitoring environmental quality at the landscape scale: Using landscape indicators to assess biotic diversity, watershed integrity, and landscape stability. BioScience 47:513-519.

O'Neill, R. V., J. R. Krummel, R. H. Gardner, G. Sugihara, B. Jackson, D. L. DeAngelis, B. T. Milne, M. G. Turner, B. Zygmunt, S. W. Christensen, V. H. Dale, R. L. Graham. 1988. Indices of landscape pattern. Landscape Ecol. 1:153-162.

Pickett, S., and M. Cadenasso. 1995. Landscape ecology: Spatial heterogeneity in ecological systems. Science 269:331-334.

Puget Sound Regional Council. 2000. Puget Sound trends.

Qi, Y., and J. Wu. 1996. Effects of changing spatial resolution on the results of landscape pattern analysis using spatial autocorrelation indices. Landscape Ecol. 11:39-49.

Riitters, K. H., R. V. O'Neill, C. T. Hunsaker, J. D. Wickham, D. H. Yankee, S. P. Timmins, K. B. Jones, and B. L. Jackson. 1995. A factor analysis of landscape pattern and structure metrics. Landscape Ecol. 100:23-39.

Stankowski, S. J. 1972. Population density as an indirect indicator of urban and suburban land-surface modifications. U.S. Geological Survey Professional Paper 800-B:B219-B224.

Turner, M., and R. Gardner [EDS..]. 1991. Quantitative methods in landscape ecology. Ecological Studies, Springer, New York, NY.

Turner, M. G. 1989. Landscape ecology: The effect of pattern on process. Annu. Rev. Ecol. Syst. 20:171-197.

Webster, C. J. 1995. Urban morphological fingerprints. Environ. Plann. B 22:279-297.

Whitehand, J. W. R. 1992. Recent advances in urban morphology. Urban Studies 29:619-636.

With, K. A., S. J. Cadaret, and C. Davis. 1999. Movement responses to patch structure in experimental fractal landscapes. Ecology 80:1340-1353.

Zheng, X.P. 1991. Metropolitan spatial structure and its determinants: A case study of Tokyo. Urban Studies 28:87-104.

Appendix 5.1. Description of study areas. URL's listed are from Microsoft Terraserver and are approximate locations. Actual study area boundaries and extents may be larger or smaller than what is displayed.

Box number: Location and description.
URL

Line 1

Box 1: Downtown Seattle, Industrial District, SoDo, Harbor Island. Primarily paved land cover, with approximately 25% of the area in single family residential around Beacon Hill, International District. I-90 and I-5 freeway bisect the area.
http://terraserver.microsoft.com/image.asp?S=13&T=1&X=344&Y=3294&Z=10&W=2

Box 2: Primarily single family residential areas in South Seattle. Includes about 20% warehouse industrial land use south of Boeing Field. Includes numerous forest patches and one large golf course.
http://terraserver.microsoft.com/image.asp?S=13&T=1&X=346&Y=3292&Z=10&W=2

Box 3: Primarily single family residential areas in South Seattle, Bryn Mawr and Renton. Includes a small amount (~10%) of warehouse industrial development along the I-90 corridor and a significant swath of I-90.
http://terraserver.microsoft.com/image.asp?S=13&T=1&X=348&Y=3288&Z=10&W=2

Box 4: Kent / Renton area, including a portion of the Southcenter Mall. Very heterogeneous mix of warehouse industrial, singly family residential, fallow and active agricultural lands, wetlands and forest patches.
http://terraserver.microsoft.com/image.asp?S=13&T=1&X=349&Y=3285&Z=10&W=2

Box 5: Eastern portion of Kent. Suburban residential area with extensive amount of highly fragmented forest lands. Includes a segment of Soos Creek.
http://terraserver.microsoft.com/image.asp?S=13&T=1&X=351&Y=3282&Z=10&W=2

Box number: Location and description. URL
Box 6: Lake Meridian and Covington area. Significant amount of what appears to be remnant agricultural land. Small areas of single family residential development interspersed with the ag lands. Large swath of Soos Creek and associated riparian area through center of area. http://terraserver.microsoft.com/image.asp?S=13&T=1&X=354&Y=3278&Z=10&W=2
Box 7: Covington and Lake Morton area. Primarily forested matrix with patches of low density and medium density single family development. Small amounts of what appears to be agricultural land / open fields. http://terraserver.microsoft.com/image.asp?S=13&T=1&X=355&Y=3276&Z=10&W=2
Box 8: Kummer area, along the Green River, which runs east-west across study area. Large unfragmented patches of forest cover within riparian area and to the north of the scene. Significant patches of agricultural land. Very small areas of rural residential development (farmhouses, small rural towns). http://terraserver.microsoft.com/image.asp?S=13&T=1&X=357&Y=3272&Z=10&W=2
Line 2 Box 1: Downtown Seattle, Industrial District, SoDo, Harbor Island. Primarily paved land cover, with approximately 25% of the area in single family residential around Beacon Hill, International District. I-90 and I-5 freeway bisect the area. http://terraserver.microsoft.com/image.asp?S=13&T=1&X=344&Y=3294&Z=10&W=2
Box 2: South Seattle / Rainier Valley / Boeing Field area. 60% of study area is in mid density residential development in South Seattle. Includes small patches of remnant forest and a large golf course. 40% of scene is warehouse residential, industrial development in the Duwamish Valley and around Boeing Field. Includes a large portion of the Boeing complex itself. http://terraserver.microsoft.com/image.asp?S=13&T=1&X=344&Y=3291&Z=10&W=2
Box 3: Atlantic City, Skyway, S Duwamish Valley area. Primarily mid density residential areas in South Seattle both along Lake Washington and within the Duwamish Valley. Significant patches of warehouse and industrial lands along the Duwamish River. Small forest patches present. http://terraserver.microsoft.com/image.asp?S=13&T=1&X=348&Y=3287&Z=10&W=2
Box 4: Highly heterogeneous area of Renton / Tukwila. Includes large paved patch in the area of Southcenter Mall and surrounding developments. Additional significant areas of office park, warehouse development. Forested, wetland, agricultural patches along Duwamish River and along freeway. Contains some residential areas. http://terraserver.microsoft.com/image.asp?S=13&T=1&X=348&Y=3285&Z=10&W=2
Box 5: Kent area. Approximately 45% of scene is warehouse and light industrial development in the valley. Remainder of scene is primarily single family residential developments. Significant forest patches between industrial and residential areas. http://terraserver.microsoft.com/image.asp?S=13&T=1&X=350&Y=3281&Z=10&W=2

Box number: Location and description.
URL

Box 6: Kent / Lake Meridian area. Matrix of highly fragmented agricultural land with large patches of suburban residential development. Two smaller paved patches, probably shopping centers.
http://terraserver.microsoft.com/image.asp?S=13&T=1&X=352&Y=3279&Z=10&W=2

Box 7: Auburn area. Agricultural and forested mix. Large patches of forest along Green River and Highway 18. Small amounts of rural residential development. Very small amount of suburban development near Auburn (southwest corner of area).
http://terraserver.microsoft.com/image.asp?S=13&T=1&X=352&Y=3275&Z=10&W=2

Box 8: Green River / Muckleshoot Indian Reservation area along Green River. Very large areas of forested and agricultural lands, mostly contiguous. Small areas of rural residential development.
http://terraserver.microsoft.com/image.asp?S=13&T=1&X=354&Y=3273&Z=10&W=2

Box 9: White River area. Large patches of forest along White River. East of forested area, primarily agricultural lands with small areas of rural residential development.
Off maps, but just south of area around:
http://terraserver.microsoft.com/image.asp?S=14&T=1&X=177&Y=1635&Z=10&W=2

Line 3
Box 1: Downtown Seattle, Industrial District, SoDo, Harbor Island. Primarily paved land cover, with approximately 25% of the area in single family residential around Beacon Hill, International District. I-90 and I-5 freeway bisect the area.
http://terraserver.microsoft.com/image.asp?S=13&T=1&X=344&Y=3294&Z=10&W=2

Box 2: Primarily single family residential areas in South Seattle. Includes about 15% warehouse industrial land use south of Boeing Field. Includes some forest patches and one large golf course.
http://terraserver.microsoft.com/image.asp?S=13&T=1&X=346&Y=3292&Z=10&W=2

Box 3: Primarily single family residential areas in South Seattle, Bryn Mawr and Renton. Includes a small amount (~10%) of warehouse industrial development along the Lake Washington waterfront in Renton. Moderate amount of remnant forest patches. Also includes small sections of south Mercer Island and Kennydale.
http://terraserver.microsoft.com/image.asp?S=13&T=1&X=348&Y=3288&Z=10&W=2

Box 4: Patchy single-family residential development to the southeast of Renton. Extensive areas of forested land along Cedar River. Approximately 15-20% paved urban land in office and light industrial development along I-405 and the Valley Freeway.
http://terraserver.microsoft.com/image.asp?S=13&T=1&X=350&Y=3286&Z=10&W=2

Box 5: Lake Youngs area, including the full extent of Lake Youngs. Highly heterogeneous mix of single-family residential developments, forested patches and a small amount of rural fields. Large, contiguous patch of forest cover around Lake Youngs.
http://terraserver.microsoft.com/image.asp?S=13&T=1&X=354&Y=3282&Z=10&W=2

Box number: Location and description. URL
Box 6: Large contiguous forested patches and many smaller fragmented forest patches in rural area. Some large bare soil patches, possibly open pit mines. Moderate amount of low density residential development in scattered patches, primarily in the southern portion of the study area. http://terraserver.microsoft.com/image.asp?S=13&T=1&X=356&Y=3279&Z=10&W=2
Box 7: Large contiguous forested patches around Lake Sawyer and forest regrowth in clearcut areas. Includes moderate amounts of low density residential development around the lake as well as the town of Morganville. http://terraserver.microsoft.com/image.asp?S=13&T=1&X=358&Y=3275&Z=10&W=2

Chapter 6

Urbanization, avian communities, and landscape ecology

James R. Miller[1], Jennifer M. Fraterrigo[1], N. Thompson Hobbs[2,3], David M. Theobald[3], and John A. Wiens[1]

[1]*Graduate Degree Program in Ecology / Department of Biology, Colorado State University, Fort Collins, CO 80523 USA, e-mail: jimmiller@facstaff.wisc.edu*

[2]*Colorado Division of Wildlife, 317 West Prospect, Fort Collins, CO 80523 USA*

[3]*Natural Resource Ecology Lab, Colorado State University, Fort Collins, CO 80523 USA*

Key words: Birds, Colorado, gradient, landscape ecology, policy, study design, urbanization

Abstract: Human settlement is a prevalent source of land-use change worldwide, but our understanding of the effects of settlement on avian communities is limited. Settlement has been characterized as a gradient that extends from urban development to exurban and rural areas. An important advantage of the gradient approach is the potential to identify thresholds in the response of birds to settlement. Although gradients have been used in studying the effects of urbanization on birds, relatively little attention has been paid to the exurban end of the spectrum, despite the potential for this type of development to affect large expanses of habitat. Studies at relatively fine scales are useful for investigating the influence of proximate factors, such as vegetation structure and composition, on birds in human-dominated areas. However, such studies must be coupled with investigations at broader scales to gain a more complete understanding of the ways that human settlement affects bird communities. Urban-wildland gradients can be quantified using remotely-sensed imagery in combination with data on the intensity and pattern of settlement. However, metrics used to describe patterns of settlement are only useful to the extent that they represent something meaningful to birds. The extent of the landscape mosaic surrounding a study site that needs to be quantified depends on the goals for a particular study. We propose a research protocol for studying the effects of settlement on birds that may be useful at multiple scales. Study sites are distributed among land-cover and settlement types, and replicated surveys are progressively aggregated from small to large scales. These data serve as the basis for interpolation, using traditional statistical tools and geostatistical methods, to areas not sampled. To be effective, conservation must be focused on landscape mosaics, not habitat patches. Areas where people live and work are important components of these mosaics.

1. INTRODUCTION

Development associated with human settlement has emerged as a prevailing source of land-use change throughout the world (Berry 1990; United Nations Centre for Human Settlements 1996; Cohen 1997; Marzluff 2001). In the U.S.A. alone, nearly 6 million ha of forest and rangeland were converted to housing and various settlement-related infrastructure between 1982 and 1992 (NRCS 1995). Such development is not evenly distributed geographically, nor is it limited to urban areas. Extensive landscape transformation in the Rocky Mountain region, for example, is the result of a population growth rate that is three times the national average. Colorado is a bellwether of change for this region, with 6 of the 10 fastest-growing counties in the nation (including the top three) - all exceeding 6% population increases per year between 1990 and 1998 (U.S. Bureau of the Census 1998). Although much of this growth is manifest in typical urban and suburban sprawl, the conversion of agricultural and forested land to low-density residential development is occurring at unprecedented rates (Knight et al. 1995, Riebsame et al. 1996).

Urbanization clearly harms some native bird species (Marzluff 2001). Our understanding of the mechanisms underlying such effects has been explored by Bolger (2001) and Marzluff (2001), but remains rudimentary, largely because many ecologists have avoided human-dominated systems as locations for study. In the relatively few studies that have focused on settled areas, researchers have usually concentrated on urban or suburban environments while areas of lower residential density have generally been ignored. Furthermore, most studies have been conducted at relatively fine scales (but also see Blair 1996, Nilon et al. 1995, Clergeau et al. 1998). For these reasons, the scientific basis for avian conservation in developed areas remains weak, particularly in areas of lower settlement densities and at landscape and regional scales.

Our goal is to contribute to the dialogue on the direction of future research on bird communities in human-dominated areas. We begin by examining gradients of human settlement with special attention to portions of such gradients that have thus far received scant attention by avian ecologists. Next, we consider issues of spatial scale and stress the importance of a multi-scale approach to examining the effects of development on birds. Finally, we discuss research needs as well as several important challenges in designing and implementing effective policies for the conservation of avian diversity in areas of human settlement.

2. GRADIENTS OF HUMAN SETTLEMENT

Ecologists often have treated "urban" as a separate type of land-cover, akin to "agriculture" or "grassland." This is similar to the way that settlement is portrayed on roadmaps, where only towns or cities above some minimum population threshold are depicted, or in land-use databases such as USGS Land-use/Land-cover (Anderson et al. 1976), where the only distinction is between urban and non-urban areas. However, the true extent of development may be vastly underestimated by such conceptualizations (Fig. 6.1).

In fact, "urban" is an endpoint on a continuum of settlement intensity – a gradient that extends from sparsely populated rural areas to large cities (Marzluff et al. 2001). When we compare densely populated urban areas to suburban development, there is a noticeable increase in average lot size (and thus a decrease in human density), the amount of greenspace, and the density of trees and shrubs. Suburban areas often give way to exurban residential development that is characterized by still larger lot sizes and greater amounts of vegetation. Ultimately, exurban development grades to rural areas, dominated by agricultural activities, or wildlands. Settlement is not a matter of "either-or" but one of degree (Alberti et al. 2001).

The "gradient paradigm" has long been used in ecology as a basis for examining environmental variation (Whittaker 1967) and has been applied to the study of bird communities at a variety of spatial scales (Terborgh 1971, Cody 1975, Wiens and Rotenberry 1981). The underlying assumption is that the structure and function of ecological systems vary in space, and understanding this variation can enhance our knowledge of system-level processes. McDonnell and Pickett (1990) have described urbanization as a "complex environmental gradient" and suggested that the gradient paradigm is therefore a "powerful organizing tool" for ecological research in human-dominated environments.

Urban gradients are indeed complex. In major metropolitan areas, transects extending from the urban core reveal an interspersion of dense residential development, parks, lower-density suburbs, and smaller city centers (Alberti et al. 2001). A design based on linear transects is useful for

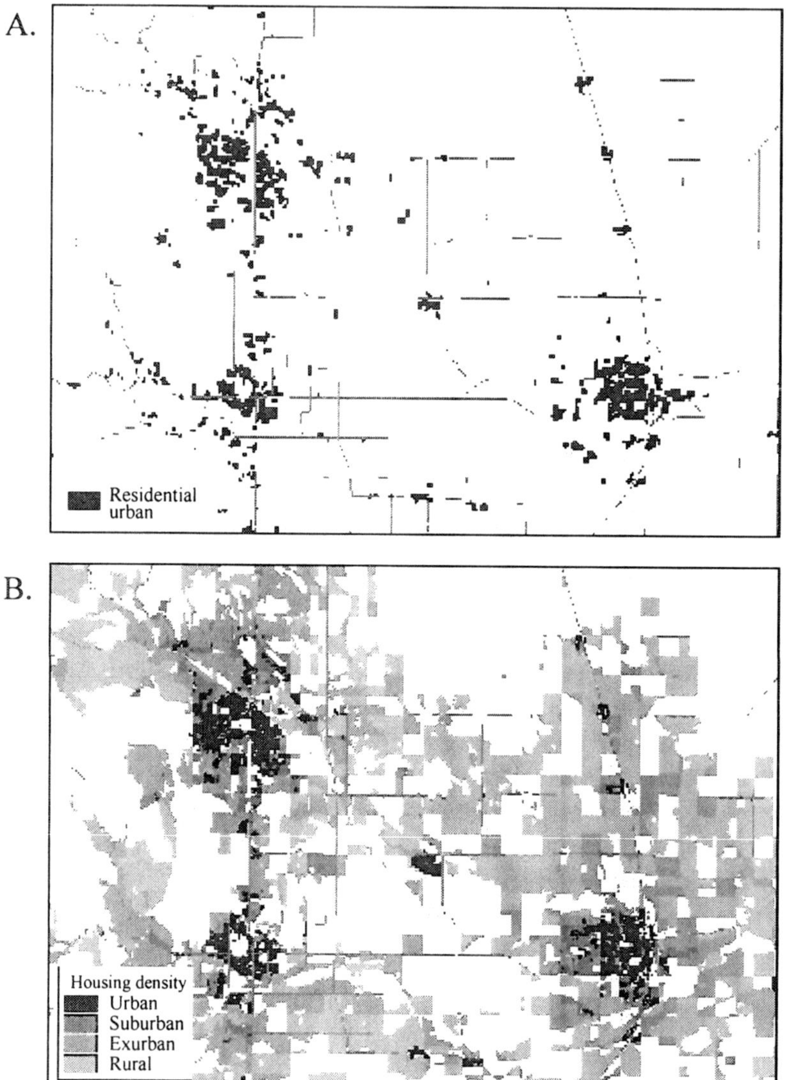

Figure 6.1. Human settlement along the northern Front Range of Colorado (both panels: Fort Collins-top, Loveland-bottom, and Greeley-right). (A) Areas classified as residential urban land-cover from the USGS 1:250,000 LU/LC map series. (B) The full gradient of housing densities from urban to rural using U.S. Census Bureau block-level data.

characterizing patterns of development (see Savard and Falls 2001), but as Alberti et al. (2001) note, urban-rural gradients are not purely geographical. A gradient approach to examining the effects of settlement on birds does not require study sites to be aligned in this fashion and other ways of representing gradients may be more informative. One might compare sites that are characteristic of different types or intensities of development (Blair 1996, 2001; Clergeau et al. 1998), or examine the effect of varying settlement intensity surrounding a particular habitat type (Friesen et al. 1995, Miller et al. 2001).

An important advantage of the gradient approach, as opposed to binary comparisons (e.g., urban vs. rural or suburban vs. wildland), lies in its potential for identifying thresholds or breakpoints where human impacts cause marked changes in biotic responses (Wiens 1989). Such thresholds might be found, for example, in measures of human density, patterns of settlement, or distance from developed areas (Bolger et al. 1997). Thresholds are important simply because they offer a clearer basis for policy, such as the determination of buffer distances for sensitive species that occur in particular habitats or appropriate zoning restrictions.

We recently surveyed 41 peer-reviewed studies of avian communities in areas of human settlement that were published between 1970 and 1999. Of these investigations, 12 utilized a gradient-based research design. In most cases the gradients were restricted in the levels of development that were sampled; only five studies covered the spectrum from urban or suburban to rural/exurban. In fact, the low-density portion of the gradient has not received much attention generally (only four of the 41 studies included exurban sites) despite the increasing pervasiveness of this type of development.

2.1 Exurban Development

In Colorado, exurban development expanded at a pace of 8.3% annually between 1960 and 1990 - nearly three times the rate of population growth (Theobald 2000). This type of low-density development is expected to continue in Colorado and elsewhere, fueled by the growth of recreation-, service-, and information-based industries; dissatisfaction stemming from various problems associated with urban and suburban living; aesthetic considerations; and recreational opportunities. Locations near federal lands such as National Parks and National Forests – places often viewed as conservation areas – appear to be particularly attractive (Nelson 1992, Howe et al. 1997, Marzluff et al. 1998).

Although individual buildings are more widely spaced in exurban developments, this does not necessarily mean that the effects of settlement are restricted to a small portion of the landscape. As rural areas are converted to exurban residential development, road density is likely to increase. Existing roads will experience greater traffic because widely dispersed development means more vehicles per household and more trips per day per vehicle (Romme 1997).

Roads are associated with an effect zone – an area of ecological impact on either side of the road itself – that extends outward from tens to hundreds of meters (Forman and Alexander 1998). Numerous species of grassland birds and forest birds have been found to avoid roads (Ferris 1979, van der Zande et al. 1980). Research in the Netherlands showed that the effect-distances (the distance from a road at which population decreases were detected) varied with traffic density and habitat; 60% of the bird species present occurred at lower densities near roads, and songbird declines were coincident with thresholds in traffic noise (Reijnen et al. 1995, Reijnen et al. 1996). Moreover, evidence suggests that mammalian predators travel along low-traffic roads (Bennett 1991, Forman 1995) and increased rates of nest predation have been found near low-traffic road edges (Small and Hunter 1988). Roads may turn source habitats into sinks by increasing mortality rates (Mumme et al. 2000) and roads may also alter avian community composition by attracting edge species, particularly in forested habitats (Hanowski and Niemi 1995).

One can imagine that buildings also are associated with an effect zone, although this phenomenon has not received as much attention from ecologists. A building-effect zone with reference to birds might be a function of human activity, noise, or free-ranging pets. Although the effect of any one building on the surrounding matrix is probably small, the cumulative effect of many buildings may be substantial (Theobald et al. 1997). Building effects might cause avoidance in some species (O'Dell 2000), whereas species that are tolerant of human presence may benefit. Given the rapid increase and dispersed nature of exurban development, this little-studied portion of the gradient of human settlement has the potential to alter the structure of bird communities over large areas in some regions.

3. SPATIAL SCALE

3.1 Local Scales

Traditionally, ecologists have focused on within-habitat diversity when conducting research on bird communities and have ascribed differences in avian diversity to variation in habitat features such as foliage-height diversity or horizontal habitat heterogeneity (MacArthur and MacArthur

1961, Karr and Roth 1971, Willson 1974). In an effort to reduce spatial variation and thereby allow "fair" comparisons, samples have usually been collected over small plots or quadrats, usually < 0.5 km^2, located in a single habitat type.

Study designs based on within-habitat diversity also have been common in research on urban birds. Several workers have examined relationships between avian community composition and attributes of urban woodlots (Tilghman 1987) or parks (Gavareski 1976, Cicero 1989) including vegetation, patch area, or the presence of human artifacts such as recreational trails. Similarly, DeGraaf and Wentworth (1986) described suburban bird assemblages in Amherst, Massachussetts, and attributed differences in avian community structure to various aspects of the built environment, particularly those relating to vegetation (see also Savard and Falls 2001).

Nest loss in developed areas also has been examined relative to local habitat features (Haskell et al. 2001). For example, Major et al. (1996) found that nest predation increased with nest height and tree density in residential backyards located in Australian urban areas. Miller and Hobbs (2000) examined nest predation in lowland riparian areas adjacent to suburban development along the Front Range of Colorado as a function of distance from recreational trails. Mammalian nest predators tended to avoid trails, whereas avian predators tended to prey more on nests near trails (Miller and Hobbs 2000), similar to the effects of buildings and human activity on nesting birds reported by Osborne and Osborne (1980).

Studies at local scales have advanced our understanding of proximate factors that affect bird distributions in areas of human settlement, but many questions remain to be investigated. Vegetation structure and floristics, for example, have been frequently cited as important determinants of habitat suitability for some species (Beissinger and Osborne 1982, Green et al. 1989, Jokimaki and Suhonen 1993, Rolando et al. 1997, Germaine et al. 1998). However, an exclusive focus on the fine-scale attributes of human-dominated areas reflects the belief that local features are synonymous with "habitat." Evidence suggests that additional factors operating at broader scales also influence avian community structure.

3.2 Landscape Context

Landscape context may influence ecological processes such as those that determine local avian diversity because birds are highly mobile organisms and likely respond to habitat features across a range of spatial scales (Hostetler 2001). Hilden (1965) suggested that habitat selection by migratory birds could best be viewed as a hierarchical decision process that starts at regional and landscape scales and proceeds to fine-scale habitat characteristics. If this is true, then we might sometimes expect similar habitats to support different bird assemblages, depending on the structure and composition of the surrounding landscape matrix.

Pearson (1993) found that much of the variation in the abundance and diversity of wintering birds could be explained solely by landscape variables. Several workers have reported a positive relationship between the width of a riparian area and avian diversity (Stauffer and Best 1980, Darveau et al. 1995, Hodges and Krementz 1996), but this relationship does not always hold when wider riparian areas are surrounded by urban or suburban development (Miller et al. 2001).

Many mechanisms may mediate the effects of landscape context on avian communities in a given habitat type. Birds nesting in forest fragments may be subjected to different nest predators and experience different rates of nest predation depending on the nature of the areas surrounding these patches (Andren 1992, Marzluff and Restani 1999). Szaro and Jakle (1985) found that birds associated with riparian habitats comprised nearly a third of avian assemblages found in desert scrub but that this percentage decreased with distance from the riparian zone. Clearly, conclusions drawn on the sole basis of local habitat features without consideration of landscape context may be misleading.

When assessing the effects of landscape context, ecologists often have focused on landscapes where the primary anthropogenic impacts involve the conversion of land-cover types such as forest and woodland to clearcuts and cropland, commonly thought of as habitat fragmentation (Harris 1988, Saunders et al. 1993, Schwartz 1997). Indeed, contemporary thinking about anthropogenic impacts on landscapes is dominated by the fragmentation paradigm, by the idea that the integrity of many ecosystems is diminished by human activities that isolate well-defined habitat patches in an ecologically compromised matrix. Although this paradigm may apply well to landscapes in which the primary effects of human land-use involve the wholesale conversion of land-cover types, including native habitat remnants embedded in high-density urban areas (e.g., Central Park in New York City), it is not especially faithful to the effects of settlement at lower densities.

In their study of the Golden-cheeked Warbler (*Dendroica chrysoparia*), Engels and Sexton (1994) noted that it is often difficult to delineate patches in urban and suburban areas in a meaningful way because the effects on birds of the defining features of built environments, such as streets and buildings, are poorly understood. Engels and Sexton (1994: p. 287) observed that the woodlands in the vicinity of Austin, Texas "...are more frequently speckled by urban structures than isolated by them into discrete islands." Generalizing this idea, we suggest that in many instances the effects of human settlement may be better represented as "perforations" of landscapes rather than agents of fragmentation (Fig. 6.2). Furthermore, data typically used by ecologists when studying fragmentation at broad scales, such as satellite images or aerial photos, may fail to capture the true extent of exurban development in forested areas. Two sites may be similar in terms of tree canopy cover, but quite different with respect to the intensity of development underneath (Brooks and Begazo 2001).

Available data, frequently used by urban planners and geographers, are well-suited to measuring both the intensity and pattern of settlement (Alberti et al. 2001). Parcel maps, for example, may be acquired from county tax assessor's offices and provide spatially-explicit information on building densities, building age, and zoning restrictions for a given parcel. Such products, in combination with remotely-sensed data, provide the means for quantifying gradients of human settlement. This being said, it should be noted that our ability to compute an ever-expanding array of landscape

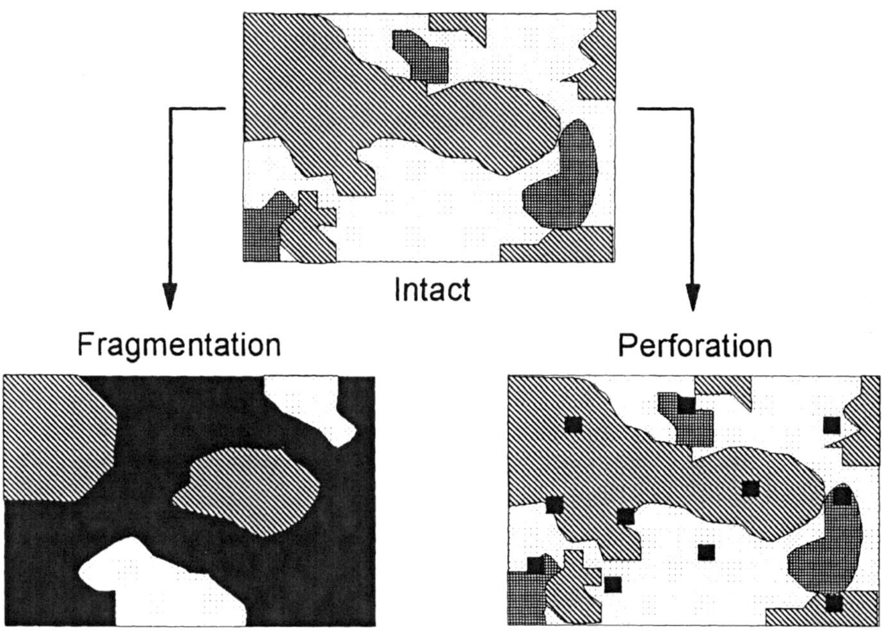

Figure 6.2. Contrasting landscape patterns associated with habitat fragmentation and habitat perforation. Black areas in the lower panels represent habitat removal. Fragmentation results in the large-scale conversion of land-cover types (by timber harvest or agriculture, for example) and discrete habitat patches surrounded by a highly-altered matrix, whereas habitat perforation results in a landscape speckled with disturbance points (such as individual houses or clustered developments).

metrics for the purpose of describing settlement patterns far exceeds our ability to relate these metrics to ecological processes in a meaningful fashion (Gustafson 1998, Turner et al. 2000). Once the metrics are computed, the more-difficult task of determining the extent to which they represent patterns that are important to birds still remains to be done.

How much of the surrounding landscape matrix should be quantified? This is a vexing question that only recently has begun to be addressed by ecologists. The answer likely depends on a suite of factors, but two important considerations are the ecological traits of the species under investigation and the portion of the urban-rural gradient that is being examined. Different species may respond to the landscape at different scales, depending on traits such as body size and trophic status (Hostetler 2001). Thus, a study of the effects of settlement on habitat use by raptors may require measurements over a much broader area than a similar study on songbirds.

The type of development that occurs in the surrounding landscape also may dictate the scale of study. The pattern or intensity of development within, say, 100 m of an urban study site may be quite different than that occurring within 500 or 1000 m. Because exurban development is, by definition, more diffuse, spatial patterns usually change more gradually compared to urban areas. One way in which investigators have tried to assess the scale at which birds are responding to their environment is by determining the amount of variation in avian community structure (e.g., richness, density, composition, species presence/absence) explained by landscape measures beyond that accounted for by local habitat features (Pearson 1993, Knick and Rotenberry 1996, Bolger et al. 1997, Saab 1999).

As a final point, we note that most research on human settlement has been conducted in forested landscapes (Marzluff et al. 2001). In this context, it is not surprising that ecologists have asserted that urbanization is synonymous with the simplification of habitat structure compared to undeveloped areas and leads to overall reductions in avian diversity (Geis 1974, Aldrich and Coffin 1980, Beissinger and Osborne 1982, DeGraaf and Wentworth 1986). However, development also can result in a more complex habitat structure compared to the presettlement landscape. Human settlement in the arid and semi-arid regions of North America often is accompanied by the planting of trees, shrubs, and expansive lawns as people try to recreate the familiar landscapes from which they came (Limerick 1987). As a result, habitat structure becomes more complex and the diversity of birds associated with the built environment in these areas is, in some cases, greater than that of the predevelopment landscape (Guthrie 1974, Vale and Vale 1976, Sodhi 1992). The same may be said of development in areas formerly dedicated to intensive agriculture, as monocultures are replaced with more diverse habitats. Again, landscape context matters.

3.3 Regional Perspectives

Most investigations of the effects of urbanization on bird communities have been somewhat restricted in terms of spatial extent. In our review of avian community studies in areas of human settlement, the majority were conducted in a single city or town ($n = 32$). Five studies compared bird communities across more than one city, but only a handful were extended to include the matrix in which the cities were embedded. Three studies examined effects at a county-wide scale and only two considered effects across more than one county. As a result, the cumulative effects of settlement at regional scales are virtually unknown.

A regional perspective is important, both in assessing the cumulative effects of settlement and in managing human population growth. For example, Marzluff (2001) noted that growth control in Portland, Oregon, has simply transferred development pressure to outlying communities. Furthermore, just as landscapes are perforated by individual buildings and small residential clusters, regions are perforated by cities, towns, and larger suburban enclaves surrounded by exurban development and linked by transportation networks. Thus, a regional perspective is a necessary component of comprehensive avian conservation strategies if they are to be effective (Knopf and Samson 1994).

4. A MULTI-SCALE RESEARCH PROTOCOL

Because we will never be able to measure processes and states for more than a tiny fraction of a region, we need sampling designs that allow us to extrapolate to unmeasured areas. We therefore propose a multi-scale research protocol for assessing the effects of development on avian communities (Fig. 6.3). This protocol could be used to compare the effect of different patterns or densities of settlement on birds within a city, between sites at different points on the urban-rural gradient, or across a region. Data obtained with this method also could be integrated with spatially-explicit models of land-use change, based on demographic and economic drivers (Theobald and Hobbs 1998), to identify important habitats that are particularly vulnerable to future residential development.

Traditional avian census techniques (point or line transects, circular plots) in combination with fine-scale habitat measurements (vegetation structure and floristics) are the foundation of this method. Sets of local surveys are

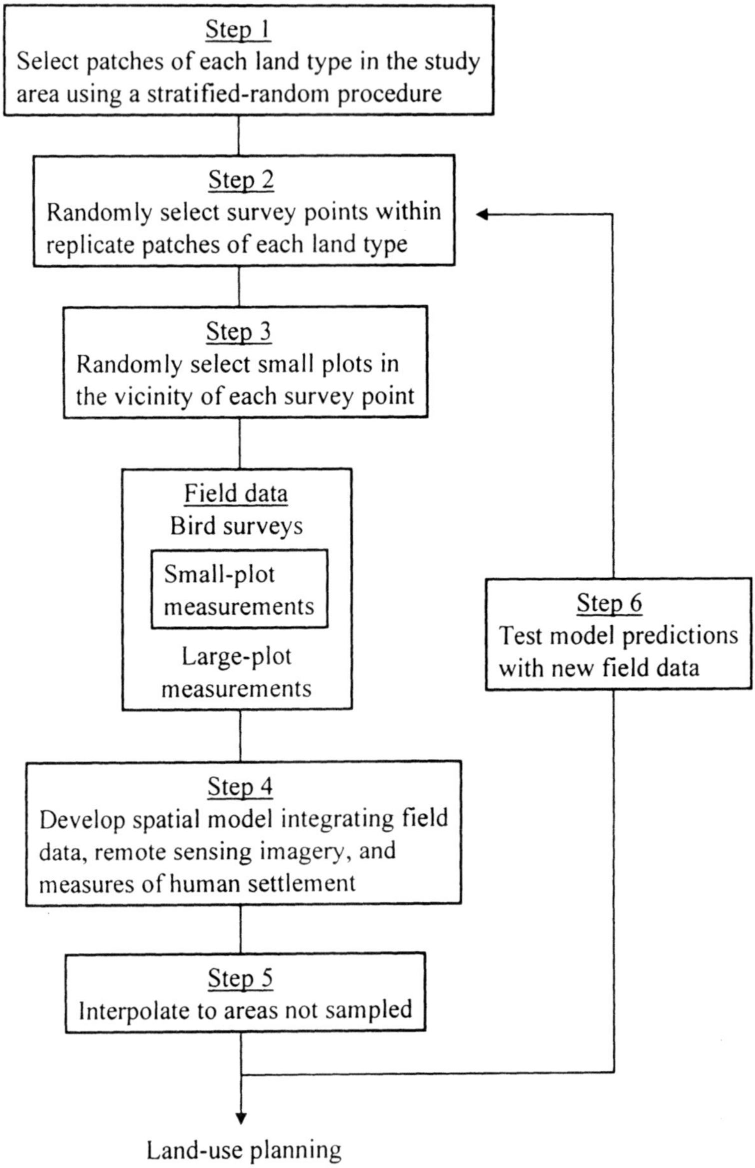

Figure 6.1. Multi-scale research protocol for assessing the effects of development on bird communities.

distributed across the spectrum of land types in the study area using a geographic information system (GIS) and a stratified random design (Fig. 6.3; step 1). For our purposes, land types are defined by measures of human settlement intensity (road density, building density, etc.) superimposed on habitat types (forest, grassland, etc.) to represent the full gradient of settlement occurring in the focal area. Undeveloped sites also could be included. Bird survey points are placed within patches of each land type (step 2) and smaller plots are placed in the vicinity of each point (step 3); the exact locations of points and plots are determined using a stratified-random procedure. Appropriate habitat measurements (canopy cover, shrub cover, etc.) are taken over the larger plots, or the area over which birds are counted at each point, and fine-scale habitat measurements (leaf litter, coarse woody debris, etc.) are taken in the smaller plots. The sampling design is hierarchically nested so that replicated surveys and habitat measures can be progressively aggregated at broader scales.

A spatial model is then developed that can be used to correlate broad-scale data derived from satellite imagery or aerial photography and measures of settlement pattern or intensity with fine-scale data on bird communities and habitat structure (step 4). To examine spatial dependence among variables, one might use traditional statistical methods such as stepwise regression in combination with geostatistical tools such as trend-surface analysis (Gittens 1968). It then becomes possible to interpolate to areas not sampled (Robertson 1987; step 5). Land-use planning often involves the application of data-based models to unsampled areas and this is preferable to planning in the absence of field data. Optimally though, a model's performance would first be assessed by testing its predictions in a new set of study sites (step 6).

In addition to their use in applied research, these methods can be used to address basic ecological questions. For example, beta diversity is generally thought to increase with an increasing number of habitat types. A recent study conducted in the Colorado Front Range of the Rocky Mountains explored the extent to which this relationship is altered by the effect of human settlement (Fraterrigo 2000). The overall study area encompassed two counties and was dominated by ponderosa pine (*Pinus ponderosa*), Douglas fir (*Pseudotsuga menziesii*), lodgepole pine (*Pinus contorta*), aspen (*Populus tremuloides*), and alpine meadows. The research focused on several small towns that lie along a north-south axis. The basic premise was that if areas of human settlement tend to be more similar to one another in terms of habitat structure than they are to the pre-development landscape (Schneider 1996), their respective bird communities should exhibit greater similarity to one another than to avian assemblages in the surrounding matrix. There is evidence to support this assertion in larger developments at lower elevations in other regions (Aldrich and Coffin 1980, Beissinger and Osborne 1982, DeGraaf and Wentworth 1986), but here the emphasis was on smaller settlements or exurban areas in western coniferous forests.

5. RESEARCH NEEDS

There is a pressing need for more research on virtually all aspects of the relationship between urbanization and bird communities. This applies to all points along the urban-wildland gradient and across scales from local to regional. Of course, no single study can address more than a few aspects of settlement's effects on birds. Particularly at broader scales, logistical constraints make it difficult to include a sufficient number of study sites to tease apart more than a few "treatment" effects, even if we relax assumptions about strict replication (Hargrove and Pickering 1992). Therefore, researchers must consider the objectives of a project carefully and realistically. This being said, we offer some thoughts on areas of research where the need for additional study is particularly acute.

Our emphasis in this paper on research conducted over relatively broad scales should not be construed to mean that investigations at local scales are unnecessary. An understanding of the interaction between development, local habitat features, and bird communities is fundamental to accurate interpretation of patterns discerned over landscapes and regions. Data collected at smaller scales also are useful in providing guidelines for habitat enhancement to individual property owners, neighborhood organizations, and municipal government agencies. For example, an urgent need exists for information on the proximate effects of streets and vehicular traffic, various sorts of human activity, and different patterns of tree and shrub planting on birds. The relative value of native versus exotic vegetation also requires further examination, particularly the insect loads that different species support (Beissinger and Osborne 1982) and their effect on nest predation rates (Schmidt and Whelan 1999) .

We believe that investigations aimed at determining the effect zone or "footprint" of individual houses or clusters of buildings should be afforded a high priority by ecologists. Such data are useful in an urban or suburban context, but they are critical in assessing the impacts of exurban development. As we have noted, very little research has been conducted on this part of the urban-wildland gradient, although this type of development potentially affects vast expanses of habitat. When examining the effects of features of the built environment at local scales, it is important that some consideration be given to the nature of the surrounding landscape.

At larger scales, we urge workers to address the entire gradient of development, from urban to wildland, in their study designs. There are logistical challenges inherent in such an approach and it is almost inevitable that a tension will exist between collecting an adequate amount of data at local scales and visiting a sufficient number of sites at different points along the gradient. Nevertheless, such studies are necessary to place smaller-scale efforts in a broader context and to make fair comparisons among different intensities of settlement on the gradient. Until there is a greater balance in the spatial scales that researchers address when conducting research, a comprehensive understanding of the effects of development on bird communities is unlikely to emerge.

6. POLICY AND MANAGEMENT IMPLICATIONS

Over the last few decades, scientists have become increasingly interested in the role that their research might play in the environmental policy process. Here, our goal is not to provide an exhaustive treatment of the intricacies of policy development regarding natural resource management – this has been done elsewhere (e.g., Ostrom 1991, Lee 1993). Rather, we wish to share a few observations that stem from our attempts (particularly by authors NTH and DMT) to provide scientific support for land-use decisions that affect human settlement (Theobald et al. 2000).

Historically, ecologists and conservation biologists have exerted their greatest impact on environmental protection by influencing decisions at the national level. The Clean Air Act, the Clean Water Act and the Endangered Species Act all are examples of policies that extend "top-down" from the federal government to influence local actions. However, most decisions regarding residential and commercial development are made at the base of the government hierarchy by county governments, municipalities, and landowners (Beatley 1994).

This creates several challenges for implementing conservation policy. The first challenge is created by the diffuse nature of land-use decisions. Many seemingly small decisions – choices made at many different times and locations – accumulate to cause large-scale impacts. The fact that these choices are diffuse in time and space means that it is difficult for experts to inform them. It follows that many local decisions are made without the benefit of scientific input regarding their ecological impacts. Second, the fact that many local jurisdictions, including individual landowners, are responsible for land-use decisions implies that conservation policy at the regional scale must include input from numerous sources. Achieving consensus or even agreement among these many jurisdictions may be impossible, and when consensus is reached it might be forged from compromises that fail to deal with fundamental conservation issues. Finally, the large degree of discretion granted to landowners by the U.S. Constitution (Smith 1993, Cullingworth 1997) means that conservation policy developed by government may have very little impact on what happens on the landscape. Unless we are content with merely cleaning up the infrastructure, it will be imperative to find ways to compensate private landowners for not developing their land.

Rising to these challenges will require the development of strategies for conserving and managing biodiversity at multiple scales and the integration of actions by state, county, and municipal governments with those of landowners. This is a daunting task, but there are encouraging examples to guide our efforts (DeGrove 1992, Lee 1993, Beatley 1994, Duerksen et al. 1997). These multi-scale strategies must achieve reasonable compromises between maintaining local and regional diversity of species and habitats.

Achieving conservation objectives in the face of rapid land-use change requires a new set of responses by scientists. We must assure that data and analyses are both readily understood and easily accessible to policy-makers and to the citizens who wish to influence decisions in established review processes. We must be willing to work locally to contribute to the solution of regional problems. Finally, we must emphasize that conservation should be focused on landscape mosaics, not habitat patches, and that developed areas are important components of these mosaics. Rather than writing off settled areas as having no conservation value, the challenge is to enhance the habitat value of these lands through better design.

ACKNOWLEDGEMENTS

We thank M. Hostetler, M. Alberti, and J. Marzluff for helpful suggestions on an earlier draft of this manuscript.

REFERENCES

Alberti, M., E. Botsford, and A. Cohen. 2001. Quantifying the urban gradient: linking urban planning and ecology, p. 89-115. *In* J. M. Marzluff, R. Bowman, and R. Donnelly [EDS.], Avian ecology and conservation in an urbanizing world. Kluwer Academic, Norwell, MA.

Aldrich, J. W., and R. W. Coffin. 1980. Breeding bird populations from forest to suburbia after thirty-seven years. Amer. Birds 34:3-7.

Anderson, J. R., E. E. Hardy, J. T. Roach, and W. E. Witmer. 1976. A land-use and land-cover classification system for use with remote sensor data. USGS Professional Paper 964. USGS, Reston, VA.

Andren, H. 1992. Corvid density and nest predation in relation to forest fragmentation: a landscape perspective. Ecology 73:794-804.

Beatley, T. 1994. Habitat conservation planning. University of Texas Press, Austin.

Beissinger, S. R., and D. R. Osborne. 1982. Effects of urbanization on avian community organization. Condor 84:75-83.

Bennett, A. F. 1991. Roads, roadsides and wildlife conservation: a review, p.99-117. *In* D. A. Saunders and R. J. Hobbs [EDS.], Nature conservation 2: the role of corridors. Surrey Beatty & Sons, Chipping Norton, NSW, Australia.

Berry, B. J. L. 1990. Urbanization, p.103-120. *In* B. L. I. Turner, W. C. Clark, R. W. Kates, J. F. Richards, J. T. Mathews, and W. B. Meyer [EDS.], The earth as transformed by human action. Cambridge University Press, England.

Blair, R. B. 1996. Land-use and avian species diversity along an urban gradient. Ecol. Appl. 6:506-519.

Blair, R. 2001. Creating a homogeneous avifauna, p. 461-488. *In* J. M. Marzluff, R. Bowman, and R. Donnelly [EDS.], Avian ecology and conservation in an urbanizing world. Kluwer Academic, Norwell, MA.

Bolger, D. 2001. Urban birds: population, community, and landscape approaches, p. 155-177. *In* J. M. Marzluff, R. Bowman, and R. Donnelly [EDS.], Avian ecology and conservation in an urbanizing world. Kluwer Academic, Norwell, MA.

Bolger, D. T., T. A. Scott, and J. T. Rotenberry. 1997. Breeding bird abundance in an urbanizing landscape in coastal Southern California. Conserv. Biol. 11:406-421.

Brooks, D. and A. Begazo. 2001. Mccaw abundance in relation to human population density, p. 429-439. *In* J. M. Marzluff, R. Bowman, and R. Donnelly [EDS.], Avian ecology and conservation in an urbanizing world. Kluwer Academic, Norwell, MA.

Cicero, C. 1989. Avian community structure in a large urban park: controls of local richness and diversity. Land. Urban Plann. 17: 221-240.

Clergeau, P., J. P. L. Savard, G. Mennechez, and G. Falardeau. 1998. Bird abundance and diversity along an urban-rural gradient: A comparative study between two cities on different continents. Condor 100:413-425.

Cody, M. L. 1975. Towards a theory of continental species diversities: bird distributions over mediterranean habitat gradients, p.214-257. *In* M. L. Cody and J. M. Diamond [EDS.], Ecology and evolution of communities. Harvard University Press, Cambridge, MA.

Cohen, J. 1997. Conservation and human population growth: what are the linkages?, p.29-42. *In* S. T. A. Pickett [ED.], The ecological basis of conservation: heterogeneity, ecosystems, and biodiversity. Chapman & Hall, New York, NY.

Cullingworth, B. 1997. Planning in the USA: policies, issues, and processes. Routledge, London, England.

Darveau, M., P. Beauchesne, L. Belanger, J. Huot, and P. Larue. 1995. Riparian forest strips as habitat for breeding birds in boreal habitat. J. Wildl. Manage. 59:67-78.

DeGraaf, R. M., and J. M. Wentworth. 1986. Avian guild structure and habitat associations in suburban bird communities. Urban Ecol. 9:399-412.

DeGrove, J. M. 1992. Planning and growth management in the States. Lincoln Institute of Land Policy, Cambridge, MA.

Duerksen, C. J., D. L. Elliot, N. T. Hobbs, E. Johnson, and J. R. Miller. 1997. Habitat protection planning: where the wild things are. Amer. Planning Assoc. Planning Advisory Serv. Rep. No. 470-471.

Engels, T. M., and C. W. Sexton. 1994. Negative correlation of Blue Jays and Golden-cheeked Warblers near an urbanizing area. Conserv. Biol. 8:286-290.

Ferris, C. R. 1979. Effects of Interstate 95 on breeding birds in northern Maine. J. Wildl. Manage. 43:421-427.

Forman, R. T. T. 1995. Land mosaics. Cambridge University Press, Cambridge, MA.

Forman, R. T. T., and L. Alexander. 1998. Roads and their major ecological effects. Ann. Rev. Ecology Syst. 29:207-231.

Fraterrigo, J. M. 2000. Low-density settlement in the Rocky Mountain West: effects on bird communities and landscape patterns. M.S. thesis, Colorado State University, Fort Collins.

Friesen, L. E., P. F. J. Eagles, and R. J. MacKay. 1995. Effects of residential development on forest-dwelling neotropical migrant songbirds. Conserv. Biol. 9:1408-1414.

Gavareski, C. A. 1976. Relation of park size and vegetation to urban bird populations in Seattle, Washington. Condor 78:375-382.

Geis, A. D. 1974. Effects of urbanization and type of urban development on bird populations, p.97-105. *In* J. H. Noyes and D. R. Progulski [EDS.], Wildlife in an Urbanizing Environment. November 27-29, 1973, Springfield, Massachusetts. University of Massachusetts, Amherst.

Germaine, S. S., S. S. Rosenstock, R. E. Schweinsburg, and W. S. Richardson. 1998. Relationships among breeding birds, habitat, and residential development in greater Tucson, Arizona. Ecol. Appl. 8:680-691.

Gittens, R. 1968. Trend-surface analysis of ecological data. J. Ecology 56:845-869.

Green, R. J., C. P. Catterall, and D. N. Jones. 1989. Foraging and other behaviour of birds in subtropical and temperate suburban habitats. Emu 89:216-222.

Gustafson, E. J. 1998. Quantifying landscape spatial pattern: what is the state of the art? Ecosys. 1:143-156.

Guthrie, D. A. 1974. Suburban bird populations in southern California. Am. Midl. Nat. 92:461-466.

Hanowski, J. M., and G. J. Niemi. 1995. A comparison of on- and off-road bird counts: Do you need to go off road to count birds accurately? J. Field Ornithology 66:469-483.

Hargrove, W. W., and J. Pickering. 1992. Pseudo-replication: a sine qua non for regional ecology. Landscape Ecol. 6:251-258.

Harris, L. D. 1988. Edge effects and conservation of biotic diversity. Conserv. Biol. 2:330-332.

Haskell, D., A. M. Knupp, and M. C. Schneider. 2001. Nest predator abundance and urbanization, p. 245-260. *In* J. M. Marzluff, R. Bowman, and R. Donnelly [EDS.], Avian ecology and conservation in an urbanizing world. Kluwer Academic, Norwell, MA.

Hilden, O. 1965. Habitat selection in birds. Ann. Zool. Fenn. 2:53-75.

Hodges, M. F. J., and D. G. Krementz. 1996. Neotropical migratory breeding bird communities in riparian forests of different widths along the Altamaha River, Georgia. Wilson Bull. 108:496-506.

Hostetler, M. 2001. The importance of multi-scale analyses in avian habitat selection studies in urban environments, p. 139-154. *In* J. M. Marzluff, R. Bowman, and R. Donnelly [EDS.], Avian ecology and conservation in an urbanizing world. Kluwer Academic, Norwell, MA.

Howe, J., E. McMahon, and L. Propst. 1997. Balancing nature and commerce in gateway communities. Island Press, Washington, DC.

Jokimaki, J., and J. Suhonen. 1993. Effects of urbanization on the breeding bird species richness in Finland: a biogeographical comparison. Ornis Fenn. 70:71-77.

Karr, J. R., and R. R. Roth. 1971. Vegetation structure and avian diversity in several new world areas. Am. Nat. 105:423-435.

Knight, R. L., G. N. Wallace, and W. E. Riebsame. 1995. Ranching the view: subdivisions versus agriculture. Conserv. Biol. 459-461.

Knick, S. T., and J. T. Rotenberry. 1996. Landscape characteristics of fragmented shrubsteppe habitats and breeding passerine birds. Conserv. Biol. 9:1059-1071.

Knopf, F. L., and F. B. Samson. 1994. Scale perspectives on avian diversity in western riparian ecosystems. Conserv. Biol. 8:669-676.

Lee, K. N. 1993. Compass and gyroscope: integrating science and politics for the environment. Island Press, Washington, DC.

Limerick, P. N. 1987. The legacy of conquest. W. W. Norton & Company, New York, NY.

MacArthur, R. H., and J. W. MacArthur. 1961. On bird species diversity. Ecology 42:594-598.

Major, R. E., G. Gowing, and C. E. Kendal. 1996. Nest predation in Australian urban environments and the role of the Pied Currawong, *Strepera graculina*. Aust. J. Ecol. 21:399-409.

Marzluff, J. M. 2001. Worldwide increase in urbanization and its effects on birds, p. 19-47. *In* J. M. Marzluff, R. Bowman, and R. Donnelly [EDS.], Avian ecology and conservation in an urbanizing world. Kluwer Academic, Norwell, MA.

Marzluff, J. M., R. Bowman, and R. Donnelly. 2001. A historical perspective on urban bird research: trends, terms and approaches, p. 1-17. *In* J. M. Marzluff, R. Bowman, and R. Donnelly [EDS.], Avian ecology and conservation in an urbanizing world. Kluwer Academic, Norwell, MA.

Marzluff, J. M., F. R. Gehlbach, and D. A. Manuwal. 1998. Urban environments: influences on avifauna and challenges for the avian conservationist. In J. M. Marzluff and R. Sallabanks [EDS.], Avian conservation: research and management. Island Press, Washington, DC.

Marzluff, J. M., and M. Restani. 1999. The effects of forest fragmentation on avian nest predation, p.155-169. In J. A. Rochelle, L. A. Lehmann, and J. Wisniewski [EDS.], Forest fragmentation. Wildlife and management implications. Brill, Leiden, the Netherlands.

McDonnell, M. J., and S. T. A. Pickett. 1990. Ecosystem structure and function along urban-rural gradients: an unexploited opportunity for ecology. Ecology 71:1232-1237.

Miller, J. R., and N. T. Hobbs. 2000. Effects of recreational trails on nest predation rates and predator assemblages. Land. Urban Plann. 50:227-236.

Miller, J. R., J. A. Wiens, and N. T. Hobbs. 2001. How does urbanization affect bird communities in riparian habitats? An approach and preliminary assessment, p.427-439. *In* J. Craig, N. Mitchell, and D. Saunders [EDS.], Nature conservation 5. Nature conservation in production environments: Managing the matrix. Surrey Beatty and Sons, Chipping-Norton, New South Wales, Australia.

Mumme, R. L., S. J. Schoech, G. E. Wooflenden, and J. W. Fitzpatrick. 2000. Life in the fast lane: Demographic consequences of road mortality in the Florida Scrub-Jay. Cons. Biol. 14: 501-512.

Nelson, A. 1992. Characterizing exurbia. J. Planning Lit. 6:350-368.

Nilon, C. H., C. N. Long, and W. C. Zipperer. 1995. Effects of wildland development on forest bird communities. Land. Urban Plann. 32:81-92.

NRCS. 1995. USDA Natural Resource Conservation Service, National Res. Inv. http://www.nhq.nrcs.usda.gov/NRI/tables/1992/table.7.html.

O'Dell, E. 2000. Wildlife communities and exurban development in Pitkin County, Colorado. M.S. Thesis, Colorado State University, Fort Collins.

Osborne, P., and L. Osborne. 1980. The contribution of nest site characteristics to breeding-success among Blackbirds *Turdus mercula*. Ibis 122:512-517.

Ostrom, E. 1991. Managing the commons. Cambridge University Press, New York.

Pearson, S. M. 1993. The spatial extent and relative influence of landscape-level factors on wintering bird populations. Landscape Ecol. 8:3-18.

Reijnen, R., R. Foppen, C. ter Braak, and J. Thissen. 1995. The effects of car traffic on breeding bird populations in woodland. III. Reduction in density in relation to the proximity of main roads. J. of Appl. Ecol. 32:187-202.

Reijnen, R., R. Foppen, and H. Meeuwsen. 1996. The effects of traffic on the density of breeding birds in Dutch agricultural grasslands. Biol. Conserv. 75:255-260.

Riebsame, W. E., H. Gosnell, and D. M. Theobald. 1996. Land-use and landscape change in the Colorado mountains I: Theory, scale, and pattern. Mount. Res. and Devel. 16: 395-405.

Robertson, G. P. 1987. Geostatistics in ecology: Interpolating with known variance. Ecology 63:744-748.

Rolando, A., G. Maffei, C. Pulcher, and A. Giuso. 1997. Avian community structure along an urbanization gradient. Italian J. of Zool. 64:341-349.

Romme, W. H. 1997. Creating pseudo-rural landscapes in the mountain west, p.139-161. *In* J. I. Nassauer [ED.], Placing nature. Island Press, Washington, DC.

Saab, V. 1999. Importance of spatial scale to habitat use by breeding birds in riparian forests: a hierarchical analysis. Ecol. Appl. 9:135-151.

Saunders, D. A., R. J. Hobbs, and P. R. Ehrlich. 1993. Nature conservation 3: Reconstruction of fragmented ecosystems, global and regional perspectives. Surrey Beatty & Sons, Chipping Norton, NSW, Australia.

Savard, J. P. and J. B. Falls. 2001. Bird-habitat relationships of breeding birds in residential areas of Toronto, Canada, p. 543-568. *In* J. M. Marzluff, R. Bowman, and R. Donnelly [EDS.], Avian ecology and conservation in an urbanizing world. Kluwer Academic, Norwell, MA.

Schmidt, K. A. and C. J. Whelan. 1999. Effects of exotic *Lonicera* and *Rhamnus* on songbird nest predation. Cons. Biol. 13: 1502-1506.

Schneider, D. W. 1996. Effects of European settlement and land-use on regional patterns of similarity among Chesapeake forests. Bull. Torrey Bot. Club 123: 223-239.

Schwartz, M. W. 1997. Conservation in highly fragmented landscapes. Chapman & Hall, New York, NY.

Small, M. F., and M. L. Hunter. 1988. Forest fragmentation and avian nest predation in forested landscapes. Oecologia 76:62-64.

Smith, H. H. 1993. The citizen's guide to planning. American Planning Association, Chicago, IL.

Sodhi, N. S. 1992. Comparison between urban and rural bird communities in prairie Saskatchewan: urbanization and short-term population trends. Can. Field-Nat. 106:210-215.

Stauffer, D. F., and L. B. Best. 1980. Habitat selection by birds of riparian communities: evaluating effects of habitat alterations. J. Wildl. Manage. 44:1-15.

Szaro, R. C., and M. D. Jakle. 1985. Avian use of a desert riparian island and its adjacent scrub habitat. Condor 87:511-519.

Terborgh, J. 1971. Distribution on environmental gradients: theory and a preliminary interpretation of distributional patterns in the avifauna of the Cordillera Vilcabamba, Peru. Ecology 52:23-40.

Theobald, D.M. 2000. Fragmentation by inholdings and exurban development, p.155-174. *In* R.L. Knight, F.W. Smith, S.W. Buskirk, W.H. Romme, and W.L. Baker, [EDS.], Forest fragmentation in the central Rocky Mountains. University Press of Colorado, Boulder.

Theobald, D.M. and N. T. Hobbs. 1998. Forecasting rural land-use change: a comparison of regression- and spatial transition-based models. Geog. Env. Mod. 2: 57-74.

Theobald, D. M., N. T. Hobbs, T. Bearly, J. A. Zack, T. Shenk, and W. E. Riebsame. 2000. Incorporating biological information in local land-use decision making: designing a system for conservation planning. Landscape Ecol. 15: 35-45.

Theobald, D. M., J. R. Miller, and N. T. Hobbs. 1997. Estimating the cumulative effects of development on wildlife habitat. Land. Urban Plann. 39:25-36.

Tilghman, N. G. 1987. Characteristics of urban woodlands affecting breeding bird diversity and abundance. Land. Urban Plann. 14:481-495.

Turner, M. G., R. H. Gardner, and R. V. O'Neill. 2000. Pattern and process: landscape ecology in theory and practice. Springer-Verlag, New York.

United Nations Centre for Human Settlements. 1996. An urbanising world: global report on human settlements, 1996. Oxford University Press for the United Nations Centre for Human Settlements (Habitat), Oxford, United Kingdom.

US Bureau of the Census. 1998. Statistical abstract of the United States. U.S. Government Printing Office, Washington, DC.

Vale, T. R., and G. R. Vale. 1976. Suburban bird populations in west-central California. J. Biogeogr. 3:157-165.

van Der Zande, A., W. Ter Keurs, and W. Van Der Weijden. 1980. The impact of roads on the densities of four bird species in an open field habitat - evidence of a long-distance effect. Biol. Conserv. 18:299-321.

Whittaker, R. H. 1967. Gradient analysis of vegetation. Biol. Rev. 42:207-264.

Wiens, J. A. 1989. Spatial scaling in ecology. Func. Ecol. 3:385-397.

Wiens, J. A., and J. T. Rotenberry. 1981. Habitat associations and community structure of birds in shrubsteppe environments. Ecol. Monogr. 51:21-41.

Willson, M. F. 1974. Avian community organization and habitat structure. Ecology 55:1017-1029.

Chapter 7

The importance of multi-scale analyses in avian habitat selection studies in urban environments

Mark Hostetler
Center for Environmental Studies and Department of Biology, Arizona State University, P.O. Box 873211, Tempe, AZ 85287-3211 USA., e-mail: hostetlerm@wec.ufl.edu

Key words: Habitat selection, hierarchy, landscape, landscape design, scale, urban

Abstract: We have a limited understanding of how birds respond to different scales of urban landscape structure. Such information is important to concerned homeowners, developers, and city planners because each need to know whether the scale at which they manipulate the landscape will appreciably reduce or attract certain avian species. Often in urban habitat selection studies, one arbitrary scale is chosen to correlate habitat structure to avian abundance, composition, and/or distribution. This scale may or may not reflect the primary scale at which species respond to landscape structure. In this paper, I (1) give a hierarchical perspective on how birds may select habitat from limited to broad scales, (2) describe a multi-scale analysis to determine the broad scales at which birds respond to landscape structure, and (3) discuss why multi-scale studies will ultimately help urban wildlife managers evaluate the scales at which a landscape design is most effective at attracting various species. A connection exists between the scales at which birds respond to landscape structure and the scales at which humans impact urban areas.

1. INTRODUCTION

Urban environments are a significant landscape type in North America. The expansion of urban cities (i.e., urban sprawl) has been quite rapid; the rate that lands are converted to urban use has been estimated to exceed population growth by a factor of six to ten (Richmond 1996). For example, metropolitan Phoenix, AZ, annexed 214 miles^2 between 1990 and 1997 (Gober 1998). Many homeowners prefer to live in the suburbs, which ultimately increases the rate of urban sprawl. Members of one noteworthy group, the "baby-boomers," have an estimated one trillion dollars to spend on retirement, and they prefer to buy land outside urban city centers (Towne 1994). The end result of urban sprawl is the loss of wildlife habitat and the fragmentation or perforation of vast regions of North American landscapes (Marzluff 2001, Miller et al. 2001).

Urban landscapes have a large impact on avian communities. Researchers have reported higher bird densities in urban areas than in natural areas; bird diversity decreased as the degree of urbanization increased; and species composition changed dramatically with a few species (e.g., Rock Doves, *Columba livia*) becoming dominant (Woolfenden and Rohwer 1969; Walcott 1974; Emlen 1974; Degraaf and Wentworth 1981; Blair 1996). Of the studies that have been conducted on birds in urban areas, many have concentrated on how habitat structure affects avian abundance, composition, and richness (e.g., Emlen 1974, Penland 1984, Horak 1986, Sexton 1987, Mills et al. 1989, Blair 1996, Germaine et al. 1998). In most cases, these studies measured habitat structure at one scale. But different species probably respond to habitat at different scales, and single-scale studies may not adequately reflect what many species are responding to in a landscape (Holling 1992, Hostetler 1999, Hostetler and Holling 2000). I define response as the ability of a bird to utilize structural objects in a landscape (e.g., tree patches); operationally, response can be measured by the presence/absence and/or variation in abundance of a species across a variety of survey areas. Areas that attract a species are assumed to contain habitat structure at a range of scales that are relevant to this species.

The range of scales relevant to animals is an important determinant in the spatial distribution and population dynamics of animals (Wiens 1989, Levin 1992, Wiens et al. 1993). In particular, landscape structure at the level of the home range and larger may be the primary factor that first limits whether a species occurs in an area (Johnson 1980, Hutto 1985). For example, the appearance of a hawk in a backyard during the breeding season first depends on whether it has a home range in the general vicinity. However, empirical studies that conduct multi-scale analyses to determine the different scales at which individual species respond to landscape structure are virtually nonexistent (Opdam 1991, Harrison 1992, Wiens 1995, Lima and Zollner 1996, although see Hostetler and Holling 2000).

If we are going to design urban landscapes to accommodate birds, we must first know the range of scales relevant to different species, particularly at the level of the home range and beyond. Homeowners, developers, and city planners who want to design landscapes to attract birds are usually constrained by the amount of area available for vegetation. Developments are planned at different scales, and different designs have the potential of affecting species differently (Nilon et al. 1995). If one knew the scales at which different birds respond to landscape structure, even the most limited amount of area could be designed as habitat for certain species. For example, the design of a homeowner's backyard may affect whether a small bird (e.g., a hummingbird) establishes a home range, but the design of a large neighborhood may affect whether a large bird (e.g., a hawk) establishes a home range in the area. Any development is defined both by its area and the sizes of structural objects (e.g., tree patches) found within the landscape. To assess the impact of a planned landscape on different avian species and perhaps to design an area for different species, two questions that need to be answered are 1) how large an area is "evaluated" by different species when selecting a habitat, and 2) what sizes and types of structural objects do species primarily respond to within these areas?

This paper presents a conceptual framework and an empirical method to address the above two questions. First, I illustrate how different birds may respond to landscape structure at different scales. Species possibly go through a hierarchy of decisions when selecting a habitat (Johnson 1980, Hutto 1985), and possibly groups of birds respond to structure within the same range of scales (Holling 1992, Hostetler and Holling 2000). Next, I describe an empirical method to determine whether different species respond to structure at different scales; here, I emphasize the importance of determining the broadest scale (i.e., home range and above) at which species respond to structure. Finally, I discuss why the results of such multi-scale analyses have implications for urban policy and management and how future research can be conducted to determine the scales at which birds respond to landscape structure.

2. HIERARCHY, SCALE, AND AVIAN HABITAT SELECTION

Ecologists have extensively used hierarchy theory to simplify studies of complex ecosystems (e.g., Allen and Starr 1982, Kent and Wong 1982, Holling 1986, O'Neill et al. 1986, Holling 1992). In hierarchy theory, complex ecosystems are decomposed into discrete hierarchical levels; each level represents a scale at which biotic and abiotic factors interact much more strongly with each other than with other factors at other levels (Allen and Starr 1982, O'Neill et al. 1986). Usually, factors at broader scales (i.e., higher hierarchical levels) constrain factors at limited scales. For example, plant growth is constrained by disturbance processes (e.g., fire) and climatic conditions that operate at broader scales. A hierarchical organization of ecosystems has been inferred from empirical studies of ecosystems and from models that captured the dynamic behavior of managed ecosystems (Allen and Starr 1982, Kent and Wong 1982, Bradbury and Reichelt 1983, O'Neill et al. 1986, Holling 1992). Such hierarchical considerations have greatly enhanced our understandings of ecosystems, and I suggest that hierarchy theory will provide a simplifying framework to study the scales at which birds respond to landscape structure.

The range of scales at which animals respond to landscape structure can be defined by the *spatial extent*, or largest area in which an organism responds to habitat structure, and the *spatial grain*, or the minimum resolution at which an organism responds to objects in a landscape (Kotliar and Wiens 1990, Wiens 1990). The structural objects in a landscape that a species responds to are dependent on natural history. For example, a bird could primarily respond to tree patches, flowering shrubs, riparian habitat, natural open spaces, and even housing density or distribution of lawns. Which objects are evaluated and in what order probably depends on the focal species (see also Johnson 1980, Hutto 1985). When a wren and a hawk fly over a landscape in search of foraging habitat, each bird probably responds to different landscape attributes as a function, in part, of its life history and body size. For example, from estimates of home range areas (Schoener 1968), a Carolina Wren (*Thyothorus ludovicianus*) probably samples small areas and selects small objects, whereas a Red-tailed Hawk (*Buteo jamaicensis*) samples large areas and large objects (Fig. 7.1). Each decision level essentially has its own spatial extent and grain. Thus, I am extending Wiens' (1990) definition of grain and extent to be specific for each hierarchical level. Within this decision hierarchy, broad-scale decisions probably constrain decisions made at limited scales (Johnson 1980, Holling 1992). Thus, whether a species appears at a specific location (e.g., in the confines of a backyard) depends not only on whether sufficient limited-scale structure exists at this location, but it first depends on whether sufficient structure exists at broad scales relevant to a species.

Because many ecological and behavioral traits are correlated to body size (e.g., Brown 1995), size may be a useful indicator of the scales at which animals respond to landscape structure (Wiens 1989, Holling 1992, Hostetler 1997, Hostetler and Holling 2000). In fact, certain body-size ranges could exist in which most of the species respond to structure at approximately the same range of scales. However, other variables, besides body size, can influence the scale at which animals primarily respond to landscape structure. For example, the spatial extent (e.g., home range) relevant to a given species can be quite variable depending on its trophic status, its diet, and the productivity of a landscape in which it resides (Harestad and Bunnell 1979). In general, birds occupying higher trophic positions and inhabiting a less productive landscape have larger spatial extents. For example, a bird of prey, such as an American Kestrel (*Falco sparverius*) has a much larger home range (142 ha) than an omnivorous species of approximately the same size, e.g., American Robin (*Turdus migratorius*, 0.121 ha; Schoener 1968). Large herbivorous birds (diet consisting of 0-10% animal food) have much smaller home ranges than comparable carnivorous birds (90-100% animal food; Table 7.1). However, as opposed to extent, grain size could be an invariant function of the size of a bird (Holling 1992). For example, the size of a bird generally limits the size of its food items.

Thus, body size may be a useful predictor of the scales relevant to a species, especially if similar-sized species were grouped according to trophic status and to diet. Such knowledge would help in evaluating the impact of a

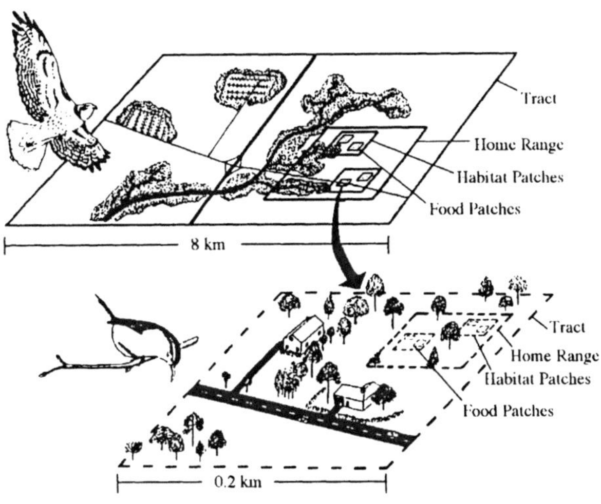

Figure 7.1. A theoretical representation of the scale-dependent landscape structures responded to by a Carolina Wren and a Red-tailed Hawk. Based on estimated home range areas, the decision hierarchy for each bird was constructed by scaling up or down an order of magnitude from one level to the next. The wren searches a tract of land (spatial extent of 0.01 - 0.04 km^2) to establish a home range (grain size of 1000.0 - 4000.0 m^2); at the next scale, the wren searches its home range (now the spatial extent) for suitable foraging patches (grain size of 100.0 - 400.0 m^2); then, within these foraging patches the wren locates areas (grain size of 10.0 - 40.0 m^2) where food is abundant; and at the smallest scale, the wren searches these areas where food is abundant for its prey (grain size of 0.1 - 1 cm^2). The hawk has a similar nested set of hierarchical decisions, but it selects much larger areas and objects at each comparable scale (e.g., home range equals a spatial extent of 2.0 - 4.0 km^2). Notice that the only overlap in the types of objects sampled by the two birds is at the food patch level for the hawk and at the tract level for the wren. (adapted from Hostetler and Holling 2000)

planned development because the scales relevant to one species could be extended to a suite of similarly sized species. Below, I outline an empirical methodology on how to conduct multi-scale analyses.

3. MULTI-SCALE ANALYSES

The goal of the following methodology is to determine the primary grain and extent, at a given hierarchical level, that is relevant to different bird species. It is important at the start of such a study to define which hierarchical level one is addressing. For example, one could measure how the distribution of vegetative features influenced the foraging locations of a hawk within its home range or one could analyze how broader scale features influence the establishment of a home range. Below, I emphasize determining the scales at which a bird selects a home range. As mentioned previously, whether a species appears at a given location first depends on whether sufficient conditions exist at broad scales relevant to that species. To simplify the following explanation of the multi-scale analyses, I refer to an example of determining the relationship between tree patches (i.e., grain) within a specific area (i.e., extent) and bird distributions during the breeding season. However, the following analyses could be extended to any season and the measured structural variable (in this case tree patches) could be any variable including lawn patches and housing densities.

First, one must select bird survey sites that where tree patch geometries vary from one scale to the next. For example, if one measured the percent cover of tree canopy at three scales (e.g., 0.2 km^2, 1.5 km^2, and 25.0 km^2) at five different sites, then the ranking of these sites *must change* in some way when measured across the three scales. If the rankings were similar from

Table 7.1. Avian body masses and home ranges of several carnivores (C) and herbivores (H)*.

Species	Diet	Body mass (g)	Home range (ha)
Colinus virginianus	H	200	1.1
Lagopus lagopus	H	510	2.6
Dendragapus obscura	H	510	1.2
Tymphanuchus cupido	H	908	5.3
Centrocercus urophasianus	H	1135	14.6
Elanus leucurus	C	270	182.2
Accipiter cooperi	C	469	225.4
Circus cyaneus	C	521	252.1
Buteo swainsoni	C	971	246.5
Buteo jamaicensis	C	1126	424.9

*Adapted from Holling (1992).

one scale to the next, then a strong correlation would exist between the percent of patch cover across the scales. In other words, if the site rankings were 5 > 4 > 3 > 2 > 1 (in terms of tree patch cover) at each of the three scales, then the tree patch geometry is similar across the three scales. Thus, three simple regressions (one at each scale) between bird abundance and percent patch cover would explain similar amounts of variation (r^2). Without this scale-dependent variance, one cannot determine the relative importance of each scale to the variance in the bird counts. I have found that correlations (say for percent tree patch cover) of 50% or less across scales are required to distinguish which scale may be most important (see Hostetler and Holling 2000).

To attain this scale-dependent variability, I recommend a stratified random sampling strategy that includes different landscape types. For example, one could randomly select sites from high-density residential areas, low-density residential areas, city parks, woodlots, industrial sites, and even golf courses. Each of these areas probably is embedded in different landscape matrices, which would mix the rankings across sites of percent cover and sizes of tree patches at several different scales. Alternatively, one could select sites that definitely contain different rankings at several different scales, and then randomly select a subset of these sites.

Determining at which scales to measure structure is probably the most difficult step in the procedure because, in general, it is unknown what scales are relevant to different species. Published home range data (e.g., Schoener 1968) will help a researcher decide which scales may be most important to a species. I suggest evaluating sites based on a scale range that is at least an order of magnitude above and below published home ranges.

Once sites are selected, the next step is to measure tree patches at several different scales (e.g., 0.2 km^2, 1.5 km^2, and 25.0 km^2). The percent cover of tree patches (independent variable) would be measured at these scales from aerial photographs that were positioned over the exact location of the survey sites (dependent variable is the bird counts). At each site, the percent cover of tree patches is the *sum* of all discernable tree patches, from small to large. Then, a simple regression between the percent cover of tree patches and bird counts would be done for each species at each scale (refer to Fig. 7.2 - Step A). From these simple regression results, if only one scale displayed a significant r^2 value, then this scale can be designated as the "best prediction area" (BPA). However, a more likely scenario is that several scales would display significant r^2 values. As explained above, this can be due to the correlation in patch cover across scales. To calculate the BPA in these situations, a uniqueness index value (SAS Institute 1994) should be calculated. Essentially, one would enter into a full, multiple regression

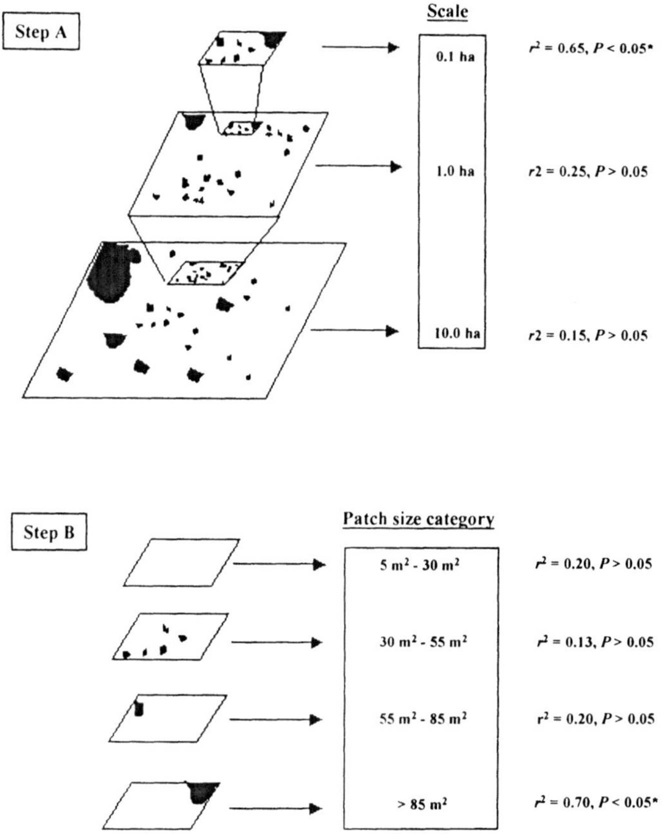

Figure 7.2. Steps A and B represent a hypothetical study that regressed the variation in bird counts of a species (across a range of sites) with the variation seen in tree patches (in black) across several different scales. The results of step A indicate that the broadest scale at which this species responded to structure was at 0.1 ha (called the Best Prediction Area or BPA). In step B, four separate regressions were run at the scale of 0.1 ha. These regressions were based on the percent of cover present across four patch-size categories. The results indicate that this species primarily responded to the > 85.0 m^2 patch size category. This category is designated as the Best Patch Size or BPS.

model the percent of patch cover at each scale that initially displayed a significant r^2 value. For example, tree patch cover at 0.1 ha and 1.0 ha could be the independent variables and bird abundance would still be the dependent variable. Then, drop one dependent variable from the full model and recalculate the r^2 value (i.e., the reduced regression). Comparing the full r^2 value and the reduced r^2 value, if a significant decrease in the r^2 value occurred ($P < 0.05$), then the scale that was removed from the full regression would be the BPA. A significant result means that the variation of bird counts was uniquely explained by one scale beyond the variance accounted for by the other scales. The result would be interpreted as the scale at which birds were primarily responding to landscape structure. This does not mean, though, that the other scales are not also important; it is just that this scale may be an extent at which a bird's occurrence is nonrandom with respect to measured habitat structure.

Now, if a BPA was calculated at a scale where regressions at broader scales were not significant by the uniqueness test, then this BPA would represent the broadest scale at which a bird responds to tree patches. If a BPA was at the broadest scale in the analyses (e.g., one did not measure tree patches at broader scales), then analyses should be extended to determine whether broader scales would have significant results. If several scales have significant r^2 values, but a uniqueness test was not significant for any of these scales, then one may have measured across several hierarchical decision levels. In this case, the broadest scale with a significant r^2 value can be considered as the broadest scale at which a species responds to tree patches. Keep in mind that this interpretation is valid only if percent cover of tree patches was *marginally* correlated for the scales that were compared. Strong correlations among the scales with significant r^2 values may not allow one to designate the broadest scale at which a species responded to tree patches. For example, a significant regression result could have been found for 0.1 ha and 1 ha, but correlation of percent cover of tree patches between the two scales was above 50%. In this case, one cannot state that 1 ha is the broadest scale at which the species responds to tree patches because it may be that this species primarily responds to tree patches at 0.1 ha. Alternatively, if correlation between the two scales is low (say 30% or below) and one still finds significant results at 0.1 ha and 1.0 ha, then perhaps this species is truly responding to structure at each scale. The 1.0 ha may be the scale at which a species responds to structure when selecting a home range (i.e., one hierarchical level) whereas at 0.1 ha, one has measured structure relevant to this species when selecting a foraging area within its home range.

Once a BPA (i.e., the primary spatial extent) is determined, one can then calculate the patch size (i.e., grain size) that is relevant to a particular species. The goal here is to determine the range of patch sizes that birds are primarily responding to in a landscape at a given BPA; this patch size range can be designated as the "best patch size" (BPS). To do this, one would construct several patch-size categories (from small to large patches) and run a separate, simple regression for each patch-size category (percent cover of tree patches is the independent variable and the abundance of birds is the dependent variable). Then, one would determine the patch-size categories that produced significant correlations (Fig. 7.2, step B). From regression analyses, if only one patch size category produced a significant correlation, then this scale can be designated as the BPS. Probably several patch sizes will have significant correlations; this means that the rankings of percent patch cover for each patch size category were similar across all sites, i.e., sites with lots of small patches also had lots of large patches. If so, conduct the uniqueness index value test for patch size categories that displayed significant correlations (i.e., compare full vs. reduced regressions). The patch size category with a significant uniqueness result ($P < 0.05$) would be the BPS.

Similar to the BPA correlation problem described above, I found that a correlation of 50% or less among patch size categories is needed to have a chance at determining BPS. If correlations are too high, an additional study should be conducted to determine the BPS. Using the extent of a BPA, the best way to determine the BPS of birds is to select several different sites where the amount of a particular structural variable is similar but the geometry is totally different. For example, one site could contain three large patches, another site could contain 30 medium patches, and a third site could contain 100 small patches. Because percent cover at each site is represented by different patch sizes (i.e., low correlation), an analysis of the bird counts among these distinct sites may reveal the BPS used by a species. In addition, recall that at each scale in the BPA analysis (Fig. 7.2, step A), the percent cover of tree patches represents the sum of small to large patch sizes. Because the sizes of tree patches are not taken into account, a non-significant correlation *across all scales measured in the study* does not necessarily mean that a bird is not responding to tree patches. Using the above example, one site could be composed of three large patches, another site could be composed of 30 medium patches, and a third site could be composed of 100 small patches; the commonality among the sites is that the percent cover is approximately the same. If birds were responding to only the three large patches (i.e., they were counted only at this site), a regression analysis would show no correlation because all of the sites would have the same amount of cover in this analysis. Thus, data should be explored to determine whether such a situation exists. To do this, one would construct several patch size categories; measure the total percent cover for each patch size category at each scale; and then run separate regressions between percent cover of tree patches and bird abundance. Any significant result would indicate both BPS and BPA.

Why not divide up the tree patches into size categories in the initial BPA regressions? The reason to sum all patch sizes at each site initially is that birds could be responding to some combination of small to large patches, and initial analyses that separated out patch sizes (and ran separate regressions) may not capture this. One is not sure of the tree patch sizes that birds are responding to in a landscape, and thus it is prudent to first look at the total percent cover of patches. If no significant correlations exist at any scale, then explore regressions with arbitrary patch size categories.

Finally, one could evaluate whether similar sizes of birds have similar BPA and BPS values (Hostetler 1997, Hostetler and Holling 2000). Here, one would search for patterns in the data. Do certain size ranges of birds have similar BPA and BPS values? Do similar-sized birds with related natural history characteristics have similar BPA and BPS values? Such data exploration may reveal that size could be used as an approximation of the scales at which birds respond to landscape structure. However, certain problems are associated with determining whether body size is a good predictor of the scales relevant to different species. One potential problem is human bias in measuring landscape structure; we may not measure what birds "see." This is especially true for correlating BPS values with the size of a bird. For example, analyses described in this paper cannot account for the possibility that large birds covering large areas on a daily basis to offset low prey density may sample clusters of patches (Holling 1992, With 1994, Hostetler and Holling 2000). Birds travelling between productive patches may respond to clusters of small, interspersed patches. Thus, the true patch size relevant to a species may be much larger than what actually is measured.

Alternatively, certain birds may be responding to microhabitats within relatively large patches. For example, this may be the case of forest-interior specialists that only breed in large forest tracts (e.g., Robbins et al. 1989, Freemark and Collins 1992). In a previous study, several small forest-interior warblers had larger BPS values than some of the large corvid species (Hostetler and Holling 2000). Thus, one has to pay particular attention to the natural histories of various species and how this may be affecting the BPS values.

I point out these difficulties not to rationalize away any irregularities in the outcome of a body-size vs. scale analyses, but to highlight possible reasons for empirical results that do not match the hypothesis that smaller species respond to structure at more limited scales than larger species. Such outcomes should be explored further. For example, if a rather large species responded to relatively small objects, a type of cluster analysis should be conducted to determine if the landscape objects were distributed in a non-random fashion. If a small species were shown to respond to relatively large objects, the possibility that it is responding to small "micro-habitat" patches within these large objects should be explored.

4. POLICY AND MANAGEMENT IMPLICATIONS

Humans alter urban landscapes at many scales. For example, the design of a homeowner's backyard (limited scale) is determined by individual choice and preference; the landscape structure of a neighborhood (medium scale) is determined by economics, local city laws, zoning, and the developer constructing a neighborhood; and the landscape structure of a large cluster of subdivisions (broad scale) is governed by the decisions of city planners and politicians that develop a city's long-range comprehensive development plan for large regions of a metropolitan area (Hostetler 1999). Thus, the character of a landscape is the result of a combination of decisions made by humans at different scales. In addition, the pre-development "state" of the landscape can play a role in structuring a landscape. For example, subdivisions can look quite different, depending on whether a subdivision was placed on an agricultural field versus a forested landscape. Further, a development can change dramatically over time. A new subdivision would initially consist of low vegetation volume (e.g., dominated by shrubs and small trees) but over the next hundred years, it may transform to high vegetation volumes as initial plantings mature. Of course decisions collectively made by homeowners will ultimately affect the future character of any landscape over time.

In a sense, decisions made by humans are hierarchical in that limited-scale decisions are nested in broad-scale decisions. A backyard design is primarily influenced by a homeowner, but this backyard is embedded in a landscape matrix that was formed from decisions made by a local developer that designed the subdivision, and in turn, the subdivision is embedded in the section of a city that was designed by a city planner. Thus, a connection may exist between the scales at which a bird primarily selects an urban area during the summer, winter, and spring/fall migration seasons and the scales at which different humans alter urban landscape structure. Small birds (e.g., hummingbirds and wrens) may be most affected by homeowners' decisions; medium-sized birds (e.g., thrushes) may be most affected by local city laws, zoning, and developers; and large birds (e.g., hawks) may be primarily affected by city planners' decisions (Fig. 7.1).

Knowing the spatial extent and grain size used by birds would be useful information to design urban areas for birds. For example, homeowners who may want to attract a thrush to the backyard may have to think not only about ways to design the yard, but talk with their neighbors to encourage the planting of vegetation at a broader scale. In fact, a study on how tree patches affected the distribution and abundance of birds in seven North American cities indicated that many species might respond to structure at broad scales (Hostetler and Holling 2000). From this study, 75% of the species during the spring (n = 84) and 93% during the summer (n = 60) displayed significant r^2 values at scales from 0.2 km^2 to 85.0 km^2.

Further, the life history stage of an organism can influence the range of scales relevant to it (Levin 1992). The migration, breeding, and overwintering season may represent stages in a bird's life where it responds to structure at different scales. Most organisms have some sort of dispersal phase and an organism's response to landscape structure may be quite different during its dispersal phase than during its sedentary phase (Wiens 1976, Levin 1992). In one previous study, some 7.0 g - 61.6 g birds had larger BPS values in the summer than in the spring (Hostetler and Holling 2000). Perhaps when species are searching for a nesting area, large tree patches are required for successful reproduction (e.g., Gibbs and Faaborg 1990, Hoover et al. 1995). During the spring, though, migrants search for stopover sites in which to rest and replenish energy resources (Moore and Simons 1992), and sites composed of small patches may serve as stopover areas (Brawn and Stotz 2001). Breeding birds are more concerned with habitat that provides cover in which to hide their nests whereas migrating individuals are procuring food as quick as possible. Such seasonal differences would indicate that urban areas with small tree patches may attract certain species during the spring but would not attract these species during the summer. Thus, studies should consider how the life history stage of birds might affect the scale at which they responds to landscape structure.

Providing a usable habitat for a species is the first step in managing cities for birds. Any urban design project is constrained by the amount of area that can be manipulated and also by the landscape matrix that it is embedded in. Understanding the scales at which different species respond to landscape structure will aid in evaluating how a given landscape could best be designed to attract bird species. Such knowledge will allow urban wildlife researchers to adequately inform city planners, developers, and homeowners how their actions influence the distribution of various avian species. I hope the approach outlined in this paper will promote future empirical studies on the scale-dependent responses of birds to urban landscape structure.

ACKNOWLEDGEMENTS

Thanks to all the people at the Arthur R. Marshall Laboratory of Ecological Science at the University of Florida who helped with this project; especially Paul Marples for his initial comments on the ideas contained in this manuscript. I thank the people at the Center for Environmental Studies, Arizona State University for the stimulating atmosphere to refine the ideas contained in this paper. This research was funded by the National Science Foundation grant DEB-9714833.

REFERENCES

Allen, T. F. H., and T. B. Starr. 1982. Hierarchy: perspectives for ecological complexity. University of Chicago Press, IL.

Blair, R. B. 1996. Land-use and avian species diversity along an urban gradient. Ecol. Appl. 6:506-519.

Bradbury, R. H., and R. E. Reichelt. 1983. Fractal dimension of a coral reef at ecological scales. Mar. Ecol. Prog. Ser. 10:169-171.

Brawn, J. and D. Stotz. 2001. The importance of the Chicago region and the Chicago Wilderness initiative for avian conservation, p. 511-524. *In* J. M. Marzluff, R. Bowman, and R. Donnelly [EDS.], Avian ecology and conservation in an urbanizing world. Kluwer Academic, Norwell, MA.

Brown, J. H. 1995. Macroecology. University of Chicago Press, IL.

Degraaf, R. M., and J. M. Wentworth. 1981. Urban bird communities and habitats in New England. Forty-sixth N. Amer. Wildl. Conf.:396-412.

Emlen, J. T. 1974. An urban bird community in Tucson Arizona: derivation, structure, regulation. Condor 76:184-197.

Freemark, K., and B. Collins. 1992. Landscape ecology of birds breeding in temperate forest fragments, p.443-454. *In* J. M. I. Hagan and D. W. Johnston [EDS.], Ecology and conservation of neotropical migrant landbirds. Smith. Inst. Press, Washington, DC.

Germaine, S. S., S. S. Rosenstock, R. E. Schweinsburg, and W. S. Richardson. 1998. Relationships among breeding birds, habitat, and residential development in greater Tucson, Arizona. Ecol. Appl. 8:680-691.

Gibbs, J. P., and J. Faaborg. 1990. Estimating the viability of ovenbird and Kentucky warbler populations in forest fragments. Conserv. Biol. 4:193-196.

Gober, P. 1998. The demographics of urban growth in Phoenix, p.30-36. *In* R. Melnick [ED.], Growth in Arizona: the machine in the garden. Morrison Institute for Public Policy, Tempe, AZ.

Harestad, A. S., and F. L. Bunnell. 1979. Home range and body weight - a reevaluation. Ecology 60:389-402.

Harrison, R. L. 1992. Toward a theory of inter-refuge corridor design. Conserv. Biol. 6: 293-295.

Holling, C. S. 1986. Resilience of ecosystems; local surprise and global change, p.292-317. *In* W. C. Clark and R. E. Munn [EDS.], Sustainable development of the biosphere. Cambridge University Press, United Kingdom.

Holling, C. S. 1992. Cross-scale morphology, geometry and dynamics of ecosystems. Ecol. Monogr. 62:447-502.

Hoover, J. P., Brittingham, M. C. and L. J. Goodrich. 1995. Effect of forest patch size on nesting success of Wood Thrushes. Auk 112:146-155.

Horak, G. C. 1986. Cumulative impacts of rapid urbanization on winter avian diversity in Northeastern Colorado. Ph.D. diss., Colorado State University, Ft. Collins.

Hostetler, M. E. 1997. Avian body-size clumps and the response of birds to scale-dependent landscape structure in suburban habitats. Ph.D. diss., University of Florida, Gainesville.

Hostetler, M. E. 1999. Scale, birds, and human decisions: a potential for integrative research in urban ecosystems. Landscape and Urban Plann. 45:15-19.

Hostetler, M. E., and C. S. Holling. 2000. Detecting the scales at which birds respond to structure in urban landscapes. Urban Ecosys. 4:25-54.

Hutto, R. L. 1985. Habitat selection by nonbreeding landbirds, p.455-476. *In* M. L. Cody [ED.], Habitat selection in birds. Academic Press Inc., New York, NY.

Johnson, D. H. 1980. The comparison of usage and availability measurements for evaluating resource preference. Ecology 61:65-71.

Kent, C., and J. Wong. 1982. An index of littoral zone complexity and measurement. Can. J. Fish. Aquat. Sci. 39: 847-853.

Kotliar, N. B., and J. A. Wiens. 1990. Multiple scales of patchiness and patch structure: a hierarchical framework for the study of heterogeneity. Oikos 59:253-260.

Levin, S. L. 1992. The problem of pattern and scale in ecology. Ecology 73:1943-1967.

Lima, S. L., and P. A. Zollner. 1996. Towards a behavioral ecology of ecological landscapes. Trends Ecol. Evol. 11:131-135.

Marzluff, J. M. 2001. Worldwide increase in urbanization and its effects on birds, p. 19-47. *In* J. M. Marzluff, R. Bowman, and R. Donnelly [EDS.], Avian ecology and conservation in an urbanizing world. Kluwer Academic, Norwell, MA.

Miller, J., J. Fraterrigo, J. Wiens, and T. Hobbs. 2001. Urbanization, avian communities, and landscape ecology, p. 117-137. *In* J. M. Marzluff, R. Bowman, and R. Donnelly [EDS.], Avian ecology and conservation in an urbanizing world. Kluwer Academic, Norwell, MA.

Mills, G. S., J. B. J. Dunning, and J. M. Bates. 1989. Effects of urbanization on breeding bird community structure in southwestern desert habitats. Condor 91:416-428.

Moore, F. R., and T. R. Simons. 1992. Habitat suitability and the stopover ecology of neotropical landbird migrants, p. 345-355. *In* J. M. Hagan III and D. W. Johnston [EDS.], Ecology and conservation of neotropical migrant landbirds. Smith. Inst. Press, Washington, DC.

Nilon, C. H., C. N. Long, and W. C. Zipperer. 1995. Effects of wildland development on forest bird communities. Landscape and Urban Plann. 32:81-92.

O'Neill, R. V., D. L. DeAngelis, J. B. Waide, and T. F. H. Allen. 1986. A hierarchical concept of ecosystems. Princeton University Press, NJ.

Opdam, P. 1991. Metapopulation theory and habitat fragmentation: a review of holarctic breeding bird studies. Landscape Ecol. 5: 93-106.

Penland, S. T. 1984. Avian responses to a gradient of urbanization in Seattle. Washington. Ph.D. diss., University of Washington, Seattle.

Richmond, H. R. 1996. Land-use policy for 21st century America: the conceptual basis of reform. Eugene V. and Clare E. Thaw Charitable Trust, Orlando, FL.

Robbins, C. S., Dawson, D. K., and B. A. Dowell. 1989. Habitat area requirements of breeding forest birds of the middle atlantic states. Wildlife Monographs 103:1-34.

SAS Institute. 1994. A step-by-step approach to using the SAS system. SAS Institute Inc., Cary, NC.

Schoener, T. W. 1968. Sizes of feeding territories among birds. Ecology 49:123-141.

Sexton, C. W. 1987. A comparative analysis of urban and native bird populations in Central Texas. Ph.D. diss., University of Texas, Austin.

Towne, M. 1994. Open space conservation in urban environments: lessons from Congo Valley, California. National symposium on urban wildlife, Bellevue, WA.

Walcott, C. 1974. Changes in bird life in Cambridge, MA, from 1860 to 1964. Auk 91:151-160.

Wiens, J. A. 1976. Population responses to patchy environments. Annu. Rev. Ecol. and Syst. 7:81-120.

Wiens, J. A. 1989. The ecology of bird communities: Foundations and patterns. Cambridge University Press, New York, NY.

Wiens, J. A. 1990. On the use of 'grain' and 'grain size' in ecology. Funct. Ecol. 3:385-397.

Wiens, J. A. 1995. Habitat fragmentation: island vs. landscape perspectives on bird conservation. Ibis Suppl. S97-S104.

Wiens, J. A., N. C. Stenseth, B. Van Horne, and R. A. Ims. 1993. Ecological mechanisms and landscape ecology. Oikos 66:369-380.

With, K. A. 1994. Using fractal analysis to assess how species perceive landscape structure. Landscape Ecol. 9:5-36.

Woolfenden, G., and S. Rohwer. 1969. Breeding birds in a Florida suburb. FL. State Mus. Bull. No. 13.

Chapter 8

Urban birds: population, community, and landscape approaches

Douglas T. Bolger
Environmental Studies Program, HB6182, Dartmouth College, Hanover, NH 03755 USA, e-mail: douglas.t.bolger@dartmouth.edu

Key words: Birds, competition, demography, edge effects, fragmentation, parasitism, predation, southern California, urbanization

Abstract: A simple conceptual framework is presented that begins with a consideration of patterns of bird abundance in urbanized landscapes. The framework includes a consideration of two levels of mechanisms which combine to produce the observed patterns: demographic and ecological. The framework is employed to review research conducted on urban-induced fragmentation and edge effects in coastal southern California.

1. INTRODUCTION

The problems posed by urban avian ecology at first often appear unique. But practitioners should not allow this to deter them from borrowing approaches from the broader fields of population, community and landscape ecology. To this end, I present a simple conceptual framework of urban bird studies and illustrate this approach with examples from the urban system in which I work, coastal southern California.

This approach begins with a consideration of patterns of bird abundance across a gradient of urbanization. These patterns are used to generate hypotheses about the demographic and ecological mechanisms that may generate the patterns. Many studies of the effects of urbanization on birds have addressed and documented patterns of abundance across an urban gradient (Vale and Vale 1976, Blair 1996, Bolger et al. 1997). However, the mechanisms generating these patterns have only infrequently been addressed directly (Marzluff and Ewing, in press). While the inferential approach can provide many valuable insights, clearly a fuller understanding will result from the investigation of mechanisms (Marzluff et al. 2000). The framework I present incorporates two levels of mechanisms: demographic and ecological. The basic demographic parameters of birth, death, emigration and immigration produce the observed patterns of abundance across a gradient of urbanization. The ecological mechanisms are the autecological and inter-specific mechanisms that influence these demographic rates. They include the use and availability of feeding and breeding resources, and the intensity of competition, predation, and parasitism along a gradient of urbanization.

2. COASTAL SOUTHERN CALIFORNIA LANDSCAPE

For the purposes of this paper I consider the coastal southern California region to include the area within 20 km of the coast from the Santa Monica Mountains in Los Angeles County, south to the border with Mexico. I concentrate on research performed by my collaborators and I in coastal San Diego County, but also cite research done in Los Angeles, Orange and Riverside Counties. Urban research in this region has focused mostly on the effects of urbanization-induced habitat fragmentation and urban edge effects (Soulé et al. 1988, Bolger et al. 1991, Lovio 1996, Bolger et al. 1997, Sauvajot et al. 1998, Crooks et al., in press, Morrison and Bolger, unpubl. data). Little effort has been devoted to the study of bird populations within the urban matrix (but see Guthrie 1974), but the framework I present applies to both the urban matrix and the natural habitat embedded in the matrix or adjacent to it.

There are five primary terrestrial habitats within the coastal zone of southern California: coastal sage scrub, chaparral, riparian woodland/scrub, oak woodland and non-native grassland (Beauchamp 1986). Most research on urban habitat fragmentation in this region has been conducted in the two predominant shrub habitat types: coastal sage scrub and chaparral (see review in Bolger, in press). These shrub habitats occur in a mosaic with non-native grassland in the coastal region (Mooney 1977, DeSimone and Burk 1992). Mixed and Chamise chaparral typically occur on mesa-tops and north-facing slopes. Coastal sage scrub (CSS) is much more common than chaparral in the coastal area and occurs on slopes and other well-drained areas. Disturbance contributes to the mosaic because CSS is an early successional community following disturbance to chaparral stands. The invasion of non-native annual plants has had a significant effect on the distribution of native shrub habitats (Mooney et al. 1986, Minnich and Dezzani 1998). Frequent or intense fires can type-convert CSS and chaparral to non-native grassland (Zedler et al. 1983). Coastal sage scrub in particular is vulnerable to conversion to non-native grassland, even in the absence of overt disturbance (Minnich and Dezzani 1998). Of the two shrub habitat types, CSS is of greater conservation concern and has been more extensively studied for urban fragmentation effects. Coastal sage scrub is notable for its restricted range within the U.S., the high diversity of endemic plants and animals, and the reported loss to development of 85-90% of its former distribution (Atwood 1993, Atwood and Noss 1994).

For the purposes of this discussion I subdivide the urban gradient of coastal San Diego County into four general elements, with decreasing exposure to the urban matrix. 1) The urban matrix. In this element, urban land-use is the predominant land-cover and is characterized by high-density (6-18 houses/ha) residential development. However, the urban matrix is actually a mosaic of urban habitat types; housing density varies from low-density exurban or "ranchette" development to dense urbanization. Non-native ornamental vegetation ranges from sparse in the higher density neighborhoods to lush in some of the older or more affluent neighborhoods. 2) Isolated fragments of native shrub habitats (1 to 1000ha). Fragments occur throughout most of the highly developed portion of the landscape. 3) The edge of large habitat blocks; habitat located within 250m of the urban edge. 250 m was chosen for the working definition of edge habitat because 1-2 typical passerine territories could occur laterally in a band of habitat of that width. 4) The interior of large undeveloped habitat blocks or wildlands (1,000-20,000 ha); this category includes only habitat greater than 250m from the urban edge. These large habitat blocks are embedded in the urban matrix but in some instances are contiguous to the east with the mountainous interior of San Diego County.

3. PATTERNS OF AVIAN ABUNDANCE ALONG A GRADIENT OF URBAN EXPOSURE IN SAN DIEGO

Patterns of bird species abundance across the urban gradient provide a useful point of departure for investigations of the effects of urbanization on bird populations and communities. Most research in southern California has focused on three discrete levels of urban exposure, isolated small fragments of native habitat surrounded by urbanization, the edge of large blocks of native habitat that abut residential development, and the interior of those blocks.

One pattern that has been observed is the relationship between the probability of bird species occurrence and the area or age (e.g. time since isolation by urban development) of urban habitat fragments. Soulé et al. (1988) found that the species richness of a group of eight shrub-habitat-requiring bird species (Bewick's Wren, Spotted Towhee, California Thrasher, Wrentit, California Quail, Greater Roadrunner, Cactus Wren, and California Gnatcatcher; see Table 8.1 for scientific names) increased with fragment area (range 0.4-103 ha) and declined with fragment age (range 2-86 years). The decline in species richness with fragment age implies relaxation or faunal collapse (Brown 1971, Soulé et al. 1979) with the rate of local extinction greater than recolonization across the urban matrix. An extinction-recolonization imbalance is supported by the observation that species richness in small fragments surrounded by urban development was significantly lower than that in similar-sized plots in unfragmented habitat (Bolger et al. 1991). Crooks et al. (in press) resurveyed the same fragments surveyed by Soulé et al. (1988) and found that, consistent with the relaxation hypothesis, there were approximately twice as many extinctions (30) as colonizations (12) between 1987 and 1997 among the original group of species studied by Soulé et al (1988). Crooks et al. (in press) also demonstrated that fragment area sensitivity is, in part, a function of fragment age in this system. Independent of area, species loss was greater in older than in younger fragments.

Patterns of bird species abundance also have been described at larger spatial scales. Bolger et al. (1997) analyzed landscape-scale patterns of bird abundance in CSS and chaparral habitat in coastal southern California as a function of the degree of urban exposure. They documented the patterns of abundance (detections per point count) of the 20 most common breeding bird species at locations that varied in exposure to the urban matrix. The study area (260 sq. km) included the interior of a large unfragmented habitat block, the edge areas of the block adjacent to the urban matrix, and fragments embedded in the urban matrix (Fig. 8.1). The fit of logistic regression models to the presence or absence of individual species was improved for 14 of the 20 species by the addition of landscape metrics to models containing variables describing local habitat conditions. These landscape metrics described the percentage of CSS and chaparral habitat versus urban land and the degree of interdigitation with urban areas in the larger landscape (buffer distances of 250 m to 3 km) around each sample point. Based on these analyses and a canonical correspondence analysis, the 20 species were characterized as edge/fragmentation-insensitive (10 species),

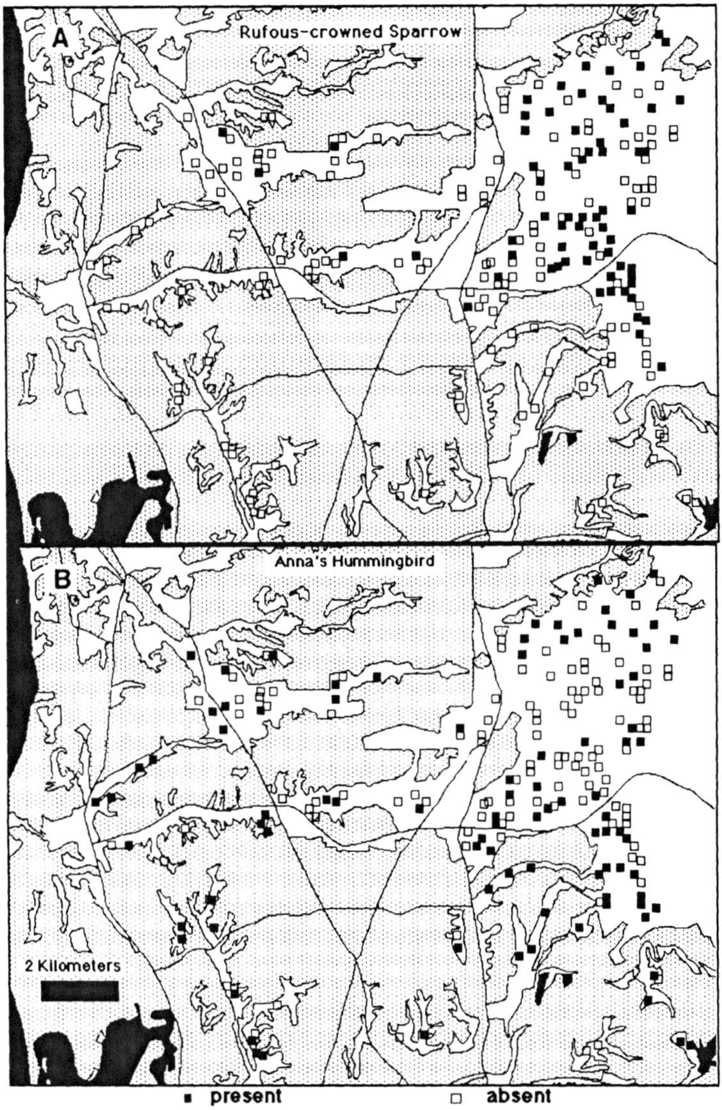

Figure 8.1. The landscape distribution patterns of (A) Rufous-crowned Sparrow and (B) Anna's Hummingbird within a 260 km^2 study area in coastal San Diego County. Presence/Absence refers to the detection or lack of detection of that species within a 100 m radius during one eight minute point count at each of 202 random locations during the spring of 1993. White areas are the undeveloped habitat mosaic of coastal sage scrub and chaparral. Stippling represents residential and commercial development. From Bolger et al. 1997.

edge/fragmentation-reduced (6 species) or edge/fragmentation-enhanced (4 species). The finding that half of the common species appear to respond to larger-scale patterns of edge and fragmentation underscores the importance of urban exposure as a determinant of bird abundance in this region.

Bolger (in press) further characterized patterns of landscape sensitivity using three descriptors of landscape pattern: 1) area sensitivity (Soulé et al. 1988, Lovio 1996, Bolger et al. 1997, Crooks et al., in press and unpubl. data); 2) edge sensitivity (Bolger et al. 1997, Bolger, in press); and 3) ability to exploit the urban matrix (Lovio 1996, Crooks et al., in press and unpubl. data). Based on a consideration of these three characteristics, species were assigned to three categories: 1) species that appear strongly negatively affected by fragmentation in the landscape, 2) species that appear moderately negatively affected by fragmentation, and 3) species that appear positively affected or neutral (Table 8.1).

Species sensitivity to fragmentation was assessed at two scales (10-20 ha or 100-200 ha fragments; Bolger, in press). However, area sensitivity in this system depends on fragment age (Crooks et al., in press) and possibly on isolation (Lovio 1996), so a simple categorization based solely on area was not possible. Thus our two categories provided a qualitative contrast of area sensitivity only for patches of CSS less that 50 years old and isolated by at least 500 m of residential development. Species categorized as sensitive to fragmentation at the scale of 10-20 ha are often found in fragments of this size, but have been shown to experience local extinction (Soulé et al. 1988, Bolger et al. 1991, Crooks et al., in press) and generally have lower abundance in fragments than in unfragmented habitat (Crooks et al., unpubl. data). Species categorized as sensitive to fragmentation at the scale of 100-200 ha are generally absent or rare in fragments smaller than that size (Lovio 1996, Bolger et al. 1997, Crooks et al., unpubl. data). Edge sensitivity was determined from a comparison of the relative abundance of species in interior habitat and habitat near the edge (<250m from urban edge) of large habitat blocks (Bolger et al. 1997, Bolger, in press).

Species in the first category, those strongly negatively affected, are generally found only in the largest habitat blocks remaining in the region. These species do not occur in the urban matrix, generally have reduced abundance near urban edges (Table 8.1; Bolger et al. 1997), and are extremely rare in smaller fragments (Crooks et al., unpubl. data). They tend to be a mixture of resident and migrant species. Although their landscape patterns suggest a high degree of sensitivity to urban exposure this is the most problematic category because the fewest data are available and alternative explanations for the landscape patterns of these species need further investigation. Many of these species are primarily grassland (Lark

Table 8.1. Patterns of landscape sensitivity in the avifauna of coastal southern California (from Bolger, in press).

Species	Area Sensitivity (ha)	Exploits Urban Matrix	Edge Response	Habitat
Strongly negative				
Lesser Nighthawk *(Chordeiles acutipennis)*	100-200	No	?	CSS/Chap.
Sage Sparrow *(Amphispiza belli)*	100-200	No	Negative	CSS/Chap.
Rufous-crowned Sparrow *(Aimophila ruficeps)*	100-200	No	Negative	CSS
Black-chinned Sparrow *(Spizella atrogularis)*	100-200	No	Negative	CSS/Chap.
Lark Sparrow *(Chondestes grammacus)*	100-200	No	Negative	Grassland
Grasshopper Sparrow *(Ammodramus savannarum)*	100-200	No	Neutral	Grassland
Western Meadowlark *(Sturnella neglecta)*	100-200	No	Negative	Grassland
Moderately negative				
California Quail *(Callipepla californica)*	10-20	No	Neutral	CSS/Chap.
Costa's Hummingbird *(Calypte costae)*	10-20	Yes	?	CSS/Chap.
Bewick's Wren *(Thryomanes bewickii)*	10-20	Yes	Neutral	CSS/Chap./ Riparian
California Thrasher *(Toxostoma redivivum)*	10-20	No	Negative	CSS/Chap.
Wrentit *(Chamaea fasciata)*	10-20	No	Negative	CSS/Chap.
California Gnatcatcher *(Polioptila californica)*	10-20	No	Neutral	CSS
Lazuli Bunting *(Passerina amoena)*	10-20	No	Negative	CSS/Chap.
Spotted Towhee *Pipilo crissalis)*	10-20	No	Negative	CSS/Chap.

Table 8.1 (cont.)

Species	Area Sensitivity (ha)	Exploits Urban Matrix	Edge Response	Habitat
Positive or neutral				
Mourning Dove *(Zenaida macroura)*	None	Yes	Neutral	Grass/CSS/ Chap./Rip.
Anna's Hummingbird *(Calypte anna)*	None	Yes	Positive	CSS/Chap.
Western Scrub Jay *(Aphelocoma californicus)*	None	Yes	Positive	CSS/Chap.
American Crow *(Corvus brachyrhynchus)*	None	Yes	Positive	CSS/Chap./ Riparian
Common Raven *(Corvus corax)*	None	Yes	Positive	CSS/Chap./ Riparian
Common Bushtit *(Psaltriparus minimus)*	None	Yes	Neutral	CSS/Chap./ Riparian
Northern Mockingbird *(Mimus polyglottus)*	None	Yes	Positive	CSS/Chap./ Riparian
Lesser Goldfinch *(Carduelis psaltria)*	None	Yes	Neutral	CSS/Chap.
House Finch *(Carpodacus mexacanus)*	None	Yes	Positive	CSS/Chap.
California Towhee *(Pipilo fuscus)*	None	Yes	Neutral	CSS/Chap.

Classifications of area sensitivity derive from Soulé et al. (1988), Lovio (1996), Bolger et al. (1997), Crooks et al. (in press) and unpublished data. Presence in the urban matrix is from Lovio (1996) and Crooks et al. (unpubl. data). Edge responses are derived from Bolger et al. (1997) and Bolger (in press). Habitat use is from Bolger et al. (1997), Unitt (1985) and pers. obs.

Sparrows, Grasshopper Sparrows and Western Meadowlarks) or chaparral species (Black-chinned Sparrow) (Table 8.1) that in the coastal habitat mosaic also occur in open CSS habitat, grassland/CSS ecotones or CSS/chaparral ecotones. Their patterns of abundance could reflect the distribution of these or other uncommon habitat elements that may be distributed non-randomly with respect to fragment size or edge proximity. For several of these species (e.g. Black-chinned Sparrow) the interpretation of their pattern of abundance is complicated by the confounding of an east-west gradient in fragmentation (Fig. 8.1) with a historical east-west gradient in abundance (Unitt 1984, Bolger, in press). The abundance gradient is probably due to the reduced cover of chaparral nearer the coast. Because a *priori* there is no reason to believe that the habitat elements required by these species would be underrepresented in fragments and near edges I suspect that the distribution of most of the species in this category are determined at least in part by patterns of fragmentation and edge. However, alternative explanations have not been thoroughly evaluated.

The second category is comprised of species that show area sensitivity in the range of 10-20 ha. Some of the species show edge sensitivity, others are neutral with regard to edge (Table 8.1; Bolger et al. 1997). These are generally resident species and are among the common and distinctive species of these habitats. They occur abundantly in both CSS and chaparral and are apparently shrub habitat generalists (Bolger et al. 1997). With the exception of Spotted Towhee, Costa's Hummingbird and Bewick's Wren, these species are generally not found within the urban matrix. Spotted Towhee and Bewick's Wren occur in the urban matrix at very low abundance while Costa's Hummingbird is reasonably common there (Crooks et al., unpubl. data).

The species categorized as neutrally or positively affected by fragmentation are all frequently observed in the urban environment. They reside and breed within developed habitats in San Diego (Unitt 1984, pers. obs.). All display positive or neutral edge responses (Table 8.1, Bolger et al. 1997). They vary in abundance in unfragmented habitat and none display obvious area sensitivity, in fact most are more abundant in fragments than in unfragmented habitat (Crooks et al., unpubl. data).

These patterns and categorizations suggest that, not surprisingly, these species differ in their responses to urbanization. The more interesting and difficult question is how do they differ, inviting the generation of hypotheses about the demographic and ecological mechanisms that generate these patterns. In the following sections I suggest categories of demographic and ecological mechanisms that could generate these patterns and evaluate the evidence for the action of these mechanism in the partially urbanized landscapes of coastal southern California.

4. DEMOGRAPHIC MECHANISMS

To understand how urbanization produces patterns of abundance across an urban gradient it is necessary to examine the rates of birth, death, immigration and emigration at points along the gradient. In non-urban systems the heterogeneity of demographic rates in different components of the landscape has received increasing attention. This has taken the form of source-sink dynamics (Pulliam 1988, Donovan et al. 1995), spatially-explicit population models (Pulliam et al. 1992, McKelvey et al. 1993) and habitat specific demography (Holmes et al. 1996). These are all manifestations of the importance of spatial variation in demographic rates to overall population dynamics. This approach is attractive for the study of urban bird populations because the gradient of urbanization creates a variety of urban habitat types. Patterns of abundance of individual species in the urban landscape are a product of the relative birth and death rates and dispersal between different landscape elements.

Morrison and Bolger (unpubl. data) carried out a demographic study of the Rufous-crowned sparrow, the species that displayed the most convincing pattern of urban fragmentation sensitivity (Table 8.1, Fig. 8.1a). The Rufous-crowned Sparrow is much less abundant in isolated fragments of habitat than in unfragmented habitat (Fig. 8.1a; Bolger et al. 1997, Crooks et al., unpubl. data) and significantly less abundant in the edge of large habitat blocks than in the interior (Bolger, in press). However, Morrison and Bolger (unpubl. data) found no difference in breeding success of Rufous-crowned Sparrows between edge and interior areas. Total seasonal reproductive output did not differ between pairs in habitat adjacent to urban development (<250 m from the urban edge) as compared to those a minimum of 500 m from urban edge. The lack of a reproductive edge effect (Morrison and Bolger, unpubl. data) suggests the landscape pattern of the Rufous-crowned Sparrow must result from differences in survival, emigration and/or immigration. This study included only two points on the gradient of urban exposure: the edge and interior of large habitat blocks. Thoroughly understanding the relationship of this species and others to urbanization will require demographic studies on a range of bird species across a wider spectrum of the urban gradient.

4.1 Immigration and Emigration Through the Urban Landscape

Immigration and emigration are fundamental demographic parameters that have been relatively ignored in avian population studies because of the inherent difficulties in quantifying these rates (Walters 1998). However, knowledge of the degree to which the urban matrix is permeable to dispersal is essential for interpreting patterns of apparent urban fragmentation sensitivity. For some fragmentation-sensitive bird species the urban matrix may be relatively impermeable to dispersal. For others, such as American Crows, it may facilitate dispersal (Marzluff et al. 2001).

Bolger (in press) suggested that the loss in species richness with increasing fragment age and decreasing area demonstrated by Soulé et al. (1988) and Crooks et al. (in press) was a result of the relative impermeability of the urban matrix to the group of species considered in those studies. However, this dispersal-limitation hypothesis needs direct testing. Bolger et al. (in press) demonstrated that fragmentation-sensitive species are more selective in their use of narrow, linear habitat features than fragmentation-tolerant species, suggesting the sensitive species have more stringent corridor requirements. This result is consistent with their movements through the urban matrix being more constrained. Dispersal studies and experimental recolonizations of empty patches could be used to test the alternative hypothesis that habitat degradation in fragments renders them unsuitable. Similarly, demographic studies on these bird species in habitat fragments and unfragmented habitat would quantify the relative habitat suitability of fragments.

Lovio (1996) studied fragments in another part of San Diego and found generally higher diversity in the same species group considered by Soulé et al. (1988) in similar-sized fragments. The differing results are probably the result of differing levels of isolation in the two sets of fragments. The Soulé et al. (1988) study area was in the western part of the city and the fragments were generally isolated canyons embedded in highly developed coastal mesas. Lovio's study area was slightly east and south in the Rancho San Diego area of San Diego and many of the fragments were portions of slopes and ridgetops that formed a fairly dense network of patches (Fig.1 in Lovio 1996). The mean interpatch distances (476 vs. 674 m) were shorter in Lovio's study area, and the intervening urban matrix was characterized by a lower density of houses and a higher cover of mature ornamental vegetation (Weser 1996, Bolger, pers. obs.). A number of the fragments were connected to other fragments by narrow habitat strips or areas of disturbed and non-native vegetation (Lovio 1996) and the set of fragments was immediately adjacent to a large unfragmented habitat block. The difference in species diversity between Lovio's and those of Soulé et al. (1988) studies may be indicative of the importance of the degree of fragment isolation and the permeability of the urban matrix. It also may suggest the existence of a isolation threshold below which species richness is sensitive to the distance between fragments. However, Lovio did not ascertain the age of individual fragments, so differing fragment ages also could be responsible for the differences between the studies.

The urban matrix is clearly not impermeable to all shrub habitat bird species. Two studies have documented dispersal of banded California gnatcatchers through fragmented landscapes. A banded juvenile was detected 1.3 km from its natal patch, having had to cross a lightly developed landscape of large wooded lots and parkland (Atwood et al. 1995 cited in Bailey and Mock 1998). Bailey and Mock (1998) also document a number of dispersal events in a heterogeneous landscape in San Diego. A number of these apparently occurred from a large block of habitat through an archipelago of fragments separated by blocks of development up to 1 km wide. This study was conducted in the same landscape as Lovio (1996) with dense ornamental vegetation and sufficient relief to often provide line-of-sight between patches of habitat. This probably facilitates inter-patch movement.

4.2 Source versus Sink Status Along the Urban Gradient

One surprising result from Bolger et al. (1997) was the abundance of urban-exploiting birds some distance into unfragmented habitat adjacent to urban areas. The abundance of the House Finch, Anna's Hummingbird (Fig. 8.1b), Northern Mockingbird and Lesser Goldfinch, species common in the urban matrix, was higher in habitat within 1km of the urban edge than further into the patch interior. These results suggest that the urban matrix could be a net source of these species, increasing densities in natural habitat adjacent to the matrix. In a similar fashion, depressed abundance of other species in natural habitat adjacent to urban areas (Table 8.1) might be caused by the urban matrix acting as a sink for some species and drawing off individuals from edge habitat. Determining the source, sink, and even sponge (Marzluf et al. 2001) status of species along the urban gradient would clarify these patterns of abundance.

5. ECOLOGICAL MECHANISMS

5.1 Breeding and Feeding Resources Across the Urban Gradient

Little research has focused directly on the autecological response of bird species to the urban gradient in coastal southern California. It is clear that species that occur in the urban matrix, such as the European Starling, House Finch, Rock Dove, Mourning Dove, Anna's Hummingbird and Common Raven, must exploit nest sites and feeding areas in these highly developed areas. House eaves, ornamental vegetation and roof tops are available for nesting sites, and feeders, refuse bins, gardens, lawns and other ornamental vegetation, can be used as foraging sites (Emlen 1974, Vale and Vale 1976, Beissinger and Osborne 1982, Mills et al. 1989). With the exception of Guthrie (1974), the particular resources exploited and the relative availability of these resources along a gradient of urbanization has not been quantified or compared to the availability of equivalent resources in natural areas in this region. And perhaps more importantly, the relative availability of these resources has never been linked to demographic rates.

5.2 Urban-induced Habitat Degradation

Fragmentation and the creation of urban edge exposes CSS and chaparral habitat to increased levels of human-induced disturbance. The effect of increasing disturbance in the form of mechanical damage, fire and exotic plant invasion on vegetation and birds in habitat fragments has not been thoroughly described. Alberts et al. (1993) found that fragments lose shrub cover through time and native plant diversity declines while exotic plant diversity increases. Disturbance opens up the vegetation in fragments by causing internal fragmentation with stands of shrubs becoming separated by non-native grasses and forbs.

The effect of disturbance-induced changes in vegetation structure on bird communities has not been well-studied in this region. Sauvajot et al. (1998) found that chaparral bird species abundance did not respond to disturbance-induced changes in vegetation structure. Bird species clearly assort along a gradient of shrub density from grassland to open CSS to dense CSS and chaparral (Cody 1975, Bolger et al. 1997). By decreasing shrub cover, disturbance should move the bird community along this gradient. The relationship of this avifauna to shrub vegetation structure needs further quantification.

5.3 Community Interactions: Top-down and Bottom-up Effects in the Urban Landscape

An active area of theoretical and empirical interest in ecology is in the relative importance of top-down (predators and parasites) versus bottom-up (food or productivity) influences on the regulation of ecosystems, communities and populations (Power 1992, Hunter and Price 1992, Polis and Strong 1996). Since the principal determinant of avian reproductive success is the rate of nest failure (Ricklefs 1969), most avian landscape studies have focused on the top-down effects of nest predation and brood parasitism by Brown-headed Cowbirds (Paton 1994, Robinson et al. 1995, Donovan et al. 1997, Hartley and Hunter 1998). Relative to top-down effects, potential bottom-up effects of fragmentation and edge on vertebrates, such as their effects on prey availability, have received much less attention (Burke and Nol 1998, Zanette et al. 2000). Edge between urban land-use and natural habitat creates the potential for important bottom-up effects. Urban land-use often involves supplements of water, energy and nutrients that can carry over into adjacent natural habitat and enhance productivity causing bottom-up changes. Edges between habitats that differ greatly in productivity can create strong bottom-up changes in the less productive habitat (Polis and Hurd 1996).

5.4 Top-down Effects - Predation and Brood Parasitism

Morrison and Bolger (unpubl. data) found no evidence to suggest that the landscape distribution of the Rufous-crowned Sparrow (Fig. 8.1a, Table 8.1) resulted from top-down effects near edges. Nest predation rates and cowbird parasitism did not differ between edge and interior areas. The predation result is surprising considering that some putative nest predators (e.g. Western Scrub-Jays and Common Ravens, Bolger, in press; California ground squirrels, *Spermophilus beechyi*, pers. obs.) are more abundant in edge areas. Video surveillance and direct observation documented ten predation events, nine by snakes, suggesting snakes are the principal predator on Rufous-crowned Sparrow nests. The rate at which snakes were encountered during field activities was equivalent in edge and interior areas. So the lack of a demographic edge effect in this bird species may be due to a lack of response to edge of its main predator. Understanding how important predators respond to edges is critical to understanding top-down effects (Marzluff and Restani 1999).

Top-down changes may be important in isolated habitat fragments. Crooks and Soulé (1999) found evidence for mesopredator release in fragments lacking coyotes (*Canis latrans*). They found that the abundance of mesopredators (grey fox, *Urocyon cinereoargenteus*; opposum, *Didelphis virgininiana*; striped skunk, *Mephitis mephitis*; and domestic cat, *Felis catus*) was negatively correlated with coyote abundance (after accounting for the effects of fragment area, age, and isolation). Moreover, they documented that mesopredator activity was higher at times when coyote activity was lower. Species richness of shrub-specialist birds was significantly positively correlated with coyote presence and showed a non-significant negative trend with increasing mesopredator abundance. Crooks and Soulé (1999) concluded that the presence of coyotes enhanced survival and reproduction of these birds through the suppression of mesopredators.

Brown-headed Cowbirds have been shown to be another important top-down influence in fragmented forest habitat (Robinson et al. 1995). However, they do not seem to be a significant influence in urban fragmented CSS vegetation. In four years in edge and interior habitat of Riverside and San Diego Counties, Morrison and Bolger (unpubl. data) found no brood parasitism by Brown-headed Cowbirds on Rufous-crowned Sparrows (n = 342 nests). Cowbirds in an urbanized region could be limited by access to agricultural feeding areas, although it is possible that cowbirds could use urban lawns for foraging (Robinson 1999). Ellison (1999) provided data suggesting that cowbirds may breed too late to parasitize resident breeding sparrows in southern California. However, in this region cowbirds are a significant problem in riparian woodland. The endangered Least Bell's Vireo is significantly affected by cowbirds (Kus 1999) as are other bird species (Laymon 1987).

5.5 Bottom-up Effects

Changes in food abundance along urban gradients, their ecological causes, and effects on bird populations have received little attention. Bolger et al. (2000) found complex relationships between arthropod diversity and abundance and urban fragment size, age, and urban edge proximity. Arthropods on foliage of California buckwheat (*Eriogonum fasiculatum*), a dominant shrub species in coastal sage scrub, generally declined in abundance and diversity with decreasing fragment size and increasing fragment age. Thus, food availability for foliage gleaners foraging on buckwheat was potentially lower in smaller and older fragments.

Reponses of ground-dwelling arthropods were more varied, but were generally similar to the shrub insects. Interestingly, spiders increased in abundance and diversity with decreasing area and increasing age. The most abundant ground arthropods in habitat fragments were non-native species: the sowbug (*Armadillidium vulgare*), the European earwig (*Forficula auriculatum*) and the Oriental cockroach (*Blatta orientalis*; Bolger et al. 2000). There were not large differences between the edge and interior of fragments in arthropod abundance and diversity.

In contrast, ground arthropods are generally less abundant adjacent to urban edge than in the interior of large habitat blocks in San Diego (Bolger, unpubl. data). Grasshoppers, mites, spiders, jumping bristletails and native ants were significantly less abundant in edge plots than in the interior plots. Beetles, bees and wasps, and flies did not differ between edge and interior plots. No arthropod order was significantly more abundant in edge plots than in interior plots. These results suggest food availability is lower in habitat adjacent to urban areas.

The arthropod taxa most vulnerable to fragmentation and edge are the native ants. In San Diego the non-native Argentine ant (*Linepithema humile*) invades CSS habitat from urban edges (Suarez et al. 1998, Bolger, unpubl. data). In isolated habitat fragments (Suarez et al. 1998) and in edge areas of large habitat blocks (Bolger, unpubl. data) the abundance and diversity of native ants is strongly negatively correlated with the abundance of the Argentine ant. Argentine ants are invasive human commensals and have become established in Mediterranean climates worldwide (Majer 1994) and have been implicated in the decline of native ants in a number of locations (Erickson 1971, Ward 1987, Majer 1994, Holway 1995, Cammell et al. 1996, Human and Gordon 1996).

The effect of the invasion of Argentine ants on bird populations is unclear. Ants frequently appear in lists of prey consumed by ground-foraging birds, but their relative importance as prey is unclear. Several studies have reported negative correlations of Argentine ants, or other exotic ants, with non-ant arthropods (Porter and Savignano 1990, Cole et al. 1992, Human and Gordon 1997, Bolger et al. 2000), while others have found no effect (Holway 1998). Bolger et al. (2000) found significant partial negative correlations between the abundance of Argentine ants and several non-ant arthropod taxa. However, the magnitude of the correlations was generally small suggesting the effect of Argentine ants on non-ant arthropods is less severe than their effect on native ants.

Urbanization could affect avian foraging patterns by altering the abundance and taxonomic composition of potential arthropod prey and by changing foraging microhabitat structure. The changes in arthropod abundance discussed above may not necessarily reflect food availability to the foraging birds in these sites (Hutto 1990). The increase in common non-native arthropods in fragmented habitat could offset losses in other species if birds use them as prey. To understand the effect of these bottom-up changes on birds it will be necessary to quantify foraging success, adult and nestling diet and their demographic implications across the urban gradient.

6. URBAN LANDSCAPE ECOLOGY

The field of landscape ecology has brought increasing attention to the role of large-scale heterogeneity in the regulation of local ecological processes (Turner 1989). Local bird abundance, demographic rates, and ecological processes may be a function of the larger urban landscape. In fragmented forest landscapes ecological interactions change as a function of the ratio of forested to unforested land area (Robinson et al. 1995, Donovan et al. 1997). In a similar fashion urban bird populations should be affected by the mix of landscape elements present. The urban landscape consists of patches of urbanization of varying intensities as well as remnant and revegetated patches of native vegetation, and patches of ornamental vegetation. This heterogeneity at the landscape scale could be a strong influence over local population dynamics and abundance of prey, predator and parasite species. Studies of urban bird populations and communities should increasingly incorporate landscape approaches. This will require partial abandonment of the urban gradient approach (McDonnell et al. 1993) for a patch-based conception of the urban landscape.

7. CONCLUSIONS

Research in southern California has focused on the effects of urbanization on bird populations residing in natural habitat adjacent to, or embedded in the urban matrix. The distribution and abundance of a number of bird species in this region suggests sensitivity to patterns of urban-induced fragmentation and edge. There is correlative evidence to support the view that the highly urbanized areas of the landscape are dispersal barriers to some fragmentation-sensitive species. The matrix is not necessarily a complete barrier to dispersal but it appears to reduce colonization rates below extinction rates for a number of species (Crooks et al., in press). Direct tests of this hypothesis in the form of dispersal studies or experimental introductions to unoccupied patches are needed. The only study of top-down effects did not find them, but nest predation must be investigated on a range of bird species to discover whether the results on the Rufous-crowned Sparrow are generalizable to other species nesting in different strata and with differing landscape sensitivities. Mesopredator release occurs in urban habitat fragments, but a direct impact on birds has yet to be demonstrated. Arthropod communities change substantially with urban exposure, but our understanding of the trophic effects of fragmentation is hindered by the lack of basic autecological data on bird foraging and diet, including adult and nestling food. The role of urbanization-induced vegetative habitat disturbance is still unclear.

To gain a fuller understanding of the effects of urbanization on this avifauna it will be necessary to consider avian population, community and landscape ecology across a gradient of urbanization. In this region that gradient includes low-density exurban or "ranchette" development, low-density affluent neighborhoods with lush landscaping, dense single family housing, high-density multi-family residential development, and commercial and light industrial areas. Studies of foraging, breeding, competition, predation and parasitism need to be carried out along this gradient. Such studies would reveal essential elements of an urban landscape conception of bird communities in this region. This is a daunting task and undoubtedly it will be some time before we compile the complete picture in this and most regions. In the meantime, we will have to continue to employ judicious interpretation of patterns of abundance, combined with a few demographic studies on carefully chosen species.

I believe the simple pattern-to-process framework I describe of documenting patterns of abundance along urban gradients and then generating and testing hypotheses about the demographic and ecological mechanisms that generate these patterns, can be helpful in advancing our understanding of urban bird populations in this and other landscapes.

8. POLICY AND MANAGEMENT IMPLICATIONS

Conservation planning in this region occurs through the state of California's Natural Communities Conservation Planning Program (NCCP). This NCCP was initiated partly in response to petitions to list the California Gnatcatcher as a state and federal endangered species (Atwood and Noss 1994). Private landowners voluntarily participate and the state coordinates subregional planning processes that prioritize lands based on conservation value. Planning occurs in 11 subregions with the purpose of designating an interconnected system of reserves which should result in no reduction in the ability of the region to maintain viable populations of target species over the long term (Atwood and Noss 1994). Several subregional plans have already been approved that include significant acreage of reserve lands (e.g. the Central-Coastal Orange County subregional plan 37,000 acres of reserve, and the MSCP subregional plan in San Diego, 170,000 acres of reserves; http://ceres.ca.gov/CRA/NCCP/updates.htm).

The reserve system that ultimately results from this effort will by necessity be set within a predominantly urban matrix. The research reviewed here points to the importance of the urban portion of the southern California landscape to the ultimate success or failure of this conservation effort. The categorization of species as to their sensitivity to the urbanized landscape (Table 8.1; Bolger et al. 1997, Bolger, in press) can help prioritize research, management, and monitoring efforts. Species whose patterns of abundance suggest sensitivity to urban exposure, all else being equal, will probably require additional efforts in these areas.

The studies reviewed here suggest that highly isolated shrub habitat patches less than 100 ha provided little conservation value for fragmentation-sensitive species over the long term. However, they do support other members of the regional fauna in abundance (Soulé et al. 1988, Crooks et al., in press). The limitations of fragmented habitat for conservation are acknowledged in the NCCP reserve selection guidelines that emphasize large, contiguous blocks of habitat (Atwood and Noss 1994.). A comparison of the results of Soulé et al. (1988) and Lovio (1996) suggests that denser archipelagos of fragments would support more inter-patch movement and higher abundance of fragmentation-sensitive species.

However, since we do not know whether fragments are sink or source habitat for most species it seems unwise to design landscapes to encourage dispersal to fragments from source habitat.

The dependence of fragmentation-sensitivity on fragment age (Crooks et al., in press) points to the dynamic nature of urban effects. It also suggests that static management concepts such as minimum area requirements and reserve buffer distances are unlikely to be sufficient. Management and monitoring efforts need to address the dynamism of urban effects. Monitoring should focus on bird populations and habitat quality in fragmented and edge habitat. Do habitat quality or bird populations decline through time and does the penetration of urban edge effects into reserves increase?

Finally, these results suggest that population segments residing within the urban matrix could affect the reserve system. Are these areas sinks for some species that could draw off reserve populations? Or, are they sources of other species that will elevate their abundances in reserves with community-level effects? Research, monitoring and management may be necessary in non-reserve urban areas to insure the success of the reserve system.

ACKNOWLEDGEMENTS

The ideas presented here grew out of discussions and collaborations with T. Case, K. Crooks, P. Mock, S. Morrison, J. Rotenberry, T. Scott, M. Soulé, and A. Suarez. I thank C. Chen, J. Faaborg, R. Bowman, and an anonymous reviewer for helpful feedback on the manuscript. I also acknowledge financial support from the National Science Foundation (DEB-9424559 and DEB-9981758) and the Metropolitan Water District of Southern California.

REFERENCES

Alberts, A. C., A. D. Richman, D. Tran, R. Sauvajot, C. McCalvin, and D. T. Bolger. 1993. Effects of habitat fragmentation on populations of native and exotic plants in southern California coastal scrub, p. 103-110. *In* J.E. Keely [ED.], Interface between ecology and land development in California. Southern California Academy of Science, Los Angeles, CA.

Atwood, J. L. 1993. California gnatcatchers and coastal sage scrub: The biological basis for endangered species listing, p. 149-169. *In* J. E. Keely [ED.], Interface between ecology and land development in California. Southern California Academy of Science, Los Angeles, CA.

Atwood, J. L., and R. F. Noss. 1994. Gnatcatchers and development: A "trainwreck" avoided? Illahee 10:123-130.

Bailey, E. A., and P. J. Mock. 1998. Dispersal capability of the California gnatcatcher: A landscape analysis of distribution data. Western Birds 29:351-360.

Beauchamp, R. M. 1986. A flora of San Diego County. Sweetwater River Press, National City, CA.

Beissinger, S. R., and D. R. Osborne. 1982. Effects of urbanization on avian community organization. Condor 84:75-83.

Blair, R. B. 1996. Land use and avian species along an urban gradient. Ecol. Appl. 6:506-519.

Bolger, D. T. In press. Habitat fragmentation effects on birds in southern California: Contrast to the paradigm. Stud. Avian Biol.

Bolger, D. T., A. C. Alberts, and M. E. Soulé. 1991. Bird species occurrence patterns in habitat fragments: sampling, extinction and nested species subsets. Am. Nat. 137:155-166.

Bolger, D. T., A. V. Suarez, K. Crooks, S. A. Morrison., and T. J. Case. 2000. Arthropods in habitat fragments; effects of area, edge and Argentine ants. Ecol. Appl. 10:1230-1248.

Bolger, D. T., T. A. Scott, and J. T. Rotenberry. 1997. Breeding bird abundance in an urbanizing landscape in coastal southern California. Conserv. Biol. 11:406-421.

Bolger, D. T., T. A. Scott, AND J. T. Rotenberry. In press. Use of corridor-like landscape structures by birds and small mammals. Biol. Conserv.

Brown, J. H. 1971. Mammals on mountaintops: Nonequilibrium insular biogeography. Am. Nat. 105:467-478.

Burke, D. M., and E. Nol. 1998. Influence of food abundance, nest-site habitat, and forest fragmentation on breeding ovenbirds. Auk 115:96-104.

Cammell, M. E., M. J. Way, and M. R. Paiva. 1996. Diversity and structure of ant communities associated with oak, pine, eucalyptus and arable habitats in Portugal. Insect Sociaux 43:37-46.

Cole, F. R., A. C. Medeiros, L. L. Loope, and W.W. Zuehlke. 1992. Effects of the Argentine ant on arthropod fauna of Hawaiian high-elevation shrubland. Ecology 73:1313-1322.

Cody, M. L. 1975. Towards a theory of continental species diversities, p. 214-257. *In* M.L. Cody and J.M. Diamond [EDS.], Ecology and Evolution of Communities. Belknap Press, Cambridge, MA.

Crooks, K. R., and M. E. Soulé. 1999. Mesopredator release and avifaunal extinctions in a fragmented system. Nature 400:563-566.

Crooks, K. R., A. V. Suarez, D. T. Bolger, and M. E. Soulé. In press. Extinction and recolonization of birds on habitat islands. Conserv. Biol.

DeSimone, S. A., and J. H. Burk. 1992. Local variation in floristics and distributional factors in Californian coastal sage scrub. Madroño 39:170-188.

Donovan T. M., F. R. Thompson, J. Faaborg, and J. R. Probst. 1995. Reproductive success of migratory birds in habitat sources and sinks. Conserv. Biol. 9:1380-1395.

Donovan, T. M., P. S. Jones, E. M. Annand, and F. R. Thompson III. 1997. Variation in local-scale edge effects: Mechanisms and landscape context. Ecology 78:2064-2075.

Ellsion, K. 1999. Importance of predation and brood parasitism on nest success in four sparrow species in southern California coastal sage scrub. Stud. Avian Biol. 18:191-199.

Emlen, J. T. 1974. An urban bird community in Tuscon, Arizona: Derivation, structure, regulation. Condor 76:184-197.

Erickson, J. M. 1971. The displacement of native ant species by the introduced Argentine ant *Iridomyrmex humilis* (Mayr). Psyche 78:257-266.

Guthrie, D. A. 1974. Suburban bird populations in southern California. Am. Midl. Nat. 92:461-466.

Hartley, M. J., and M. L. Hunter, Jr. 1998. A meta-analysis of forest cover, edge effects, and artificial nest predation rates. Conserv. Biol. 12:465-469.

Holmes, R. T., P. P. Marra, and T.. Sherry. 1996. Habitat-specific demography of breeding black-throated warblers (Dendroica caerulescens): implications for population dynamics. J. Anim. Ecol. 65:183-195.

Holway, D. A. 1995. The distribution of the Argentine ant (*Linepithema humile*) in central California: a twenty year record of invasion. Conserv. Biol. 9:1634-1637.

Holway, D. A. 1998. Effect of Argentine ant invasions on ground-dwelling arthropods in northern Califonia riparian woodlands. Oecologia 116:252-258.

Human, K. G., and D. M. Gordon. 1996. Exploitation and interference competition between the invasive Argentine ant, *Linepithema humile*, and native ant species. Oecologia 105:405-412.

Human, K. G., and D. M. Gordon. 1997. Effects of Argentine ants on invertebrate biodiversity in northern California. Conserv. Biol. 11:1242-1248.

Hunter, M. D., and P. W. Price. 1992. Playing chutes and ladders: heterogeneity and the relative roles of bottom-up and top-down forces in natural communities. Ecology 73:724-732.

Hutto, R. L. 1990. Measuring the availability of food resources. Stud. Avian Biol. 13:20-28.

Kus, B. E. 1999. Impacts of Brown-headed Cowbird parasitism on productivity of the endangered Least Bell's Vireo. Stud. Avian Biol. 18:160-166.

Laymon, S. A. 1987. Brown-headed cowbirds in California: historical perspectives and management opportunities in riparian habitats. Western Birds 18:63-70.

Lovio, J. C. 1996. The effects of habitat fragmentation on the breeding bird assemblage in California coastal sage scrub. M.S. thesis, San Diego State University.

Majer, J. D. 1994. Spread of Argentine ants (*Linepithema humile*), with special reference to Western Australia, p. 163-173. *In* D.F. Williams [ED.], Exotic ants: biology, impact, and control of introduced species. Westview, Boulder, CO.

Marzluff, J. M., and K. Ewing. In press. Restoration of fragmented landscapes for the conservation of birds: a general framework and specific recommendations for urbanizing landscapes. Restor. Ecol.

Marzluff, J. M., K. McGowan, R. Donnelly, and R. Knight. 2001. Causes and consequences of expanding American Crow populations, p. 333-365. *In* J. M. Marzluff, R. Bowman, and R. Donnelly [EDS.], Avian ecology and conservation in an urbanizing world. Kluwer Academic, Norwell, MA.

Marzluff, J. M., M. G. Raphael, and R. Sallabanks. 2000. Understanding the effects of forest management on avian species. Wildl. Soc. Bull. 28:1-12.

Marzluff, J. M., and M. Restani. 1999. The effects of forest fragmentation on avian nest predation, p. 155-169. *In* J. A. Rochelle, L. A. Lehmann, and J. Wisniewski [EDS.], Forest fragmentation: wildlife and management implications. Brill, Boston, MA.

McDonnell, M. J., S. T. A. Pickett and R. V. Pouyat. 1993. The application of the ecological gradient paradigm to the study of urban effects, p. 175-189. *In* M. J. McDonnell, S. T. A. Pickett [EDS.], Humans as components of ecosystems: The ecology of subtle human effects and populated areas. Springer-Verlag, New York, NY.

McKelvey, K., B. R. Noon, and R. H. Lamberson. 1993. Conservation planning for species occupying fragmented landscapes: The case of the Northern Spotted Owl, p. 424-450. *In* P.M. Kareiva, J. G. Kingsolver, and R. B. Huey [EDS.], Biotic interactions and global climate change. Sinauer Associates, Sunderlund, MA.

Mills, G. S., J. B. Dunning, Jr., and J. M. Bates. 1989. Effects of urbanization on breeding bird community structure in southwestern desert habitats. Condor 91:416-428.

Minnich, R. A., AND R. J. Dezzani. 1998. Historical decline of coastal sage scrub in the Riverside-Perris Plain, California. Western Birds 29:366-391.

Mooney, H. A. 1977. Southern coastal scrub, p. 471-489. *In* M. G. Barbour and J. Major [EDS.], Terrestrial vegetation of California. Wiley, New York, NY.

Mooney, H. A., S. P. Hamburg, and J. A. Drake. 1986. The invasions of plants and animals into California, p. 250-272. *In* H. A. Mooney and J. A. Drake [EDS.], Ecology of the biological invasions of North America and Hawaii. Springer-Verlag, New York, NY.

Paton, P. W. 1994. The effect of edge on avian nest success: how strong is the evidence? Conserv. Biol. 8:17-26.

Polis, G. A., and S. D. Hurd. 1996. Linking marine and terrestrial food webs: Allochthonous input from the ocean supports high secondary productivity on small islands and coastal land communities. Am. Nat. 147:396-423.

Polis, G. A., and D. R. Strong. 1996. Food web complexity and community dynamics. Am. Nat. 147:813-846.

Porter, S. D., and D. A. Savignano. 1990. Invasion of polygyne fire ants decimates native ants and disrupts arthropod community. Ecology 71:2095-2106.

Power, M. 1992. Top-down and bottom-up forces in foodwebs: Do plants have primacy? Ecology 73:733-746.

Pulliam, R. H. 1988. Sources, sinks, and population regulation. Am. Nat. 132: 652-661.

Pulliam, H. R., J. B. Dunning, and J. G. Liu. 1992. Population dynamics in complex landscapes – a case study. Ecol. Appl. 2:165-177.

Ricklefs, R. E. 1969. An analysis of nesting mortality in birds. Smithson. Contrib. Zool. 9:1-48.

Robinson, S. K. 1999. Section I: Cowbird ecology: Factors affecting the abundance and distribution of cowbirds. Stud. Avian Biol. 18:4-9.

Robinson, S. K., F. R. Thompson III, T. M. Donovan, D. R. Whitehead, and J. Faaborg. 1995. Regional forest fragmentation and the success of migratory birds. Science 267:1987-1990.

Sauvajot, R. M., M. Buechner, D. A. Kamradt, and C. M. Schonewald. 1998. Patterns of human disturbance and response by small mammals and birds in chaparral near urban development. Urban Ecosystems 2:279-297.

Soulé, M. E., B. A. Wilcox, and C. Holtby. 1979. Benign neglect: a model of faunal collapse in the game reserves of East Africa. Biol. Conserv. 15:259-272.

Soulé, M. E., D. T. Bolger, A. C. Alberts, R. Sauvajot, J. Wright, M. Sorice, and S. Hill. 1988. Reconstructed dynamics of rapid extinction of chaparral requiring birds in urban habitat islands. Conserv. Biol. 2:75-92.

Suarez, A. V., D. T. Bolger, and T. J. Case. 1998. The effects of fragmentation and invasion on the native ant community in coastal southern California. Ecology 79: 2041-2056.

Turner, M. Q. 1989. Landscape ecology: The effect of pattern on process. Ann. Rev. Ecol. Syst. 20:171-197.

Unitt, P. 1984. The birds of San Diego County. San Diego Society of Natural History, San Diego, CA.

Vale, T. R., and G. R. Vale. 1976. Suburban bird populations in west-Central California. J. Biogeogr. 3:157-165.

Ward, P. S. 1987. Distribution of the introduced Argentine ant (*Iridomyrmex humilis*) in natural habitats of the lower Sacramento Valley and its effects on the indigenous ant fauna. Hilgardia 55:1-16.

Walters, J. R. 1998. The ecological basis of avian sensitivity to habitat fragmentation, p. 181-192. *In* J. M. Marzluff and R. Sallabanks [EDS.], Avian conservation: research and management. Island Press, Washington, DC.

Weser, O. E. 1996. San Diego County 1995-96 aerial foto-map book. Aerial Fotobank, Inc., San Diego, CA.

Zanette, L., P. Doyle, and S. M. Trémont. 2000. Food shortage in small fragments: evidence from an area-sensitive passerine. Ecology 81:1654-1666.

Zedler, P. H., C. R. Gautier, and G. S. McMaster. 1983. Vegetation change in response to extreme events: the effect of a short interval between fires in California chaparral and coastal sage scrub. Ecology 64:809-818.

Chapter 9

Interactions among non-native plants and birds

Sarah Hayden Reichard, Linda Chalker-Scott, and Solai Buchanan
Center for Urban Horticulture, University of Washington, Box 354115, Seattle, WA 98195 USA, e-mail: reichard@u.washington.edu

Key words: Dispersal, exotic plants, frugivory, horticulture, invasive species, landscaping, ornamental, urbanization

Abstract: Humans have introduced many non-native plants into urbanizing landscapes. In numerous cases, the dispersal and establishment of non-native plants are facilitated by birds. We have reviewed documented relationships between birds and non-native plants with specific attention to the urbanizing environment. Birds consume fruits and disperse seeds of non-native plants. They may also increase seed production or plant/seed survival by pollinating non-native plant species or consuming insect predators, respectively. Some avian frugivores facilitate the spread of, and benefit from, non-native plant species. In contrast, some avian nectarivores and insectivores have evolved mutualisms with specific plants and have been negatively influenced by the spread of competing, non-native plants. We enumerate the characteristics of successful, avian-dispersed fruit displays, and then highlight conditions in the urbanizing landscape that influence avian community structure and the spread of non-native plants. Finally, we discuss how these characteristics should influence management decisions and future research directions.

1. INTRODUCTION

Horticultural introductions bring plants across physical barriers to areas of potentially suitable habitat. Having thus overcome geographical barriers to invasion, horticultural plants need only an effective means of dispersal to become naturalized and possibly invasive (Kruger et al. 1986). Birds play an important role in the dispersal of many non-native horticultural plant species from human-altered environments (White and Stiles 1992). In general, successful avian dispersal of plants is characterized by the movement of diaspores over large distances, and by a high probability that diaspores will be deposited in microenvironments suitable for germination and survival (Proctor 1968, Howe and Smallwood 1982). Some birds enhance non-native plant dispersion and may, in turn, be affected by the resulting plant community depending on their feeding guild and habitat use. Frugivores and omnivores may benefit widely from and, in some cases, facilitate the spread of non-native plant species. Frugivorous and omnivorous birds are often generalists who prove quite able to incorporate new fruit types into their diet (MacDonald 1986). Relationships between granivorous or nectarivorous birds and plants often evolve more closely; thus, non-native plants that rely on avian dispersal of seeds or pollen appear to be less likely to find suitable dispersers in non-native habitat (Stiles 1989). Similarly, non-native plants that displace native vegetation are more likely to negatively affect granivores and nectarivores than frugivorous birds (Breytenbach 1986). There have been notable exceptions, however, in which non-native horticultural plants with showy flowers have been utilized by nectarivorous birds (Waterhouse 1997). Insectivorous birds also can influence the spread of non-native plants through predation on herbivorous insects, but because their effect is mediated by other organisms, the influences of insectivorous birds on plants is usually not as ecologically salient as that of the other feeding guilds (Bock et al. 1986, Bossard 1991). But there are many notable instances in which non-native invasive plants profoundly alter native insect communities and, in this way, exert strong negative effects on insectivorous bird species, many of which have specific dietary and structural vegetation needs (Cohan et al. 1978). Birds also spread plants used for perching and nesting habitat when they inadvertently pick up propagules via adhesion (Sorensen 1986). Despite these obvious relationships, in many instances the impact that a non-native plant species will exert on bird species and *vice versa* is not straightforward.

The human-altered environments into which horticultural plants are introduced often present favorable conditions for effective avian dispersal. Fruiting horticultural ornamentals, selected for their aesthetic appeal to humans, often have characteristics such as abundant brightly colored fruits that make them particularly attractive to avian dispersers (Stiles 1982). Landscape plantings with a high proportion of edge area, and highly visible, well-spaced plantings are amenable to birds' visually-based food search strategies (White and Stiles 1992).

Though birds are often important dispersers of non-native plants, this phenomenon is probably not as widespread as casual observation of foraging birds would suggest. To illustrate this point, we present some original research on the incidence of avian dispersal of non-native plants in North

America and Hawaii. We note some habitat types in which avian dispersal of plants is common.

Bird community responses to urbanization are discussed in detail elsewhere in this volume (Section IV). Our purpose in writing this chapter is to elucidate the interactions between birds and non-native plants in human-altered landscapes and to discuss predictive models of such interactions. It is our hope that critically reviewing diverse sources will help direct future research on non-native plants and their avian dispersers in urbanizing environments. Despite many anecdotal observations, rigorously quantified studies in this area are scarce. We draw on a range of information, much of it collected in natural environments, to make inferences about bird interactions with non-native plants in human-altered environments. While remaining cognizant of the lack of causally linked relationships, small sample size, and a high degree of methodological incompatibility in the existing body of studies, we have tried to construct generalities that can aid managers faced with the all too real ramifications of non-native plant invasions.

This review is organized into nine major sections. We start by clarifying terminology and describing the general processes of plant invasion and avian plant dispersal. We then survey the literature describing the relationships between birds of different feeding guilds and non-native plants. Because the dispersal of non-native plants by frugivorous birds has been shown to constitute an efficient and widespread pathway of non-native plant dispersal, we highlight characteristics of fruit displays that make them especially attractive to birds. We then discuss the importance of avian dispersal in non-native plant invasions and attributes of urban environments that may increase its importance. Finally, we conclude by highlighting some of the information needs and management implications of the discussed interactions between birds and non-native plants.

2. TERMINOLOGY ASSOCIATED WITH PLANT PROPAGULES AND INVASIONS

We use the term *fruit* to describe the matured ovary of a flowering plant. The *pericarp* is the matured ovary wall of a fruit and the outer layer of the pericarp is the *exocarp*. Fruits can be either fleshy or dry, and while these terms necessarily stem from the particulars of a fruit's developmental process, they may be adequately understood here on an intuitive level. The term *seed* is used to describe the fertilized matured ovule(s) housed in the ovary of flowering plants. The outer coating of the seed is the seed *coat*. In some species, the seed coat can develop fleshy outgrowths called *arils* that make the seed desirable to dispersers. Still, not all seed plants bear fruits. The seeds of conifers are housed in *cones*, which may appear fruit-like but are scale-derived structures.

A *naturalized* plant is a plant of foreign geographic origin that is able to establish and reproduce as a native (Reichard 1997). *Establishment* describes the period in which a seed germinates and fully develops into a seedling. The term *invasive* describes a non-native species that can establish self-sustaining populations in environments with little to no disturbance (Bazzaz 1986).

In this review, we have tried to classify landscapes of the cited studies in a manner consistent with other articles in this volume (Marzluff et al. 2001). We recognize the value of using standardized land-use classifications, but must use these terms rather liberally as accurate landscape classification was often unspecified in the reviewed literature. We have listed the probable landscape classification of cited sources in the fourth column of Appendix 9.1.

3. INTRODUCTION AND INVASION OF NON-NATIVE PLANTS

Human modification of landscapes has served as a conduit for the introduction of non-native plants into urban and suburban ecosystems and the subsequent dispersal of these non-natives into natural ecosystems (Henderson 1995, Reichard 1997). Invasive non-native plants may compete with native species for resources (Huenneke and Thomson 1994), alter nutrient and hydrologic cycles (Neill 1983, Vitousek et al. 1987), and change disturbance frequency and intensity (Hughes et al. 1991, D'Antonio and Vitousek 1992). Biological invasions are second only to habitat loss and degradation as causes of endangerment to rare species in the United States, accounting for 49% of all threats to species and 57% of threats to rare plant species (Wilcove et al. 1998). Historically, the most frequently employed modes of invasive plant remediation have been directed at controlling existing populations. While existing population control efforts remain essential to preserving ecosystem integrity, increasing attention has been directed at preventing the dispersal of non-native plant species (Reichard 1997).

Various physical and biological factors act as barriers to the movement of invasive species (Kruger et al. 1986). To become invasive, a species must (1) cross a geographic barrier (such as that between continents), (2) find suitable growing conditions (such as those that match its water and temperature tolerances), and (3) establish and spread independently of cultivation (via vegetative, abiotic, and/or biotic means). Species of horticultural or agricultural value overcome the barriers associated with geography and habitat suitability when they are introduced as functional plants. Whether a non-native species overcomes the third barrier to invasion and is able to spread outside of cultivation depends on whether the species' means of establishment and dispersal is amenable to its new environment (Andersen 1995). In many cases, species require biotic relationships with pollinators or seed dispersers to spread successfully (Stiles 1989).

4. AVIAN DISPERSAL OF PLANTS

Birds are often efficient dispersers of plants. Many birds travel long distances, carrying propagules with them (Proctor 1968). Unlike abiotic dispersal, such as wind or water, birds select microhabitat and are more likely to transport propagules to areas that will be favorable to seedling establishment (Howe and Smallwood 1982). Also, many plants have evolved a requirement for chemical and/or physical treatment by the avian gastrointestinal tract and can germinate only after being consumed by birds (Glyphis et al. 1981).

4.1 Long Distance Seed Dispersal by Birds

Birds are able to transport seeds considerable distances away from parent plants. The distance of seed dispersal by birds is variable; seed may be held in the digestive tract of a bird for up to 100 hrs (Proctor 1968). While dispersal is usually less than 100 m from the maternal plant, medium to large birds may disperse seeds ≥1 km (Debussche and Isenmann 1994). Generally, there is a negative exponential decrease in propagule density as distance from the parent plant increases (Howe and Primack 1975). Many species travel considerable distances between foraging habitat and roosting or nesting areas every day (Oatley 1984, Moulton 1993, Kerley et al. 1996, Nakamura and Miyazawa 1997, Waterhouse 1997). Fruits are heavily utilized before and during migration periods when birds travel great distances, dispersing seed on the way (Willson 1986, White and Stiles 1992).

Avian dispersal of non-native plants leads to the formation of numerous satellite populations, increasing the rate of spread of the species and making control more difficult. Consumption of even a single fruit may lead to a clumped deposition of seeds, allowing several seeds to be "planted" simultaneously (Willson 1986). Models have demonstrated that an avian dispersal strategy is hypothetically the most effective way for a plant invasion to progress rapidly (Moody and Mack 1988). The numerous small populations that arise from bird dispersal may be difficult to detect because of their scattered distribution. As a plant species continues to spread from scattered foci and create more widespread infestations, the distribution of the plant species can quickly become so diffuse that eradication or even control becomes logistically and economically impossible (Reichard 1997).

4.2 Seed Deposition into Favorable Microhabitat

Unlike random dispersal by wind or water, bird dispersal is more likely to transport many terrestrial seeds to locations that are favorable to germination. This increased probability of reaching favorable habitat primarily occurs because birds select the habitat where they forage, perch, and nest (Howe and Smallwood 1982). Thus, because of active habitat selection, the distribution of areas a bird occupies are likely to be much more homogeneous than the distribution of areas a wind might deposit a given seed. An example of favorable differential seed deposition is that of wild olive (*Olea africana*), a riparian species consumed in large quantities by the Pied Currawong (*Strepera graculina*). The wild olive's penetration of riparian areas has been aided by the currawongs' tendency to regurgitate seed pellets after drinking water, thus differentially depositing the olive seeds near streams (Buchanan 1989).

Evidence of differential avian use of habitat and the consequent patterns of seed deposition have been documented in several studies that combine behavioral observations with examinations of the distribution of avian-dispersed seed depositions. Studies such as those conducted in patchy Mediterranean vegetation (Debussche and Isenmann 1994) and in abandoned German orchards (Milton et al. 1997) have found that avian seed deposition is disproportionately located under the canopies of isolated trees. Such patterns of deposition would not benefit shade intolerant plant species, but in hot dry environments the microclimate created by isolated trees is important for seed germination and seedling establishment (Debussche et al. 1982). Seed deposition density by birds has also been found to increase with vegetation height because higher perching sites often attract more birds (Debussche and Isenmann 1994), with fruit display abundance (Sallabanks 1993), and tree crown accessibility (Denslow and Moermond 1982).

The deposition of seeds by birds may favor successful germination in other ways. Seeds of pinyon pine (*Pinus edulis*) (Christensen et al. 1991) and limber pine (*Pinus flexilis*) (Feldman et al. 1999) are primarily dispersed by the seed predator Clark's Nutcracker (*Nucifraga columbiana*). Nutcrackers ensure against winter forage scarcity by subterranean caching of cone seeds while cones are plentiful in the fall. In a typical winter, about one-fifth of the cached seeds are not recovered (Feldman et al. 1999). Some fruits such as European hawthorn (*Crataegus monogyna*), and holly (*Ilex aquifolium*), contain secondary compounds that are mildly toxic to many avian consumers. These secondary compounds may limit the amount of fruit eaten at one time by any given individual, effectively ensuring seed dispersal through time and by many individuals (Barnea et al. 1993). Also, while the seeds inside fruits are sometimes regurgitated by birds, they are more commonly defecated. Invasions into nutrient-poor environments may be facilitated by nutrient enrichment from fecal matter (Kerley et al. 1996, Auge and Brandl 1997).

4.3 Seed Treatment for Germination by Birds

Many plant species have evolved strong protective seed coats that require passage through the avian gastrointestinal tract to facilitate seed germination (Glyphis et al. 1981). Removal of the exocarp and/or a break in the seed coat in the avian digestive tract may allow for the absorption of water. One such example would be that of a cultivated non-native species, *Michelia champaca* (Sapu), in Brazil. De Figueiredo (1997) found that seeds of *M. champaca* ingested by Swallow Tanagers (*Tersina viridis*) had a higher germination rate than uningested seeds. Even closely related species may differ in their requirements for treatment by birds. A comparative study of two weedy *Solanum* species found that *S. nigrum*, an invader of disturbed areas, did not exhibit bird-enhanced germination but *S. luteum*, an invader of more natural, arid sites, did (Barnea et al. 1990).

5. USES OF NON-NATIVE PLANTS BY BIRDS OF DIFFERENT FEEDING GUILDS

In this review, we have chosen to emphasize trophic relationships among non-native plant species and bird species. Birds also rely upon plants to provide habitat structure for nests and perches from which to forage, attract mates, hide from predators, preen, and roost. Certainly, drawing a distinction between trophic interactions and other kinds of interactions is somewhat synthetic as selection simultaneously acts on all activities, forage related or otherwise, influencing avian survivorship. Suitable breeding habitat must at once provide ample access to food resources, protection from predators, and suitable nesting substrates. Thus, in addition to the discussion on trophic interactions, we will highlight some of the relationships among bird species and non-native plants that are specifically related to changes in birds' nesting and perching habitat needs. We will discuss passive avian dispersal of plants, such as that which occurs due to seed adhesion, in this context.

5.1 Consumption by Frugivorous and Omnivorous Birds

Frequently, invasion by fruiting plant species has been facilitated by frugivorous and omnivorous birds. The many examples listed in Appendix 9.1 indicate a probable positive correlation between the presence of fruit-bearing invasive species and the occurrence of the birds that consume them (Table 9.1). In general, avian frugivory may be more the result of the ecological congruence of the historical, environmental, and geographical factors that have shaped a given area rather than a strict co-evolution among particular plant species and bird species (Herrera 1984). Certainly, a frugivorous relationship that has evolved between a given plant and bird can benefit several bird or plant species, regardless of whether they were historically in proximity to one another or not (Howe and Smallwood 1982). The positive feedback loop between birds and fruit plants might theoretically be expected to be stronger for frugivores than omnivores, as frugivores are more dependent on fruiting species. In actuality, however, the diets of omnivores usually develop with a greater degree of plasticity than those of frugivores, so that omnivores may be the feeding guild that most frequently benefits from the presence of non-native fruiting plant species (Middlemiss 1963, Stiles 1982, Baldwin and Lovvorn 1994, La Rue 1994). One notable example of an omnivore that has effectively incorporated many species of non-native plants into its diet is the Pied Currawong abundant in suburban Australia (Buchanan 1989, Humphries et al. 1991). In currawong populations of suburban Sydney, Buchanan (1989) demonstrated that most of the dietary intake was composed of non-native plant species. During the

Table 9.1. Probability of change in bird abundance in relation to change in invasive plant abundance. A positive correlation indicates that an increase in the abundance of a bird or plant type is likely to be followed by an increase in the associated group. Similarly, a decrease in a bird or plant group is likely to lead to a decrease in the associated group. The +/- correlation or invasive plants and insectivores reflects the ambiguous relationship between these groups. Insectivores might increase or decrease with invasive plants depending on the effects of plants on native and introduced insect faunas.

Invasive plant type	Bird's trophic position	Sign of correlation between plant and bird abundance
Bird-dispersed	Frugivore	+
Bird-dispersed	Omnivore	+
Bird-pollinated	Nectarivore	+
All	Insectivore	+/-

winter, its diet was largely wild olive, an invasive species with a prolific winter crop. The currawong is also a significant disperser of the non-native tree European hawthorn, non-native privets (*Ligustrum*), and firethorns (*Pyracantha*) (Buchanan 1989, Bass 1990, Bass 1995). The pattern of establishment of sweet briar (*Rosa rubiginosa*) around large trees has been attributed to roosting currawongs (Hatton 1989). Another omnivorous species that has successfully utilized non-native species is the European Starling (*Sturnus vulgaris*). The starling has become widespread in many human-altered landscapes around the globe. It has been implicated in the spread of *Hakea serica* (Middlemiss 1963, Breytenbach 1986) and *Acacia cyclops* (Fraser and Crowe 1990) in South Africa, the Russian olive (*Elegaganus angustifolia*) in the southwest U.S. (La Rue 1994), common elder (*Sambucus nigra*) in New Zealand (Williams and Karl 1996), and still other non-native plants in Australia (Green 1984).

Some non-native fruits may not be successfully ingested by frugivorous birds because the birds have gapes too small to effectively ingest the fruits (Howe and Smallwood 1982). Large-fruited species, more common in tropical environments, usually depend on mammalian dispersers (Balasubramanian 1996). The behavior of some birds, however, enables them to effectively forage on large fruits. Many generalist birds that consume fruits are now ubiquitous in suburban landscapes; fruits used by generalists in one area are probably utilized by birds in areas of comparable land-use intensities and environmental parameters.

Although there does not appear to be a lack of suitable avian frugivorous dispersers in urbanizing environments (Buchanan 1989, Calley et al. 1993), there are instances in certain areas where the severe decline of avian dispersers has limited plant dispersal. The Hawaiian avifauna has undergone severe transformations in response to land-use change and non-native plant introduction. Land birds like the Nene (*Nesochen sandvicensis*), a disperser of many native plant fruit species, have become extinct or rare (Cole et al. 1995). Introduced game birds including Ring-necked Pheasants (*Phasianus colchicus*) and Chukars (*Alectoris chukar*) are now the dominant avifauna in high shrublands on Haleakala, Maui. These birds primarily eat the fruit of native woody dicots and have been beneficial to restoring degraded ecosystems. They efficiently spread the native pukiawe (*Styphelia tameiameiae*), which has been very difficult to germinate under artificial conditions (Cole et al. 1995). It should be noted that even though Cole et al. concluded that pheasants and chukars did not significantly compete with native bird species for food and other resources, they might transmit bird parasites and diseases to vulnerable native birds, function as disease reservoirs, or increase numbers of avian predators. Thus, the net effect of these game birds on native avifauna may be detrimental, despite benefiting native plant species.

There are instances in which specialist avian frugivores have been negatively affected by the introduction of new fruiting plant species that either displace native fruiting species (Cohan et al. 1978, Breytenbach 1986, Williams and Karl 1996) or support population growth of a competing avian species (MacDonald 1986, Schmidt and Whelan 1999). In most of the cases that we reviewed, however, frugivorous birds have benefited from the presence of new non-native fruiting species (Appendix 9.1). The characteristics that make fruits particularly attractive to avian dispersers are discussed in detail in section six.

Sometimes frugivorous birds also digest the seeds within the fruit, suppressing plant dispersal. Even in cases of seed predation, however, birds may aid dispersal if they only digest a portion of the seeds in the fruits they consume. Many fruits contain vast numbers of seeds and rely upon such a dispersal mechanism. For example, Kalij Pheasant (*Lophura leucomelana*), an omnivorous non-native of Hawaii, eats fruits and seeds of the highly invasive vine banana poka (*Passiflora mollisima*) and the invasive shrub thimbleberry (*Rubus rosaefolius*). Lewin and Lewin (1984) found that *P. mollisima* was present in the gut of 82% of the examined Kalij Pheasants. From one gizzard alone, more than one hundred *P. mollisima* seeds and several thousand *R. rosaefolius* seeds were extracted. Many of the seeds were destroyed, but the sheer number of seeds recovered suggests that pheasants may be important dispersers of these species.

Frugivorous birds may consume fruits found on a variety of plant types ranging from vines to shrubs to trees and including native and invasive species. For example, in the northeastern United States Cedar Waxwings (*Bombycilla cedorum*) have incorporated fruits of the invasive shrub Morrow's honeysuckle (*Lonicera morrowii*) into their diet. The honeysuckle has become such a large part of waxwings' dietary intake that, due to certain compounds in the fruits, an increasing number of birds develop orange tips on their tail feathers rather than the usual yellow (Witmer 1996).

Often, the presence of non-native fruits may allow frugivorous and omnivorous birds to substantially expand their ranges (Oatley 1984, Binggeli and Hamilton 1993, Williams and Karl 1996) and increase survivorship (Stiles 1982, La Rue 1994). For instance, MacDonald (1986), working in South Africa, found that the Pied Barbet (*Lybius leucomelas*) utilized fruits of non-native *Acacia cyclops*, allowing the birds to expand beyond their former range. Stiles (1982) examined Christmas Bird Count data and the increasing popularity of multiflora rose (*Rosa multiflora*) as a multi-use ornamental in suburban and urban areas. He concluded that there was a causal link between increased planting of *R. multiflora* and the increased range of Northern Mockingbirds (*Mimus polyglottos*) in the northeastern U.S. (Stiles 1982).There is also evidence that longer temporal availability of non-native fruits has encouraged migratory birds to postpone migration (Debussche and Isenmann 1990).

5.2 Consumption by Granivorous Birds

Granivorous birds have facilitated the spread of non-native plant species, but these instances are much less widespread than those that arise with frugivorous birds (Richardson and Higgins 1998). Successful plant dispersal by avian granivores is often a more specifically evolved bird-plant interaction and in most cases is not transferable from one location to another (Feldman et al. 1999). Although granivorous birds digest seed, mechanisms do exist by which granivorous birds disperse seeds. Seed-containing structures, such as cones and winged arils, are often removed and carried some distance by granivorous birds before they are ingested (Debussche et al 1982). While many such seeds are digested, some typically are lost as the birds transport and dismantle cones and the other seed containing structures. For example, the cones of slash pine (*Pinus elliottii*) and Monterey pine (*P. radiata*), invasive species in Australia, are spread by Glossy Black Cockatoos (*Calyptorhynchus lathami)* (Richardson and Higgins 1998). Similarly, Carnaby's Cockatoo (*Calyptorhynchus funereus*) carries away the cones of the invasive maritime pine (*P. pinaster*). While the seeds of *Pinus* species are generally considered to be wind-dispersed, the role of avian seed dispersal appears to be quite significant (Richardson and Higgins 1998, Feldman et al. 1999).

Most seeds ingested by granivores are ground to a pulp in the avian digestive tract, but those stored in the crop remain intact. These viable seeds can escape digestion via incidental regurgitation or regurgitation that is stimulated when a bird has consumed enough of a fruit that threshold levels of mildly toxic secondary compounds are reached (Barnea et al. 1993), or even through the activity of raptors that prey on granivorous bird species. Raptors can act as secondary vectors for seed dispersal when they regurgitate or defecate the seeds from the crop of granivorous avian prey or the cheek caches of rodent prey. In a study by Dean and Milton (1988) in southern Africa, numerous raptor pellets contained viable seeds of several plant species. Although the actual numbers of seeds spread in this manner are likely to be small, this pathway of seed dispersal is of note because the sites in which seeds are deposited by raptors are often remote from the parent plant.

5.3 Pollination by Nectarivorous Birds

The majority of flowering plants, especially perennial plants, are genetically facultative or obligate cross-fertilizers, although the percentages may vary by latitude and other factors (Lloyd 1980). Some cross-fertilizing species are wind pollinated, but many depend on animals for pollination. The majority of zoophilous plant species are insect pollinated but others are adapted for bird pollination (Baker and Hurd 1968). Pollination by only a single bird species is rare, as nectarivorous birds are usually attracted to a number of plant species (Calley et al. 1993). Still, there are significant evolutionary pressures for the relationship between plant and pollinator to develop some degree of specificity (Faegri and van der Pijl 1979). Nectarivorous species are generally less common and occur only in limited geographic regions (Faegri and van der Pijl 1979). Thus, avian pollinated plants will often not find suitable pollinators in non-native environments, particularly if the plant has a flower morphology that requires specific traits of its pollinators, such as a long curved beak. Suitable species substitution for an evolved relationship is less likely in isolated areas, such as islands like Hawaii. This is illustrated by the example of the many now-extinct avian nectarivores of Hawaii, and the endemic Lobelioid species with which they evolved highly specific mutualisms (Faegri and van der Pijl 1979, Wagner et al. 1990). Success of non-native plants that utilize avian pollinators will also depend on the particular geographical context into which the plant is introduced (White and Stiles 1992). Certainly, when suitable nectarivorous birds are present, the probability that bird-pollinated plants are able to invade increases because suitable nectarivores are likely to increase seed set in a plant (Robertson et al. 1999). And when nectar-bearing plants are present, nectarivorous birds may be expected to increase as well (Table 9.1).

Whether pollination is a significant factor in limiting seed set is a somewhat controversial issue (Bawa and Beach 1981), but pollination limitation does appear to occur in natural populations, as well as introduced ones (Parker 1997). For example, Levin (1970) found that having abundant native and introduced pollinators can increase seed set in an introduced species. In New Zealand, human changes to the landscape and invasion by non-native plants have reduced populations of endemic mistletoes (*Peraxilla* spp.). Robertson et al. (1999) linked the decline in native bird pollinators and consequent reduction in pollinator visitations to a decline in the rates of fruit set in two species of *Peraxilla*.

Conversely, non-bird-pollinated plants may replace native bird-pollinated species to the detriment of native bird species (Fraser and Crowe 1990). In central Australia, tamarisk (*Tamarix aphylla*) often forms dense stands along riparian corridors and negatively impacts nectarivorous avian populations

(Griffin et al. 1989). Similar results were seen in the South African fynbos following the replacement of native nectar-bearing shrub species with introduced *Acacia cyclops*. As *A. cyclops* has supplanted native plant species, the density and species richness of native nectarivorous birds has decreased (Fraser and Crowe 1990).

Some horticultural species with large, showy flowers can offer native nectarivores rich food resources. In some notable instances, avian nectarivores that experienced population declines due to land conversion and food plant loss have been able to rebound because of increased nectar supplies provided by horticultural species. The Rainbow Lorikeet (*Trichoglossus haematodus*) nearly disappeared from urban areas around Sydney, Australia, as the sclerophyll forest, its major foraging habitat, was cleared. Due to the planting of non-native coral tree (*Erythrina variegata*) in urban parks and reserves, however, the Rainbow Lorikeet is increasingly abundant in this area (Waterhouse 1997).

5.4 Birds that Consume Insects and Other Invertebrates

The relationship between non-native plants and insectivorous birds is less direct than that of the other feeding guilds, and prediction of the impact that a non-native plant will have on insectivorous birds must take into account the effects on other species. Non-native vegetation may attract fewer insects (Southwood 1961, Green et al. 1989, Griffin et al. 1989), resulting in lower densities of insectivorous birds than those maintained by native vegetation (Beissinger and Osborne 1982, Griffin et al. 1989, Mills et al. 1989). Rottenborn (1999) and Griffin et al. (1989) found lower densities of insectivorous birds in non-native riparian vegetation than in native riparian vegetation.

Even if non-native plant species depress insect populations, the resulting impact on insectivorous birds may not be clear-cut. For example, the non-native saltcedar (*Tamarix chinensis*) has spread along water courses in the southwest U.S., usually forming dense stands, replacing native *Populus* spp. (cottonwood) and *Salix* spp. (willow) populations, and altering native insect populations (Carothers and Brown 1991). [Note: many botanists consider the plants invading the southwest to be *T. ramosissima*, but the article cited used the alternate nomenclature.] Saltcedar has been implicated in the decline of the Southwestern Willow Flycatcher (*Empidonax trailii extimus*), a federally listed endangered species. Many plants are pollinated by generalist insects that can be depressed by insectivorous birds, reducing seed production (Kerley et al. 1996). We found little published research on the effects of insectivorous birds on pollinators of invasive species and do not know if the lack of research implies a lack of effect. Indeed, avian insectivores have been found to significantly decrease insect herbivores that suppress the spread of non-native plants (Gardner and Thompson 1998). Many studies have documented effective avian reduction of insect pests of agricultural plant species (MacLellan 1958, Solomon et al. 1976, Quiring and Timmins 1988, Witt et al. 1995).

Like invasive terrestrial species, non-native aquatic plants often grow in dense populations. When generalist invertebrates flourish in response to these infestations, the non-natives support additional food resources available to waterbirds. There are well-documented cases in which dense populations of non-native aquatic plants have benefited waterbirds. Invertebrates important to the diet of waterbirds are more prevalent in areas with non-native Japanese eelgrass (*Zostera japonica*) than comparable non-vegetated or natively vegetated areas invasion in the Pacific Northwest. In Boundary Bay, British Columbia, a 17-fold increase in the *Z. japonica* population occurred between 1970-1991, resulting in an additional 2161 ha of habitat for several species of waterfowl (Baldwin and Lovvorn 1994). Similarly, the highly invasive freshwater aquatic plant, *Hydrilla verticillata*, increases horizontal vegetation diversity, which in turn increases invertebrate abundance and richness (as well as substantial grazing fodder), ultimately increasing the energetic return for foraging waterbirds. Experimental removal of *H. verticillata* in Texas led to a decrease in bird abundance (Esler 1990).

5.5 Role of Non-native Plants on Avian Nesting and Other Habitat Use Activities

Some birds are generalist foragers but have specialized nesting substrate needs (Lynn et al. 1998). For example, native birds in eastern Australian cities preferred native plants for non-foraging type activities but showed no foraging preference for native plants. (Green et al. 1989). Birds that select native nesting substrates can lose nesting substrate altogether or shift to sub-optimal substrates when invasions occur (Lynn et al. 1998). In contrast, nesting generalists may benefit from additional nesting substrate provided by non-native plants (MacDonald 1986). Structurally, whether or not nesting habitat is suitable for a given avian species often depends on the particular growth form of a given non-native and the density of its shoots relative to bird size and habitat utilization tendencies (Braithwaite et al. 1989). Non-natives that grow in dense stands discourage larger birds' use because they are difficult to navigate and forage within, but, for similar reasons, they may offer small birds protection from predators (Braithwaite et al. 1989, Benoit and Askins 1999). Structural features such as thorns and spikes may provide protection from mammalian predators. In northern Australia, *Mimosa pigra* has replaced thousands of hectares of sedge wetland. Overall, bird abundance is diminished in areas that have been invaded by *M. pigra,* but some small birds appear to nest in it. *Mimosa pigra* may be preferred as a nest plant by these species because it has thorns that offer protection from larger predators, and because their dense tangles of adventitious roots provide sites free from flooding (Braithwaite et al. 1989). Similarly, thorns of the non-native multiflora rose may benefit Northern Mockingbirds by helping to protect their nests and recent fledglings from predation (Stiles 1982).

Non-native plants may possess characteristics that are consistent with a bird's evolved search image of suitable nesting habitat but also possess attributes that increase rates of nest predation or brood parasitism. Because birds interacting with non-natives have not evolved with these species, they may not have the capacity to select against such negative attributes; in this way, non-native species may become significant sources of avian population sink. In Illinois, Whelan and Dilger (1992) observed avian species disproportionately selecting non-native species including amur honeysuckle (*Lonicera mackii*), tartarian honeysuckle (*Lonicera tatarica*), and common buckthorn (*Rhamnus cathartica*) for nesting habitat. These non-native species added structural characters such as extensive branching within the mid-layer of the altered arboretum landscape of their study. Among native plant species, such structural features had not been encouraged by managers. Schmidt and Whelan (1999) observed similar disproportionate nesting preferences by native thrushes for non-native honeysuckles and buckthorns. Unfortunately for avian nesters, higher rates of predation occurred on nests within these non-native plants than among nests located in native species. In the western U.S., brood parasitism rates appear to be higher among birds nesting in *Tamarix ramosissima* stands than those using native riparian vegetation. Brown-headed Cowbirds (*Molothrus ater*) often increase in abundance in areas that have been disturbed, as does *T. ramosissima.* The combination of these factors may make *T. ramosissima*, though structurally suitable for nest sites, a nest plant that indirectly decreases nesting success (Anderson et al. 1977, Lynn et al. 1998).

5.6 Passive Dispersal of Non-native Plants

Published literature on bird-dispersed seeds focuses primarily on those bird species that ingest and release viable seeds. Invasive plants can, however, spread their seeds passively via attachment to feet, feathers, or other appendages of birds (Stiles 1989). Adhesion can occur through barbed, hooked, or viscid plant structures surrounding the seed; because these structures are carried passively, they may be retained by a bird for quite a long time depending upon specific seed characteristics and avian preening behavior. Adhesive seed types are prevalent in disturbed urban and roadside environments where a high proportion of non-native plant species occur (Sorenson 1986). Such areas are not only affected by human activities that have been implicated in seed dispersal (Lonsdale and Lane 1994, Hodkinson and Thompson 1997) but also are submitted to high levels of bird use. Thus, the role of birds passively dispersing seeds cannot be discounted (Sorenson 1986). Invasive plants may attract birds as perching, roosting, or nesting sites, whereupon birds may inadvertently pick up seeds of that species (Glyphis et al. 1981) or disperse seeds of another invasive (Debussche et al. 1982). This mechanism of seed dispersal has not received much quantitative investigation among introduced plants. Still, anecdotal information on passive dispersal exists in the literature for certain plant genera. In Ridley's *Dispersal of Plants Throughout the World* (1933) an entire chapter is devoted to "dispersal by adhesion due to viscid exudation." *Pisonia, Boerhaavia*, and several composite genera were identified as having particularly sticky fruits that readily cling to the feathers and feet of various bird species, often to the detriment of the birds themselves (Ridley 1933).

6. FRUIT DISPLAYS OF BIRD-DISPERSED PLANTS

Most bird-dispersed, non-native plants use fruits, as opposed to seeds or vegetative parts as dispersal propagules (Appendix 9.1). In surveying the characteristics correlated with avian frugivory some common patterns emerge (Willson 1986). The traits we highlight here increase the probability that a fruit display will attract avian foragers. We caution that relative fruit preference by birds can vary with location, season, and annual weather patterns (Whelan and Willson 1994). This information may serve to predict which non-native species are likely to have significant relationships with frugivorous avian dispersers.

Bird-dispersed fruits are usually red and/or black (Willson 1986, Balasubramanian 1996) or have red arils (Winterbottom 1970) or pseudoarils; these species are preferred over those with a white epidermis (Knight and MacDonald 1991). The average length of fruits eaten by birds in eastern North America is around 8 mm (Willson 1986). Small fruits appear to attract a wider array of dispersal species than larger ones (Balasubramanian 1996). Similarly, species with small-seeded fruits are more likely to have more potential dispersers than large-seeded fruits (Herrera 1984). American Robins (*Turdus migratorius*) appear to prefer fruits of an invasive hawthorn (*Crataegus*) over a native congener. This preference is likely to stem from the invasive species offering fruit with twice as much pulp as that found in the native species (Sallabanks 1993). An unpublished study reported by Willson (1986) found that birds foraged preferentially on fruits with large amounts of pulp. Willson reported that nutrient content and seed size had no effect on fruit choices. The water content of fruits must also be considered as one of the potentially desirable aspects of fruit resources, especially in drier habitats (Stiles 1989). Given otherwise comparable characteristics, fruits in which the seeds are easier to remove from surrounding tissues are preferred over those that require more energy for removal (Stiles 1989).

The timing of a fruit display and the extent of the display appear to have a strong influence on which fruits are eaten. For the purposes of discussion, we will describe the two extremes of fruiting display strategies adopted by fruiting plants (as diagrammed in Figure 9.1). The first achieves high visibility and increases foraging efficiency by producing large displays of high quality, high lipid fruits. These large, often visually distinct displays are generally intense but short in duration (Stiles 1989). Invasive plant

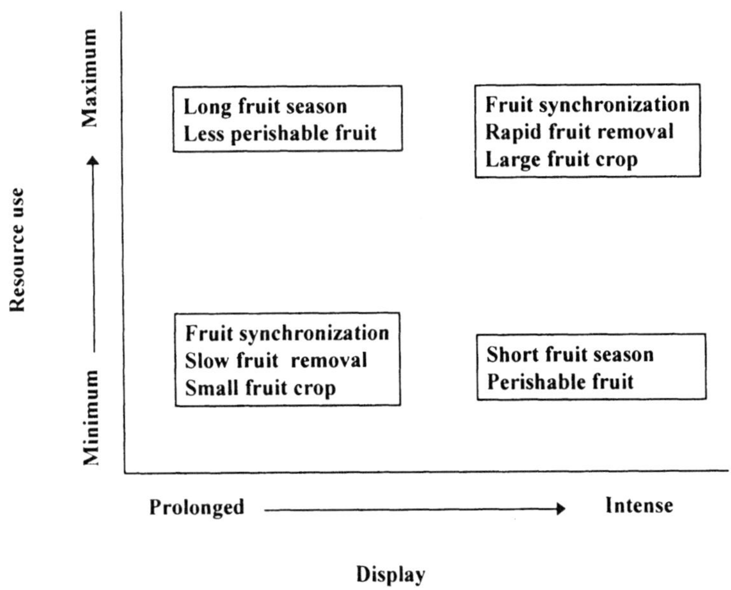

Figure 9.1. Fruit display strategies that attract birds (after Knight 1986).

species that exhibit this short and intense display strategy in the southwest Cape region of South Africa also have high synchronization of fruit display with local conspecifics, effectively increasing the forage efficiency for avian visitors and the attractiveness of the fruits of less abundant species (Knight 1986). Sallabanks (1992) found that fruit abundance strongly affected the probability of American Robin *(Turdus migratorius)* visitation and dispersal success of European hawthorn (*Crataegus monogyna*). Also, some studies comparing congeners have found that highly invasive species have larger fruit displays and consequently higher fruit removal rates than native or less invasive sister species (Sallabanks 1993, Vilà and D'Antonio 1998).

The other major fruiting display strategy is characterized by the production of fruit over an extended time period, producing lower quality but persistent fruit. These displays may achieve competitive advantage in attracting foragers by outlasting richer fruit displays or by presenting themselves at an alternative time of year. This method may ensure dispersal and would likely require a less intense use of resources by the plant (Knight 1986). Some bird-dispersed non-native species may succeed because they use a prolonged fruiting display strategy that allows the plants to present fruit when native species have completed fruiting. White and Stiles (1992) found that of 45 bird-dispersed plant species studied in New Jersey, 33% were non-native. These species generally had a fall fruit maturation that ultimately comprised half of the winter fruit biomass, with use by birds increasing from fall through winter. These fruits had lower lipid content than natives, but provided food after the depletion of native sources. In another study, Amur honeysuckle, an invasive shrub in the eastern United States, was found to produce low-quality fruit that were dispersed in late autumn and winter, after higher quality fruit types had been depleted (Hutchinson and Vankat 1998). Honeysuckle fruit is low in both protein and lipids and thus a poor energy source for birds, but the fruit was abundant enough that birds did not have to expend much energy in foraging (Ingold and Craycraft 1983). It should be noted in the latter study that the analyses of fruit content were done in the fall and some fruit types increase lipid content throughout the season. Therefore it is possible that late in the season, when the fruit were actually consumed, they might have offered a greater nutritional reward (Willson 1986).

Non-native plants may also display fruit during episodes when a native population fails. This situation could arise if a non-native species happens to be more resistant than the native species to herbivore activity, cold spells, or other disruptions. By withstanding the event that eliminates a native fruit, non-native species might lure bird dispersers (Oatley 1984). Researchers have long observed a lag phase between the time that a non-native species is introduced and when it begins to invade (Kowarik 1995). One hypothesis is that dispersal may not begin immediately because birds adopt new resources gradually (Kruger et al. 1986), perhaps initially failing to recognize a potential food source. A non-native plant species may begin to spread when the native food crop fails and birds are forced to use secondary resources (Oatley 1984, Hardt 1986).

Finally, birds are visual foragers. Therefore, it is not surprising that they prefer fruit that bears a strong resemblance to familiar dietary components and that non-native invasive fruiting species closely resemble native displays in ecosystems with a limited variety of native fruiting species (Knight 1986).

7. INCIDENCE OF BIRD DISPERSAL OF NON-NATIVE PLANT SPECIES

Although birds are effective dispersers of invasive plant seeds, the phenomenon is probably not as pervasive as might be expected. Nearly all avian dispersal of plants aids in *sexual* reproduction but many plant species invade predominantly or exclusively through *vegetative* reproduction (Calley et al. 1993, Andersen 1995). Among 56% of the plant species that invade seminatural areas in Denmark, vegetative reproduction is more important than sexual reproduction (Andersen 1995). And although some believe that the worst plant invaders are bird-dispersed (Cronk and Fuller 1995), many invasive species in the U.S. and elsewhere that most profoundly alter ecosystems are wind-dispersed (e.g., *Casuarina equisetifolia, Ailanthus altissima,* and *T. ramosissima*). In disturbed habitats, seed dispersal by ants has been found to be more instrumental in dispersing plants than other biotic vectors (Andersen 1995).

While there are many invasive species with bird and otherwise biotically dispersed seeds, there is often no difference in invasion success between biotically and abiotically dispersed species (Reichard 1997). As a part of two larger studies comparing traits of introduced plants, both invasive and non-invasive, seed dispersal mechanisms were compared. The frequency of dispersal by biotic agents (predominantly birds) among 141 invasive and 94 non-invasive woody species in North America is not significantly different, but some mechanisms appear to be more common in particular community types. Biotic dispersal (again, predominantly birds) is more common in forests where, presumably, there are adequate numbers of frugivorous or omnivorous birds (Table 9.2). In other ecosystem types, such as grasslands and riparian areas, fruits tend to be water and wind dispersed. In Hawaii, like North America, bird-dispersal does not vary significantly among 94 invaders and 63 non-invaders when examined across all habitat types (Table 9.3). Although bird dispersal is more common among invasive plants in forests, in low altitude grasslands, along montane roadsides, and on the wet, forested, windward sides of islands; it is less common in montane grasslands and on the drier, more sparsely vegetated leeward sides of islands (Table 9.3). It appears that bird-dispersed invasive plants are not necessarily most damaging to ecosystems, but may be of particular concern in certain community types.

Evidence from the broader literature supports these conclusions. Few herbaceous species have bird-dispersed fruits; instead, most plant species that rely on birds for propagule dispersal are woody and exist primarily in forested ecosystems rather than grasslands (Willson 1986, Andersen 1995, Reichard 1997). Debussche and Isenmann (1994), working in France, found 19 non-native plant species that were dispersed by 15 different species of birds. Even though some of the non-natives in the Debussche and Isenmann study were invasive in other regions, only six had been naturalized in France and none had become invasive. They attributed the lack of invasion to edaphic and climactic constraints and to competition for dispersers by 65

Table 9.2. Dispersal methods of woody invasive plants in North America by community type (species invade more than one community type).

Community type	Biotic dispersal	Abiotic dispersal
Disturbed	102 (62%)	62 (38%)
Forest	80 (70%)	34 (30%)
Wetland	17 (46%)	20 (54%)
Coastal	9 (64%)	5 (36%)
Grassland	6 (40%)	9 (60%)
Riparian	13 (34%)	25 (65%)

$*P < 0.01$, $**P < 0.001$.

Table 9.3. Dispersal methods of woody invasive plants in Hawaii by community type and island side (species invade more than one community type or island side).

Community type	Biotic dispersal	Abiotic dispersal
Low altitude roadside	29 (59%)	20 (41%)
Montane roadside	13 (72%)	5 (28%)
Low altitude forest	32 (63%)	19 (37%)
Montane forest	25 (61%)	16 (39%)
Coastal**	2 (17%)	10 (83%)
Low altitude grassland**	30 (83%)	6 (17%)
Montane grassland	6 (55%)	5 (45%)
Leeward side (all communities)	23 (55%)	19 (45%)
Windward side (all communities)*	67 (61%)	43 (39%)

$*P < 0.01$, $**P < 0.001$.

native fruit-producing species. In some instances, explanations of a non-native species' invasive success over that of congeners can definitively be linked to dispersal by birds. For example, the invasive success of *Acacia cyclops* in South Africa over other acacia species can be attributed to its production of fleshy pseudoarils not found on other introduced acacias (Glyphis et al. 1981). But whether a plant invades depends upon a combination of factors related to the species' traits and environmental factors including dispersal agents, plant competition, nutrient availability, and disturbance regimes into which a species is introduced (Andersen 1995).

8. THE HUMAN ALTERED LANDSCAPE AND THE ROLE OF BIRDS IN THE DISPERSAL OF ITS NON-NATIVE PLANTS

Urban and suburban areas are highly invasible because they have low vegetative cover and are frequently disturbed (Crawley 1987, Cook 1990). Urbanized areas often serve as centers of invasive plant introduction and establishment from which plants are dispersed into more natural areas (Sukopp and Werner 1983). Species of horticultural value are constantly being introduced into human-altered environments. In the U.S. alone, horticulture is an \$11.2 billion dollar industry annually (D. Johnson pers. comm.). This value has been steadily increasing as suburban land-use continues to grow in the U.S. (Butterfield 1997). Worldwide, the majority of non-native naturalized and invasive woody plants have been intentionally introduced (Henderson 1995). In Smith's (1985) review of the 86 most invasive non-native plants of Hawaii, over half were introduced for horticultural purposes. In South Africa, 17 out of the 20 most serious invasive plant species were intentionally introduced to supply ornamentation, erosion control, timber, and crops (Kruger et al. 1986).

The mosaic of urban greenspace, comprised of diverse cultivated plant communities, results in a rich assemblage of fleshy-fruited species not available in most native temperate environments (Beissinger and Osborne 1982). Suburban lawns have a higher concentration of ground bird food resources and offer greater efficiency in foraging due to the lack of structural complexity and increased view of predators than do other grassland habitats (Beissinger and Osborne 1982). Often, horticultural landscape plantings include non-native woody ornamentals selected specifically for their brightly colored fruits (Date et al. 1996, Panetta and McKee 1997). Some of these species produce fruits during different seasons than native species and are considered highly desirable sources of winter color in ornamental landscapes (Hutchinson and Vankat 1998, White and Stiles 1992). Other species are valued as ornamentals because they produce flowers for extended periods (Morton 1976, Calley et al. 1993). A number of authors have noted birds in the suburban environment responding to the habitats, fruits, and flowers of specific plantings of non-native ornamentals. Obviously, such plants represent a food source to native and non-native birds when other food is scarce or unavailable (Siegfried 1968, Debussche et al. 1982, Debussche and Isenmann 1990, White and Stiles 1992).

Plants in urban environments occur as patches of ornamental landscapes, urban forests, and greenbelts. Those plants occupying edge areas and gaps produce more fruit than species growing under a closed canopy, even among shade tolerant individuals. As one would expect of visual foragers, avian fruit discovery and removal rates are significantly greater among plants occupying edge and light gap sites than those in closed canopy sites (Thompson and Willson 1978, Beissinger and Osborne 1982, Denslow et al. 1986). These natural phenomena are magnified in urban environments, where greenspace patchiness results in significantly greater edge habitat.

A number of studies have found that urban areas support bird communities of low species diversity but high abundance of a few dominant species (Beissinger and Osborne 1982, Lenz 1990, Wood 1993, Clergeau et al. 1998, Sewell and Catterall 1998, Vosshenrich and Mattes 1998, Rottenborn 1999, Marzluff 2001). Ground foraging granivores and omnivores are most successful in urban environments (Beissinger and Osborne 1982, Konstantinov 1996, Vosshenrich and Mattes 1998), perhaps because urban landscapes are more open at ground level than natural systems (Beissinger and Osborne 1982). Even higher total avian abundances are found in suburban environments (Degraaf 1985, Dulisz and Nowakowski 1996, Vosshenrich and Mattes 1998). Suburban areas support those feeding guilds successful in urban areas, as well as frugivores and some nectarivores, but they generally do not support many species of insectivores compared to natural environments (Wolcott 1974, Beissinger and Osborne 1982, Jones 1983, DeGraaf 1991, Zalewski 1994, Konstantinov 1996, Rolando et al. 1997, Clergeau et al. 1998, Sewell and Catterall 1998). These observations hold true across a wide variety of geographic locations including northeast France, Italy, Poland, Russia, the U.S, and Canada. Insect and invertebrate diversity may decrease in response to urban development (Beissinger and Osborne 1982, Bolger 2001) and use of non-native landscaping trees (Southwood 1961). The fact that human settlements are accompanied by purposeful and incidental reductions in insect populations and a simplification of habitat structure in forest zones most likely plays a large role in the resulting reductions in insectivorous species (Sukopp and Werner 1983) in most urban environments.

Avian diversity and/or insectivore abundance can increase, however, in response to resource augmentation through human activities. For example, suburban areas in the arid southwest U.S. have greater insect availability than natural desert areas, perhaps due to additional water and tree resources (Emlen 1974, Guthrie 1974). Bird community richness may increase under light suburban development in which ornamental vegetation has been planted (Blair 1996). In urbanizing areas around a city in northeast Australia, total bird abundance and species richness were higher in suburban areas than large and small forest remnants (Sewell and Catterall 1998). The authors found that the suburbs were planted with a broad array of native and non-native ornamentals, many of which flowered prolifically. The suburban areas supported a great abundance of species whose composition varied substantially from those of remnant forests of original habitat. In general, greater avian species richness has been found in older suburbs (Vale and Vale 1976, Munyenyembe et al. 1989, DeGraaf 1991, Dulisz and Nowakowski 1996) or lower density developments (Tweit and Tweit 1986). Furthermore, those suburbs that maintain a high percentage of green space (DeGraaf 1991, Dulisz and Nowakowski 1996) containing a diverse assemblage of groundcover (Dulisz and Nowakowski 1996), shrubs (Munyenyembe et al. 1989), and trees (DeGraaf 1985, DeGraaf 1991, Mistrukova 1998), especially native vegetation (Guthrie 1974, DeGraaf 1991, Sewell and Catterall 1998, Rottenborn 1999), support the greatest diversity of bird species (Savard and Falls 2001). Finally, diversity in floristic and vegetative structure (Vale and Vale 1976, DeGraaf 1985, Munyenyembe et al. 1989, Wood 1993), nectar and fruit resources, (Recher and Serventy 1991), and water resources and wetlands (Cicero 1989, Recher and Serventy 1991, Dulisz and Nowakowski 1996) will support more avian species in suburbs.

Birds in the urban environment utilize invasive species and often travel between suburban areas and natural areas (Green 1984, Catterall et al. 1989, Calley et al. 1993, Waterhouse 1997), thus enhancing the spread of non-native species (Smith 1985, Zalewski 1994, Hutchinson and Vankat 1998). In a study of an urbanizing area in Poland, Zalewski (1994) found 75% of the birds observed in urban areas foraged outside of their breeding sites, while only a small minority of species breeding in forests foraged outside their territories. Zalewski's study demonstrated that birds in urban environments cover large areas and often utilize some resources outside of the urban environment even if they primarily inhabit urban areas. Some species have particular reproductive strategies allowing them to move back and forth between natural and suburban environments. For example, Nakamura and Miyazawa (1997) found that during the breeding season Common Cuckoos (*Caculus canorus*) spent half their time in natural areas where brood hosts were abundant and half their time in neighboring suburbs. Migrants may also disperse propagules across and among suburban landscapes if they are attracted to islands of remnant vegetation and artificial bodies of water (Whelan and Willson 1994).

While suburban species may utilize forest remnant edges, breeding forest birds do not appear to utilize suburban habitats (Catterall et al. 1991, Waterhouse 1997). The presence of insectivores in forest remnants has been positively related to tree cover (DeGraaf 1991) and to distance from neighboring suburban habitat (Askins and Philbrick 1987, Catterall et al. 1991). Friesen et al. (1995) found that 4 ha woodlots with no surrounding development supported greater neotropical migrant species richness and abundance than 25 ha lots with surrounding housing developments. Similarly, Rottenborn (1999) found that development of land adjacent to native riparian vegetation exerts a significant negative impact on avian communities. In the southwest U.S., Tweit and Tweit (1986) found that low development density (≤ 2 houses ha^{-1}) where native vegetation was maintained did not diminish natural levels of avian diversity, but that denser developments or the removal of most of the native vegetation substantially altered avian diversity.

9. POLICY AND MANAGEMENT IMPLICATIONS

To predict which non-native plant species will establish in new areas, negatively impact native plants, or otherwise compromise the ecological integrity of ecosystems, we need sufficient data to recognize patterns and make generalizations. The study of the relationships between birds and invasive plant species has been limited. Isolated studies tend to highlight differences, not commonalities. In reviewing this literature, we were struck by how much on these topics had been supported only anecdotally. Where information does exist, there are sometimes conflicting results that confound efforts to generalize and extrapolate.

The spread of non-native plants by bird species occurs over spatial and temporal scales. When dealing with phenomena that occur over large scales and in which several contributing variables are simultaneously changing, it is often hard to decisively identify causal relationships. Despite that, more effort needs to be directed at studying the spread of non-natives by birds over large landscapes and over time. Certainly, studies need to be carried out over the range of land-use types used by a given bird population of interest. For some species, this is may be a very large area. Also, studies may benefit from increased attention to how birds utilize the landscape. van Ruremonde and Kalkhoven (1991) pointed out that some species primarily forage over the landscape via flight while others forage mostly on the ground. Thus, birds profoundly differ in the ways that habitat fragmentation and connectivity would affect their foraging activities. While agricultural expanses may serve as a barrier to dispersal of non-native plant species by some avian species (Hutchinson and Vankat 1998), they may present no such barrier to the spread of non-natives by other species (Middlemiss 1963). After reviewing the existing literature, we still found ourselves with many unanswered questions about how land-use and connectivity affect the movement of invasive plants into natural areas.

Schmidt and Whelan (1999) found that nest site competition exerted by a bird species that increases with suburban development reduced nesting success of a species that declines with suburban development. The role landscape structure has in mediating the way that a bird dispersed non-native plant spreads and the resultant nature of bird-bird competitive interactions for non-native plant resources or affected native plant resources is still largely unexplored. While there is usually a bias in management that natural habitats are better buffers for interior forest species, there may also be advantages to establishing unnatural buffers that discourage plant establishment near forest remnants and thus, reduce the immigration of non-native plant species into remaining "natural" areas (Hutchinson and Vankat 1998). Further investigation of this matter and the relative detriment of abrupt habitat edges versus non-native plant species penetration into an area must be more widely assessed and investigated.

Besides a general lack of large-scale studies, we found several specific holes in the information base on the topics discussed. For example, we found no published data on birds contributing to the spread of hollies, although they are widely believed to be bird-dispersed. Also, we found no quantified studies of passive avian dispersal of non-native plant species, although this method of dispersal is often referred to in general discussions of historical plant distributions and dispersal strategies. Additionally, there seemed to be little research directed at the effect that insectivorous birds have on non-native plant insect pollinators and insect herbivores, even though this topic has many important implications for integrated pest management and biological control programs' viability.

In the coming years, the movement of invasive plants out of urban areas is likely to accelerate as the landscape becomes increasingly urbanized and landscaped, and as more plant species are introduced into the horticultural trade. Before introducing new plants for horticultural applications, appropriate research on how a plant might be expected to interact with bird communities directly and indirectly could be examined. When selecting species as potential landscaping plants, it would be advised to conduct mixed aviary and field trials to determine the potential for avian uptake and spread of proposed plant introductions. Such trials could also be applied to plants that have already been introduced to help determine which non-native species may invade in the future.

We also suggest that, on a more localized basis, managers and horticulturists rigorously examine likely implications for the avian community of their management areas before implementing new plantings. It appears that non-native plants that have attractive fruit displays are the plants most likely to strongly change dynamics within avian communities because they often present significant amounts of new food resources. We found that most urbanizing environments already have a ready population of avian frugivorous dispersers. Thus, a lack of avian frugivores will usually not be a limiting factor in the spread of non-native plants, although some bird communities may not be well adapted to spread large fruits. We have outlined some of the fruit traits that are particularly likely to attract avian dispersers to highlight some of the characters of fruit bearing plants that must be assessed in relation to avian frugivore preference. Still, the physical foundations for avian preference can vary greatly depending on plant habit, planting density, climate, conditions, and avian species present (Whelan and Willson 1994). Mixed aviary and field trials would also be useful in determining native species that should be more heavily planted in restoration efforts, especially efforts aimed at avian community restoration.

Better understanding of the relationships between birds and non-native plants may also allow managers to understand how to mitigate problems created by large flocks of pest species like the European Starling. For example, review of national roost surveys of wintering starlings in the U.S. found that the plant species with the most desirable roosting attributes was non-native bamboo (*Pyhllostachys* spp.) (Glahn et al. 1994). Such information could allow land managers to establish plantings that could potentially attract birds away from problem areas. Similarly, when undertaking efforts to control non-native species, managers must consider how their efforts will affect the bird community. There are many instances of non-native plants substituting for or augmenting nesting and foraging resources, so eradication must not proceed blindly (MacDonald 1986, Whelan and Dilger 1992, Sallabanks 1993, Date et al. 1996). For example, before removing an invasive shrub utilized by birds from an area, managers may want to plant native shrubs that could serve as an alternative resource for birds. Many very straightforward measures such as removing favored nesting shrubs during the winter rather than the breeding season, and removing food resource plants during periods when there is ample alternative food supplies can lessen the impact that non-native control efforts exert on avian species.

It is also important that managers consider the broader landscape and temporal implications that suburban non-native plantings have for populations in nearby natural areas. Non-native plantings can displace native plants in remnant natural areas, thereby converting critical habitat for some birds (Breytenbach 1986, La Rue 1994); in many cases the conversion may be quite subtle to the human observer. Alternatively, introduced plants can alter the outcome of competitive or predator/prey relationships by subsidizing aggressive and predatory bird species (MacDonald 1986, Askins and Philbrick 1987, Catterall et al. 1991, Friesen et al. 1995, Major et al. 1996). Identifying which plants may cause problems can prove difficult, as the impact of a non-native plant species may depend on the bird species or season (La Rue 1994). In the absence of complete information, it is wise to mix non-native plantings with native vegetation (Cohan et al. 1978). This approach will minimize any severe, negative impacts on birds that depend on natural areas.

Beyond choices that managers make about the species of plantings, managers must take into account structural aspects of how urban and suburban landscapes are maintained. In our section on urban areas, we pointed out several characteristics that make human altered areas more favorable to bird species richness and diversity. Managers should consider the broad urban system of structural change. New housing developments can maintain many elements of an area's original vegetation and new development projects can be locally staggered so that birds that rely on mature suburban vegetation within an area are not suddenly deprived of all habitat.

Though only a small percentage of invasive plant species have a significant relationship with birds, a greater awareness of these relationships could be vital to the effective management and conservation of the landscape as human land-use needs and values continue to dictate its transformation.

ACKNOWLEDGEMENTS

The authors would like to thank John Marzluff, Roarke Donnelly, and two anonymous reviewers for their helpful comments on an earlier draft of this manuscript.

REFERENCES

Andersen, A. V. 1995. Comparison of dispersal strategies of alien and native species in the Danish flora, p.61-70. *In* P. Pyšek, K. Prach, M Rejmánek, and M. Wade [EDS.], Plant invasions - general aspects and special problems. SPB Academic Publishing, Amsterdam, The Netherlands.

Anderson, B. W., A. Higgins, and R. D. Ohmart. 1979. Avian use of saltcedar communities in the lower Colorado River Valley, p.128-136. *In* Importance, preservation, and management of riparian habitat: a symposium. USDA Forest Service General Tech. Report RM 143.

Askins, R. A., and M. J. Philbrick. 1989. Effect of changes in regional forest abundance on the decline and recovery of a forest bird community. Wilson Bull. 99:7-21.

Auge, H., and R. Brandl. 1999. Seedling recruitment in the invasive clonal shrub, *Mahonia aquifolium* Pursh (Nutt.). Oecologia 110:205-211.

Baker, H. G. and P. D. Hurd. 1968. Intrafloral ecology. Ann. Rev. of Entom.: 385-414.

Balasubramanian, P. 1996. Interactions between fruit-eating birds and bird-dispersed plants in the tropical dry evergreen forest of Point Calimere, South India. J. Bombay Nat. Hist. Soc. 93:428-441.

Baldwin, J. R., and J. R. Lovvorn. 1994. Expansion of seagrass habitat by the exotic *Zostera japonica*, and its use by dabbling ducks and Brant in Boundary Bay, British Columbia. Mar. Ecol. Prog. Ser. 103:119-129.

Barnea, A. Y., J. B. Harborne, and C. Pannell. 1993. What parts of fleshy fruits contain secondary compounds toxic to birds and why. Biochem. Syst. Ecol.. 21: 421-429.

Barnea, A., Y. Yom-Tov, and J. Friedman. 1990. Differential germination of two closely related species of *Solanum* in response to bird ingestion. Oikos 57:222-228.

Bass, D. A. 1990. Pied currawongs and seed dispersal. Corella 14:24-9.

Bass, D. A. 1995. Contribution of introduced fruits to the winter diet of pied currawongs in Armidale, New South Wales. Corella 19:127-32.

Bawa, K. S., and J. H. Beach. 1981. Evolution of sexual systems in flowering plants. Ann. Missouri Bot. Gard. 68:254-274.

Bazzaz, F. A. 1986. Life history of colonizing plants: some demographic, genetic, and physiological features. *In* H.A. Mooney and J.A. Drake [EDS.], Ecology of biological invasions of North America and Hawaii. Springer-Verlag, New York, NY.

Beissinger, S. R., and D. R. Osborne. 1982. Effects of urbanization on avian community organization. Condor 84:75-83.

Benoit, L. K., and A. Askins. 1999. Impact of the spread of *Phragmites* on the distribution of birds in Connecticut tidal marshes. Wetlands 19:194-208

Binggeli, P., and A. C. Hamilton. 1993. Biological invasion by *Maesopsis eminii* in the East Usambara forests, Tanzania. Opera Bot. 121:229-235.

Blair, R. B. 1996. Land-use and avian species diversity along an urban gradient. Ecol. Appl. 6:506-519.

Bock, C. E., J. H. Bock, K. L. Jepson, and J. C. Ortega. 1986. Ecological effects of planting African lovegrasses in Arizona. National Geographic Research 2:456-463.

Bolger, Douglas. 2001. Urban birds: population, community, and landscape approaches, 155-177. *In* J. M. Marzluff, R. Bowman, and R. Donnelly [EDS.], Avian ecology and conservation in an urbanizing world. Kluwer Academic, Norwell, MA.

Bossard, C. C. 1991. The role of habitat disturbance, seed predation, ant dispersal on establishment of the exotic shrub *Cytisus scoparius* in California. Am. Midland Nat. 126:1-13.

Braithwaite, R. W., W. M. Lonsdale, and J. A. Estbergs. 1989. Alien vegetation and native biota in tropical Australia: the impact of *Mimosa pigra*. Biol. Conserv. 48:189-210.

Breytenbach, G. J. 1986. Impacts of alien organisms on terrestrial communities with emphasis on communities of the south-western Cape, p.229-238. *In* I. A. W. MacDonald.

F. J. Kruger, and A. A. Ferrar [EDS.], The ecology and management of biological invasions in southern Africa. Oxford University Press, Cape Town, South Africa.

Buchanan, R. A. 1989. Pied currawongs (*Strepera graculina*): their diet and role in weed dispersal in suburban Sydney, New South Wales. Proc. Linn. Soc. N.S.W. 111:241-255.

Butterfield, B. W. 1999. National gardening survey 1996-1999. National Gardening Association, Burlington, VT.

Calley, M., R. W. Braithwaite, and P. G. Ladd. 1993. Reproductive biology of *Ravenala madagascariensis* Gmel. as an alien species. Biotropica 25:61-72

Carothers, S. W. and B. T. Brown. 1991. The Colorado River through Grand Canyon: natural history and human change. University of Arizona Press, Tempe.

Catterall, C. P., R. J. Green, and D. N. Jones. 1989. Occurrence of birds in relation to plants in a subtropical city. Aust. Wildl. Res. 16:289-305.

Catterall, C. P., R. J. Green, and D. N. Jones. 1991. Habitat use by birds across a forest-scrub interface in Brisbane: implications for corridors. *In* D.A. Saunders and R.J. Hobbs [EDS.], Nature conservation 2: the role of corridors. Surrey Beatty and Sons, Australia.

Christensen, K. M., T. G. Whitham, and R. P. Balda. 1991. Discriminating among pinyon pine trees by Clarks' nutcrackers: effects of cone crop size and cone characters. Oecologia 86:402-409.

Cicero, C. 1989. Avian community structure in a large urban park: controls of local richness and diversity. Landscape and Urban Plann. 17:221-240.

Clergeau, P., J. L. Savard, G. Menneciez, and G. Falardeau. 1998. Bird abundance and diversity along an urban-rural gradient: A comparative study between two cities on different continents. Condor 100:413-425.

Cohan, D. R., B. W. Anderson, and R. D. Ohmart. 1978. Avian population response to salt cedar along the Lower Colorado River, p. 371-382. *In* Strategies for protection and management of flood plain wetlands and other riparian ecosystems: proceedings of the symposium Dec. 11-13, 1978. USDA Forest Service General Tech. Report WO-12.

Cook, C. D. K. 1990. Origin, autoecology, and spread of some of the world's most troublesome weeds, p.31-38. *In* A.H. Pieterse and K.J. Murphy [EDS.], The ecology and management of nuisance aquatic vegetation. Oxford University Press, England.

Cole, F. R., L. L. Loope, A. C. Medeiros, J. A. Raikes, and C. S. Wood. 1995. Conservation implications of introduced game birds in high-elevation Hawaiian shrubland. Conserv. Biol. 9:306-313.

Crawley, M.J. 1989. What makes a community invasible?, p. 429-451. *In* A.J. Gray, M.J. Crawley, and P.J. Edwards [EDS.], Colonization, succession, and stability. Blackwell Scientific Publications, Oxford, England.

Cronk, Q. B., and J. L. Fuller. 1995. Plant invaders: the threat to natural ecosystems. Chapman and Hall, London, England.

D'Antonio, C., and P. M. Vitousek. 1992. Biological invasions by exotic grasses, the grass/fire cycle and global change. Ann. Rev. Ecol. System. 23:63-89.

Date, E. M., H. F. Reicher, H. A. Ford, and D. A. Stewart. 1996. The conservation and ecology of rainforest pigeons in northeastern New South Wales. Pacific Conserv. Biol. 2:299-308.

Dean, W. R. J., and S. J. Milton. 1988. Dispersal of seeds by raptors. Afr. J. Ecol. 26:173-176.

Debussche, M., J. Escarre, and J. Lepart. 1982. Ornithochory and plant succession in Mediterranean abandoned orchards. Vegetatio 48:255-266.

Debussche, M., and P. Isenmann. 1990. Introduced and cultivated fleshy-fruited plants: consequences of a mutualistic Mediterranean plant-bird system, p.399-416. *In* F. Di

Castri, A. J. Hansen and M. Debussche [EDS.], Biological invasions in Europe and the Mediterranean Basin. Kluwer Academic Publishers, Dordrecht, The Netherlands.

Debussche, M., and P. Isenmann. 1994. Bird-dispersed seed rain and seedling establishment in patchy Mediterranean vegetation. Oikos 69:414-426.

De Figueiredo, R. A. 1999. Testing a biological model of adaptation for the exotic tree *Michelia champaca* L. (Magnoliaceae) in Brazil. Ci. Cult. Braz. Assoc. Adv. Sci. 49:278-280.

DeGraaf, R. M. 1985. Residential forest structure in urban and suburban environments: some wildlife implications in New England. J. Arboriculture 11:236-241.

DeGraaf, R. M. 1991. Winter foraging guild structure and habitat associations in suburban bird communities. Landscape and Urban Planning 21:173-180.

Denslow, J. S., T. C. Moermond. 1982. The effect of accessibility on rates of fruit removal from tropical shrubs: an experimental study. Oecologia 54:170-176.

Denslow, J. S., T. C. Moermond, and D. J. Levy. 1986. Spatial components of fruit display in understory trees and shrubs, p.37-44. *In* A. Estrada and T. H. Fleming [EDS.], Frugivores and seed dispersal. Dr. W. Junk, Publishers, Dordrecht, The Netherlands.

Dulisz, B., and J.J. Nowakowski. 1996. The species diversity of avifauna in built-up areas in the city of Olsztyn (NE Poland). Acta Ornithologica 31:33-38.

Emlen, J. T. 1974. An urban bird community in Tucson, Arizona: derivation, structure, regulation. Condor 76:184-199.

Esler, D. 1990. Avian community responses to *Hydrilla* invasion. Wilson Bull. 102:427-440.

Faegri, K., and L. Van der Pilj. 1979. The principles of pollination ecology. Pergamon Press, Inc., Oxford, England.

Feldman, R., D. F. Tomback, and J. Koehler. 1999. Cost of mutualism: Competition, tree morphology, and pollen production in limber pine clusters. Ecology 80:324-329

Fraser, M. W., and T. M. Crowe. 1990. Effects of alien woody plant invasion on the birds of Mountain Fynbos in the Cape of Good Hope Nature Reserve. S. Afr. J. Zool. 25:97-108.

Friesen, L. E., P. F. J. Eagles, and R. J. Mackay. 1995. Effects of residential development on forest dwelling neotropical migrant songbirds. Conserv. Biol. 9:1408-1414.

Gardner, K. T., and D. C. Thompson. 1998. Influence of avian predation on a grasshopper (Orthoptera: Acrididae) assemblage that feeds on threadleaf snakeweed. Env. Entom. 27:110-116.

Glahn, J. F., R. D. Flynt, and E. P. Hill. 1994. Historical use of bamboo/cane as blackbird and starling roosting habitat: implications for roost management. J. Field Ornith. 65:237-246.

Glyphis, J. P., S. J. Milton, and W. R. Siegfried. 1981. Dispersal of *Acacia cyclops* by birds. Oecologia 48:138-141.

Green, R. J. 1984. Native and exotic birds in a suburban habitat. Aust. Wildl. Res. 11:191-190.

Green, R. J, C. P. Catterall, D. N. Jones. 1989. Foraging and other behavior of birds in subtropical and temperate suburban habitats. Emu 89:216-222.

Griffin, G. F., D. M. Stafford Smith, S. R. Morton, G. E. Allan, and K. A. Masters. 1989. Status and implications of the invasion of tamarisk (*Tamarix aphylla*) on the Finke River, Northern Territory, Australia. J. Environ. Manag. 29:297-315.

Guthrie, D. A. 1974. Suburban bird populations in southern California. Am. Midl. Nat. 92:461-466.

Hardt, R. A. 1986. Japanese Honeysuckle: from "one of the best" to ruthless pest. Arnoldia 46:27-34.

Hatton, T. J. 1989. Spatial patterning of sweet briar (*Rosa rubiginosa*) by two vertebrate species. Aust. J. Ecol. 14:199-205.

Henderson, L. 1995. Plant invaders of southern Africa. Plant Protection Institute, Pretoria.

Herrera, C. M. 1984. A study of avian frugivores, bird-dispersed plants, and their interaction in Mediterranean scrublands. Ecol. Monog. 54:1-23.

Hodkinson, D. J., and Thompson, K. 1999. Plant dispersal: the role of man. J. Appl. Ecol. 34:1484-1496.

Howe, H. F., and R. B. Primack. 1975. Differential seed dispersal by birds of the tree *Cascaria nitida* (Flacourtiaceae). Biotropica 7:278-283.

Howe, H. F., and J. Smallwood. 1982. Ecology of seed dispersal. Ann. Rev. Ecol. Syst.. 13:201-28.

Huenneke, L. F., and J. K. Thompson. 1994. Potential interference between a threatened endemic thistle and an invasive non-native plant. Conserv. Biol. 9:416-425.

Hughes, F., P. M. Vitousek, and T. Tunison. 1991. Alien grass invasion and fire in the seasonal submontane zone of Hawaii. Ecology 72:743-746.

Humphries, S. E., R. H. Groves, D. S. Mitchell. 1991. Plant invasions of Australian ecosystems. A status review and land management directions. Kowari 2:1-134.

Hutchinson, T. F., and J. L. Vankat. 1998. Landscape structure and spread of the exotic shrub *Lonicera maackii* (amur honeysuckle) in southwestern Ohio forests. Am. Midl. Nat. 139:383-390.

Ingold, J. L., and M. J. Craycraft. 1983. Avian frugivory on honeysuckle (*Lonicera*) in southwestern Ohio in fall. Ohio J. Sci. 83:256-258.

Jones, D. N. 1983. The suburban bird community of Townsville, a tropical city. Emu 83:12-18.

Kerley, G. I. H., A. McLachlan, and J. G. Castley. 1996. Diversity and dynamics of bushpockets in the Alexandria coastal dunefield, South Africa. Landscape Urban Plann. 34:255-266.

Knight, K. S. 1986. Fruit displays of indigenous and invasive alien plants in the south-western Cape. S. Afr. J. Bot. 52:249-255.

Knight, R. S., and I. A. W. MacDonald. 1991. Acacias and korhaans: an artificially assembled seed dispersal system. S. Afr. J. Bot. 57:220-225.

Konstantinov, V. M. 1996. Anthropogenic transformations of bird communities in the forest zone of the Russian Plain. Acta Ornithologica 31:53-58.

Kowarik I. 1995. Time lags in biological invasions with regard to the success and failture of alien species, p.15-38. *In* Pyšek P, Prach K, Rejmánek M, and Wade M [eds], Plant invasion – general aspects and special problems. Academic Publishing, Amsterdam, the Netherlands.

Kruger, F. J., D. M. Richardson, and B. W. van Wilges. 1986. Processes of invasions by alien plants, p.145-155. *In* I. A. W. MacDonald, F. J. Kruger, and A. A. Feuar [EDS.], The ecology and management of biological invasions in southern Africa. Oxford University Press, Cape Town, South Africa.

La Rue, C. T. 1994. Birds of northern Black Mesa, Navajo County, Arizona. Great Basin Nat. 54:1-32.

Lenz, M. 1990. The breeding bird communities of three Canberra suburbs. Emu 90:145-153.

Levin, P. A. 1970. Reinforcement of reproductive isolation: plants versus animals. Am. Nat. 104:571-581.

Lewin, V., and G. Lewin. 1984. The Kalij pheasant, a newly established game bird on the island of Hawaii. Wilson Bull. 96:634-646.

Lloyd, D. G. 1980. Demographic factors and mating patterns in angiosperms, p.67-88. *In* Solbrig, O.T. [ED.], Demography and evolution in plant populations. The Alden Press, Oxford, Great Britain.

Lonsdale, W. M., and Lane, A. M. 1994. Tourist vehicles as vectors of weed seeds in Kakadu National Park, Northern Australia. Biol. Conserv. 69:277-283.

Lynn, S., M. L. Morrison, A. J. Kuenzi, J. C. C. Neale, B. N. Sacks, R. Hamlin, and L. S. Hall.1998. Bird use of riparian vegetation along the Truckee River, California and Nevada. Great Basin Nat. 58:328-343.

MacDonald, I. A. W. 1986. Range expansion of the pied barbet and the spread of alien tree species in Southern Africa. Ostrich 57:75-94.

MacLellan, C. R. 1958. Role of woodpeckers in control of the codling moth in Nova Scotia. Can. Entom. 90:18-22.

Major, R. E., G. Gowing, and C. E. Kendal. 1996. Nest predation in Australian urban environments and the role of the Pied Currawong. Aust. J. Ecol. 21:399-409.

Marzluff, J. M. 2001. Worldwide increase in urbanization and its effects on birds. p. 19-47. *In* J. M. Marzluff, R. Bowman, and R. Donnelly [EDS.], Avian ecology and conservation in an urbanizing world. Kluwer Academic, Norwell, MA.

Marzluff, John M., Kevin McGowan, Roarke Donnelly, and Richard Knight. 2001. Causes and consequences of expanding American Crow populations. p. 333-365. *In* J. M. Marzluff, R. Bowman, and R. Donnelly [EDS.], Avian ecology and conservation in an urbanizing world. Kluwer Academic, Norwell, MA.

Middlemiss, E. 1963. The distribution of *Acacia cyclops* in the Cape Peninsula area by birds and other animals. S. Afr. J. Sci. 59:419-420.

Mills, G. S., J. B. Dunning Jr., and J. M. Bates. 1989. Effects of urbanization on breeding bird community structure in southwestern desert habitats. Condor 91:416-429.

Milton, S. J., W. R. J Dean, and S. Klotz. 1999. Thicket formation in abandoned fruit orchards: processes and implications for the conservation of semi-dry grasslands in central Germany. Biodiv. Conserv. 6:275-290.

Mistrukova, L. M. 1998. Study of the ornithofauna of the urban and suburban green plantations in the central Cis Dnieper Forest Steppe of Ukraine. Vestnik Zoologii 32:107-113.

Moody, M. E., and R. N. Mack. 1988. Controlling the spread of plant invasions: the importance of recent foci. J. Appl. Ecol. 25:1009-1021.

Morton, J. F. 1976. Pestiferous spread of many ornamental and fruit species in south Florida. Proc. Fl. State Hort. Soc. 89:348-358.

Moulton, M. P. 1993. The all-or-none pattern in introduced Hawaiian passeriforms: the role of competition sustained. Am. Nat. 141:105-119.

Munyenyembe, F. J. Harris, J. Hone, and H. Nix. 1989. Determinants of bird populations in an urban area. Aust. J. Ecol. 14:549-559.

Nakamura, H., and Y. Miyazawa. 1999. Movements, space use and social organization of radio-tracked common cuckoos during the breeding season in Japan. Jap. J. Ornith. 46:23-54.

Neill, W. 1983. The tamarisk invasion of desert riparian areas. Educ. Bull. Desert Protective Council Vol., p. 83-84. Ed. Found. Desert Protective Council, Spring Valley, CA.

Oatley, T. B. 1984. Exploitation of a new niche by the Rameron Pigeon *Columba arquatrix* in Natal, p.323-330. *In* J. Ledger [ED.], Proceedings of the Fifth Pan African Ornithological Congress. S. Afr. Ornith. Soc., Johannesburg, South Africa.

Panetta, F. D., and J. McKee. 1999. Recruitment of the invasive ornamental, *Schinus terebinthifolius*, is dependent upon frugivores. Aust. J. Ecol. 22:432-438.

Parker, I. M. 1999. Pollinator limitation of *Cytisus scoparius* (Scotch broom), an invasive exotic shrub. Ecology 78:1457-1470.

Proctor, V. W. 1968. Long-distance dispersal of seeds by retention in the digestive tract of birds. Science 160:321-322.

Quiring, D. T., and P. R Timmins. 1988. Predation by American crows reduced overwintering European corn borer populations in southwestern Ontario. Can. J. Zool. 66:2143-2154.

Recher, H. F., and D. L. Serventy. 1991. Long term changes in the relative abundances of birds in Kings Park, Perth, Western Australia. Conserv. Biol. 5:90-102.

Reichard, S. 1999. Prevention of invasive plant introductions on national and local levels, p.215-219. *In* J. O. Luken and J. W. Thieret [EDS.], Assessment and management of plant invasions. Springer-Verlag, New York, NY.

Richardson, D. M., and S. I. Higgins. 1998. Pines as invaders in the southern hemisphere, p.450-473. *In* D. M. Richardson [ED.], Ecology and biogeography of pines. Cambridge University Press, England.

Ridley, H. N. 1933. The dispersal of plants throughout the world. Otto Koeltz Science Publishers, Koenigstein West Germany.

Robertson, A. W., D. Kelly, J.J. Ladley, and A.D. Sparrow. 1999. Effects of pollinators on endemic New Zealand mistletoes (Loranthaceae). Conserv. Biol. 13:499-508.

Rolando, A., G. Maffei, C. Pulcher, and A. Giuso. 1999. Avian community structure along an urbanization gradient. Ital. J. Zool. 64:341-349.

Rottenborn, S. C. 1999. Predicting the impacts of urbanization on riparian bird communities. Biol. Conserv. 88:289-299.

Sallabanks, R. 1992. Fruit fate, frugivory, and fruit characteristics: a study of hawthorn, *Crataegus monogyna* (Rosaceae). Oecologia 91:296-304.

Sallabanks, R. 1993. Fruiting plant attractiveness to avian seed dispersers: native vs. invasive *Crataegus* in western Oregon. Madroño 40:108-116.

Savard, Jean-Pierre, and J. Bruce Falls. 2001. Bird-habitat relationships of breeding birds in residential areas of Toronto, Canada, p. 543-568. *In* J. M. Marzluff, R. Bowman, and R. Donnelly [EDS.], Avian ecology and conservation in an urbanizing world. Kluwer Academic, Norwell, MA.

Schmidt, K. A., and C. J. Whelan. 1999. Effects of exotic *Lonicera* and *Rhamnus* on songbird nest predation. Conserv. Biol. 13:1502-1506.

Sewell, S. R., and C. P. Catterall. 1998. Bushland modification and styles of urban development: their effects on birds in south-east Queensland. Wildlife Res. 25:41-63.

Siegfried, W. R. 1968. Ecological composition of the avifaunal community in a Stellenbosch suburb. Ostrich 105-129.

Smith, C. W. 1985. Impact of alien plants on Hawaii's native biota, p.180-250. *In* C. P. Stone and J. M. Scott [EDS.], Hawaii's terrestrial ecosystems: preservation and management. University of Hawaii, Honolulu.

Solomon, M. E., D. M. Glen, D. A. Kendall, and N. F. Milsom. 1976. Predation of overwintering larvae of codling moth (*Cydia poponella* L.) by birds. J. Appl. Ecol. 13:341-352.

Sorensen, A. E. 1986. Seed dispersal by adhesion. Ann. Rev. Ecol. Syst. 17:443-63.

Southwood, T. R. E. 1961. The number of species of insect associated with various trees. J. Anim. Ecol. 30:1-8.

Stiles, E. W. 1982. Expansions of mockingbird and multiflora rose in the northeastern United States and Canada. Am. Birds 36:358-364.

Stiles, E. W. 1989. Fruits, seeds, and dispersal agents. *In* W.G. Abrahamson [ED.], Plant-animal interactions. McGraw-Hill Book Co., NY.

Sukopp, H., and P. Werner. 1983. Urban environment and vegetation, p.247-260. *In* W. Holzner, M. J. A. Werger and I. Ikisima [EDS.], Man's impact on vegetation,. Dr. W. Junk Publishers, The Netherlands.

Thompson, J. N., and M. F. Willson. 1978. Disturbance and the dispersal of fleshy fruits. Science 200:1161-1163.

Tweit, R. C., and J. C. Tweit. 1986. Urban development effects on the abundance of some common resident birds of the Tucson area of Arizona. Am. Birds 40:431-436.

Vale, T. R., and Vale, G. R. 1976. Suburban bird populations in west-central California. J. Biog. 3:157-165.

van Ruremonde, R. H. A. C., and J. T. R. Kalkhoven. 1991. Effects of woodlot isolation on the dispersion of plants with fleshy fruits. J. Veg. Sci. 2:377-384.

Vilà, M., and C. M. D'Antonio. 1998. Fruit choice and seed dispersal of invasive and non-invasive *Carpobrotus* (Aizoaceae) in coastal California. Ecology 79:1053-1060.

Vitousek, P. M., L. R. Walker, L. D. Whiteaker, D. Mueller-Dombois, and P. Matson. 1989. Biological invasion by *Myrica faya* alters ecosystem development in Hawaii. Science 238:802-804.

Vosshenrich, A., and H. Mattes. 1998. Avifauna in built-up areas of Muenster, NW Germany in winter 1995/96. Vogelwelt 119:311-321.

Wagner, W. L., D. R. Herbst, and S. H. Sohmer. 1990. Manual of the flowering plants of Hawaii. University of Hawaii Press and Bishop Museum Press, Honolulu.

Waterhouse, R. D. 1999. Some observations on the ecology of the rainbow lorikeet *Trichoglossus haematodus* in Oatley, South Sydney. Corella 21:17-24.

Whelan, C. J., and M. L. Dilger. 1992. Invasive, exotic shrubs: a paradox for natural area managers. Nat. Areas J. 12:109-110.

Whelan, C. J., and M. F. Willson. 1994. Fruit choice in migrating North American birds: field and aviary experiments. Oikos 71:137-151.

White, D. W., and E. W. Stiles. 1992. Bird dispersal of fruits of species introduced into eastern North America. Can. J. Bot. 70:1689-1696.

Wilcove, D. S., D. Rothsteen, J. Dubow, A. Phillips, and E. Losos. 1998. Quantifying threats to imperiled species in the United States. BioSci. 48:607-615.

Williams, P. A., and B. J. Karl. 1996. Fleshy fruits of indigenous and adventitive plants in the diet of birds in forest remnants, Nelson, New Zealand. N.Z. J. Ecol. 20:127-145.

Willson, M. F. 1986. Avian frugivory and seed dispersal in eastern North America, p.223-279. *In* R. F. Johnston [ED.], Current ornithology, Vol. 3. Plenum Press, New York, NY.

Winterbottom, J. M. 1970. The birds of the alien *Acacia* thickets of the south western Cape. Zool. Afr. 5:49-59.

Witmer, M. C. 1996. Consequences of an alien shrub on the plumage coloration and ecology of cedar waxwings. Auk 113:735-743.

Witt, A. B. R., R. M. Little, T. M. Crowe. 1995. The effectiveness of helmeted guineafowl *Numida meleagris* (Linnaeus 1776) in controlling the banded fruit weevil *Phlyctinus callosus* (Schonherr 1826), and their impact on other invertebrates in apple orchards in Western Cape Province, South Africa. Agriculture, Ecosystems, and Environment 55:169-179.

Wolcott, C. F. 1974. Changes in bird life in Cambridge, Massachusetts from 1860 to 1964. Auk 91:151-160.

Wood, K. A. 1993. The avian population of an urban bushland reserve at Wollongong, New South Wales: implications for management. Landscape and Urban Plann. 23:81-95.

Zalewski, A. 1994. A comparative study of breeding bird populations and associated landscape character, Torum, Poland. Landscape and Urban Plann. 29:31-41.

Appendix 7.1. Survey of the existing literature on how non-native plants interact with birds of different feeding guilds. Plant species are listed in the first column, and unless otherwise noted, it should be assumed that plants are non-native and invading or invasive threats in the region where each study was conducted. In the second column are the major species of birds that have been ecologically linked with the non-native plant(s) listed in the first column of the corresponding row. Non-native species of birds are indicated by an asterisk. Many studies concerned a large number of different bird species and did not include substantial data on any one species. In these cases, groupings of birds utilized in the articles, such as "neotropical migrants" are listed. The ecological links between birds and non-native plants included in the table are quite broad in nature. For example, the non-natives may be dispersed by birds eating their fruits, they may form dense stands that discourage nesting, they may be aided by avian insectivores predating limiting herbivores, or they may relate to birds in several other ways. We have indicated whether associations between bird species and non-native plants are negative (indicated by a "-"), positive (indicated by a "+"), neutral (indicated by a "="), or not specifically assessed in the study (indicated by the lack of any sign prefix). The choice of placement of signs beside bird species, plant species, or both for a given study reflects the perspective that the authors took in the study. For instance, if there is a "+" in front of a bird species and no sign in front of a plant species, the authors concluded that the given bird species was benefiting from the spread of the non-native plant. Signs have been indicated within parentheses if we felt that the study had not unequivocally linked effects of the listed birds and plants. We opted to list some of the secondary sources of information to maximize the potential usefulness and accessibility of known information on this topic. This table should be seen as a broad citation overview of the subject matter but not an exhaustive list of available sources.

Non-native plant species list ***Birds found to have*** + positive impact - negative impact = no impact **no sign** not assessed ( ) correlation implied but not rigorously assessed	**Bird species list** ***Non-native plants found to have*** + positive impact - negative impact = no impact **no sign** not assessed ( ) correlation implied but not rigorously assessed * indicates bird species is non-native	**Landscape Class & Location of Study Site(s)** **W** = wildland **R** = rural **S** = suburban **U** = urban **?** = unclear	**Citation**
(+)*Berberis darwinii* (barberry) **Naturalized**	*Turdus merula* (European blackbird)* *Turdus philomelos* (song thrush)* *Zosterops lateralis* (silvereye) *Hemiphaga novaeseelandiae* (New Zealand pigeon)	**?** Lab, New Zealand	Allen & Wilson 1992
Tamarix chinensis (saltcedar)	-cavity nesters; neotropical migrants; insectivores -species richness	**?** SW U.S.	Anderson et al. 1977

	+*Molothrus ater* (brown-headed cowbird) +*Zenaida* spp. (dove) -*Dendroica petechia* (yellow warbler) -*Vireo bellii* (Bell's vireo)		
Zostera japonica (Japanese eelgrass)	+*Anas americana* (American widgeon) +*Anas platyrhynchos* (mallard) +*Anas acuta* (northern pintail) +*Anas crecca* (green-winged teal) +*Branta bernicla* (brant)	Ocean bay next to **W** Canada (British Columbia)	Baldwin & Lovvorn 1994
(+)*Solanum nigrum* (black nightshade) (+)*Solanum luteum* (hairy nightshade)	*Pycnonotus xanthopygos* (white spectacled bulbul) *Turdus merula* (European blackbird)	**S?** N. Israel, Lab	Barnea et al. 1990
(+)*Crataegus monogyna* (European hawthorn)	(+)*Strepera graculina* (pied currawong)	? Australia (New S. Wales)	Bass 1990
Pyracantha spp. (firethorns) *Ligustrum* spp. (privets)	(+)*Strepera graculina* (pied currawong)	**S?** Australia (New S. Wales)	Bass 1995
Phragmites australis (common reed)	-species diversity; state listed species -*Ammodramus maritima* (seaside sparrow) -*Ammodramus caudacutus* (sharp-tailed sparrow) -*Catoptrophorus sentipalmatus* (willet) +*Cistothorus palustris* (marsh wren) +*Melospiza georgiana* (swamp sparrow)	**W/U** marshes U.S. (Connecticut)	Benoit & Askins 1999
+*Maesopsis eminii* (musizi)	+*Bycanisies bucinator* (trumpeter hornbill) +*Bycanistes brevis* (silvery-cheeked hornbill)	**W?** Tanzania	Binggeli & Hamilton 1993
Bromus tectorum (cheatgrass) *Agropyron cristatum* (crested wheatgrass)	(-)*Sturnella magna* (eastern meadowlark) (-)*Ammodramus savannarum* (grasshopper sparrow) (-)*Aimophila cassinii* (Cassin's sparrow) (-)*Aimophila botterii* (Botteri's sparrow) (-)*Pooecetes gramineus* (vesper sparrow) (-)*Lanius ludovicianus* (loggerhead shrike)	**W-R?** U.S. (Arizona)	Bock et al. 1986

-*Cytisus scoparius* (Scotch broom)	*Dendragapus obscurus* (blue grouse)	**?** U.S. (California)	Bossard 1991
Mimosa pigra (black mimosa)	-most terrestrial and waterbirds -*Conopophila albogularis* (rufous-banded honeyeater) - *Geopelia striata* (zebra dove) + *Rhipidura leucophrys* (willy wagtail)	**R?** N. Australia	Braithwaite et al. 1989
Hakea sericea (needlebush)	-nectivores +*Sturnis vulgaris* (European starling)* +*Prinia maculosa* (spotted prinia) -*Promerops cafer* (Cape sugarbird)	**?** S. South Africa	Breytenbach 1986 (review)
(+)*Acacia* spp.	-nectivore, insectivore, and frugivore densities +*Lybius leucomelas* (pied barbet) +*Bostrychia hagedash* (hadada ibis)	**?** S. South Africa	Breytenbach 1986 (review)
Acacia cyclops (rooikrans)	(+)*Numida meleagris* (helmeted guineafowl) (+)*Streptopelia semitorquata* (redeyed dove)	**?** S. South Africa	Brooke 1984
+*Rubus procerus* [*R. discolor*] (Himalaya blackberry)	*Dromaius novaehollandiae* (emu)	**W** Australia (Victoria)	Brunner et al. 1976
+*Ligustrum lucidum* (glossy privet) +*Ligustrum sinense* (Chinese privet) +*Olea africana* (wild olive) +*Morus alba* (Asian mulberry) +*Ochna atropurpurea* (orange firethorn) +*Solanum pseudocapsicum* (winter cherry nightshade)	(+)*Strepera graculina* (pied currawong)	**S-U** Australia (New South Wales)	Buchanan 1989
= *Ravenala madagascariensis* (traveller's tree)	*Meliphagidae* spp. (honeyeaters)	**S-U** Australia (Northern Territory)	Calley et al. 1993
(+) *Spathodea* spp. (African tulip tree) (+) *Caesalpinia* spp. (leopard tree)	Various native birds	**S-U**	Catterall et al. 1989
Tamarix chinensis (saltcedar)	+Old World genera -insectivores; frugivores	**?** SW U.S.	Cohan et al. 1978

(+) native shrubs and sedges (+) *Styphelia tameiameiae* (pukiawe)	*Phasianus colchicus* (ring-necked pheasant)* *Alectoris chukar* (chukar)*	**W** U.S. (Hawai'i)	Cole et al. 1995
(+)*Acacia cyclops* (rooikrans)	*Columba arquatrix* (African olive pigeon) *Andropadus importunus* (sombre bulbul) *Colius striatus* (speckled mousebird) *Pycnonotus capensis* (Cape bulbul) *Cossypha caffra* (Cape robin) *Onychognathus morio* (African red-winged starling) *Zosterops pallidus* (Cape white-eye)	**?** South Africa	Cowling et al. 1997
Lantana camara *Cinnamonum camphora* (camphor laurel) *Ligustrum lucidum* (glossy privet) *Ligustrum sinense* (Chinese privet) *Solanum mauritianum* (bugweed)	(+)*Ptilinopus magnificus* (wompoo pigeon) (+)*Ptilinopus regina* (rose-crowned pigeon) (+)*Ptilinopus superbus* (suburb pigeon) (+)*Lopholaimus antarcticus* (topknot pigeon) (+)*Columba leucomela* (white-headed pigeon) (+)*Macropygia amboinensis* (brown pigeon) (+)*Leucosarcia cyanoleuca* (wonga pigeon) (+)*Chalcophaps indica* (emerald dove)	**W-R?** Australia (New South Wales)	Date et al. 1996
(+)*Acacia cyclops* (rooikrans) (+)*Sorghum* spp. (+)*Datura innoxia* (thorn-apple)	raptors and owls that eat granivorous birds and rodents	**?** Zimbabwe, South Africa, Others?	Dean & Milton 1988
(+)*Vitis vinifera* (grape) (+)*Phytolacca americana* (pokeweed) (+)*Pyracantha coccinea* (firethorn) (+)*Ligustrum ovalifolium* (California privet) (+)*Parthenocissus quinquefolia* (Virginia creeper) (+)*Lonicera japonica* (Japanese honeysuckle) (+)*Lycium barbarum* (matrimony vine)	(+)*Turdus philomelos* (song thrush) (+)*Turdus iliacus* (redwing) (+)*Sylvia atricapilla* (blackcap) *Sylvia melanocephala* (Sardinian warbler) *Erithacus rubecula* (European robin) others	**R-S?** France	Debussche & Isenmann 1990, 1994

All spp. naturalized, not invasive			
(+)*Celastrus orbiculatus* (oriental bittersweet)	birds	**?** U.S. (Connecticut)	Dreyer et al. 1987
Tamarix chinensis (saltcedar)	-neotropical migrants -*Piranga rubra* (summer tanager) -*Sitta carolinensis* (white-breasted nuthatch) +*Icteria virens* (yellow-breasted chat) +*Guiraca caerulea* (blue grosbeak) +*Geothlypis trichas* (common yellowthroat) many others	**W?** SW U.S.	Ellis 1995
Parkinsonia aculeata (Mexican palo verde) non-native ornamentals	(+) *Auriparus flaviceps* (verdin) (+)*Carpodacus mexicanus* (house finch)* (+)*Cardinalis cardinalis* (northern cardinal)	**S-U, W** U.S. (Arizona)	Emlen 1974
Hydrilla verticillata	+*Fulica americana* (American coot) +*Podilymbus podiceps* (pied-billed grebe) *Anas clypeata* (northern shoveler) +other *Anas* spp. (dabbling ducks)	Reservoir, **W?** U.S. (Texas)	Esler 1990
(+)*Schinus terebinthifolius* (Brazialian pepper tree)	*Turdus migratorius* (American robin) others	**W** U.S. (Florida)	Ewel 1986
+*Cinnamomum zeylanicum* (Ceylon cinnamon)	*Acridotheres tristis* (common myna)*	**?** Seychelles	Fleischmann 1997
(+)*Acacia cyclops* (rooikrans)	+frugivores; granivores	**W?**	Fraser &

	-nectivores +*Pycnonotus capensis* (Cape bulbul) +*Sturnus vulgaris* (European starling)* +*Zosterops pallidus* (Cape white-eye) -*Promerops cafer* (Cape sugarbird) -*Nectarinidae* spp. (sunbird)	South Africa	Crowe 1990
(+)*Gutierrezia microcephala* (threadleaf snakeweed)	*Sturnella* spp. (meadowlark) *Mimus polyglottis* (northern mockingbird) *Aimophila cassinii* (Cassin's sparrow)	**W** U.S. (Arizona)	Gardner & Thompson 1998
+*Pittosporum undulatum* (Australian daphne) **Native but invasive**	*Turdus merula* (European blackbird)*	**W-S?** Australia (Victoria)	Gleadow 1982
(+)*Acacia cyclops* (rooikrans)	*Streptopelia* spp. (dove) *Colius colius* (whitebacked mousebird) *Pycnonotus capensis* (Cape bulbul) *Sturnus vulgaris* (European starling)* *Onychognathus morio* (African red-winged starling) others	**W?** South Africa	Glyphis et al. 1981
non-native plants	native birds *Anthochaera carunculata* (red wattlebird) *Turdus merula* (European blackbird)* *Passer domesticus* (house sparrow)* *Sturnus vulgaris* (European starling)* *Acridotheres tristis* (common myna)*	**S-U** E. Australia	Green et al. 1989
non-native plants	*Turdus merula* (European blackbird)* *Passer domesticus* (house sparrow)* *Sturnus vulgaris* (European starling)* *Acridotheres tristis* (common myna)* *Carduelis tristis* (American goldfinch)*	**S-U** E. Australia	Green 1984
+*Rosa rubiginosa* (sweetbriar)	(+)*Strepera graculina* (pied currawong)	**R-S** Australia (New South Wales)	Hatton 1989
Tamarix chinensis (saltcedar)	-*Vireo bellii* (Bell's vireo)	**R?**	Hunter et al.

	-Piranga rubra (summer tanager) *-Coccyzus americanus* (yellow-billed cuckoo)	SW U.S.	1988
+Lonicera maackii (Amur honeysuckle)	birds	**R-S** Mid-west U.S.	Hutchinson & Vankat 1998
(+)*Lonicera maackii* (Amur honeysuckle) (+)*Lonicera xylosteum* (European fly honeysuckle)	*Catharus minimus* (gray-cheeked thrush) *Bombycilla cedorum* (cedar waxwing) *Carpodacus purpureus* (purple finch) *Turdus migratorius* (American robin) others	**W** (altered) U.S. (Ohio)	Ingold & Craycroft 1983
(+) *Sapium sebiferum* (Chinese tallow tree) **Naturalized, possibly invasive**	*Dryocopus pileatus* (pileated woodpecker) *Quiscalus major* (boat-tailed grackle)	**W-U?** U.S. (Florida)	Jubinsky & Anderson 1996
+Ficus microcarpa (laurel fig)	birds	**?** U.S. (Florida)	Kaufman et al. 1991
(+)*Acacia cyclops* (rooikrans) (+)*Opuntia ficus-indica* (Indian fig)	birds	**?** South Africa	Kerley et al. 1996
(+)*Acacia cyclops* (rooikrans) (+)*Lantana camara* (+)*Myoporum serratum* [*M. tetrandum*] (Tasmanian Ngais) (+)*Pittosporum undulatum* (Australian daphne) (+)*Rubus fruticosus* (blackberry) (+)*Solanum mauritianum* (bugweed)	*Onychognathus morio* (red-winged starling) *Pycnonotus barbatus* (blackeyed bulbul) *Pycnonotus capensis* (cape bulbul) *Sturnus vulgaris* (European starling)* *Oriolus larvatus* (African black-headed oriole)	**?** South Africa	Knight 1986
+Acacia cyclops (rooikrans) *+Acacia longifolia* (Sydney golden wattle)	(+)*Eupodotis afra* (black korhaan)	**?** South Africa	Knight & Macdonald 1991
+Myrica faya (firetree)	*Phaeornis obscurus* (Lanai thrush) *Carpodacus mexicanus* (house finch)* *Hemignathus virens* (Hawai'ian amakihi)	**W** U.S. (Hawai'i)	La Rosa et al. 1985

Tamarix chinensis (saltcedar)	(+)foraging insectivores in fall (+)*Junco hyemalis* (dark-eyed juncos) in winter (-)*Tyrannus verticalis* (western kingbirds) (-)*Icteria virens* (yellow-breasted chat) (-)*Passerina* spp. (bunting) (-)*Icterus* spp. (oriole)	**W-R** U.S. (New Mexico)	LaRue 1994
Salsola iberica (Russian thistle) *Elaeagnus angustifolia* (Russian olive) *Kochia scorparia* (summer cypress)	(+)*Junco hyemalis* (dark-eyed junco) (+)*Eremophila alpestris* (horned lark) (+)*Picoides pubescens* (downy woodpecker) (+)*Colaptes auratus* (northern flicker) (+)*Sialia currucoides* (mountain bluebird) (+)*Turdus migratorius* (American robin) (+)*Sturnis vulgaris* (European starling)* (+)*Bombycilla cedorum* (cedar waxwing) (+)*Dendroica coronata* (yellow-rumped warbler) (+)*Zonotrichia leucophrys* (white-crowned sparrow) (+)*Coccothraustes vespertinus* (evening grosbeak)	**W-R** U.S. (New Mexico)	LaRue 1994
(+)*Passiflora mollissima* (banana passionfruit) (+)*Rubus rosaefolius* (thimbleberry) (+)*Psidium guajava* (guava) Others	*Lophura leucomelana* (kalij pheasant)*	**W-S?** U.S. (Hawai'i)	Lewin & Lewin 1984
Cardaria draba (whitetop)	-bird species richness -riparian obligates =*Molothrus ater* (brown-headed cowbird)	**R?** U.S. (California, Nevada)	Lynn et al. 1998
+*Acacia* spp.	+*Lybius leucomelas* (pied barbet)	**?** S. South Africa	Macdonald 1986
(+)*Miconia calvescens* (velvet tree)	*Zosterops lateralis* (silvereye)* *Pycnonotus cafer* (red-vented bulbul)* *Zosterops japonicus* (Japanese White-eye)* *Leiothrix lutea* (red-billed leiothrix)*	**W-U** U.S. (Hawai'i)	Medeiros et al. 1997

	Acridotheres tristis (common myna)*		
(+)*Acacia cyclops* (rooikrans)	*Gallinula chloropus* (African moorhen) 3 *Streptopelia* spp. (dove) *Pycnonotus capensis* (Cape bulbul) *Cossypha caffra* (Cape robin) *Sturnus vulgaris* (European starling)* *Onychognathus morio* (African red-winged starling) *Zosterops pallidus* (Cape white-eye)	**W?** South Africa	Middlemiss 1963
(+)*Opuntia ficus-indica* (Indian fig)	*Corvus capensis* (black crow) *Corvus alba* (pied crow)	**?** South Africa	Milton & Dean 1987
(+)*Adenanthera pavonina* (red sandalwood) (+)*Bischofia javanica* (bishopwood) (+)*Cestrum diurnum* (day jasmine) (+)*Jasminum fluminense* (Brazilian jasmine) others	birds	**S?** U.S. (Florida)	Morton 1976
+*Solanum mauritianum* (bugweed) *Morus nigra* (African mulberry) *Phytolacca octandra* (inkweed)	+ *Columba arquatrix* (African olive pigeon)	**R-S?** Natal	Oatley 1984
(+)*Schinus terebinthifolius* (Brazialian pepper tree)	*Zosterops lateralis* (silvereye)	**W?** and Lab NE Australia	Panetta & McKee 1997
Pittosporum undulatum (Australian daphne) *Cryptomeria japonica* (sugi)	-*Pyrrhula murina* (Azores bullfinch)	**W-S?** Azores	Ramos 1996
Pinus elliotti (slash pine) *P. radiata* (Monterey pine) *P. pinaster* (maritime pine)	*Calyptorhynchus lathami* (glossy black cockatoos) *Calyptorhynchus funereus latirostris* (Carnaby's cockatoo)	**W?** Australia	Richardson & Higgins 1998
+*Peraxilla* spp. (mistletoe) (Natives)	+ *Anthornis melanura* (bellbird) + *Prosthemadera novaeseelandiae* (tui)	**W-R** New Zealand	Robertson et al. 1999
+*Crataegus monogyna* (European hawthorn)	*Turdus migratorius* (American robin)	**W?** U.S. (Oregon)	Sallabanks 1993
Lonicera maackii (amur honeysuckle)	- *Turdus migratorius* (American robin)	**W?**	Schmidt &

Rhamnus cathartica (European buckthorn)	-*Hylochichla mustelina* (wood thrush)	U.S. (Illinois)	Whelan 1999
(+)*Lantana camara* (+)*Clidemia hirta* (Koster's curse) (+)*Myrica faya* (firetree) (+)*Passiflora mollissima* (banana passionfruit) (+)*Psidium cattleianum* (strawberry guava) (+)*Rubus argutus* (sharp-toothed blackberry) (+)*Schinus terebinthifolius* (Brazialian pepper tree) (+)24 horticultural non-natives	*Pycnonotus cafer* (red-vented bulbul)* Non-native frugivores* Non-native granivores* Native frugivores	**W-U?** U.S. (Hawai'i)	Smith 1985 (review)
Rosa multiflora (multiflora rose) *Berberis thunbergii* (Japanese barberry) *Lonicera japonica* (Japanese honeysuckle) *Ligustrum vulgare* (privet)	+*Mimus polyglottis* (northern mockingbird)	**W-U?** Eastern U.S. & SE Canada	Stiles 1982
Myrica faya (firetree)	*Zosterops japonicus* (Japanese white-eye)*	**W?** U.S. (Hawai'i)	Vitousek & Walker 1989
Erythrina variegata (coral tree) *Ficus microcarpa* (laurel fig) *Schefflera actinophylla* (umbrella tree) Queensland native Others	+*Trichoglossus haematodus* (rainbow lorikeet)	**S** Australia (New South Wales)	Waterhouse 1997
Lythrum salicaria (purple loosestrife)	+avian density -avian diversity +*Melospiza georgiana* (swamp sparrow) *Cistothorus palustris* (marsh wren) Others	**R-S?** U.S. (Michigan)	Whitt et al. 1999
Lonicera mackii (Amur honeysuckle) *Lonicera tatarica* (Tartarian honeysuckle) *Rhamnus cathartica* (European buckthorn)	(+)Breeding birds nest site needs *Hylochichla mustelina* (wood thrush) *Dumetella carolinensis* (gray catbird) *Turdus migratorius* (American robin)	**W?** U.S. (Illinois)	Whelan & Dilger 1992

	Pheucticus ludovicianus (rose-breasted grosbeak)		
Berberis thunbergii (Japanese barberry) *Rosa multiflora* (multiflora rose) *Pyrus* spp. (pear) *Euonymus alatus* (burning bush) *Lonicera japonica* (Japanese honeysuckle) *Lonicera tatarica* (Tartarian honeysuckle) *Celastrus orbiculatus* (oriental bittersweet)	(=) fall migrants (+) spring migrants (+) 10 fruit-eating species *Turdus migratorius* (American robin) *Bombycilla cedrorum* (cedar waxwing) *Mimus polyglottis* (northern mockingbird) *Dumetella carolinensis* (gray catbird) *Catharus* spp. (thrush) *Colaptes auratus* (northern flicker)	**W-S?** U.S. (New Jersey)	White & Stiles 1992
Arbutus unedo (strawberry tree) *Berberis glaucocarpa* (barberry) *Crateagus monogyna* (European hawthorn) *Leycesteria formosa* (Himalayan honeysuckle) *Ligustrum sinense* (Chinese privet) *Rubus fruticosus* (blackberry) *Sambucus nigra* (elderberry) *Solanum nigrum* (black nightshade) 19 non-native plants	(+)*Turdus merula* (European blackbird)* (+)*Sturnis vulgaris* (European starling)* (+)*Zosterops lateralis* (silvereye) (-)*Prosthemadera* (tui) (-)*Anthornis melanura* (bellbird) (-)native honeyeaters; some native pigeons others	**W-R?** New Zealand	Williams & Karl 1996
Acacia cyclops (rooikrans)	-species diversity; insectivores +granivores -*Erythropygia coryphaeus* (karoo robin) -*Cisticola subruficapilla* (gray-backed cisticola) -*Nectarinia chalybea* (lesser doublecollared sunbird) others	**W?** S. South Africa	Winterbottom 1970
Acacia cyclops (rooikrans)	+*Sturnis vulgaris* (European starling)*	**W?** S. South Africa	Winterbottom & Liversidge 1954
Lonicera morrowii (Morrow's honeysuckle)	(+) *Bombycilla cedorum* (cedar waxwing)	**S-U** Lab, U.S. (New York)	Witmer 1996

Chapter 10

Urban sprawl and juniper encroachment effects on abundance of wintering passerines in Oklahoma

Bryan R. Coppedge, David M. Engle, Samuel D. Fuhlendorf, Ronald E. Masters, and Mark S. Gregory

Div. Agricultural Sciences and Natural Resources, Oklahoma State University, Stillwater, OK 74078 USA e-mail: bcoppedg@tulsa.cc.ok.us

Key words: Christmas Bird Count, fire suppression, grasslands, juniper, Oklahoma, passerines, urban sprawl

Abstract: Oklahoma landscapes have experienced significant alteration from low-density urban development. Correlated with this development is the encroachment of woody plants such as juniper (*Juniperus* spp.) into grassland habitat remnants as a result of disruption of historic fire regimes. Because junipers produce prolific fruits that are utilized by frugivorous passerines in winter, we examined how increases in human population density (an index of urban development) and juniper encroachment affected long-term abundance patterns of common passerines using data from the Audubon Society Christmas Bird Counts (CBC). Abundance of seven species was significantly related to human population density. Five of these species exhibited unimodal response patterns, indicating a preference for intermediate levels of habitat alteration by humans. Abundance of nine species was related to regional juniper encroachment levels; three had positive linear trends with juniper levels, two exhibited unimodal trends, while four species were negatively related to juniper encroachment levels. Juniper is projected to invade one-half of the grassland remnants in the state by the next century, so the winter abundance of many frugivorous species and species adaptable to urbanized environments will likely increase in Oklahoma. Conversely, continued development and juniper encroachment will hinder conservation efforts for many grassland birds already in decline.

1. INTRODUCTION

In Oklahoma, as in many western and plains states, significant landscape alteration has occurred as large rural areas are converted to sprawling, low-density human developments (Alig and Healy 1987, Huntsinger and Hopkinson 1996). One of the most serious impacts of human development on ecosystems is the disruption of historic, recurrent fires (Baker 1992, Leach and Givnish 1996). Fire is a critical process that developed and maintained most North American grasslands (Bragg and Hulbert 1976, Axelrod 1985). Furthermore, the decreasing size of grassland remnants has led to changes in management regimes, especially in the use of prescribed fire for control of invasive woody plants (Bernardo et al. 1988, Engle et al. 1999). Thus, the suppression of fire that accompanies human settlement and landscape fragmentation has resulted in significant woody plant encroachment in Great Plains grasslands (Bragg and Hulbert 1976, Archer 1994). In Oklahoma, juniper (*Juniperus* spp.) has expanded exponentially into suburban and rural habitat fragments since 1950 (Engle et al. 1995). Juniper increase has paralleled a steady rise in human population density for this same time period (Fig. 10.1). Thus, low-density urban sprawl, fire suppression, and juniper expansion appear to be concurrent and connected processes within Oklahoma landscapes.

Junipers produce prolific fruits that are highly desired by frugivorous passerines (Holthuijzen and Sharik 1985), and subsequent seed dispersal by birds plays a key role in juniper expansion (Holthuijzen and Sharik 1984, Chavez-Ramirez and Slack 1994). Juniper fruit ripens in late fall and winter and provides a critical food for many overwintering passerines (Holthuijzen and Sharik 1984). These facts led us to hypothesize that the abundance of birds that utilize juniper fruit will show a temporal increase concomitant with temporal patterns of juniper encroachment and the development of rural areas. We utilized long-term data from the Audubon Society Christmas Bird Counts (Butcher 1990) to examine the relationship of urban sprawl and associated juniper encroachment to the abundance patterns of common passerines that winter in Oklahoma.

2. METHODS

2.1 Christmas Bird Count Data

The Christmas Bird Count (CBC) is the largest and oldest wildlife survey in existence (Drennan 1981). The CBC began in 1900 in the eastern U.S. and by 1986 included over 1500 sites in North America (Butcher 1990).

CBC locations are 24 km diameter circles within which numerous participants count all birds seen or heard in a 24 hr period designated within two weeks of Christmas (Bock and Root 1981). Some controversy exists as to the utility of the data, including how to standardize species counts, determining appropriate spatial and temporal scales for analysis, and potential biases in habitat coverage (Butcher 1990). The analysis of CBC data for certain species also is complicated by flocking behavior (Bock and Root 1981). Accurate counts of large flocks are difficult and recorded abundance can be altered by chance flock movements (Root 1981). Analysis of very rare species also is difficult because of small sample sizes. Thus, the use of multiple CBC counts from multiple years (i.e. large sample sizes) is generally recommended for analysis of CBC data, especially to alleviate sample-size problems (Butcher 1990). We followed the methods of Root (1981) to standardize species totals for each CBC count by pooling counts over adjacent years and calculating an average abundance as:

$$X = \left(\sum_{i=1}^{Y} (I / H)_i\right) / Y$$

where X = the average species abundance; I = total number of individuals counted of species i; H = total party hours, an index of observer counting effort; and Y = number of count years (Root 1981). We calculated average abundance using data from a 5-yr period centered on the dates of juniper surveys (detailed below). Thus, CBC data from 1948 to 1952 were used to derive a mean species abundance for 1950, 1963-1967 CBC data for 1965, etc.

2.2 Selection of Species and Study Area

Because of the variety of habitats covered during the CBC survey, only broad generalizations of bird-habitat relationships are possible (Arbib 1981). To minimize sources of variation in species occurrences due to major habitat differences, we limited our data to include only 19 CBC circles located within the Osage Plains-Cross Timbers physiographic province (Bystrak 1981). This province includes most of Oklahoma but excludes the high plains and mesa country of the panhandle region and Ozark-Ouachita plateau to the east (Fig. 10.2). By limiting our data to counts within this province, we intentionally avoided counts west of the 100th Meridian, a major biogeographic transition zone in the southern plains (Bock et al. 1977, Root et al. 1981). Our study therefore focused on grassland areas that have experienced various degrees of juniper encroachment. We examined the abundance patterns of 22 common passerines (Appendix 10.1) that winter across the Osage Plains-Cross Timbers province (Root 1981).

2.3 Urban Metrics

As of 1990, Oklahoma had two major urban centers with human populations greater than 250,000; Oklahoma City, in the center of the state, and Tulsa, in the northeast, as well as 12 other cities with human populations greater than 25,000 (Fig. 10.2). While most CBC circles were near at least one urban area, some CBC circles were located in predominately rural areas (Fig. 10.2). We used human population density (HPD) as a general measure of low-density urban sprawl and human development. We obtained U.S. Census records from 1950, 1960, 1970, 1980, and 1990 to calculate HPD as the number of people/km^2 for each of the 77 Oklahoma counties.

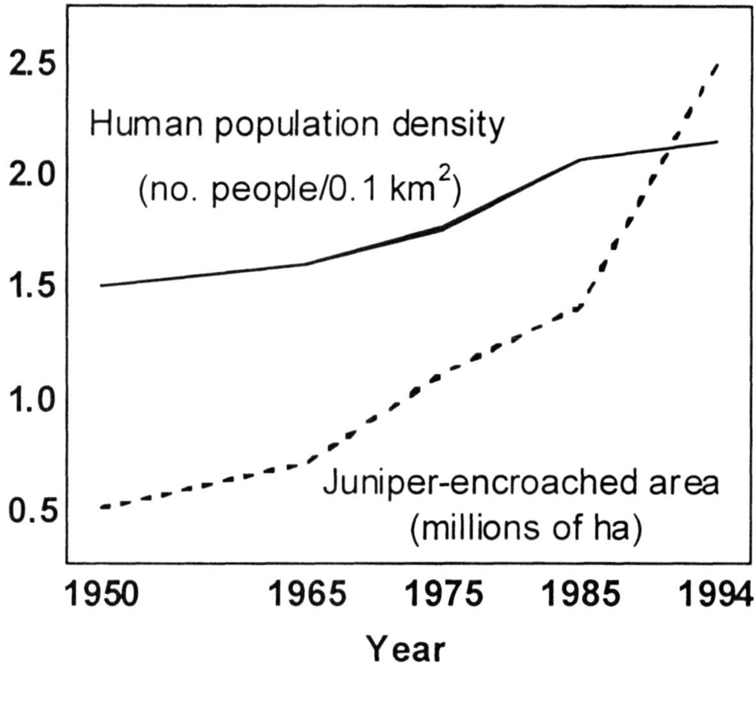

Figure 10.1. Human population density and juniper-encroached area in Oklahoma from 1950 to 1994.

2.4 Juniper Encroachment Data

We obtained data from earlier statewide surveys (Snook 1985, Engle et al. 1995) that quantified juniper encroachment levels by county in 1950, 1965, 1975, 1985, and 1994. The survey data were collected by all Natural Resources Conservation Service field offices in Oklahoma by having field personnel identify on maps of their county the areas invaded by juniper with an average minimum density of 127 trees per ha. Figure 10.2 was derived from the summarization of this county-level data by Engle et al. (1995) and illustrates the extent of juniper in 1950 and subsequent encroachment levels as of 1994. A typical landscape in the study area is shown in Fig. 10.3.

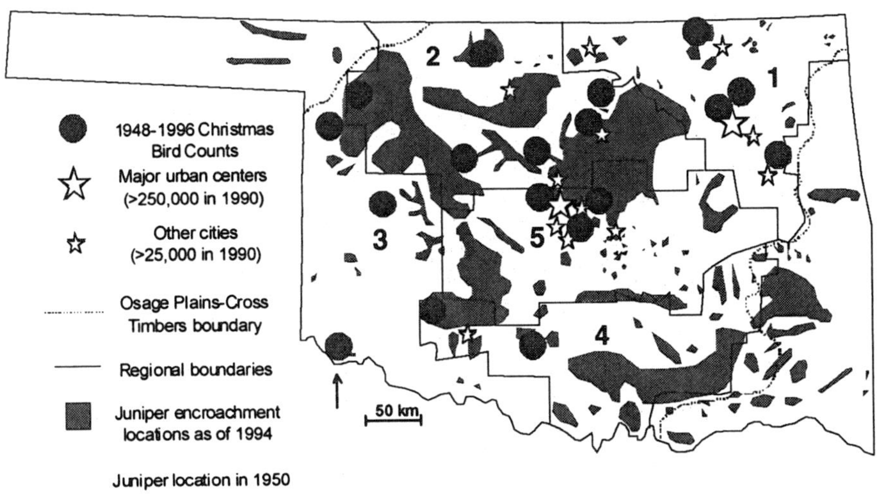

Figure 10.2. Geographic features of Oklahoma, including the locations of Christmas Bird Counts, cities, and the Cross Timbers-Osage Plains physiographic province boundaries and areas of juniper encroachment and study analysis regions.

Figure 10.3. Urban development (modular homes) and juniper encroachment (dark green vegetation) into grasslands near Stillwater, OK.

2.5 Scale Issues and Data Analysis

Because we were using three datasets collected at various spatial scales with different objectives, a common scale was needed at which data would be comparable. Thus, we scaled up by pooling habitat data from adjacent counties, thereby dividing the Osage Plains-Cross Timbers province into five regions (Fig. 10.2). Each region contained several CBC circles and possessed a range of urban and juniper encroachment levels. With this larger scale approach it also was necessary to pool the bird data. Although this reduced total sample size somewhat, this approach did not violate general recommendations for pooling CBC counts to derive average abundances (Butcher 1990). Bird abundance data were then calculated for the 25 spatio-temporal replicates (five regions x five time periods, $n = 25$). The percentage of each region invaded by juniper and HPD data also were summarized by region for each survey date and used in analyzing bird-habitat relationships. Bird data were then modeled against juniper and urbanization variables with simple linear and non-linear regression by transforming independent data into polynomial terms (Meents et al. 1983). We modeled HPD and juniper separately because of the collinear nature of these variables (Fig. 10.1), so these data should be interpreted cautiously. Because of sample size we limited our models to second-order polynomials, which are nonetheless sufficient to detect unimodal (Gaussian) response curves (Meents et al. 1983). Furthermore, our use of spatio-temporal replicates allowed a wider sampling of the juniper encroachment and HPD gradients (Wiens and Rotenberry 1981). For comparative purposes, we also obtained survey-wide and statewide population trends from the CBC website (Sauer et al. 1996) for the bird species studied.

3. RESULTS

3.1 Urbanization and Species Abundance Patterns

Seven of 22 species used in this study showed significant relationships to regional HPD in Oklahoma. Interestingly, five of these species exhibited unimodal patterns (Fig. 10.4), indicative of maximum abundance at intermediate levels of human development and urban sprawl. The Eastern Bluebird showed a similar but insignificant trend with HPD. Two species, the Northern Mockingbird and Northern Cardinal, had strong positive relationships with HPD.

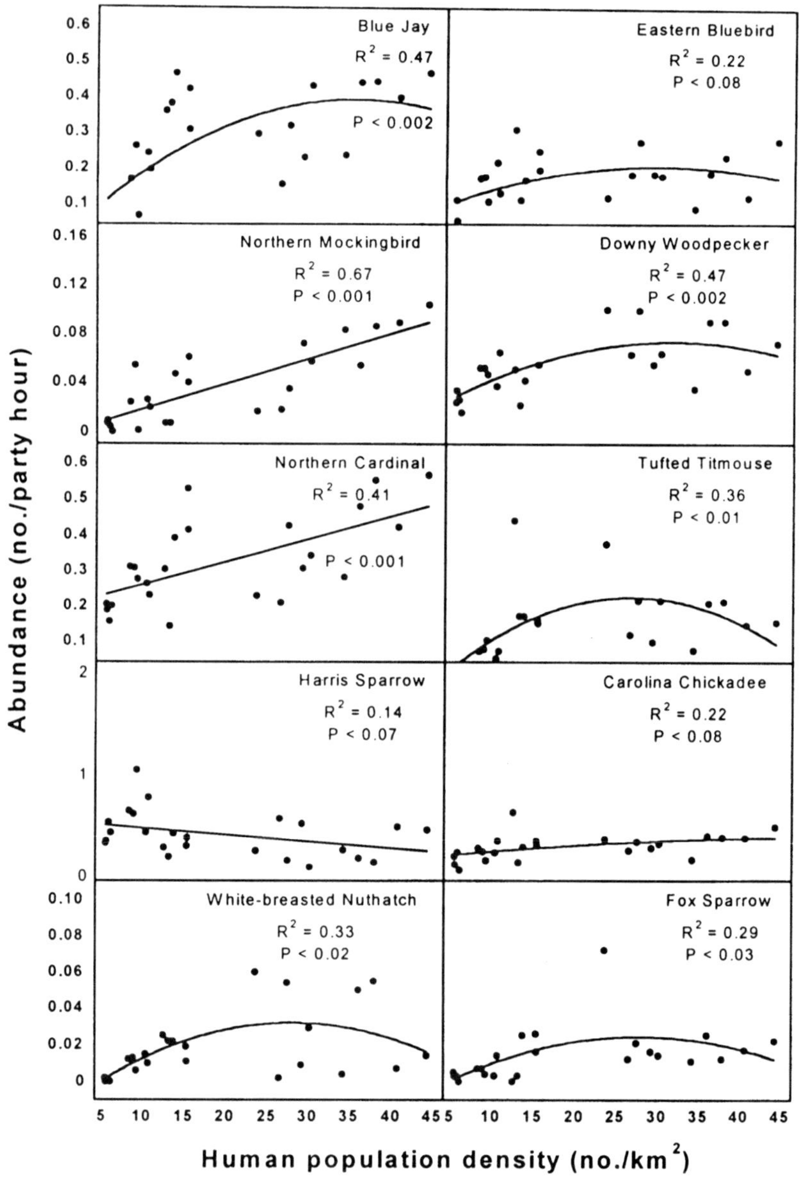

Figure 10.4. Results of regression models relating mean species abundance of passerines wintering in Oklahoma to regional human population density for the 1950-1994 time period.

3.2 Juniper Encroachment and Species Abundance Patterns

The winter abundance of nine of 22 passerine species was related to regional juniper encroachment levels (Fig. 10.5). Two species, the American Robin and Eastern Bluebird, exhibited unimodal trends with juniper levels, indicating maximum abundance at moderate levels of habitat modification by the invasive woody junipers. The Blue Jay showed similar but insignificant trends with juniper. The Cedar Waxwing, Ruby-crowned Kinglet, and Yellow-rumped Warbler, all potential juniper feeders (Appendix 10.1), had positive linear trends with regional juniper levels. The abundance of four species, the Song Sparrow, White-crowned Sparrow, House Sparrow, and American Goldfinch, were negatively related to juniper (Fig. 10.5).

3.3 Comparison of CBC Trends

Only wintering American Robins, Cedar Waxwings, Carolina Chickadees, and White-crowned Sparrows are increasing in Oklahoma (Appendix 10.1). But Cedar Waxwings and Carolina Chickadees also are increasing across the North American continent, so increasing populations of these two species in Oklahoma may simply reflect a wider continental trend. Interestingly, two species that potentially feed on juniper and have declining continental trends, the Northern Mockingbird and Dark-eyed Junco, are stable in Oklahoma. Two species with declining continental populations, the White-crowned Sparrow and House Sparrow, also were negatively related to juniper levels in Oklahoma. The American Goldfinch is increasing continentally but negatively related to juniper levels and declining in Oklahoma (Appendix 10.1). One other species, the Field Sparrow, is declining in Oklahoma, but this seems unrelated to human or juniper encroachment and is likely a reflection of wider continental trends (Appendix 10.1).

4. DISCUSSION

Our results revealed unimodal relationships of several species to regional juniper and HPD levels, indicating that our analyses did not measure simple temporal trends in bird abundance, but instead revealed regional-scale winter patterns in bird-habitat relationships in Oklahoma.

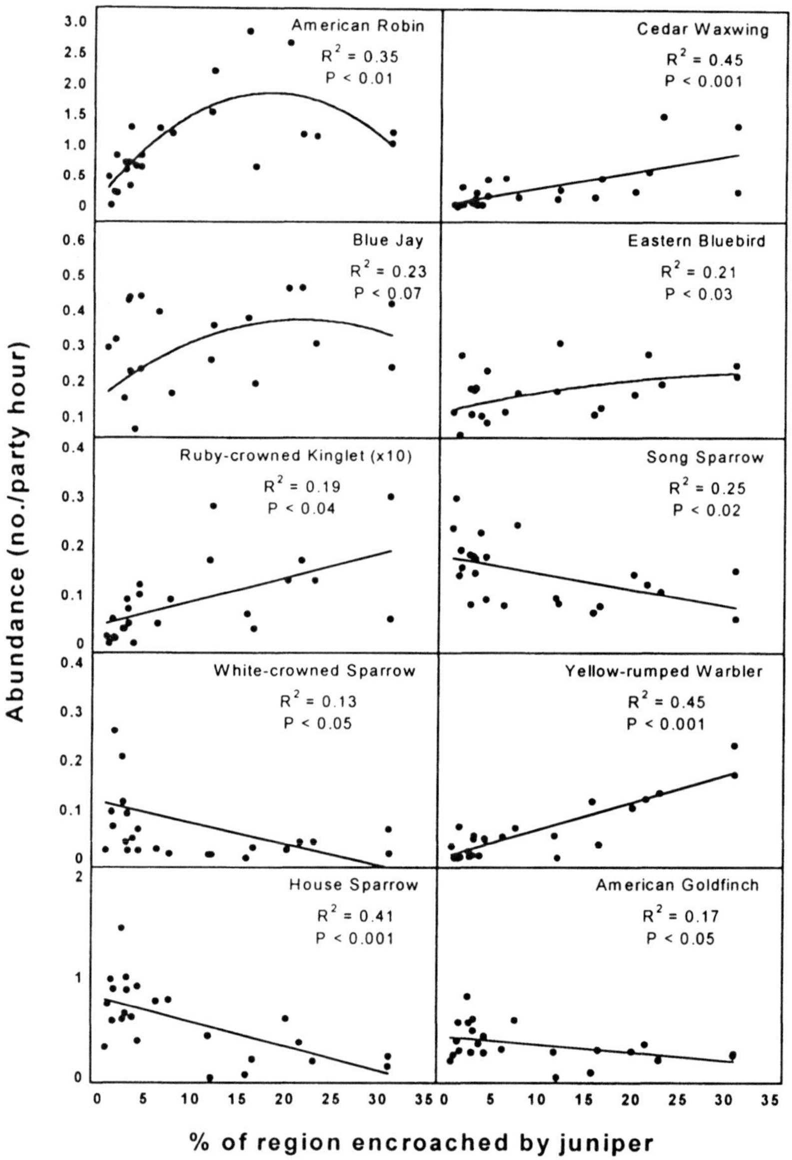

Figure 10.5. Results of regression models relating mean species abundance of passerines wintering in Oklahoma to regional juniper encroachment levels for the 1950-1994 time period.

Interestingly, intermediate levels of habitat disturbance are known to maximize avian species diversity and abundance by increasing the breadth of resources available to the avian community (Blair 1996). Blair (1996) attributed these increasing resources, at least partially, to an increase in woody vegetation, as the relative abundance of exotic species and habitat generalists whose feeding depends on woody vegetation usually increase, while endemic species and habitat specialists usually decline (see also Mills et al. 1989). Similar changes in the avian community are seen in woody plant invasions into grasslands, which attracts shrubland or woodland birds while the abundance of open grassland species decrease (Lloyd et al. 1998). Although only a few studies have been done in the Great Plains, results have shown that juniper woodlands have high numbers of forest and edge species and habitat generalists but decreased numbers of grassland specialists (Seig 1991).

4.1 Effects of Urbanization

Our results suggest urbanization in Great Plains grasslands has a two-fold effect. The first is an increase in woody vegetation that often occurs within residential and suburban areas due to planting for windbreaks, shelterbelts, or in landscaping. This woody vegetation is known to attract exotic and generalist avian frugivores (Blair 1996). In our study, all of the species that related to HPD have affinities with woody habitats (Fig. 10.4). Another specific and relevant example was reported by McPherson (1987), who found that Cedar Waxwings fed almost exclusively on the fruits of exotic woody vegetation within an urbanized area of central Oklahoma.

A second, indirect effect of urbanization is landscape fragmentation. Fragmentation, at least in Oklahoma, leads to fire suppression and allows woody plant expansion into grassland remnants, further altering avian community structure. Studies have shown that many eastern North American woodland species (Knopf 1994) have moved westward via woodlands that have developed along major plains rivers (Johnson 1994). We see in Oklahoma a similar pattern as most of the birds responding to juniper increases in Oklahoma were forest or open woodland species (Fig. 10.5). Our study demonstrates an important regional corollary to the continental pattern of habitat modification that results in the mixing of regional avifaunas, a process that can result in loss of local community distinctiveness, hybridizations, and ultimately in the loss of overall avian diversity (Knopf 1986, 1994).

In Oklahoma, urban sprawl and woody plant encroachment are confounded processes that result in similar effects on the avian community.

The winter abundance of five species that commonly visit bird feeders (Dunn 1995) showed unimodal responses to HPD. Thus, the popular activity of winter bird feeding that would intuitively accompany moderate levels of urbanization such as the development of residential areas may affect bird abundance in Oklahoma at a regional scale, much as it is presumed to have done at larger scales in other regions of North America (Ehrlich et al. 1988).

4.2 Juniper Effects

The abundance patterns of the American Robin, Cedar Waxwing, Eastern Bluebird, Blue Jay, and Yellow-rumped Warbler, all species known to consume juniper fruit during winter, exhibited significant relationships with regional juniper levels. Four of these species also had stable or positive population trends in Oklahoma according to the CBC (Appendix 10.1). Thus, it may be that the abundance of these species is simply coincidental with juniper increases in Oklahoma. However, we offer an alternate explanation. We propose that the abundance of juniper fruit may significantly affect both the overwinter survival and subsequently the populations of these species. For example, the Northern Mockingbird is declining continentally, but increasing in Oklahoma. Although not significantly related to juniper levels at the scale of this study, this urban-adapted bird is positively associated with increased HPD, which is likely an important factor behind juniper increases. Thus, the abundance of potential juniper dispersers is likely both a product of and cyclic cause of increasing juniper (McDonnell and Stiles 1983; Jordano 1993, Santos et al. 1999).

During daily foraging bouts, the American Robin is known to feed on juniper fruit, then switch to ground foraging for invertebrates (Chavez-Ramirez and Slack 1994). This feeding pattern makes the Robin one of the most important dispersers of juniper seeds into grasslands in the southern plains (Chavez-Ramirez and Slack 1994), and the unimodal peak of this species' abundance with regional juniper level verifies this locally distinct pattern. Initially, robin abundance increased with juniper to the point at which the species requirements for open grassland for ground foraging became limiting, thereafter abundance declined (Fig. 10.5).

Because urban sprawl and the presence of abundant seed sources will likely continue, juniper invasion is projected to increase exponentially in Oklahoma (Engle et al. 1995). Thus, the winter abundance of many frugivorous species, juniper dispersers, and species adaptable to woody, urban environments will likely increase in Oklahoma as well. However, of most concern are the changes in the composition of the avian community that accompany juniper invasion. While this study demonstrated decreasing abundance of some grassland species with rising levels of urbanization and juniper encroachment, most grassland birds of conservation concern are neotropical migrants that are present in Oklahoma only in summer months (Askins 1993). Thus, this study could not fully evaluate juniper effects on grassland birds that migrate to other locations in winter.

Grassland birds are already one of the fastest declining avifaunas in North America (Askins 1993), and the projected influx of juniper will likely detract from conservation efforts for many species in two ways. First, it is known that woody vegetation attracts nest predators and parasites in prairie environments (Johnson and Temple 1990). Thus, the nesting success of ground-nesting grassland birds will likely be lowered by an influx of juniper, although specific research into this area is needed. Secondly, studies on the negative effects of woody vegetation on grassland birds also have demonstrated the area-sensitivity of many species (Herkert 1994, Helzer and Jelinski 1999). Woody plants will negatively affect habitat suitability for grassland birds with large areas requirements. For example, we have found that the abundance of several breeding grassland birds such as the Western Meadowlark (*Sturnella neglecta*), Cassin's Sparrow (*Aimophila cassinii*), and Horned Lark *(Eremophila alpestris)* were negatively related to woody vegetation (primarily juniper) and grassland patch size in areas of northwestern Oklahoma invaded by juniper (Coppedge et al. 2001). Thus, our studies clearly demonstrate the effects juniper is already having on both breeding and overwintering avian communities in southern Great Plains grasslands.

4.3 Scale Effects

Juniper and population density may affect winter bird abundance at scales larger or smaller than regional because birds respond to habitat at a variety of spatial scales (Wiens 1981, Wiens et al. 1987, Hostetler 2001). Our regional approach was based on county-level data because that was the scale at which juniper and human population data were collected. Smaller plot-level studies have shown that local juniper levels do influence avian frugivore abundance in winter in the southern plains (Chavez-Ramirez and Slack 1996). Similarly, continent-wide studies have concluded that the abundance of some species, especially the Cedar Waxwing, are closely associated with juniper distribution (Root 1981). Thus, we attribute the lack of significant relationships between some common wintering birds in Oklahoma and regional variables to species-specific scale effects (Wiens 1981). Meaningful analyses of CBC data for certain highly gregarious species, such as the European Starling, are simply not possible (Bock and Root 1981). Therefore, our regional approach simply may not have been appropriate for all species. Nonetheless, our results show that urbanization and juniper are important, long-term influences on the winter abundance patterns of several passerine species in Oklahoma.

4.4 Policy and Management Implications

The potential decline of grassland birds implores the preservation and management of grasslands and their integrity (Knopf 1994). These habitat remnants need protection from both further urban development and juniper invasion by a comprehensive landscape management plan. A juniper control program also is needed in the southern plains, though one has yet to be proposed. Funds are available for landowners interested in improving their land for wildlife through the Environmental Quality Incentive Program (EQIP) administered by the U.S. Department of Agriculture Natural Resource Conservation Service. But EQIP funds are limited, especially those earmarked for brush control. Presently, the decision to control encroaching juniper is at the discretion of individual landowners. Economic incentives exist for cattle producers to control juniper on open rangeland because of the negative effect heavy infestations have on forage production (Bernardo et al. 1988). But fear of liability due to the lack of modern prescribed burning laws currently precludes the widespread use of fire, the most effective juniper control method, by most landowners with smaller management units.

Thus, we conclude with several policy and management suggestions. First, juniper needs to be designated a noxious weed so that a comprehensive control plan with significant supporting funds can be implemented. Secondly, negative connotations surrounding the use of prescribed fire as a land management tool should be removed. Essential to this process would be the passage of legislation removing the severe penalties of liability for landowners who properly use fire to control juniper but nonetheless commit accidental property damage. Third, some efforts must be made to curb the spread of low-density human developments into prairies and grasslands. Creating development easements or setting aside grassland preserves will help provide habitat for declining grassland birds, but only if properly managed and protected from juniper invasion.

ACKNOWLEDGMENTS

This work was supported by the USDA National Research Initiative Competitive Grants Program (no. 9600853) and the Oklahoma Agricultural Experiment Station. We thank two anonymous reviewers for helpful suggestions and comments on an earlier manuscript draft. This article is published with the approval of the director, Oklahoma Agricultural Experiment Station. We also thank all of the volunteers over the years who have labored to build the CBC database.

REFERENCES

Alig, R. J. and R. G. Healy. 1987. Urban and built-up land area changes in the United States: an empirical investigation. Land Economics 63: 215-226.

Arbib, R. S. 1981. The Christmas Bird Count: constructing an ideal model. Stud. Avian Biol. 6: 30-33.

Archer, S. 1994. Woody plant expansion into southwestern grasslands and savannas: rates, patterns, and proximate causes, p. 13-68. *In* M. Vavra, W. Laycock, and R. Pieper [EDS.], Ecological implications of livestock herbivory in the West. Society for Range Management, Denver, CO.

Askins, R. A. 1993. Population trends in grassland, shrubland, and forest birds in eastern North America. Current Ornithol. 11: 1-34.

Axelrod, D. I. 1985. Rise of the grassland biome, central North America. Bot. Rev. 51: 163-202.

Baker, W. L. 1992. Effects of settlement and fire suppression on landscape structure. Ecology 73: 1879-1887.

Bernardo, D. J., D. M. Engle and E. T. McCollum. 1988. An economic assessment of risk and returns from prescribed burning on tallgrass prairie. J. Range Manage. 41: 178-183.

Blair, R. B. 1996. Land use and avian species diversity along an urban gradient. Ecol. Appl. 6: 506-519.

Bock, C. E., J. H. Bock and L. E. Lepthien. 1977. Abundance patterns of some bird species wintering on the Great Plains of the U.S.A. J. Biogeogr. 4: 101-110.

Bock, C. E. and T. L. Root. 1981. The Christmas Bird Count and avian ecology. Stud. Avian. Biol. 6: 17-23.

Bragg, T. B. and L. C. Hulbert. 1976. Woody plant invasion of unburned Kansas bluestem prairie. J. Range Manage. 29: 19-24.

Butcher, G. S. 1990. Audubon Christmas Bird Counts, p. 5-13. *In* J. R. Sauer and S. Droege [EDS.]. Survey designs and statistical methods for the estimation of avian population trends. Biological Rep. 90, U.S. Dept. Int., Washington, D.C.

Bystrak, D. 1981. The North American Breeding Bird Survey. Stud. Avian Biol. 6: 34-41.

Chavez-Ramirez, F. and R. D. Slack. 1994. Effects of avian foraging and post-foraging behavior on seed dispersal patterns of Ashe juniper. Oikos 71: 40-46.

Chavez-Ramirez, F. and R. D. Slack. 1996. Winter phenology and frugivory of American Robins and Cedar Waxwings on the Edwards Plateau of central Texas. Tex. J. Sci. 48: 129-136.

Coppedge, B. R., D. M. Engle, R. E. Masters and M. S. Gregory. 2001. Avian response to landscape change in fragmented southern Great Plains grasslands. Ecol. Appl. 11:47-59.

Drennan, S. R. 1981. The Christmas Bird Count: an overlooked and underused sample. Stud. Avian Biol. 6: 24-29.

Dunn, E. H. 1995. Bias in Christmas Bird Counts for species that visit feeders. Wilson Bull. 107: 122-130.

Engle, D. M., T. G. Bidwell and M. E. Moseley. 1995. Invasion of Oklahoma rangelands and forests by eastern redcedar and ashe juniper. Oklahoma St. Univ., Okla. Coop. Ext. Serv. E-947, Stillwater, OK.

Engle, D. M., T. L. Criner, J. C. Boren, R. E. Masters and M. S. Gregory. 1999. Response of breeding birds in the Great Plains to low density urban sprawl. Great Plains Res. 9: 55-73.

Ehrlich, P. R., D. S. Dobkin and D. Wheye. 1988. The birder's handbook: a field guide to the natural history of North American birds. Simon and Schuster, New York.

Helzer, C. J. and D. E. Jelinski. 1999. The relative importance of patch area and perimeter-area ratio to grassland breeding birds. Ecol. Appl. 9: 1448-1458.

Herkert, J. R. 1994. The effects of habitat fragmentation on midwestern grassland bird communities. Ecol. Appl. 4: 461-471.

Holthuijzen, A. M. A. and T. L. Sharik. 1984. Seed longevity and mechanisms of regeneration of eastern red cedar (*Juniperus virginiana* L.). Bull. Torr. Bot. Club 111: 153-158.

Holthuijzen, A. M. A. and T. L. Sharik. 1985. The avian seed dispersal system of eastern red cedar (*Juniperus virginiana*). Can. J. Bot. 63: 1508-1515.

Hostetler, Mark. 2001. The importance of multi-scale analyses in avian habitat selection studies in urban environments, p. 139-154. *In* J. M. Marzluff, R. Bowman, and R. Donnelly [EDS.], Avian ecology and conservation in an urbanizing world. Kluwer Academic, Norwell, MA.

Huntsinger, L. and P. Hopkinson. 1996. Viewpoint: Sustaining rangeland landscapes: a social and ecological process. J. Range Manage. 49: 167-173.

Johnson, W. C. 1994. Woodland expansion in the Platte River, Nebraska: Patterns and causes. Ecol. Monogr. 64: 45-84.

Johnson, R. G. and S. A. Temple. 1990. Nest predation and brood parasitism of tallgrass prairie birds. J. Wildl. Manage. 54: 106-111.

Jordano, P. 1993. Geographical ecology and variation of plant-seed disperser interactions: southern Spanish junipers and frugivorous thrushes. Vegetatio 107/108: 85-104.

Knopf, F. L. 1986. Changing landscapes and the cosmopolitism of the eastern Colorado avifauna. Wildl. Soc. Bull. 14: 132-142.

Knopf, F. L. 1994. Avian assemblages on altered grasslands. Stud. Avian Biol. 15: 247-257.

Leach, M. K. and T. J. Givnish. 1996. Ecological determinants of species loss in remnant prairies. Science 273: 1555-1558.

Lloyd, J., R. W. Mannan, S. Destefano and C. Kirkpatrick. 1998. The effects of mesquite invasion on a southeastern Arizona grassland bird community. Wilson Bull. 110: 403-408.

McDonnell, M. J. and E. W. Stiles. 1983. The structural complexity of old field vegetation and the recruitment of bird-dispersed plant species. Oecologia 56: 109-116.

McPherson, J. M. 1987. A field study of winter fruit preferences of Cedar Waxwings. Condor 89: 293-306.

Meents, J. K., J. Rice, B. W. Anderson and R. D. Ohmart. 1983. Nonlinear relationships between birds and vegetation. Ecology 64: 1022-1027.

Mills, G. S., J. B. Dunning and J. M. Bates. 1989. Effects of urbanization on breeding bird community structure in southwestern desert habitats. Condor 91: 416-428.

Root, T. L. 1981. Atlas of wintering North American birds: an analysis of Christmas Bird Count data. Univ. Chicago Press, Chicago, IL.

Root, T. L., M. A. Holmgren and R. W. Andrews. 1981. Winter abundance patterns of some songbirds near the 100th Meridian in the southern United States. Southwest. Natural. 26: 95-100.

Santos, T., J. L. Telleria and E. Virgos. 1999. Dispersal of Spanish juniper *Juniperus thurifera* by birds and mammals in a fragmented landscape. Ecography 22: 193-204.

Sauer, J. R., S. Schwartz and B. Hoover. 1996. The Christmas Bird Count homepage. Ver. 95.1. (http://www.mbr.nbs.gov/bbs/cbc.html),.Patuxent Wildlife Research Center, Laurel, MD.

Seig, C. H. 1991. Geographic affinity of bird species associated with Rocky Mountain juniper woodlands and adjacent grasslands in southwestern South Dakota. Prairie Natural. 23: 25-33.

Snook, E. C. 1985. Distribution of eastern redcedar on Oklahoma rangelands, p. 45-52. *In* R. F. Wittwer and D. M. Engle [EDS.], Conference Proceedings: Eastern redcedar in Oklahoma. Oklahoma St. Univ., Okla. Coop. Ext. Serv. E-349, Stillwater, OK.

Wiens, J. A. 1981. Scale problems in avian censusing. Stud. Avian. Biol. 6: 513-521.

Wiens, J. A. and J. T. Rotenberry. 1981. Habitat associations and community structure of birds in shrubsteppe environments. Ecol. Monogr. 51: 21-41.

Wiens, J. A., J. T. Rotenberry and B. Van Horne. 1987. Habitat occupancy patterns of North American shrubsteppe birds: the effects of spatial scale. Oikos 48: 132-147.

Appendix 10.1. Passerine species examined to determine the influence of low-density urban sprawl and juniper encroachment on winter abundance patterns in Oklahoma.

Species	Scientific name	Juniper berry feeder?	CBC trend 1959-88	
			Survey-wide	Oklahoma
American Robin	*Turdus migratorius*	Y	0	+
Cedar Waxwing	*Bombycilla cedrorum*	Y	+	+
European Starling	*Sturnus vulgaris*	Y	0	0
Northern Mockingbird	*Mimus polyglottos*	Y	-	0
Eastern Bluebird	*Sialia sialis*	Y	+	0
Downy Woodpecker	*Picoides pubescens*	Y	+	0
Dark-eyed Junco	*Junco hyemalis*	Y	-	0
Blue Jay	*Cyanocitta cristata*	Y	+	0
Carolina Chickadee	*Parus canadensis*	?	+	+
Tufted Titmouse	*P. bicolor*	?	0	0
Ruby-crowned Kinglet	*Regulus calendula*	?	+	0
Harris Sparrow	*Zonotrichia querula*	?	-	0
White-crowned Sparrow	*Z. leucophrys*	?	-	+
House Sparrow	*Passer domesticus*	D	-	0
Song Sparrow	*Melospiza melodia*	?	0	0
Field Sparrow	*Spizella pusilla*	D	-	-
Fox Sparrow	*Passerella iliaca*	D	0	0
White-breasted Nuthatch	*Sitta canadensis*	Y	+	0
American Goldfinch	*Carduelis tristis*	?	+	-
Yellow-rumped Warbler	*Dendroica coronata*	Y	*	*

Species	Scientific name	Juniper berry feeder?	CBC trend 1959-88	
			Survey-wide	Oklahoma
Nortern Cardinal	*Cardinalis cardinalis*	Y	0	0
Horned Lark	*Eremophila alpestris*	D	0	0

Y = documented feeder on juniper fruit.

? = at least partially frugivorous, but not documented to consume juniper fruit.

D = unlikely to consume juniper fruit.

+ = statistically significant ($P < 0.10$) positive trend.

- = statistically significant ($P < 0.10$) negative trend.

0 = statistically insignificant ($P > 0.10$) trend.

* = trend data not available.

Chapter 11

Nest predator abundance and urbanization

David G. Haskell, A. M. Knupp, and M. C. Schneider
Department of Biology, 735 University Ave., University of the South, Sewanee, TN 37383 USA, e-mail: dhaskell@sewanee.edu

Key words: Artificial nest, domestic cat, nest predation, quail eggs, urban gradient

Abstract: We used infra-red triggered cameras, point counts, footprint monitors, and artificial nests baited with quail eggs to investigate the abundance of potential nest predators at 75 sites spread across a gradient of housing density in Franklin County, Tennessee. The total number of predators detected at each site showed a shallow positive rise with increasing housing density. Blue Jays (*Cyanocitta cristata*), domestic cats (*Catus silvestris*), raccoons (*Procyon lotor*), and opossums (*Didelphis marsupialis*) all increased in abundance, American Crows (*Corvus brachyrhynchos*) decreased in abundance, and the abundance of domestic dogs (*Canis familiaris*) was not related to housing density. Thus, in our study area, urbanization may increase populations of some nest predators, but not of others. The probability of predation on artificial nests containing a quail egg did not change with housing density, and quail egg predation was correlated only with the presence of raccoons and American Crows. These data suggest that quail eggs may not be an appropriate method for studying rates of predation on eggs or nestlings in urban settings, but may be helpful for assessing the abundance of American Crows and raccoons. More information is needed about how changes in the abundance of predators translates into changes in rates of predation on bird nests, and for studies of how urbanization affects populations of nest predators in regions other than North America and Europe. We also outline two parallel studies of economics and policy that may provide ways to connect conservation biology with changes in land-use planning.

1. INTRODUCTION

Many studies have shown that adding houses, commercial developments and roads to a landscape can change the abundance and diversity of avian communities (Emlen 1974, Beissinger and Osborne 1982, Blair 1996). A number of mechanisms might underlie these changes. Urbanization changes the taxonomic diversity and structural complexity of vegetation (Rottenborn 1999, Savard and Falls 2001), which may affect the availability of nest sites and foraging microhabitats. Edges created by roads and other disturbances can reduce food availability for insectivorous birds (Burke and Nol 1998, Haskell 2000). Urbanization also has been suggested to increase rates of predation on nestlings and eggs (e.g., Terborgh 1989, Wilcove 1999).

The few studies that have examined rates of nest predation across a range of levels of urbanization have generally done so by using artificial nests baited with quail or chicken eggs. These previous studies have come to differing conclusions. Wilcove (1985), Wong et al. (1998) and Hogrefe et al. (1998) found that rates of predation on artificial nests were higher in urban or suburban areas. In contrast, Melampy et al. (1999) found no difference in rates of artificial nest predation between suburban, semi-rural and rural sites and Danielson et al. (1997) found lower rates of predation on artificial nests at suburban edges compared to rural edges.

The interpretation of artificial nest experiments is, unfortunately, fraught with problems. Although these artificial nests are baited with eggs, their results have been used to make inferences about overall rates of predation on real nests, including predation on both the egg and nestling stages of these real nests. This may be problematic if some predators have different responses to eggs and nestlings, and the results of artificial nest experiments might better be interpreted as being applicable only to the egg stage of real nests. In addition, some predators cannot prey upon the quail and chicken eggs often used to bait nests (Roper 1992, Haskell 1995a, DeGraaf and Maier 1996). If the abundance of these predators varies across the treatment groups in the experiment, the results of the experiment will have a directional bias. This bias can produce spurious correlations between measures of landscape change (e.g., fragmentation) and rates of predation on artificial nests (Haskell 1995b). Because we do not have a complete understanding of how the relative proportions of predators vary across urbanization gradients, and because we do not know the extent to which some predators in urban areas will prey on quail eggs (e.g., domestic cats and dogs [*Catus silvestris* , *Canis familiaris*]), we cannot yet draw any well-supported conclusions from artificial nest studies conducted in urban areas.

Another approach to understanding how urbanization affects nest predation is to study the effects of urbanization on the abundance and behavior of animals that are potential predators of bird nests. This approach requires either the assumption that there is a positive correlation between the abundance of nest predators and rates of nest predation, or, better, concurrent studies of the nest searching behavior of different predator species. The former assumption underlies many studies of nest predator abundance, but has not been tested (see Discussion). In addition, these studies assume that we have a good knowledge about which animals are predators of bird nests. This information is readily available for some species (e.g., raccoons, *Procyon lotor*), but not for others (e.g., we lack information about the extent to which many snakes, small rodents and domestic animals will prey on bird nests during the egg or nestling stages).

Our knowledge of how urbanization affects the abundance of nest predators is patchy. Most of this information comes from North America and Europe, but even in these areas we lack information for many potential predators. Raccoons have been extensively studied in North America because they are vectors of rabies. Riley et al. (1998) reviewed previous studies of raccoon densities in urban and rural areas. Although the highest density reported in the literature (250 per km^2) comes from a rural area, there is a trend for urban areas to have higher raccoon densities (rural sites: median density = 12.5 per km^2, $n = 22$; suburban/urban sites: median = 61.2 per km^2, $n = 6$). Studies of bird nest predators indicate that in North America the density of American Crows (*Corvus brachyrhynchos*) increases with increasing urbanization, whereas Blue Jays (*Cyanocitta cristata*) are most associated with suburban areas (Crosby and Blair 2001, Marzluff et al. 2001, McGowan 2001). In Australia, Major et al. (1996) found that Pied Currawongs (*Strepera graculina*) were very abundant in urban areas and accounted for 52% of predation of artificial nests in these areas.

Red foxes (*Vulpes vulpes*) are found in higher densities in suburban and urban parts of Europe (Macdonald 1987), but may be more associated with suburban fringes in North America (Adkins and Stott 1998). Opossums (*Didelphis marsupialis*) and skunks (*Mephitis mephitis*) may be less abundant in urban areas compared to rural or suburban areas (Matthiae and Stearns 1981).

Williamson (1983) quantified the abundance of gray squirrels (*Sciurus carolinensis*) at 75 sites across an urbanization gradient in Massachusetts. He found squirrels at highest densities in park-like areas (1.10 squirrels ha^{-1}), and at lowest densities in urban areas with low numbers of trees (0.15 squirrels ha^{-1}). Natural forests, forested areas with scattered residential housing and suburban areas with higher densities of housing did not differ in the density of squirrels (0.58 squirrels ha^{-1}). Other rodents that may prey on birds' nests include chipmunks (*Tamias striatus*) and some mice (*Peromyscus* sp.). Matthiae and Stearns (1981) found similar abundances of both chipmunks and *Peromyscus leucopus* in urban and rural forest fragments, although Nupp and Swihart (1996) found that the abundance of *Peromyscus leucopus* was increased by forest fragmentation.

The domestic cat has been singled out as a predator whose effects are most pronounced in urban or suburban areas (e.g., Stallcup 1991, Luoma 1997). Cats certainly kill many birds (Churcher and Lawton 1987), but few studies have examined how cat densities change with increasing urbanization, nor if cats prey on bird eggs and nestlings in addition to adults. Crooks and Soulé (1999) showed that cats can reach high densities in urban areas and that these densities increase with increasing fragmentation of remnant habitat patches. Other studies suggest that cats also may reach high densities in some rural areas. Coleman and Temple (1993) found that 86% of farms in Wisconsin have cats, and that the mean number of cats per farm was 8.6. Other studies of cat densities (reviewed in Liberg and Sandell 1988) have high densities of cats in urban areas, rural farms and fishing villages. We know of no studies that have examined cat abundance across a gradient of urbanization in one habitat.

Thus, neither studies of the abundance of predator populations nor artificial nest studies provide unambiguous evidence that urbanization increases rates of nest predation. We lack data on the effects of urbanization on populations of many predators and our ability to interpret the results of artificial nest experiments is hampered by our lack of knowledge about the responses of predators to these nests. The objectives of this study were to test for one county in Tennessee: (1) Whether the abundance of domestic cats, other medium-sized mammals and corvids increases with the density of residential housing, and (2) Whether the abundance of these potential nest predators is related to rates of nest predation on artificial nests baited with quail eggs.

2. METHODS

The study was conducted on the Cumberland Plateau of Franklin County, Tennessee (study area covered 581 km^2, centered on 35°12'30"N 85°52'30"W, mean elevation 560 m) in June and early July, 1998, during the peak and end of the nesting season for most local birds. The Cumberland Plateau has, until recently, been covered with extensive mixed hardwood forests (typically oak-hickory forests) with scattered urban and suburban developments (Fig. 11.1). The rate of housing development in this region has been rapid in the past decade. For example, the number of houses in Franklin County, which includes part of the Cumberland Plateau, increased by 43% from 1991 to 1998 (data from county 911 emergency services database).

We collected data from 75 sites spread across the Plateau in Franklin County. All sites were at least 400 m apart (mean distance between sites =

1.4 km). All sites were located in forested habitat at least 50 m from forest edges. We chose sites according to which landowners gave us permission to work on their land, and to ensure that sites encompassed the full range of housing densities in the study area. At each site, we conducted one ten-minute, 75 m radius point count for Blue Jays and American Crows within one hour of dawn. In addition, at each site, we placed an infra-red triggered camera (Buckshot 35, Foresite, Denham Springs, LA) at a height of 0.5m on a tree. This camera system takes a photograph when triggered by the motion of an animal passing within range of its infra-red detector. We placed a 0.5m by 0.5m board 2m in front of the camera. We covered the board with 1 cm deep fine, damp sand to record footprints. We also placed a Japanese quail egg in a small depression in the ground within 5m of each camera to make an artificial nest. We quantified the local vegetation density at each

Figure 11.1. Aerial photo taken from digital-ortho quarter quad of a part of the study area in Sewanee, Tennessee. The image shows one of the areas of higher housing density in the study (140 houses per square kilometer).

site by measuring the percent of a 30 cm by 30 cm white card that was visible from 5m. We held the card 75 cm above the ground and repeated the measure from the north, south, east and west. We used the mean of these percentages as our measure of vegetation density. We checked the camera system, footprint monitors and quail eggs daily for three 24-hour periods. For the last 24-hour period, we baited the camera by placing the contents of a 156 g can of cat food (Special Kitty, Wal-Mart Stores, Inc.) in the center of the sand.

We used a Geographic Information Systems (GIS) database to calculate housing density at each site. We used the county's latest 911 emergency services database of street addresses to establish the spatial location of each house. We then calculated the number of houses within a circular area of 1 km^2 around each of our sampling sites. These densities ranged from wildland (0 houses/km^2) to suburban (140 houses/km^2).

To further assess the effectiveness of quail egg experiments, we used the University of the South's "classified advertisement" email list-server to solicit volunteers for a quail egg feeding trial with pet cats. We gave each volunteer a quail egg and a datasheet with instructions. We asked the volunteers to place the quail egg on the ground in front of their pet cat, then observe the cat's behavior until the cat left the vicinity of the egg. If the cat did not prey upon the egg, the volunteers were asked to place the quail egg in the cat's food dish, then to observe the cat's behavior.

2.1 Statistical Analyses

Data from the cameras and footprint monitors were converted into presence/absence scores for each site. This avoided potential over-counting of individuals (e.g., the same raccoon walking past the camera on consecutive nights is counted only once), and is thus a conservative measure of abundance. We excluded from the analysis photographs of deer and rabbits, and included only those species we judged to be potential nest predators. Data for cats (n = 13 detections), dogs (n = 15 detections) raccoons (n = 23 detections) and opossums (n = 13 detections) were analyzed by logistic regression. We illustrated trends by breaking the 75 sites into three groups (High density, > 10 houses per km^2; medium density, 0 - 10 houses per km^2; low density, 0 houses per km^2), each group containing 25 sites.

The perceived predation risk varied among species (e.g., dogs may occasionally prey on bird nests, but likely much less frequently than raccoons); we discuss the implications of this bias in the discussion. We also omitted foxes (n = 2 detections) and squirrels (n = 3 detections) from the analysis because of the small number of detections. No mice, chipmunks, skunks, or snakes were detected.

After checking for normality and homogeneity of variance, the relationship between housing density and the overall abundance of all predators, the abundance of American Crows, the abundance of Blue Jays, and vegetation density were analyzed by linear regression; the relationship between housing density and predation on quail eggs, and the presence or absence of dogs, cats, raccoons and opossums were analyzed by logistic regression. All statistical tests used an alpha value of 0.05 to determine significance.

3. RESULTS

The total number of predators detected at each site (summing across all predator species) showed a shallow positive rise with increasing housing density ($F_{1,73} = 5.94$, adj. $R^2 = 0.06$, $P = 0.02$). Individual species showed different responses to housing density. Blue Jays increased in abundance ($F_{1,73} = 31.13$, adj. $R^2 = 0.29$, $P < 0.001$) whereas American Crows decreased in abundance ($F_{1,73} = 5.09$, adj. $R^2 = 0.06$, $P = 0.03$) as housing density increased. Data for both these species came from point counts only. The presence of dogs was not related to housing density ($P = 0.28$), but cats, raccoons and opossums were all more likely to be detected in areas of higher housing density (cats, $P < 0.001$; raccoons, $P = 0.03$; opossums, $P = 0.03$; Fig. 11.2).

The probability of predation on the artificial nest containing the quail egg did not change with housing density ($P = 0.49$). Different predators showed very different correlations with quail egg predation. Sites at which raccoons and American Crows were detected were more likely to have quail eggs preyed upon than were sites where these predators were not detected (Fig 11.3; raccoon: $\chi^2 = 4.80$, $P < 0.05$; American Crow: $\chi^2 = 3.90$, $P < 0.05$). Sites at which cats, dogs, Blue Jays, or opossums were detected showed no association with quail egg predation (Fig 11.3; cat: $\chi^2 = 1.02$, $P > 0.1$; dog: $\chi^2 = 1.71$, $P > 0.1$; Blue Jay: $\chi^2 = 1.24$, $P > 0.1$; opossum: $\chi^2 = 0.10$, $P > 0.1$). No association existed between the housing density and the measure of vegetation density at our sites ($F_{1,73} = 0.29$, adj. $R^2 = 0.004$, $P = 0.59$).

Thirty-six cats participated in the volunteer quail egg experiment. Thirty-four cats sniffed the quail egg, 13 pawed the quail egg and two broke the egg. Both these breakages occurred after the cat "played with" the egg, batting it with their paws. Three cats attempted to bite the egg, but none of these attempts resulted in the egg breaking.

4. DISCUSSION

4.1 Relationship Between Urbanization and Nest Predator Abundance

The abundance of domestic cats, opossums and raccoons increased with increasing housing density (Fig. 11.2). Cats were almost entirely absent from wildlands and exurban areas of low housing density, whereas raccoons and opossums were still found, albeit at lower densities, in these areas. This is in accord with previous studies of raccoons (Riley et al. 1998). Matthiae and Stearns (1981) found higher opossum densities in rural compared to urban areas of Wisconsin, although these densities varied considerably among both rural and urban sites. This is, to our knowledge, the first study to examine cat populations across a gradient of housing densities.

The near absence of cats from exurban areas is intriguing. It may be that cats in these areas are themselves subject to predation by coyotes, domestic

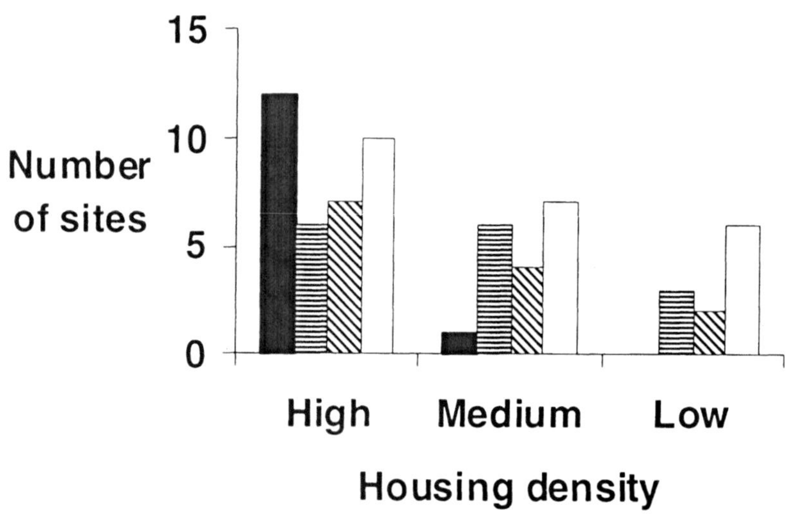

Figure 11.2. Number of sites at which cats (filled), dogs (horizontal stripes), opossums (diagonal stripes), and raccoons (open) were detected with infra-red triggered cameras and footprint monitors. High density, > 10 houses per km^2; medium density, 0 - 10 houses per km^2; low density, 0 houses per km^2.

dogs or owls that might prevent feral cat populations from establishing themselves and may cause owners of pet cats to keep their cats indoors. For example, Crooks and Soulé (1999) reported an inverse relationship between the presence of coyotes and the abundance of cats and Quinn (1997) found that coyote scat often contained the remains of cats. Anecdotal evidence from our study area suggests that roaming house cats in these areas are often killed within days of their departure, but we have not tested whether this risk of predation varies across the gradient of housing densities. Coyotes are occasionally seen in our study area, although we detected none with our camera systems. We did detect roaming domestic dogs, which also prey on domestic cats (pers. obs.). If dog predation was causing the absence of cats from some of our samples, we would expect a negative association to exist between the presence of dogs and the presence of cats. However, there was a positive association between the presence of cats and dogs in our samples (χ^2 = 6.72, $P < 0.01$). Also, the abundance of dogs did not differ across housing densities, so differential predation by dogs is unlikely to explain the absence of cats from exurban areas.

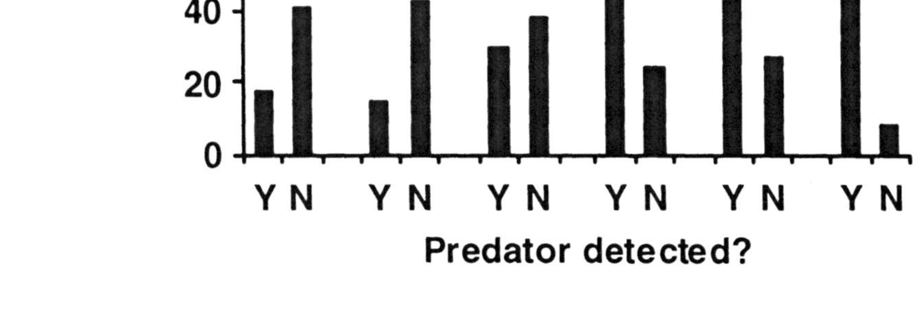

Figure 11.3. The relationship between predation on quail eggs in artificial nests and the presence of six potential nest predators. For each predator, all sites (n = 75) were broken into those at which the predator was detected (Y) and those which did not detect the predator (N).

We found that Blue Jays were more abundant in areas with higher housing densities. These data accord with other studies which have found a high abundance of Blue Jays in suburban areas (e.g., Crosby and Blair 2001). The areas with highest housing densities in this study retained significant numbers of mature white oak trees and the housing densities at these sites (50-150 houses per km^2) are comparable to suburban areas elsewhere. We found a negative correlation between housing density and the abundance of American Crows. This contrasts with studies from most other regions of North America, which have reported high abundance of crows in urban and suburban areas (Marzluff et al. 2001, McGowan 2001). It is hard to explain these results without a study of the demography and behavior of crows in this region.

4.2 Issues in the Interpretation of Nest Predator Abundance

Overall, the total abundance of potential nest predators in this study did increase with increasing housing density, although rates of predation on quail eggs did not differ across housing densities. Can we use these data to draw definitive conclusions about the effects of urbanization on rates of nest predation? Three factors suggest that we cannot.

First, this study looked at one part of the United States. Until studies are replicated across a variety of locations, the generality of our findings remains unknown. Studies conducted in landscapes dominated by agriculture have found that significant regional differences exist in the effects of agriculturally-induced forest fragmentation (Trine 1998, Friesen et al. 1999). It may be that the effects of urbanization on rates of nest predation also differ geographically.

Second, this study was not designed to gather data on the relative abundance of snakes or rodents, both of which may prey on bird nests. We know of no studies examining rates of nest predation by snakes across urban gradients. The abundance of rodents may be high in some urban settings (e.g., parks), but relatively low in more heavily urbanized areas (see Introduction).

The third reason that the results of this study cannot be used to make definitive judgements about the effects of urbanization on rates of nest predation is that it is unclear how the abundance of predators translates into actual rates of predation on bird nests. For example, our data found correlations between rates of predation on artificial nests containing quail eggs and the abundance of some predators (American Crows and raccoons), but not others (Blue Jays, opossums, cats). In addition, the correlation between predator abundance and rates of nest predation may vary across habitats. If the availability of alternative food sources varies across urbanization gradients, predators may switch their searching behavior to these alternative sources, resulting in lower rates of predation on bird nests (see also Schmidt 1999). For example, although more raccoons may occur in urban areas than in rural areas, if urban raccoons spend their time foraging in garbage or parking lots instead of woodlands, rates of raccoon predation on bird nests may be lower in urban areas. This is a speculative example, but some studies do suggest that potential nest predators may alter their behavior in urban environments (Doncaster and Macdonald 1997).

Even if a strong positive correlation exists between predator abundance and rates of nest predation (such as the correlation between American Crow abundance and quail egg predation reported here), one more complicating factor limits our ability to interpret these data. Because we have a limited knowledge of the nest searching behavior of predators, we do not know which species of nest predators represent the most danger to bird nests. This would not be a significant problem if the abundance of all predators showed positive correlations with urbanization. This, however, was not the case. For example, raccoons were more abundant in areas of high housing density, whereas crows were less abundant. Is the gain of two raccoons offset by the loss of two crows? This problem becomes particularly interesting when we consider interactions between predators. For example, our data suggest that most cats do not prey on quail eggs, whereas raccoons and American Crows did prey on these eggs. If the same is true for smaller bird eggs (this is unknown), then some native predators might represent more of a danger to bird eggs that the equivalent numbers of cats (e.g., 10 cats per hectare might result in fewer egg losses than 10 raccoons per hectare). If this were the case, low densities of cats could decrease rates of egg predation by displacing native predators (see Triggs et al. (1984) for preliminary evidence that cats may compete with native predators). This hypothetical example rests on a number of untested assumptions, but illustrates the difficulties of drawing conclusions about bird conservation from data on predator abundance in urban areas.

4.3 Suitability of Artificial Nest Studies in Urban Environments

Rates of nest predation are often estimated by using artificial nests baited with quail, chicken, or clay eggs (reviewed in Major and Kendal 1996). We found a positive relationship between predation on artificial ground nests baited with one quail egg and the presence of raccoons and American Crows. Therefore, quail egg experiments may be useful tools for assessing rates of egg predation from American Crows and raccoons. We found no significant relationship between the presence of cats, dogs or Blue Jays and predation on these artificial nests. Domestic cats presented with quail eggs did not prey on these eggs, even when the eggs were presented in the cats' food bowls. If cats are significant predators of nestling birds or of small bird eggs (there have been no tests of whether this is the case), then artificial nests may not be suitable for estimating actual rates of egg or nestling predation in urban environments. Interestingly, Bowman and Woolfenden (2001) found that rates of predation on eggs were lower in suburban areas than rural sites, but that rates of predation on nestlings showed the opposite trend. This again suggests that artificial nests baited with eggs may not provide reliable information about overall rates of nest predation in urbanized areas.

4.4 Policy and Management Implications

We conducted this study of nest predation in parallel with economic and political studies in the same region of Tennessee. This provides a number of opportunities to study the interaction between biology, economics and policy. We will discuss two such interactions: the effectiveness of a tax-incentive program and the relationship between the economics of development and conservation biology.

The Tennessee Greenbelt program (Agricultural, Forest, and Open-space Act, Tennessee Legal Code 67-5-1001) provides a property tax reduction for forest owners who agree not to develop their land. Owners who do develop pay a rollback tax penalty. More than a third of forested land on the Cumberland Plateau in Franklin County is enrolled in this program (144,000 acres). This law seems, on the surface, to offer a promising method for preventing, or slowing, conversion of forested lands into residential or commercial uses. However, Brockett and Gebhard (1999) found that the incentive program is, at best, a weak tool. Their analysis of the land-use practices of Greenbelt participants in Franklin County showed that this program does little to change these practices and therefore likely has little effect slowing the rate of development. Thus, although the Greenbelt program results in a significant loss of tax revenues for local governments, it offers little in the way of benefits in the form of biodiversity protection.

Horchem and Gottfried's (1998) study provides an insight into the potential parallels between avian conservation and increases in property taxes. Under most scenarios of growth in the county, they found that property tax rates increased as growth proceeded. This increase in taxation was particularly pronounced if growth was sprawling and required large investments in infrastructure. Sprawling growth also may have a negative effect on biodiversity by fragmenting forests to a greater extent than clustered growth. For example, we performed a GIS-simulation in which we used different zoning regulation scenarios to place 300 new houses at random along roads on the Cumberland Plateau in Franklin County. We then assumed that one third of these households brought with them domestic cats with home ranges of 1 km^2 (based on the low end of cat densities presented in Liberg and Sandell 1988). In this landscape, 3% of the total area fell within the home ranges of these cats if the houses were built under zoning regulations that restricted new development to within exiting township boundaries. This increased to 15% if zoning regulations allowed housing development along any county road. These two scenarios gave all roads within the zoning boundaries the same probability of receiving a new house, so they omit many subtleties of housing development in the real world. The important point is that sprawling development may have negative effects on both bird populations and on human taxpayers and that this can provide conservation biologists with a way to encourage planned, unsprawled growth. Residents of Franklin County care about taxes -- telephone interviews of county residents showed that 39% of those surveyed (n = 187) would not support an increase in taxes, even if this increase were to solve what respondents considered was "the county's most pressing problem" (Brockett, unpub. data). Thirty-seven percent perceived excessive property taxes a threat to the quality of outdoor life in Franklin County. No data exist on the level of concern about the plight of neotropical migrant birds, but we suspect that they would rank below taxes.

We present our management recommendations in Table 11.1. These recommendations are based on the assumption that urbanization does increase rates of nest predation. As the literature review in this paper has shown, this is a tentative conclusion, and our field data only support this conclusion for some species of predators. Thus, all these management recommendations should be interpreted cautiously. The first recommendation stems from political reality; bird conservation is likely to proceed fastest if it is linked to issues important to most taxpayers. The second recommendation arises from this and many other studies; the effects of urbanization are small when urban areas are also small. We assign the

Table 11.1. Priority rankings for suggested prescriptions.

Rank	Prescription
1	Find connections between bird conservation and human "quality of life issues"
1	Minimize incursions of housing development into unfragmented forest
2	Control availability of human refuse and other food sources for urban predators
2	Encourage pet owners to keep cats indoors and limit feral cat population

last two recommendations a lower rank because of the very limited nature of our knowledge. While it seems likely that the availability of human refuse might increase populations of urban predators, this might not always result in higher rates of nest predation. Refuse might provide alternative food sources that move predators away from bird nests (see above). Likewise, cats may have a negative effect on nesting birds, but we have scant data to support this claim and we do not understand how changes in cat abundance affect the abundance of native nest predators.

ACKNOWLEDGEMENTS

We thank the Appalachian College Association and the University of the South for funding this research, the private landowners who allowed us to conduct surveys on their property, and the Tennessee Wildlife Resources Agency GAP Analysis Lab for generously providing GIS coverages and technical support. We thank R. Bowman, J. Marzluff, R. Mumme, S. Vance, and one anonymous reviewer for helpful comments on the manuscript. Our colleagues at the Cumberland Center for Conservation and Sustainable Development (C. Brockett, J. Evans, R. Gottfried, K. McGlothlin, and J. Peters) provided a stimulating environment in which to conduct this research, and gave helpful comments on the manuscript.

REFERENCES

Adkins, C. A., and P. Stott. 1998. Home ranges, movements and habitat associations of red foxes (*Vulpes vulpes*) in suburban Toronto, Ontario, Canada. J. Zool. (Lond.) 244:335-346.

Beissinger, S. R., and D. R. Osborne. 1982. Effects of urbanization on avian community organization. Condor 84:75-83.

Blair, R. B. 1996. Land use and avian species diversity along an urban gradient. Ecol. Appl. 6:506-519.

Bowman, R., and G. Woolfenden. 2001. Nest success and the timing of nest failure of Florida Scrub-Jays in suburban and wildland habitats, p. 385-404. *In* J. M. Marzluff, R. Bowman, and R. Donnelly [EDS.], Avian ecology and conservation in an urbanizing world. Kluwer Academic, Norwell, MA.

Brockett, C. D., and L. Gebhard. 1999. NIPF tax incentives: do they make a difference? J. Forestry 97:16-21.

Burke, D. M., and E. Nol. 1998. Influence of food abundance, nest-site habitat, and forest fragmentation on breeding ovenbirds. Auk 115:96-104.

Churcher, P. B., and J. H. Lawton. 1987. Predation by domestic cats in an English village. J. Zool. (Lond.) 212:439-455.

Coleman, J. S., and S. A. Temple. 1993. Rural residents' free-ranging domestic cats: a survey. Wildl. Soc. Bull. 21:381-390.

Crooks, K. R., and M. E. Soulé. 1999. Mesopredator release and avifaunal extinctions in a fragmented system. Nature 400:563-566.

Crosby, N., and R. Blair. 2001. Do temporal trends in Christmas Bird Counts reflect the spatial trends of urbanization in southwestern Ohio, p. 525-542. *In* J. M. Marzluff, R. Bowman, and R. Donnelly [EDS.], Avian ecology and conservation in an urbanizing world. Kluwer Academic, Norwell, MA.

Danielson, W. R., R. M. DeGraaf, and T. K. Fuller. 1997. Rural and suburban forest edges: effect on egg predators and nest predation rates. Landsc. Urb. Plan. 38:25-36.

DeGraaf, R. M., and T. J. Maier. 1996. Effect of egg size on predation by white-footed mice. Wilson Bull. 108:535-539.

Doncaster, C. P., and D. W. Macdonald. 1997. Activity patterns and interactions of red foxes (*Vulpes vulpes*) in Oxford city. J. Zool. (Lond.) 241:73-87.

Emlen, J. T. 1974. An urban bird community in Tuscon, Arizona: derivation, structure, regulation. Condor 76:184-197.

Friesen, L., M. D. Cadman, and R. J. MacKay. 1999. Nesting success of neotropical migrant songbirds in a highly fragmented landcape. Conserv. Biol. 13:338-346.

Haskell, D. G. 1995a. Forest fragmentation and nest-predation: Are experiments with Japanese Quail eggs misleading? Auk 112:767-770.

Haskell, D. G. 1995b. A reevaluation of the effects of forest fragmentation on rates of bird-nest predation. Conserv. Biol. 9:1316-1318.

Haskell, D. G. 2000. Effects of forest roads on macroinvertebrate soil fauna of the Southern Appalachians mountains. Conserv. Biol. 14:57-63

Hogrefe T. C., R. H. Yahner, and N. H. Piergallini. 1998. Depredation of artificial ground nests in a suburban versus a rural landscape. J. Penn. Acad. Sci. 72:3-6.

Horchem, J. D., and R. R. Gottfried. 1998. Growth: boon or bane? A report of the Cumberland Center for Conservation and Sustainable Development, Sewanee, Tennessee.

Liberg, O., and M. Sandell. 1988. Spatial organisation and reproductive tactics in the domestic cat and other felids, p. 83-98. *In* D.C. Turner and P. Bateson (EDS.), The domestic cat. Cambridge University Press, Cambridge, UK.

Luoma, J. R. 1997. Catfight. Audubon 99:85-91.

Macdonald, D. 1987. Running with the fox. Unwin Hyman, London-Sydney.

Major, R. E., G. Gowing, and C. E. Kendal. 1996. Nest predation in Autralian urban environments and the role of the pied currawong, *Strepera graculina*. Australian J. Ecol. 21:399-409.

Major, R. E., and C. E. Kendal. 1996. The contribution of artificial nest experiments to understanding avian reproductive success: a review of methods and conclusions. Ibis 138:298-307.

Marzluff, J. M., K. McGowan, R. Donnelly, and R. Knight. 2001. Causes and consequences of expanding American Crow populations, p. 333-365. *In* J. M. Marzluff, R. Bowman, and R. Donnelly [EDS.], Avian ecology and conservation in an urbanizing world. Kluwer Academic, Norwell, MA.

Matthiae, P. E., and F. Stearns. 1981. Mammals in forest islands in southeastern Wisconsin. p. 55-66. *In* R.L. Burgess and D.M. Sharpe (EDS.), Forest island dynamics in man-dominated landscapes. Springer-Verlag, New York.

McGowan, K. 2001. Demographic and behavioral comparisons of suburban and rural American Crows, p. 367-383. *In* J. M. Marzluff, R. Bowman, and R. Donnelly [EDS.], Avian ecology and conservation in an urbanizing world. Kluwer Academic, Norwell, MA.

Melampy, M. N., E. L. Kershner, and M. A. Jones. 1999. Nest predation in suburban and rural woodlots of northern Ohio. Am. Midland Nat. 141:284-292.

Nupp, T. E., and R. K. Swihart. 1996. Effect of forest patch area on population attributes of white-footed mice (*Peromyscus leucopus*) in fragmented landscapes. Can. J. Zool. 74:467-472.

Quinn, T. 1997. Coyote (*Canis latrans*) food habits in three urban habitat types of western Washington. Northwest Sci. 71:1-5.

Riley, S. P. D., J. Hadidain, and D. A. Manski. 1998. Population density, survival, and rabies in raccoons in an urban national park. Can. J. Zool. 76:1153-1164.

Roper, J. J. 1992. Nest predation experiments with quail eggs: too much to swallow? Oikos 65:528-530.

Rottenborn, S. C. 1999. Predicting the impacts of urbanization on riparian bird communities. Biol. Conserv. 88:289-299.

Savard, J., and J. B. Falls. 2001. Bird-habitat relationships of breeding birds in residential areas of Toronto, Canada, p. 543-568. *In* J. M. Marzluff, R. Bowman, and R. Donnelly [EDS.], Avian ecology and conservation in an urbanizing world. Kluwer Academic, Norwell, MA.

Schmidt, K. A. 1999. Foraging theory as a conceptual framework for studying nest predation. Oikos 85:151-160.

Stallcup, R. 1991. A reversible catastrophe. Observer 91:8-9.

Terborgh, J. 1989. Where have all the birds gone? Princeton University Press, NJ.

Triggs, B., H. Brunner, and J. M. Cullen. 1984. The food of fox, dog, and cat in Croajingalong National Park, south-eastern Victoria. Australian Wildl. Res. 11:491-499.

Trine, C. L. 1998. Wood thrush population sinks and implications for the scale of regional conservation strategies. Conserv. Biol. 12:576-585.

Wilcove, D. S. 1985. Nest predation in forest tracts and the decline of migratory songbirds. Ecology 66:1211-1214.

Wilcove, D. S. 1999. The condor's shadow: The loss and recovery of wildlife in America. W. H. Freeman and Co., New York, NY.

Williamson, R. D. 1983. Identification of urban habitat components which affect eastern gray squirrel abundance. Urban Ecol. 7:345-356.

Wong, T. C. M., N. S. Sodhi, and I. M. Turner. 1998. Artificial nest and seed predation experiments in tropical lowland rainforest remnants of Singapore. Biol. Conserv. 85:97-104.

Chapter 12

Bird tolerance to human disturbance in urban parks of Madrid (Spain): Management implications

Esteban Fernández-Juricic, Maria Dolores Jimenez, and Elena Lucas
Depto de Biologia Animal I. Facultad de Biologia, Universidad Complutense de Madrid, Madrid E-28040, Spain, e-mail:estebanfj@yahoo.com

Key words: Birds, flush distances, human disturbance, tolerance, urban park design

Abstract: The effects of human disturbance in urban parks of Madrid (Spain) on bird tolerance were studied to derive recommendations for urban park planning. We intended to determine (1) how habitat structure influenced flush distances, (2) whether flush distance increased with body-size, and (3) whether flush distances varied with the amount of people visiting the park. In four city parks in Madrid with different levels of human visitation and habitat structure, we recorded flush distances of four bird species of different sizes (overall body length: *Passer domesticus*, 15 cm; *Turdus merula*, 24-25 cm; *Columba palumbus*, 40-42 cm; *Pica pica*, 44-48 cm). Humans approached individual birds when they were foraging on the ground. Habitat structure (shrub cover, shrub and tree height) influenced flush distances of the four species. After controlling for the effects of different microhabitat use, averaged flush distances varied significantly among species; large species were less tolerant of human disturbance than small ones. Birds were more tolerant of our intrusions in parks with more human visitors. To enhance the suitability of urban parks for bird species, the following recommendations may be considered: (a) flush distances could be a first indicator of appropriate set-back distances (minimal distance that a pedestrian may approach a bird) for pathways, (b) flush distances of large bird species could be used to determine minimum area requirements of resource patches separated by pathways, (c) habitat complexity could increase the availability of escape cover for native bird species, and (d) highly variable visitation rates may be less harmful in popular parks, because bird tolerance to human disturbance appears to increase with the overall amount of visitors per park.

1. INTRODUCTION

Urban parks may serve as reservoirs for native species in densely populated areas (Baker and Graf 1989, Goode 1991). One of the main features of these natural settings within cities is that they are heavily visited by people, who may potentially interact with wildlife in different ways (Dickman 1987, Blair 1996, Fernández-Juricic and Tellería 2000). Studies conducted in protected areas demonstrate that human presence can disturb bird species, with negative effects at the individual, population and community levels (Skagen et al. 1991, Klein 1993, Rodgers and Smith 1995, Gill et al. 1996, Riffell et al. 1996, Gutzwiller et al. 1998). However, little empirical evidence exists as to how bird species in urban parks tolerate human presence and the environmental factors associated with their response. Understanding bird tolerance to humans may ultimately be useful for improving the design of urban parks and reducing the negative impacts of bird-human interactions.

Bird tolerance to human presence may depend on several factors. As habitat structure increases in complexity, birds show more tolerance to people since they have more available cover to hide from visitors (Knight and Temple 1995). Generally, species vary in tolerance levels; large species seem less tolerant of humans than small ones (Cooke 1980, Humphrey et al. 1987, Holmes et al. 1993, but see Gutzwiller et al. 1998). Habituation to human presence may explain some variation in tolerance levels; for instance, migrants are less tolerant than resident species because of less contact with people throughout the year (Burger and Gochfeld 1991).

Bird tolerance can be estimated by flush distance, the distance at which a species escapes from a visitor; the larger the flush distance, the less the tolerance (Cooke 1980, Humphrey et al. 1987, Gutzwiller et al. 1998). Such flush distances may be used to develop appropriate estimates of disturbance buffer zones or set-back distances to reduce the effects of human disturbance on wildlife (Knight and Knight 1984, Rodgers and Smith 1995). The purpose of this paper is to present flush distances of four different-sized bird species inhabiting four urban parks with different habitat structure and visitor loads in the city of Madrid (Spain). We (1) determine the influence of foraging microhabitat use on flush distances; (2) analyze the variation in flush distances within and among species relative to visitor load and species size (considering total body length); and (3) make recommendations to apply bird tolerance levels to urban planning and the conservation of urban bird species. We expected flush distances to decrease with increases in the amount of available habitat and the number of visitors, but decrease with species size.

2. METHODS

2.1 Study Area

The study was conducted from May to July 1998 in four large parks of Madrid (40.25 N; 03.43 W): Moro (18 ha), Dehesa de la Villa (25 ha), Oeste (98 ha), and Retiro (110 ha) (see Fig. 12.1 for an overview of Madrid and Retiro, Moro, and Oeste parks). Created more than 50 years ago, these parks are representative of the urban parks of this city, with many deciduous and coniferous trees and extensive areas of watered grass. All parks include recreational areas (< 1% built, 0 buildings/ha) divided into patches defined by walking paths. Although people are used to walking within paths, they sometimes encroach in patches for picnicking or resting.

Aerial photography (scale 1:10000) was used to quantify the level of urbanization surrounding parks. We measured the number of buildings in 10 ha squares throughout the fringes of each park. We averaged these figures, and expressed final estimates as number of buildings/ha. To determine the number of humans/ha surrounding parks, we quantify the number of people walking in five minutes intervals in 25-m radius circular plots at the edges of each park. These figures were averaged and transformed to humans/ha.

The matrix surrounding parks differed in several ways. The Moro park is located near Casa de Campo (a 1722 ha forest fragment immediately NW of Madrid) and is surrounded by a suburban matrix (0.6 buildings/ha), with large historic buildings and few people (10-200 humans/ha). Dehesa de la Villa is adjacent to a large natural area (Monte del Pardo) and a University campus. The matrix surrounding Dehesa de la Villa can be considered exurban, with a few sparsely distributed buildings (<0.5 buildings /ha), and relatively few people (10-150 humans/ha). The Oeste park is fringed by a more developed suburban matrix (1.5 buildings/ha), with sporting areas, roads, and office buildings and about 400-600 humans/ha. Retiro is the largest urban park, located near the city center. It is surrounded by an urban matrix, with large commercial buildings (5 buildings/ha), and a dense human population (>600 humans/ha).

2.2 Field Methods

House Sparrow (*Passer domesticus*), Blackbird (*Turdus merula*), Woodpigeon (*Columba palumbus*), and Magpie (*Pica pica*) are representative urban models for investigating tolerance levels. These species are common in urban areas of the Palaearctic, and their selection was based mainly on their considerable abundance, similar foraging habits (ground foragers), and size differences to compare specific tolerance levels to human disturbance. Following Cooke (1980), we used overall body length as an indicator of species size: House Sparrow (15 cm), Blackbird (24-25 cm), Woodpigeon (40-42 cm), and Magpie (44-48 cm) (Tellería et al. 1999).

Samples of vegetation structure were obtained independently from those taken when approaching birds (see below). We were interested only in comparing habitat structure among parks rather than analyzing specific habitat features within each park. Habitat structure was measured in 25-m radius circular plots distributed at 30-m intervals (Tellería and Santos 1997). Variables visually estimated (Prodon and Lebreton 1981) were: grass cover,

Figure 12.1. Spin-2 satellite image of downtown Madrid and three urban parks: Retiro Park, Moro Park, and Oeste Park. The image encompasses approximately 8 mi^2 and was taken in 1987.

paved ground cover, shrub cover, number of shrub species, shrub height, coniferous cover, deciduous cover, number of tree species, and tree height.

In determining human visitation in each park, a team of observers recorded simultaneously the number of pedestrians observed at five minute intervals in 10 different 25-m radius circular plots in those sectors of the parks in which birds were experimentally approached. We made 10 counts at morning (7:30-11:00) and 10 at midday (12:00-14:30). Preliminary sampling yielded significant differences in visitor load between, but not within, weekend and workdays. Hence, each park was sampled two days during workdays and one day during the weekend to compare visitor loads among parks. Final figures were averaged in each park as pedestrians/minute. Although visitation samples were not taken at the same time as an experimental approach to birds, our estimates could be considered reliable because visitation showed little variation within days. Following a similar sampling procedure, we also recorded dogs in each park, since they also could modify tolerance levels of urban birds.

Two observers (authors M.D.J. and E.L.) made all observations. They were previously trained in approaching birds in a 20-ha park different from those used in the analyses to avoid conditioning birds. Preliminary analysis showed that ambient temperature did not affect flush distances of any species (Fernández-Juricic et al. unpubl. data). All samples were taken during sunny or partially cloudy conditions, but never when raining. Samples were gathered between 7:30 and 18:00, totaling 347 observation hours. An individual bird was approached by a single observer at a steady speed (1 step/sec) to record its flush distance, which was defined as the distance from the observer to the bird at the moment it started to flush (either walking, running, or flying towards escape cover) (Rodgers and Smith 1997). All birds approached were on the ground. The approach was linear with unobstructed visual contact between the observer and the bird. The bird to be approached was selected following these criteria: the focal bird was not within 21 ± 1.7 m of any person, it was not feeding although it could be searching for food, and it did not show any alert behaviors (Fernández-Juricic and Tellería 2000). We excluded samples that involved birds looking for nest material. Most birds we approached were alone or with only one or two conspecifics. In the latter case, the observer focused their measurements on a single individual in the group chosen before the approach. When birds were disturbed by any kind of sound during our observations, we stopped our approach. Once the focal bird flew away, the observer continued the approach up to the point from which the bird left. At this point they recorded the following microhabitat variables from a 25-m circular plot: grass cover, paved ground cover, shrub cover, shrub height, coniferous cover, deciduous cover, and tree height. These data were used to analyze whether microhabitat complexity influenced flush distances.

Observers conducted flush distance samples at the same time but in different parks. They visited parks at random with at least one-day interval between visits to the same park. We collected 20-23 samples per species per park. For each species and park, samples were gathered from different locations to reduce pseudoreplication by disturbing the same individual. Before sampling we mapped species territories by means of thorough searches in the areas that were later used for approaches. The observers never collected two samples from the same species in a row nor from contiguous territories to circumvent any dependence between the first and subsequent approaches.

2.3 Statistical Analyses

We employed a discriminant function analysis (DFA) to identify the main habitat characteristics that differentiate the four parks. Significant variables were selected by forward stepwise procedures (StatSoft 1996). The results of the DFA were interpreted considering the significance of the discriminant function, and the accuracy of the classification of cases into groups. With the variables identified in the DFA we calculated a structural diversity index (Shannon-Wiener index) for each park. Visitor and dog rates were compared among parks by means of ANOVA tests. To determine the relationship between microhabitat use and flush distances in each park, we used Pearson correlations. We used ANCOVA to analyze inter and intra-specific differences in flush distances in the four studied parks. Those microhabitat structure measures that influenced flush distances were included as covariates to control for the effects of habitat use on tolerance levels. We checked for normality and homocedasticity of variables and of residuals before and after the analyses, respectively.

3. RESULTS

3.1 Park Structure, Visitor And Dog Rates

Parks differed in the degree of habitat complexity (Fig. 12.2). The discriminant function analysis identified five variables that significantly discriminated among parks: grass and shrub cover, coniferous and deciduous cover, and tree height (Fig. 12.2, Wilks' Lambda: 0.19, $F_{15,237} = 13.06$, $P < 0.001$). Seventy one percent of cases were correctly assigned to their proper parks. The squared Mahalanobis distances among parks were significant

(SMD ranged from 3.28 to 21.22, $F_{7,84}$ ranged from 5.62 to 35.87, $P < 0.001$), which indicates that all parks differed in the aforementioned variables. Because cover variables provided the greatest discrimination among parks, we calculated diversity indices relative to vegetation cover (grass, shrub, coniferous and deciduous). The Moro park was the most diverse ($H' = 1.76$) followed by the Oeste ($H' = 1.71$), Retiro ($H' = 1.6$), and Dehesa de la Villa ($H' = 1.47$) parks.

We found significantly different human visitation among parks; Moro was the least and Retiro the most visited parks, and Oeste and Dehesa de la Villa having similar visitor rates (Fig. 12.3; ANOVA, $F_{3,316} = 17.8$, $P < 0.001$). Dog visitation also differed among parks (ANOVA test, $F_{3,316} = 18.2$, $P < 0.001$). Dogs were more common in Retiro park (0.73 ± 1.15 dogs/m), followed by the Oeste (0.21 ± 0.81 dogs/m) and Dehesa de la Villa parks (0.03 ± .16 dogs/m). We did not observe any dogs in Moro park.

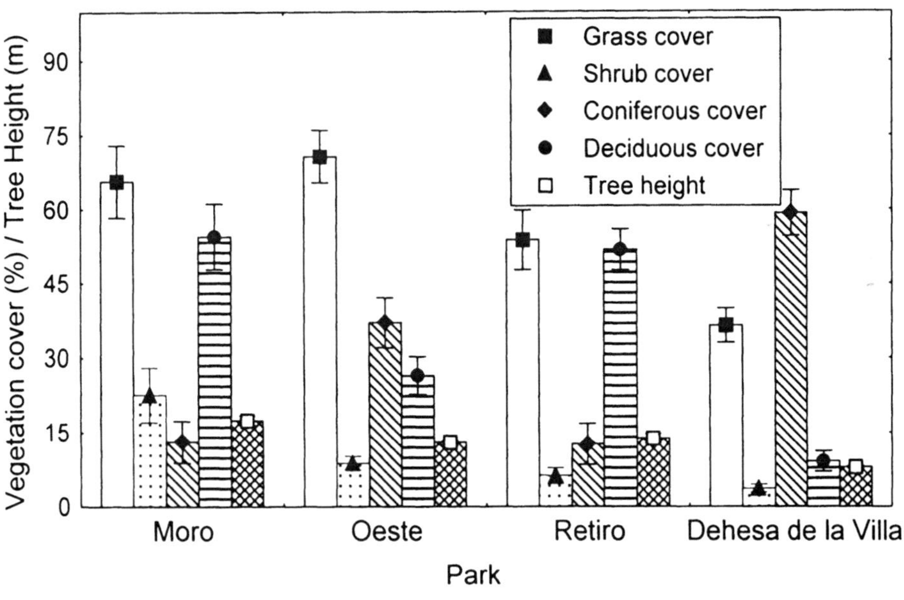

Figure 12.2. Habitat structure differences between four parks in the city of Madrid (Spain). Variables were selected from a discriminant function analysis, and differed significantly among parks.

3.2 Flush Distances

Microhabitat at the observation point influenced flush distance of all species. Individuals fled at shorter distances from the observer as the amount of shrub cover increased. This pattern was true for each species in at least one park: House Sparrows (Dehesa de la Villa), Blackbirds (Oeste and Retiro), Woodpigeons (Oeste), and Magpies (Retiro) (Table 12.1). Magpie flush distances in Dehesa de la Villa and Oeste parks were significantly shorter as shrub height increased (Table 12.1). In Retiro park, Magpies also reduced their flush distance with higher tree heights (Table 12.1) and in Moro park, Magpies reduced flush distance with increased grass cover (Table 12.1). In Dehesa de la Villa and Moro parks, Woodpigeons and Magpies, respectively, reduced flush distances with high coniferous tree cover (Table 12.1). In Oeste park, Blackbirds increased flush distances with increased paved ground (Table 12.1).

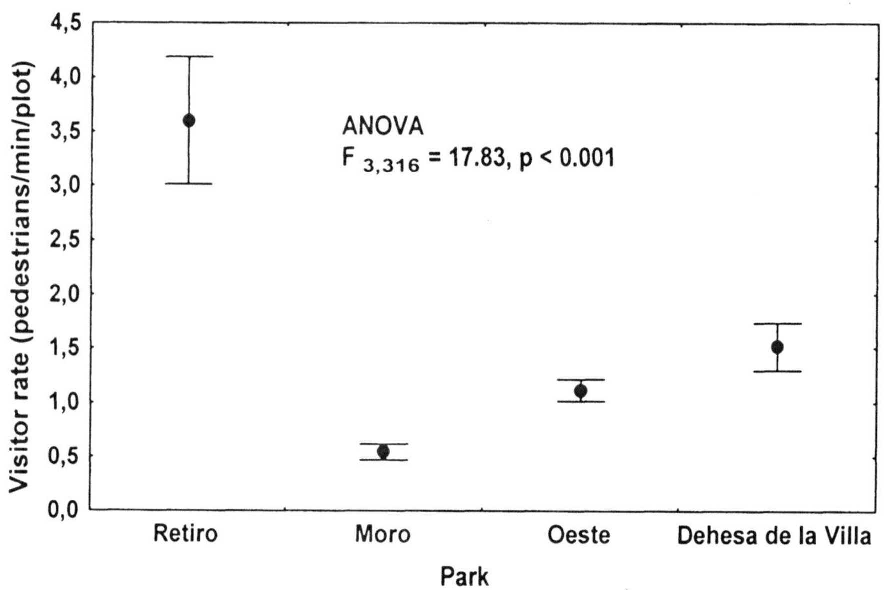

Figure 12.3. Rate of human visitors to four urban parks in the city of Madrid (Spain).

Controlling for microhabitat use, flush distances differed significantly among species and parks, and no interaction was detected (Species, $F_{3,327} = 11.59$, $P < 0.001$; Park, $F_{3,327} = 10.77$, $P < 0.001$; Interaction, $F_{9,327} = 0.71$, $P = 0.703$) (Fig. 12.4). For all parks combined, the smaller species fled at shorter distances than the larger species: House Sparrows (7.27 m) <Blackbird (8.62 m) < Woodpigeon (9.91 m) < Magpie (10.63 m). We also found a park effect, whereby flush distances differed significantly among parks (considering all species): Retiro (7.43 m) < Oeste (7.96 m) < Dehesa de la Villa (10.08 m) < Moro (10.96 m).

4. DISCUSSION

4.1 Flushing Behavior

All species flushed from observers, but varied their flush distances. Individual birds may modify flush distances to optimize escape probabilities based on a balance between the costs and benefits of fleeing (Ydenberg and Dill 1986, Henson and Grant 1991, Skagen et al. 1991). Such balance could be affected by habitat structure, with different variables affecting different species based on their escape tactics (Lima 1993). House Sparrows and

Table 12.1. Microhabitat variables that significantly correlated with flush distances of four bird species in urban parks of Madrid. Abbreviations: SHC, shrub cover; PGC, paved ground cover; CNC, coniferous cover; SHH, shrub height, TRH, tree height; GRC, grass cover.

Park	House Sparrow	Blackbird	Woodpigeon	Magpie
Dehesa de la Villa	SHC (r=-0.53, P<0.05)	-----	CNC (r=-0.43, P<0.044)	SHH (r = -0.50, P < 0.02)
Oeste	-----	PGC (r=0.65, P<0.01); SHC (r=-0.45, P<0.04)	SHC (r= -0.54, P < 0.01)	SHH (r = -0.44, P < 0.05)
Retiro	-----	SHC (r=-0.65, P<0.01)	-----	SHC (r = -0.46, P < 0.04); TRH (r = 0.47, P < 0.03)
Moro	-----	-----	-----	GRSC (r = -0.49, P < 0.02); CNC (r = -0.55, P < 0.01)

Blackbirds use shrubby vegetation as escape cover; Woodpigeons and Magpies use trees as escape cover.

The amount of shrub cover appeared to be the principal microhabitat characteristic that increased birds tolerance to people in urban parks. This may relate to a greater availability of escape cover and consequently a lower perceived risk of predation (Ydenberg and Dill 1986, Martín and López 1995, Kramer and Bonenfant 1997). This is particularly true for House Sparrows and Blackbirds, although Woodpigeons and Magpies sometimes use shrubs as escape cover. In much the same way, shrub height and coniferous cover may allow some birds to reduce the costs of not fleeing when people approach. Shrub height also increases the availability of cover, probably reducing the perception of risk for Magpies. Coniferous cover may act similarly for Woodpigeon and Magpies, since coniferous trees were their preferred escape cover. Increased tree height, on the other hand, increased Magpie flush distances. Magpies may perceive an increased risk of predation with tall trees (Ydenberg and Dill 1986), since individuals must fly longer distances to reach the tree canopy, their preferred escape cover. Although we did not measure food availability, it may vary with microhabitats and may

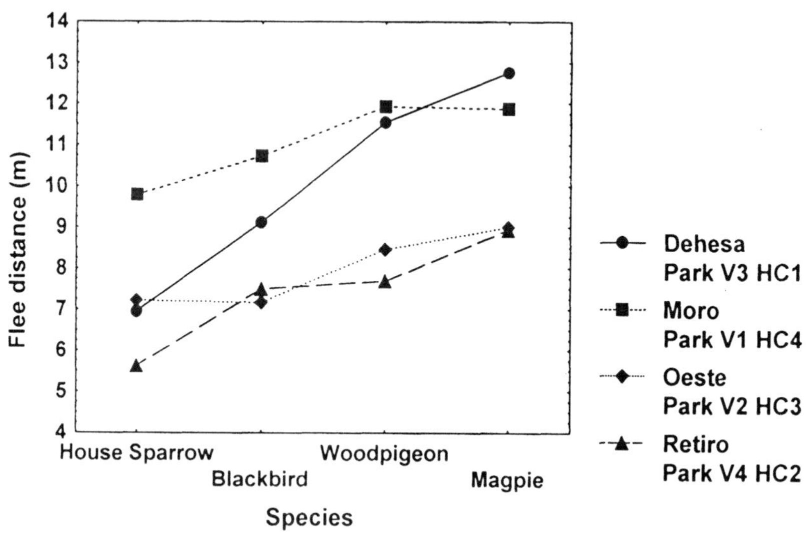

Figure 12.4. Mean flush distances of four bird species in four parks in the city of Madrid (Spain). Also indicated is the visitor load (V) and the degree of habitat complexity (HC) of each park in a scale 1-4 (1 denoting low, and 4, high).

increase the probability of individual birds finding food (Fernández-Juricic and Tellería 1999) and, as a result, delay the flushing response of individuals. Increased foraging success may mean some birds tolerate an increased risk of capture (Ydenberg and Dill 1986). This trade-off may account for the negative relationship between grass cover and Magpie flush distances. Magpies usually feed on worms and small insects on the grass, and increased foraging success could increase their tolerance to people. The positive relationship between paved ground and Blackbird flush distances may have to do with increased risk of detection by predators or with the lack of food resources.

Even though flush distances varied within and among species, they could not be directly attributed to any specific cause due to our sampling design. We can suggest that variation in park visitation, park habitat structure, and species size affects flush distance. Within species, we found a park effect; individuals in less visited parks generally showed longer flush distances (Fig. 12.3 and 12.4). Habituation to people might influence such a pattern. Visitors to urban parks are a relatively benign source of disturbance in that generally the birds are not killed or pursued (Cooke 1980). Therefore, it is likely that individuals have learned to tolerate human intrusions (Rodgers and Smith 1995, Riffell et al. 1996). For instance, House Sparrows seem to be more tolerant to people in parks with high human visitation (Retiro Park) and less tolerant in parks with low visitation (Moro). The same may hold for dogs in urban parks; birds became habituated to dogs in highly used parks.

Another relevant factor that may influence flush distances in more subtle ways is habitat structure. Parks with complex habitat structure may have a greater availability of escape covers, increasing tolerance levels to visitors (Knight and Temple 1995). For instance, Magpie flush distances were high in Dehesa de la Villa park, where tolerance may be low because few potential refuges are available (shrubs, trees, etc., Fig. 12.2). However, flush distances were also high in Moro park where habitat complexity was greater. Even though no interaction effects were detected, the influence of visitation and habitat structure in urban parks may be interacting with each other or with other factors (e.g. food availability, territoriality, etc.) to give rise to more complicated patterns of tolerance (for example, the Woodpigeon, Fig. 12.4).

We found larger species to be less tolerant of human presence (Fig. 12.4), a conclusion supported by other studies (Cooke 1980, Humphrey et al. 1987, Skagen et al. 1991, Holmes et al. 1993). Little empirical consideration has been devoted to this relationship. Perhaps small species have higher energy needs (greater area/body mass ratios) than large ones, such that the former would be more tolerant of human presence to reduce energy costs associated with fleeing (Holmes et al. 1993). Alternatively, larger species may have higher perceptual ranges (ability to perceive landscape elements, Lima and Zollner 1996) than smaller ones, so they would detect and react to human presence at greater distances.

4.2 Policy And Management Implications

Previous studies in Madrid suggest that pedestrians can reduce feeding rates and breeding densities and change selection of feeding areas of urban birds (Fernández-Juricic and Tellería 2000, Fernández-Juricic 2000). Although the four species studied here are abundant in Madrid and forage mainly on the ground, they could be used as models upon which management can be planned to improve the conservation of urban wildlife. Of interest is the design of urban parks with a network of pathways that minimize the conflicts between visitors and wildlife. In Madrid, the arrangement of pathways in parks has rarely followed biological considerations nor have specific policies regulated the use of city parks by pedestrians. The flush distances of different species could be a reliable indicator of tolerance that may serve to establish set-back distances that guide the spatial arrangement of pathways to minimize disturbance to birds. A similar approach has been applied to colonial waterbirds (Rodgers and Smith 1995, 1997).

Because pathways within parks separate patches that may be used by different bird species for foraging and breeding, set back distances refer to the minimal distance that a pedestrian may approach a bird before it is disturbed. For instance, if patches within parks were circular-shaped, it would be advisable that the radius of these patches were greater than flush distances. Indeed, some authors argue that alert distances would be more useful thresholds for set-back distances (Rodgers and Smith 1995, Fernández-Juricic et al., unpubl. data). Bird diversity in urban parks may be increased if set-back distances were greater than the tolerance levels of the most sensitive species.

To obtain good estimates of set-back distances for different birds, one alternative is to use tolerance levels of the largest species. The implementation of set-back distances based on large species automatically allows smaller ones to use parks for feeding or breeding without being disturbed.

Habitat complexity could increase tolerance levels of many urban birds by increasing the availability of cover used for escape. Management plans aimed at increasing vegetation structure, especially shrub cover, could be a relatively easy means to reduce human disturbance in urban parks and to potentially increase bird tolerance levels.

Tolerance among urban species appears to be related to their degree of habituation to people. Therefore, if visitor loads are to be increased temporarily in certain urban parks (e.g., a public act, a concert, etc.), it is deemed advisable that such increases be implemented in those parks with high pedestrian rates to minimize the harmful effects towards urban birds.

4.3 Research Needs

Future research on the effects of human disturbance on urban birds should encompass the following topics: (1) analyses of the relationships between human disturbance and tolerance levels for less tolerant and less abundant species, which are mainly the target of conservation efforts; (2) evidence on the effect of different spatial arrangements of pathways within parks on the tolerance levels and densities of urban birds; (3) assessment of the relationship between visitation and demography (breeding success and survivorship), particularly for the least tolerant species; (4) determination of visitation thresholds that may maintain breeding density levels of urban birds; and (5) investigation on how the temporal dynamics of visitors to urban parks (daily and seasonally) may influence tolerance levels of bird species.

ACKNOWLEDGMENTS

We are indebted to C.L. Alonso, F.R. Gehlbach, and Reed Bowman whose insights and comments greatly improved the quality of earlier versions of the manuscript. EFJ was funded by the Agencia Española de Cooperación Internacional (MUTIS Fellowship, Spain).

REFERENCES

Barker, G., and A. Graf. 1989. Urban wildlife now: principles for nature conservation in towns and cities. Nature Conservancy Council, Number 3.

Blair, R. B. 1996. Land use and avian species diversity along an urban gradient. Ecol. Appl. 6:506-519.

Burger, J., and M. Gochfield. 1991. Human distance and birds: tolerance and response distances of resident and migrant species in India. Environ. Conserv. 18:158-165.

Cooke, A. S. 1980. Observations on how close certain passerine species will tolerate an approaching human in rural and suburban areas. Biol. Conserv. 18:85-88.

Dickman, C. R. 1987. Habitat fragmentation and vertebrate species richness in an urban environment. J. Appl. Ecol. 24: 337-351.

Fernández-Juricic, E., and J. L. Tellería. 1999. Recruitment patterns of Blackbirds (*Turdus merula*) in urban fragmented populations. Ardeola 46:61-70.

Fernández-Juricic, E., and J. L. Tellería. 2000. Effects of human disturbance on Blackbird *Turdus merula* spatial and temporal feeding patterns in urban parks of Madrid, Spain. Bird Study 47:13-21.

Fernández-Juricic, E. 2000. Local and regional effects of human disturbance in a fragmented landscape. Condor 102:247-255.

Gill, J. A., W. J. Sutherland, and A. R. Watkinson. 1996. A method to quantify the effects of human disturbance on animal populations. J. Appl. Ecol. 33:786-792.

Goode, D. A. 1991. Wildlife in cities, p. 353-393. *In* E. A. Webb and S. Q. Foster [EDS.], Perspectives in urban ecology. Denver Museum of Natural History, Denver, CO.

Gutzwiller, K. J., H. A. Marcum, H. B. Harvey, J. Roth, and S. H. Anderson. 1998. Bird tolerance to human intrusion in Wyoming montane forests. Condor 100:519-527.

Henson, P., and T. A. Grant. 1991. The effects of human disturbance on trumpeter swan breeding behavior. Wildl. Soc. Bull. 19:248-257.

Holmes, T. L., R. L. Knight, L. Stegall, and G. R. Craig. 1993. Responses of wintering grassland raptors to human disturbance. Wildl. Soc. Bull. 21:461-468.

Humphrey, P. S., B. C. Livezey, and D. Siegel-Causey. 1987. Tameness of birds of the Falkland Islands: an index of preliminary results. Bird Behav. 7:67-72.

Klein, M. L. 1993. Waterbird behavioral responses to human disturbances. Wildl. Soc. Bull. 21:31-39.

Knight, R. L., and S. A. Temple. 1995. Wildlife and recreationists: coexistence through management, p. 327-333. *In* R. L. Knight and K. J. Gutzwiller [EDS.], Wildlife and recreationists: coexistence through management and research. Island Press, Washington, DC.

Knight, R. L., and S. N. Knight. 1984. Responses of wintering bald eagles to boating activity. J. Wildl. Manage. 48:999-1004.

Kramer, D. L., and M. Bonenfant. 1997. Direction of predator approach and the decision to flee to a refugee. Anim. Behav. 54:289-295.

Lima, S. L. 1993. Ecological and evolutionary perspectives on escape from predatory attack: a survey of North American birds. Wilson Bull. 105:1-47.

Lima, S. L., and P. A. Zollner. 1996. Towards a behavioural ecology of ecological landscapes. Trends Ecol. Evol. 11:131-135.

Martín, J., and P. López. 1995. Influence of habitat structure on the escape tactics of the lizard *Psammodromus algirus*. Can. J. Zool. 73:129-132.

Prodon, R., and J. D. Lebreton. 1981. Breeding avifauna of a Mediterranean succession: the holm oak and cork oak series in the eastern Pyrenees, 1. Analysis and modelling of the structure gradient. Oikos 37:21-38.

Riffell, S. K., K. J. Gutzwiller, and S. H. Anderson. 1996. Does repeated human intrusion cause cumulative declines in avian richness and abundance? Ecol. Appl. 6:492-505.

Rodgers, J. A., and H. T. Smith. 1995. Set-back distances to protect nesting bird colonies from human disturbance in Florida. Conserv. Biol. 9:89-99.

Rodgers, J. A., and H. T. Smith. 1997. Buffer zone distances to protect foraging and loafing waterbirds from human disturbance in Florida. Wildl. Soc. Bull. 25:139-145.

Skagen, S. K., R. L. Knight, and G. Orians. 1991. Human disturbance of an avian scavenging guild. Ecol. Appl. 1:215-225.

StatSoft, Inc. 1996. Statistica for Windows user's guide, version 4.5. StatSoft, Inc. Tulsa, Oklahoma.

Tellería, J. L., and T. Santos. 1997. Seasonal and interannual occupation of a forest archipielago by insectivorous passerines. Oikos 78:239-248.

Tellería, J. L., B. Asensio, and M. Díaz. 1999. Aves Ibéricas. II. Paseriformes. J. M. Reyero, Madrid.

Ydenberg, R. C., and L. M. Dill. 1986. The economics of fleeing from predators. Advances in the Study of Behaviour 16:229-249.

Chapter 13

Settlement of breeding European Starlings in urban areas: Importance of lawns vs. anthropogenic wastes

Gwénaëlle Mennechez[1] and Philippe Clergeau[2]
[1]UMR CNRS 6553 « Ecobio », CNRS-Université Rennes1, Avenue du Général Leclerc, 35042 Rennes cedex, France

[2]INRA Faune sauvage & UMR CNRS 6553 « Ecobio », Avenue du Général Leclerc, 35042 Rennes cedex, France, e-mail:clergeau@univ.rennes.fr

Key words: Diet, European Starling, France, nestling, *Sturnus vulgaris*, urban ecology

Abstract: We examined shifts in the feeding habits of European Starlings along an urban-rural gradient in northwestern France by comparing the diet of nestlings in three habitats: urban, suburban and rural areas. We sampled 99 broods using the neck-collar method. Broods were adjusted to four nestlings each to exclude brood size effects on diet. We have previously shown that parents foraged within 800 m from the nest in each habitat. In all three habitats, most food items brought to nestlings were invertebrates living on and in the ground or on low vegetation. However, diet composition changed along the urbanization gradient: (1) During first nest attempts nestlings reared in the urban area received less Lepidoptera and more Coleoptera than those located in the rural area; (2) Urban and suburban nestlings received less animal food than rural nestlings and more plant food and human refuse; (3) Diet varied with nest attempt and with nestling age. We conclude that, in towns, European Starlings feeding nestlings depend heavily on grasslands despite their ability to use human refuse. We suggest that recent urban planning in France, which encourages lawn plantings, may, in part, explain the increase of urban starling populations.

1. INTRODUCTION

Occupancy of birds in towns depends on the capacity of urban habitats to provide adequate nest sites, roosts, and food resources (Tatner 1983, Mills et al. 1989). Because it results in important modifications of the natural environment, urbanization is associated with acquisition of new behaviors or with behavioral shifts (Diamond 1986). Indeed, changes in the diet of birds have been described in urban environments. For example, Kestrels (*Falco tinnunculus*) feed mainly on House Sparrows (*Passer domesticus*) in urban areas and on small mammals in rural areas (Yaldon 1980). Great Tits (*Parus major*) and Blue Tits (*Parus caerulus*) bring fewer Lepidoptera to their young when breeding in gardens and use abundantly available artificial foods (Cowie 1988). The availability of foods provided by humans in urban areas is known to favor colonization by opportunistic feeders such as corvids and gulls (Erz 1966, Adams 1994, Brousseau et al. 1996).

The European Starling, *Sturnus vulgaris*, is one of the most common urban bird species (Gilbert 1989). In addition, densities of breeding pairs in urban areas often exceed those reported in nearby rural habitats (Beissinger and Osborne 1982, Mennechez 1999). The successful colonization of urban areas by starlings is often attributed to their behavioral adaptability and dietary flexibility (Erz 1966, Taitt 1973, Feare 1984). Indeed, they are able to successfully exploit anthropogenic sources of foods, such as cattle foods or human refuse from rubbish dumps (Dunnet 1955, Clergeau 1980, Feare and MacGinnity 1986). Numerous studies have documented the diet of starlings in agricultural landscapes (France; Madon 1930, Great Britain; Dunnet 1955, Poland; Gromadzki 1969), but little is known in urban habitats. We found only one study of the diet of urban starlings in Europe (Gromadzka and Luniak 1978) and it concerned a migratory population in eastern Europe. Comparable studies of the diet in a resident, breeding population are unknown, especially studies of nestling diets. We have previously shown that parents foraged within 800 m from the nest in both urban and rural habitats (Clergeau 1981), thus nestling foods are likely representative of the foods used in each habitat type.

We studied the relative importance of anthropogenic foods and natural foods in the diet of urban-breeding starlings. The urban gradient varied from a highly developed city core to a less developed residential suburb to rural farmland (Erz 1966, Adams 1994).

2. METHODS

2.1 Study Areas

This study was conducted in northwestern France, in a 400 km^2 area including the city of Rennes (48°7' N, 1°41' W), Brittany. With a human population of 200.000, Rennes is a middle-sized city similar to other towns of France, with a pattern of concentric rings of decreasing urbanization out from the city center (see Clergeau et al. 2001). We subdivided the study area into three types of land-use (Fig. 13.1): (A) the urban area which included both the commercial downtown area, i.e., historic center of the city with less than 15% of vegetated area and buildings with 3-5 stories, and the residential sector adjacent to the commercial area where buildings have 1-3 stories and vegetated areas covered between 20% and 40% of the area. Here, starlings forage mainly in gardens and parks. (B) The suburban area which consisted of residential areas of single-family detached homes, some large apartment building complexes, parks and cemeteries, and large office buildings or supermarkets with large lawns and parking. The height and density of houses and buildings in this area varied considerably and vegetated areas covered up to 70 % of this sector. (C) The rural countryside with scattered houses and dominated by farmland. For quantitative descriptions of landscape elements near Rennes see Clergeau et al. (1998).

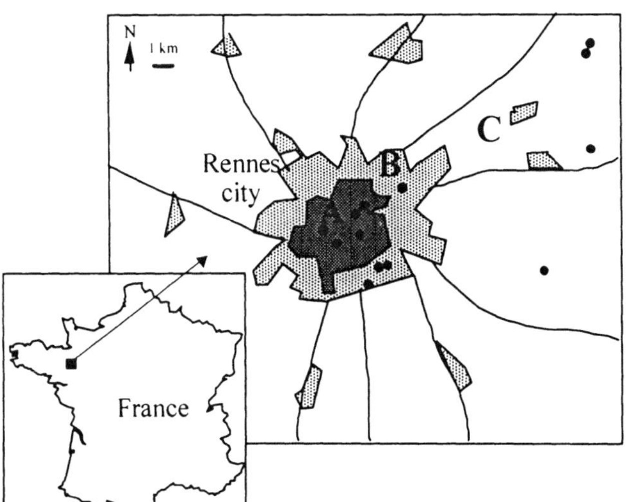

Figure 13.1. Study area and the location of European Starling nestbox clusters in (A) urban, (B) suburban and (C) rural countryside areas, near Rennes, France.

One year prior to the study, we placed wooden nestboxes in each zone: 42 in A, 22 in B and 17 in C. In urban and suburban areas, nestboxes were placed in typical sites, such as gardens, parks, on trees and on building walls. Rural nestboxes were on trees and buildings and were 15-20 km from the city centre. We collected data during the breeding seasons of 1995 and 1996, both of which appeared to be typical years (see more details in Mennechez 1999).

2.2 Sampling of the Nestling Food

European Starlings have three distinct peaks in nesting activity, representing first nests, renests, and second nests (Cramp and Perrins 1994). In Brittany, first clutches are initiated in the last days of March. About 70% of pairs (Clergeau 1985) that have already reared one brood attempt a second brood 40-50 days after the first clutches. To study temporal variation in nestling diet, we sampled first broods in April (period 1) and second broods in late May (period 2). We excluded clutches initiated between these dates because these included replacement clutches or are first clutches laid by immature birds (Clergeau 1985). All nestboxes were inspected at least twice a week, from mid-March until late June to establish the breeding phenology and nestling age. Because diet varies with brood size (Tinbergen 1981, Biermann and Sealy 1982), we adjusted all broods to contain four nestlings, the mean brood size of starlings in Brittany (Clergeau 1985).

We used the neck-collar method to collect food samples (Kluijver 1933, Henry 1982). Neck-collars were applied once a day to all nestlings in each nest for 50 min. Each brood was sampled twice when chicks were 5-7 days old and twice when they were 10-12 days old. Samples were collected throughout the day except during the first and last feeding hours. Food samples were immediately placed into 70% alcohol and prey were identified to family (when possible) in the laboratory. Food items were classified into eight categories: Annelida, Arachnida, Diptera, Coleoptera, Lepidoptera, other animals, plant materials, and human refuse (pieces of bread, meat...). Species typical of grasslands and lawns were identified from Ricou (1967) and Clergeau (1981), and from a study of 300 soil plots in lawns and grasslands (Mennechez, unpubl. data).

Because data from the two study years did not differ (Appendix 13.1), we summed the number of food items per period for both years combined. Results are expressed as the percentage of each food type in the total diet, i.e., number of items of a food type divided by the total number of items. We used contingency tests to assess significant differences among food types (Statview 1992). Shannon diversity and similarity indices were calculated following Jongman et al. (1995).

3. RESULTS

In total, we collected 5,075 food items from 366 samples, 129 in rural, 172 in suburban and 65 in urban areas (Appendix 13.1). Starling nestlings were fed primarily arthropods (90.7% of the diet), especially insects (81% of the diet). Diptera (such as Tipulidae, Stratiomyidae, Dolichopodidae, and Bibionidae), Lepidoptera (such as Noctuidae), and Coleoptera (such as Carabidae, Staphylinidae, Scarabeidae, Elateridae, and Curculionidae) dominated the diet (Table 13.1). Other invertebrates were in decreasing order of importance: Annelida (Lumbricidae), Isopoda (such as Oniscidae,

Table 13.1: Percent composition of diet of European Starling nestlings in first broods (a) and second broods (b) along an urbanization gradient. () = number of food items. Probabilities *P* from χ^2 tests are given; for calculations plant materials and human refuse were gathered in one category, non-animal; ns = not significant.

(a) PERIOD 1: first broods

	RURAL (n = 529)	SUBURB (n = 917)	URBAN (n = 293)	P 1 vs 2	P 2 vs 3	P 1 vs 3
Annelida	11 (56)	10 (92)	7 (21)	Ns	ns	ns
Arachnida	5 (28)	2 (23)	3 (10)	< 0.010	ns	ns
Lepidoptera	15 (79)	9 (81)	6 (17)	< 0.001	ns	< 0.001
Diptera	50 (266)	54 (496)	48 (141)	Ns	ns	ns
Coleoptera	8 (44)	13 (121)	15 (44)	< 0.010	ns	< 0.005
Other animals	11 (56)	10 (92)	14 (40)	Ns	ns	ns
Plant materials	0	0 (1)	0			
Human refuse	0	1 (11)	7 (20)	Ns	< 0.001	< 0.001
Shannon index	2.12	2.06	2.25			

(b) PERIOD 2: second broods

	RURAL (n = 1080)	SUBURB (n = 2101)	URBAN (n = 155)	P 1 vs 2	P 2 vs 3	P 1 vs 3
Annelida	6 (68)	5 (104)	2 (3)	Ns	ns	< 0.050
Arachnida	4 (44)	2 (40)	4 (7)	< 0.001	ns	ns
Lepidoptera	3 (33)	3 (54)	5(8)	Ns	ns	ns
Diptera	54 (589)	51 (1078)	39 (61)	Ns	< 0.001	< 0.001
Coleoptera	25 (272)	11 (236)	17 (26)	< 0.001	ns	< 0.050
Other animals	5 (53)	26 (541)	18 (28)	< 0.001	< 0.050	< 0.001
Plant materials	2 (21)	2 (33)	6 (10)			
Human refuse	0 (0)	1 (15)	8 (12)	Ns	< 0.001	< 0.001
Shannon index	1.90	1.96	2.48			

Armadillidae), Arachnida (such as Lycosidae, Pisauridae, Thomisidae), Gasteropoda (such as Cochilicopae, Zonitidae) and Myriapoda (such as Chilopodes, Diplopodes). Food was essentially taken on or just under the soil surface (Tipulidae, Noctuidae, Elateridae, Carabeidae) or around tall vegetation (Stratiomyidae, Cantharidae). In the three habitats, most food samples contained prey captured by starlings from lawn or grassland areas; earthworms and leatherjackets alone, species typical of lawn, represented more than 75% of prey (Table 13.2). Plant foods (exclusively cherries) often represented less than 5% of the diet (Table 13.1).

In the three habitats, the composition of the nestling diet varied significantly between first and second broods (rural; $\chi^2 = 161.5$, DF = 6, $P < 0.001$; suburban; $\chi^2 = 174.6$, DF = 7, $P < 0.001$; urban; $\chi^2 = 4.7$, DF = 6, $P < 0.05$). The diet of first broods was characterized by a high proportion of insect larvae (42 to 68%), whereas insect adults dominated the diet of second broods (89 to 73%) (rural; $\chi^2 = 199.8$, DF = 1, $P < 0.001$; suburban; $\chi^2 = 269.9$, DF = 1, $P < 0.001$; urban; $\chi^2 = 36.2$, DF = 1, $P < 0.001$). In general, proportions of Lepidoptera (rural; $\chi^2 = 75.5$, DF = 1, $P < 0.001$; suburban; $\chi^2 = 57.1$, DF = 1, $P < 0.001$; urban; ns) and Annelida (rural; $\chi^2 = 8.6$, DF = 1, $P < 0.005$; suburban; $\chi^2 = 26.3$, DF = 1, $P < 0.001$; urban; $\chi^2 = 4.5$, DF = 1, $P < 0.05$) decreased from first to second broods. In addition, a small proportion of cherries was brought to second-brood nestlings but not to first-brood nestlings, but only in urban habitats. Some brood differences were specific to each habitat. In the rural habitat, Coleoptera (mostly Cantharidae) were brought in greater numbers to second- brood nestlings (25 versus 8%) ($\chi^2 = 62.9$, DF = 1, $P < 0.001$) and became the second most important prey item. In the suburban habitat, the proportion of other animals (mainly Formicidae) increased from 10 to 26% ($\chi^2 = 94.2$, DF = 1, $P < 0.001$) from first to second brood.

The proportions of food items in the diet differed significantly along the urbanization gradient for both first and second broods ($\chi^2 = 95.6$, DF = 11, P

Table 13.2: Percent occurrence of the two principal prey captured in grassland soil, earthworms and leatherjackets, in European Starling food samples collected along an urbanization gradient during two periods of the breeding cycle. () = number of samples.

Breeding cycle	Rural	Suburban	Urban
Period 1	91.0 (73)	85.5 (90)	83.7 (49)
Period 2	87.9 (56)	85.4 (82)	56.3 (16)
Total	89.7 (129)	85.5 (172)	76.9 (65)

< 0.001; χ^2 = 359.3, DF = 11, *P* < 0.001, respectively). For first broods (Table 11.1a), the proportion of Lepidoptera in the diet decreased from rural to urban habitats, whereas the proportion of Coleoptera increased. In addition, human refuse often was fed to urban nestlings (8% of bread, meat, cheese,...) but was found only in a few suburban samples and was absent from rural samples. For second broods (Table 13.1b), the proportion of animal food was lower in urban (86%) than in suburban (97%) or rural (98%) habitats. Urban parents brought more human refuse and cherries and less Diptera. Rural nestlings were fed more Coleoptera (Cantharidae) and less other animals than urban or suburban nestlings.

The diversity of food items in the diet (Shannon diversity index) increased from rural to urban habitat (Table 13.1). The main difference in the diet of nestlings was that urban birds consumed human refuse. All other types of food items were consumed by birds in all habitats although some specific food items were restricted to certain habitats. For example, Cantharidae, abundant in the rural habitats, were absent from the urban habitat. Suburban habitats were characterized by a high proportion of Formicidae in the diet. In fact, the composition of the diet varied sequentially along the urban gradient; the suburban diet was intermediate of the rural and urban diet as shown by similarity indices: first broods: urban/suburban s = 88, suburban/rural s = 90, urban/rural s = 83; second broods: urban/suburban s = 77, suburban/rural s = 80, urban/ rural s = 71.

Diet also changed with nestling age (All tests of comparison between 5-7 days and 10-12 days were significant for all habitats and for both broods). During the first brood in all habitats, 5-7 days old nestlings were fed more Arachnida (χ^2 = 7.2, DF = 1, *P* < 0.01) and Lepidoptera (χ^2 = 27.9, DF = 1, *P* < 0.001), but less Annelida (χ^2 = 10, DF = 1, *P* < 0.01) and other animals (χ^2 = 6.16, DF = 1, *P* < 0.05) than older nestlings. Other differences in diets of 5-7 and 10-12 days old nestlings were found but did not apply to all habitats across the urban gradient. For example, in the urban habitat, human refuse was more commonly fed to older chicks (first brood; χ^2 = 6.2, DF = 1, *P* < 0.05; second brood; χ^2 = 9.3, DF = 1, *P* < 0.01). In the suburban habitat, the proportion of other animals increased with nestling age (first brood; 7% versus 12%, *P* < 0.05; second brood; 21% versus 30%, *P* < 0.005).

4. DISCUSSION

An opportunistic diet is often mentioned as a factor for the range expansion of certain bird species and for their settlement in urban areas (Taitt 1973, Tatner 1983, Brousseau et al. 1996). European Starlings feed on a wide variety of foods and can be regarded as opportunistic feeders with a generalist diet (Cramp and Perrins 1994). Our results support the observations (Kluijver 1933, Dunnet 1955, Bogucki 1974) that starlings have a broad diet, including 16 orders of invertebrates. The main food categories found in our study were similar to those reported in the literature: Diptera, Coleoptera, and Lepidoptera (Feare 1984, Cramp and Perrins 1994). The proportion of food categories also varied between different broods and habitats along an urban gradient. Prey was taken mainly at or just below the soil surface or in tall vegetation.

In rural landscapes, grasslands are the "preferred" foraging sites of the European Starling (Dunnet 1955, Whitehead et al. 1995). During the breeding season, 95% of food was taken underground (Dunnet 1955). Despite the variety of food items consumed in Rennes, our results support the idea that the main foraging sites of urban breeding starlings remain lawns. In France, more lawns and private gardens occur in suburban and urban areas since 1960. Lawns and gardens can provide an important source of food for starlings, facilitating their increasing abundance in suburban and urban habitats (Clergeau 1994).

Other studies of starling diet in agricultural landscapes showed a predominance of animal foods, especially in spring when parents feed their young (Havlin and Folk 1965, Gromadzki 1969). Similarly, in the rural countryside of Rennes, nestlings were fed mainly with animal foods. However, non-animal items (fruits and human refuse) represented an important portion of the diet of 10-12 day old urban nestlings (11 to 20%). Fruit exploitation was more important in the city than in the rural site. A possible explanation for this difference may be differences in the fruiting phenology because of the city microclimate (temperatures are 0.5-1.5 higher than in the surrounding rural areas, Gilbert 1989). In Rennes, cherries mature near the end of the starling breeding season and 10 days earlier than in the rural countryside (personal observation). Because of this phenological difference, they were available to urban nestlings, but not to rural nestlings. Anthropogenic foods were not found in any rural samples but were the second most common food items in the diet of older urban nestlings, after Diptera. Similarly, the diets of starling nestlings in Warsaw showed a high proportion (35%) of anthropogenic foods (Gromadzka and Luniak 1978). The use of anthropogenic food by other bird species during the rearing of nestlings is widely reported especially in urban areas and can be important to nestling survival since traditional foraging areas are scarce in urban centers (Cowie 1988, Shnack 1991). The starling's ability to feed on a wide variety of food enables it to exploit new sources of food. As a consequence it is highly adaptable to anthropogenic habitats (Feare 1984). This adaptability of starlings is probably important in its use of urban habitats, particularly during the breeding season (Adams 1994).

For tits (Cowie 1988) and gulls (Brousseau et al. 1996), the proportion of anthropogenic food in the diet increased with nestling age. Similarly in Rennes, more anthropogenic food was brought to older starling nestlings. However, the availability of these items is likely constant during the breeding season. Therefore, we suggest that starling parents avoid feeding human foods to young nestlings. Anthropogenic food may be less preferred because of low energetic content, poor nutritional value (Pierotti and Annett 2001), or because it is less easily handled and digested by younger birds. This shift to increased anthropogenic food occurred when nestling feeding demands reached their maximum (between day 9 and day 16, Tinbergen 1981), and suggests that urban parents sacrificed food quality for a more readily obtained food (Cowie 1988). In addition, the diet of urban nestlings also was characterized by a low amount of Lepidoptera (high energetic prey; Tinbergen 1981) during first broods and a high amount of other invertebrates during second broods. Our results show a lower consumption of favored items and energetic items in urban than in rural areas. As a consequence, urban habitats could be a suboptimal habitat for breeding starlings.

4.1 Policy and Management Implications

Our study of the diet of nestling starlings confirms their high degree of adaptability and their capacity to breed in urban areas by exploiting a wide variety of food. However, along the entire urbanization gradient, prey living on and under the ground or on low vegetation of grassy areas remained an important component of the diet. Lawns and gardens appear to be the major feeding areas for urban and suburban breeding starlings. We have shown that the density of urban starling nests with young was positively correlated with the amount of lawn (Mennechez 1999), and that large flocks of starlings feed in wide urban lawns and roosted near these sites in winter (Clergeau 1981). We suggest that recent urban planning in France, which has encouraged lawn plantings since 1960, probably explains the increase of urban starling populations. The evolution of landscape structures in urban areas can explain new species or new population dynamics in cities (Clergeau et al. 1998, Mennechez 1999). From an urban bird diversity perspective (including species management), we have to consider not only the positive effects of increasing green spaces in urban environments, such as biodiversity restoration, but also the negative potential of such management, such as increasing the probability of increasing urban starling populations.

ACKNOWLEDGEMENTS

Thanks to E. Simmonet and G. Place for field assistance. We are grateful to J.P.L. Savard, S. Petit, and R. Bowman for improving the English. This work was supported by Centre National de la Recherche Scientifique (PIRVilles Paris).

REFERENCES

Adams, L. W. 1994. Urban wildlife habitat: a landscape perspective. University of Minnesota, Minneapolis Press.

Beissinger, S. R., and D.R. Osborne. 1982. Effects of urbanization on avian community organization. Condor 84: 75-83.

Biermann, G.C., and S.G. Sealy. 1982. Parental feeding of nestlings Yellow Warblers in relation to brood size and prey availability. Auk 99: 332-341.

Bogucki, Z. 1974. Studies on the diet of Starling nestlings. Acta. Zool. Crac. 19: 357-390.

Brousseau, P., J. Lefebvre, and J. F. Giroux. 1996. Diet of Ring-billed Gull chicks in urban and non-urban colonies in Quebec. Colonial Waterbirds 19: 22-30.

Clergeau, P. 1980. Fréquentation d'une décharge d'ordures ménagères de Rennes par l'avifaune. Penn ar Bed 12 : 297-312.

Clergeau, P. 1981. Comportements liés à l'alimentation de l'Etourneau *Sturnus vulgaris* en Bretagne ; rôle joué par certaines variables environnementales et sociales. Thèse de doctorat, université de Rennes.

Clergeau, P. 1985. Production en jeunes et croissance chez l'étourneau *Sturnus vulgaris* en Bretagne. Caractéristiques bretonnes et signification écologique. Acta Oecol. 6: 135-159.

Clergeau, P. 1994. L'Etourneau Sansonnet, p. 670-673. *In* D.Yeatman-Berthelot and G. Jarry [EDS.], Nouvel atlas des oiseaux nicheurs de France, Paris.

Clergeau, P., G. Mennechez, A. Sauvage, and A. Lemoine. 2001. Human perception and appreciation of birds: A motivation for wildlife conservation in urban environments in France, p. 69-88. *In* J. M. Marzluff, R. Bowman, and R. Donnelly [EDS.], Avian ecology and conservation in an urbanizing world. Kluwer Academic, Norwell, MA.

Clergeau, P., J. P. Savard, G. Mennechez, and G. Falardeau. 1998. Bird abundance and diversity along an urban-rural gradient: a comparative study between two cities on different continents. Condor 100: 413-425.

Cowie, R. J. 1988. Feeding ecology of Great Tits (*Parus major*) and Blue Tits (*Parus caeruleus*), breeding in suburban gardens. J. Anim. Ecol. 57: 611-626.

Cramp, S., and C. M. Perrins. 1994. Birds of Europe, the Middle East and North Africa, p. 228-269. *In* The birds of the Western Paleartic, Oxford University Press, Oxford.

Diamond, J. M. 1986. Rapid evolution of urban birds. Nature 324: 107-108.

Dunnet, G. M. 1955. The breeding of the Starling *Sturnus vulgaris* in relation to its food supply. Ibis 97: 619-662.

Erz, W. 1966. Ecological principles in the urbanization of birds. Ostrich suppl. 6: 358-363.

Feare, C. J. 1984. The Starling. Oxford University Press, Oxford.

Feare, C.J., and N. MacGinnity 1986. The relative importance of invertebrates and barley in the diet of Starlings *Sturnus vulgaris*. Bird Study 33: 164-167.

Gilbert, L. O. 1989. The ecology of urban habitats. Chapman and Hall, London.

Gromadzka, J., and M. Luniak. 1978. Food composition of the Starling nestlings, *Sturnus vulgaris L.*, in Warsaw. Acta Orn. 16: 275-284.

Gromadzki, M. 1969. Composition of food of the starling, *Sturnus vulgaris L.*, in agrocenoses. Ekol. Pol. 17: 287-310.

Havlin, J., and C. Folk 1965. Food and economic importance of the Starling, *Sturnus vulgaris.* L. Zool. Listy 14: 193-208.

Henry, C. 1982. Etude du régime alimentaire des passereaux par la méthode des colliers. Alauda 50: 92-107.

Jongman, R. H. G., C. J. F. Ter Braak, and O. F. R Tongeren. 1995. Data analysis in community and landscape ecology. Cambridge University Press, Cambridge, UK.

Kluijver, H. N. 1933. Contribution to the biology and the ecology of the Starling (*Sturnus vulgaris vulgaris L.*) during its reproductive period. Versl. PIZiekt. Dienst Wageningen 69: 1-145.

Madon, M. 1930. L'étourneau et son régime. Alauda 11: 283-317.

Mennechez G. 1999. Urbanisation et espèces habitats-généralistes : biologie comparée de l'Etourneau *Sturnus vulgaris* lors de la période de reproduction sur un gradient rural-urbain. Thèse de doctorat, université de Rennes.

Mills, G. S., J. B. Dunning, and J. M. Bates 1989. Effects of urbanization on breeding bird community structure in southwestern desert habitat. Condor 91: 416-428.

Pierotti, R., and C. Annett. 2001. The ecology of Western Gulls in habitats varying in degree of urban influence, p. 309-331. *In* J. M. Marzluff, R. Bowman, and R. Donnelly [EDS.], Avian ecology and conservation in an urbanizing world. Kluwer Academic, Norwell, MA.

Ricou G. 1967. Etude biocénotique d'un milieu naturel, la prairie permanent pâturée. Thèse de doctorat, INRA - Université de Paris.

Shnack, S. 1991. The breeding biology and nestling diet of the Blackbirds *Turdus merula* L. and the Song Trush *Turdus philomelos* in Vienna and in an adjacent wood. Acta Orn. 26: 85-106.

Statview. 1992. Abacus Concept, Inc, Berkeley, CA.

Taitt, M. J. 1973. Winter food and feeding requirements of the Starling. Bird Study 20: 226-236.

Tatner, P. 1983. The diet of urban Magpies *Pica pica*. Ibis 125: 90-107.

Tinbergen. J. M. 1981. Foraging decisions in Starlings (*Sturnus vulgaris* L.). Ardea 69: 1-67.

Yaldon, D. W. 1980. Notes on the diet of urban Kestrel. Bird Study 27: 235-238.

Whitehead, S. C., J. Wright, AND P. A. Cotton. 1995. Winter field use by the European Starling *Sturnus vulgaris*: habitat preferences and the availability of prey. J. Avian Biol. 26: 193-202.

Appendix 13.1. Composition of European Starling nestling diet (in occurrence of prey items) along an urbanization gradient for each study year. Orders and families are given for the main prey; n = number of food samples; habitat type: s = preys captured on or just under the soil surface, g = on the ground or on grass, and o = other types such as in air, in trees.

			PERIOD 1							PERIOD 2						
			RURAL		SUBURBAN		URBAN			RURAL		SUBURBAN		URBAN		
			1995	1996	1995	1996	1995	1996		1995	1996	1995	1996	1995	1996	
		type	n=23	n=50	n=47	n=43	n=12	n=37	Total	n=20	n=36	n=44	n=38	n=8	n=8	Total
ANNELIDA	*Lumbricidae*	s	25	31	44	48	4	17	169	32	36	13	91		3	175
DIPTERA ad.	*Tipulidae*	g-o		6	35	36		2	79		4	2	1			7
	Bibionidae	g-o			129	19	26	7	181		21					21
	Stratiomyidae	o			2				2	4	223	155	424		18	824
	Empididae	g-o			14	26		4	44			66	72		17	155
	Dolichopodidae	o									4	94	21		2	121
	Scatophagidae	o	24	30	3	6			63	9	46		6			61
	Other	g-o	2	7	17	8	5	8	47	13	99	47	38		1	198
DIPTERA larv.	*Tipulidae*	s	72	99	97	101		63	432	63	64	80	64	4	6	281
	Other	s-g	2	26		4		1	33	12	27	7	1		13	60
LEPIDOPTERA ad.	*Noctuidae*	o								1	2	7	14		4	28
	Other	o									7	14	5		2	28
LEPIDOPTERA larv	*Noctuidae*	s	12	43	7	55	1	11	129	1	18	4	5			28
	Other	s-g	3	21	7	12		5	48		4	3	2	2		11
COLEOPTERA ad.	*Carabidae*	g	1	11	9	8	3	24	56	10	37	8	39		2	96
	Staphylinidae	g		1	8	3	3	3	18	4	4	12	11	4	7	42
	Scarabeidae	s-g			14				14	3	14	2	3			22
	Cantharidae	g-o			1	1			2	5	116	1	3			125
	Elateridae	s-g	1	6	21	6		1	35	1	8	35	72		6	122
	Curculionidae	g-o		4	27	2	1	2	36	4	6	6	5		1	22
	Other	g-o		1	10	3	3	1	18	2	5	8	16	2	2	35
COLEOPTERA larv.	*Staphylinidae*	s	2	6			1	4	13	41	7	4	10		2	64
	Scarabeidae	s-g	4	3	4	1			12							
	Elateridae	g		4	1	2		1	8							

			PERIOD 1							PERIOD 2						
			RURAL		SUBURBAN		URBAN			RURAL		SUBURBAN		URBAN		
			1995	1996	1995	1996	1995	1996		1995	1996	1995	1996	1995	1996	
		type	n=23	n=50	n=47	n=43	n=12	n=37	Total	n=20	n=36	n=44	n=38	n=8	n=8	Total
	Other	s-g-o						1	1	1	4	1				6
HYMENOPTERA	*Formicidae*	g			25	7		3	35			179	102			281
	Other	g-o		1	4			3	8		1	2	2		2	7
ODONATA		o										1	1		1	3
NEVROPTERA		g											1			1
HETEROPTERA		g-o			4	4	1	2	11	1	8	2	19		4	34
HOMOPTERA		o			10				10	2	3	43	23	7	1	79
DERMAPTERA		g-o			1				1				1			1
ORTHOPTERA		g		2					2		3					3
THYSANOPTERA		o		1					1						1	1
MYRIAPODA		s-g		18	3	2	1	1	25	6	13	2	1			22
GASTEROPODA	*Helicidae*	g	2	10	11	7	2	3	35	4	1	8	3	2		18
	Limacidae	g			3	1		3	7		1	1	9	1	2	14
ARACHNIDA	*Arachnidae*	g-o	6	22	5	18	4	6	61	7	31	10	23	2	3	76
	Opilionidae	g								3	3	3	4	1	1	15
ISOPODA		g		20	7	3	3	14	47	1	8	85	55		7	156
Human refuse					8	3	7	13	31			4	11	9	3	27
Plant materials										8	13	19	14	2	8	64
Indeterminate				2		1			3		1	1				2
TOTAL			156	375	531	387	89	204	1742	238	842	929	1172	36	119	3336

Chapter 14

Variation in the timing of breeding between suburban and wildland Florida Scrub-Jays: Do physiologic measures reflect different environments?

Stephan J. Schoech[1] and Reed Bowman[2]
[1]*Department of Biology, University of Memphis, TN 38152 USA, e-mail: sschoech@memphis.edu*

[2]*Archbold Biological Station, P.O. Box 2057, Lake Placid, FL 33862 USA*

Key words: Florida Scrub-Jays, physiology, time of breeding, urbanization

Abstract: Evidence exists that access to anthropogenic food by birds in urban areas can result in earlier laying dates and larger clutch sizes, both of which have the potential to increase the relative fecundity of urban bird populations. Several mechanisms exist by which supplemental food can advance laying. Supplemental food may provide additional energy, essential nutrients, or an environmental cue that the resources necessary for breeding are available. We examined several physiologic measures of pre-breeding female Florida Scrub-Jays (*Aphelocoma coerulescens*) from a suburban and a wildland population to assess these alternative hypotheses. Plasma protein levels were higher in females from the suburban scrub-jay population, but the difference was absent when we controlled for the number of days prior to laying that we sampled the bird. However, plasma protein levels during the earliest sampling period, long before protein mobilization as a result of vitellogenesis should be occurring in either population, were higher in the suburban population. We found no differences in plasma calcium levels or in the total body lipids between suburban and wildland populations. In both populations, total body lipids declined as laying neared, suggesting that Florida Scrub-Jays may use some endogenous resources to fuel reproduction. From other studies, we know that suburban scrub-jays increase food intake rates relative to jays in wildland populations. However, this strategy does not appear to result in improved body condition. We suggest that the readily available proteins in anthropogenic foods may provide essential nutrients and high intake rates may serve as an environmental cue, both resulting in earlier laying in suburban jay populations. However, evidence exists suggesting that this may not increase, but rather decrease fecundity because breeding is out of synch with the seasonally fluctuating foods, such as arthropods, necessary to rear nestlings.

1. INTRODUCTION

The impacts of urbanization on avian demography and community structure often have been considered but, with the notable exception of the effects of industrial contaminants such as DDT and its metabolites, relatively little is known about the effects of human settlement on the physiology of individuals. Consideration at this level of analysis may reveal mechanisms producing negative or positive affects on demographic measures. For example, work by Wasser et al. (1997) revealed that endangered Northern Spotted Owls (*Strix occidentalis*) that live close to roads have higher plasma levels of corticosterone than owls farther from roads. High levels of adrenal glucocorticoids (corticosterone and cortisol) have been shown to negatively influence reproduction in reptiles (Moore and Miller 1984, Moore and Zoeller 1985), birds (Wilson and Follett 1975, Siegel 1980, Wingfield 1988, 1994), and mammals (review in Sapolsky 1992). It is clear that an analysis at a physiologic level may reveal mechanisms that effect demographic patterns that may not be readily evident through standard techniques of assessing human impact.

The effects of urbanization on animals can vary and are likely species- and development-specific. Urbanization is likely to decrease essential resources for some species and increase resources for others. This process affects individuals as they compete for these essential and sometimes limited resources, which in turn affects species-specific demographic rates, and ultimately community structure. One resource essential to birds that may vary in different directions with urbanization is food. Natural foods important to birds, such as arthropods, often decline with urbanization (Blair and Launer 1997, Deny and Schmidt 1998, Bolger et al. 2000); however, anthropogenic sources of food often are plentiful and may be readily available to birds (see Pierotti and Annett 2001). Bottom-up (food and productivity) influences on population regulation have received far less attention than top-down (predation and parasitism) influences (see Bolger 2001), and yet food has important implications to species-specific demographic rates.

For many bird species, access to supplemental food often increases clutch size and advances laying date (for a review see Boutin 1991). Early-laying birds often enjoy a reproductive advantage, thus anthropogenic foods may directly benefit some species by increasing their potential fecundity. In

Florida Scrub-Jays (*Aphelocoma coerulescens*) experimentally provisioned supplemental food increased clutch size and advanced laying date (Schoech 1996) and in a suburban population of scrub-jays, where virtually all birds have access to anthropogenic sources of food, the jays lay earlier and have larger clutches than jays in natural habitats without access to extra food (Bowman et al. 1998).

Supplemental food may advance laying via several different mechanisms. If food is limiting, supplemental food may provide additional energy that improves the relative condition of birds so that additional endogenous resources are available for breeding. Alternatively, supplemental foods may provide essential nutrients, such as protein and calcium, necessary for oogenesis, which also may be limited in natural habitats. The relatively acidic scrub-oak habitats frequented by the Florida Scrub-Jay (see Abrahamson et al. 1984) might limit uptake of these nutrients. Finally, birds may use the relative availability of food or intake rates as a cue that resources are adequate for breeding.

Each of these hypotheses generates different predictions regarding the physiological state of adult females during the weeks preceding breeding. To assess each of these hypotheses, we compared (1) total body lipids as a measure of overall body condition and (2) plasma levels of total protein and calcium in breeding female Florida Scrub-Jays during the several weeks prior to breeding between a population in wildlands without access to anthropogenic foods with a population in a suburban matrix where all birds have such access (see Bowman and Woolfenden 2001 for a description and aerial photograph of the two study sites).

2. METHODS

2.1 Study Population

The two populations of Florida Scrub-Jays reside within Highlands County, Florida. Scrub-jays at Archbold Biological Station (27°10'N, 81°21'W, elevation 38 - 68 m) and a nearby suburban development were studied from mid-January through April of 1998. The suburban population is located approximately 10 km NNW of Archbold (see Bowman and Woolfenden 2001). All jays in both populations were banded with a unique combination of United States Fish and Wildlife Service aluminum bands and plastic color bands. The sex, breeding status (breeder or non-breeding helper), and group association of all jays in both populations was known from ongoing long-term studies. All nests were located during the nest building stage or incubation and, therefore, clutch initiation dates were known or could be estimated accurately (see Schoech et al. 1991, 1996; Mumme 1992; and Schoech 1996 for further details on the Archbold population).

2.2 Capture and Blood Sampling

Beginning in mid-January, six to eight weeks before the typical onset of breeding, all jays were captured in Potter traps that were baited with peanuts. Only female breeders were trapped and individuals were sampled only once. To assure that we were measuring baseline values of blood chemistry, each trap was monitored continuously, the jay was removed within seconds of capture, and a blood sample collected immediately. To control for possible diel fluctuations in plasma protein and calcium levels, all jays were captured between 0700 and 1100 h. Blood samples were collected in heparinized micro-haematocrit tubes from the ulnar vein following venipuncture with a 26 g needle, stored on ice until return to the laboratory (within 1 - 3 hours), centrifuged, and the plasma harvested for assay. A small volume (20 μl) of plasma was used for immediate evaluation of plasma protein (see below). The remaining volume was frozen and stored at -20°C until shipped to Indiana University for assay of plasma calcium levels.

Immediately following collection of a blood sample, jays were anaesthetized and total body lipids were determined (see below for details). While recovering from the anaesthetic, individuals were weighed and a standard set of morphometric data were taken, e. g., body mass, tail and wing-chord length, a series of bill measurements, head-breadth, and overall head plus bill length. From the time of capture until release at the capture site birds were held for approximately 0.5 h, thus allowing sufficient time for full recovery from the anaesthetic. All procedures were sanctioned by the Bloomington Institutional Animal Care Use Committee of Indiana University.

2.3 Total Body Lipids

Immediately following collection of a blood sample, body lipids were assessed using total body electrical conductivity (TOBEC). This allows determination of an animal's lean mass and from this, one can estimate total body lipids (Walsberg 1988, Kenagy and Barnes 1988, Roby 1991, Schoech 1996). Accurate use of the instrument requires that subjects are positioned uniformly and kept in the same position for the duration of the scanning procedure. Therefore, all jays were anaesthetized with Metophane (inhaled) and held in a nylon stocking during the procedure.

When calibrated for a species this method is useful for within species comparisons of relative body lipid content, although the technique has come under some criticism (see Morton et al. 1991, Asch and Roby 1995). Calibration of the instrument for a given species necessitates that on one occasion a number of individuals be scanned, killed, and total body fat content measured directly with a lipid extraction technique (e. g., chloroform in a Soxhlet apparatus). For this study, the previously derived linear regression equation (Schoech 1996) from the congeneric Western Scrub-Jay (*A. californica*) was used to estimate total body lipid in Florida Scrub-Jays (for calibration, validation, and methodological details see Schoech 1996).

2.4 Plasma Protein and Calcium

Plasma protein concentration was measured with a hand-held clinical refractometer (Atago Company, Ltd., Model A 300 CL). This allows determination of dissolved solute in a small volume of plasma (20 µl), based upon the degree to which light passing through the sample is 'bent' or refracted.

Plasma calcium concentration was determined with a kit obtained from Sigma Diagnostics (Procedure No. 587). Calcium reacts in a dose-dependent fashion with o-cresolphtalein complexone to produce a red complex, the intensity of which is directly proportional to the calcium concentration in the plasma sample. Color intensity was subsequently determined with a spectrophotometer set at 575 nm absorbency .

2.5 Data Analyses

Because birds mobilize endogenous reserves of certain nutrients when preparing to breed, we compared physiological attributes between populations relative to calendar date (# of days from the onset of sampling) and relative to laying date (# of days sampling occurred prior to the laying of the first egg of the first clutch). To analyze the data by calendar date, beginning on 21 January, the date sampling began, samples were grouped into 14-day periods. For example, all samples collected between 21 January and 3 February or between 4 and 17 February were combined. Therefore, there were 5 collection periods between 21 January and the termination of data collection on 31 March. To compare females from the two populations that were sampled at equivalent points in their reproductive cycle, samples were grouped by the number of days prior to clutch initiation that they were collected. The number of days within each category was 10, e. g., 10 to 1 days or 40 to 31 days before clutch initiation date. We predicted that if physiological differences were the cause of differences in the timing of breeding, those differences should exist for both calendar date and laying date analyses. If the differences were the result rather than the cause of differences in the timing of breeding, then we predicted those differences would be significant only for calendar date analyses and not for laying date analyses.

Because not all females that were paired with a male and occupying a territory initiated clutches, sample sizes are less for the analyses that are based upon clutch initiation date. However, for the analyses conducted by calendar date, we found no difference in the results whether these females were included or excluded, thus for the analyses based upon calendar date, we report data from all females sampled whether they subsequently initiated a clutch or not.

3. RESULTS

3.1 Timing of Reproduction

As in previous years, suburban jays nested earlier than wildland jays (t = -6.52, DF = 66, $P < 0.001$, Fig. 14.1). The mean laying date of first eggs by the 32 females in the suburbs that laid clutches was 24 March (± 2.3 days SE) and clutch initiation date ranged from 5 March through 23 April. In contrast, the 36 wildland females that initiated clutches had a mean clutch initiation date of 13 April (± 1.95 days SE). The earliest clutch in the wildland population was initiated on 25 March and the latest on 11 May.

3.2 Total Body Lipids

We found no differences between populations in total body lipids by calendar date ($F_{1,81}$ = 0.34, P = 0.56; Fig. 14.2a) nor did total body lipids vary with calendar date ($F_{4,81}$ = 1.84, P = 0.13), nor was there an interaction between habitat and calendar date ($F_{1,4}$ = 0.77, P = 0.55). We found no difference between populations in total body lipids by laying date ($F_{1,57}$ = 0.43, P = 0.51); however, when the difference in timing of clutch initiation between the two populations was controlled, total body lipids declined as females approached laying ($F_{5,57}$ = 5.20, P = 0.001; Fig. 14.2b). No interaction existed between habitat and laying date ($F_{1,5}$ = 0.51, P = 0.77). Post hoc pairwise comparisons found that total body lipids of females sampled within 10 days of clutch initiation were lower than for females sampled 60 - 51 days and 30 - 21 days before laying.

3.3 Plasma Protein

When compared by calendar date, females in the suburban population had higher plasma levels of protein than did females in the wildland population (ANOVA: $F_{1,81}$ = 6.38, P = 0.014; Fig. 14.3a). In both populations, plasma protein increased with calendar date ($F_{4,81}$ = 11.06, P < 0.001), but no

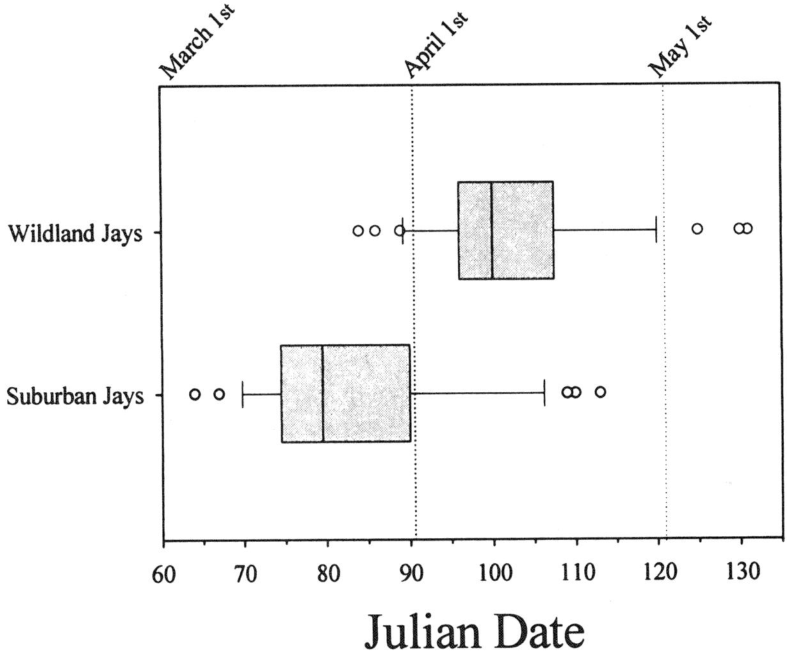

Figure 14.1. A comparison of first clutch initiation dates of Florida Scrub-Jays in a wildland and a suburban population. Suburban females lay significantly earlier than females in natural scrub habitat at Archbold Biological Station. In this horizontal box plot the shaded box represents the middle 50% of each population, the vertical lines within the boxes represent the median laying date, the error bars represent the first 10% and latest 90% of the population, and the circles represent statistical outliers.

interaction existed between date and habitat ($F_{1,4}$ = 1.04, P = 0.39). Post hoc pairwise comparisons using the Bonferroni adjustment found that plasma protein levels in females from the suburban population during the last sampling period, 18 - 31 March, were significantly higher than suburban females sampled during all three periods between 21 January and 3 March. Additionally, suburban females during this sampling period had levels that were higher than wildland females over the four sampling periods beginning 21 January and ending 17 March. Wildland females sampled during the last period (18 - 31 March) had higher plasma protein levels than females in the wildland population sampled during the first two periods (i.e., from 21 January - 13 February).

Based on our *a priori* hypothesis that suburban jays initiate reproduction earlier that wildland jays because they are in better condition and an expectation that this might be best reflected early in the sampling period, we also compared plasma protein levels during the earliest sampling period (i.e., during the first two periods, between 21 January and 13 February). At this time, suburban females had significantly higher plasma protein levels than wildland females (F = 4.90, P = 0.034).

When compared by laying date, plasma protein levels increased as clutch initiation date neared ($F_{5,58}$ = 8.82, P < 0.001; Fig. 14.3b) and females in the two populations exhibited the same pattern. However, we found no differences in plasma protein between the two populations ($F_{1,58}$ = 0.024, P = 0.88) when we controlled for time before laying and no interaction occurred between these factors ($F_{1,5}$ = 0.73, P = 0.60). Because we found no differences with respect to urbanization, we combined habitats for post hoc pairwise comparisons by days before clutch initiation. Plasma protein levels were highest during the sampling period immediately prior to clutch initiation, i. e., from 10 - 1 days before a female laid her first egg.

3.4 Plasma Calcium

Plasma calcium levels increased with calendar date ($F_{4,80}$ = 3.29, P = 0.02; Fig. 14.4a), but no differences existed between populations ($F_{1,80}$ = 1.92, P = 0.17) nor was there any interaction between habitat and calendar date ($F_{1,4}$ = 0.30, P = 0.88). Calcium in females sampled from 18 - 31 March was higher than that of those females sampled between 4 - 17 February.

Plasma calcium levels increased as laying date neared ($F_{5,58}$ = 3.89, P = 0.005), but there were no differences between populations ($F_{1,58}$ = 0.57, P = 0.46; Fig. 14.4b) nor was there any interaction between habitat and laying date ($F_{1,5}$ = 0.91, P = 0.48). Plasma calcium levels during the period immediately preceding laying (1 - 10 days) were significantly higher than all

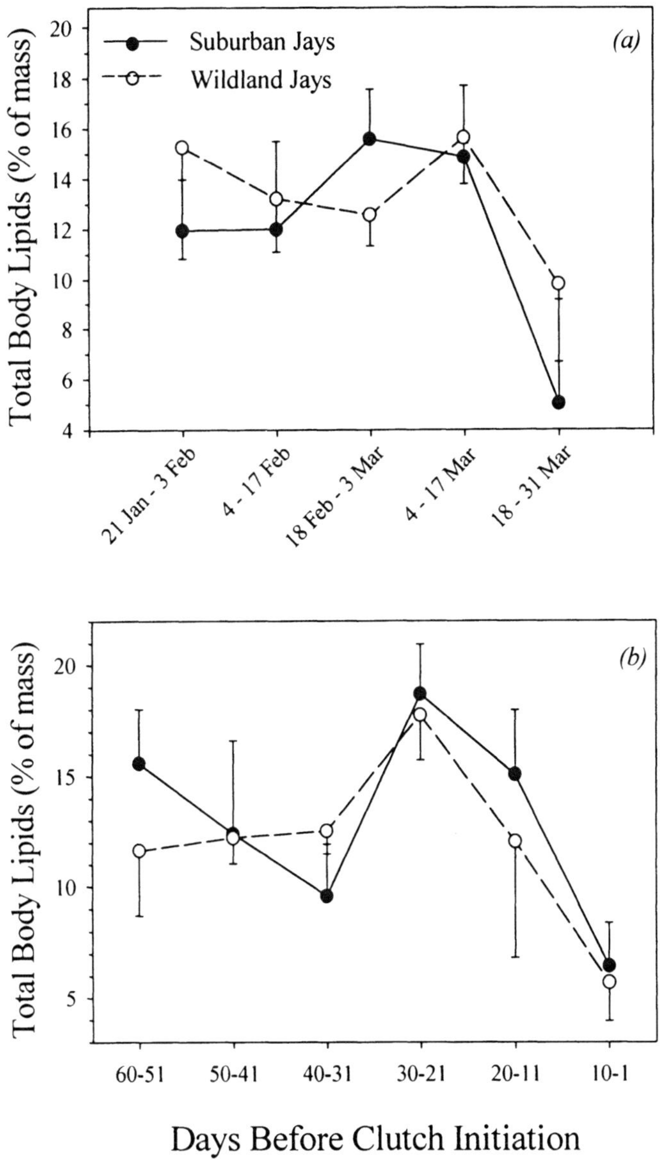

Figure 14.2. A comparison of mean (± SE) total body lipid levels in female breeders in two populations of Florida Scrub-Jays, relative to calendar date (a) and relative to time of sampling prior to laying date (b).

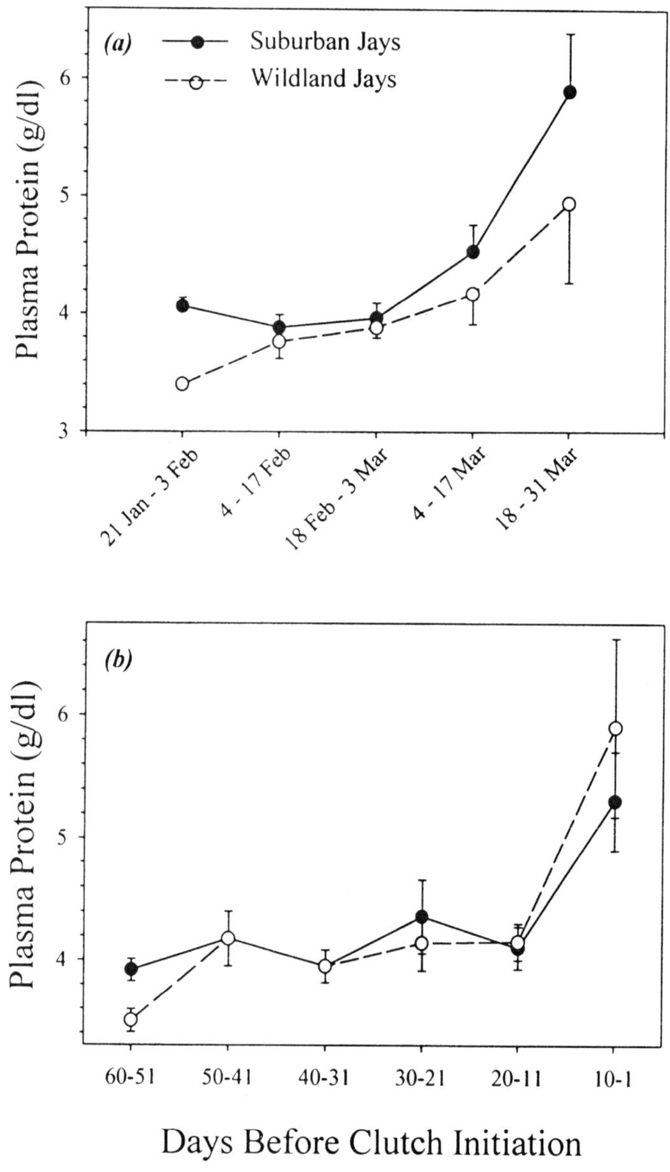

Figure 14.3. A comparison of mean (± SE) plasma protein levels in female breeders in two populations of Florida Scrub-Jays, relative to calendar date (a) and relative to time of sampling prior to laying date (b).

periods with the exception of the sampling period that was next closest to laying (i. e., 11 - 20 days before clutch commencement, $P = 0.056$).

4. DISCUSSION

Our results do not appear to support strongly any of the three hypotheses. Although plasma protein levels were higher among suburban scrub-jays, we found no population difference when we controlled for days prior to laying. The pattern in which plasma protein levels become elevated as laying approaches in many avian species has been shown to reflect an increase in blood levels of vitellogenin (White 1991). This protein which is produced in the liver in response to endocrine signals (primarily estradiol and progesterone) and transported to the ovary via the blood stream, is a major component in avian yolk (Jackson et al. 1977, Deeley et al. 1993, Ho 1991, Carey 1996). Therefore, the increases seen after 24 February in the suburban females and 9 March in the wildland females (see Fig. 2a) probably reflect the onset of vitellogenesis. However, we found a significant between-population difference in protein levels in the females that were sampled very early in our study, prior to 13 February (see Fig. 14.3b). This sampling occurred 39 days before most suburban females and nearly 60 days before most wildland females laid eggs, well before the time when jays from either population would be yolking follicles.

We suggest that this difference reflects the protein-rich and anthropogenically supplemented diets of females in the suburbs. Two of the most common food items in the suburbs are shelled peanuts and pet foods, both of which have relatively high amounts of protein (Fleischer 2000). Acorns, a common dietary item of wildland scrub-jays during the winter also are relatively high in proteins, however, they have high concentrations of tannins which tend to decrease the ability of animals to effectively metabolize proteins. Arthropods are relatively scarce in natural scrub during the winter, thus it seems as if the habitat-specific differences in protein availability may influence differences in the timing of breeding between habitats. The readily available proteins (or the essential amino acids that they contain) in supplemental food may allow females to come into full reproductive state earlier or serve as a signal that the time is appropriate to initiate reproduction.

Calcium is essential for somatic growth in young animals, and later, in homeostasis of bodily functions; in birds calcium also is necessary for egg shell deposition. Wildland scrub-jays may obtain calcium by taking small vertebrate prey, such as small reptiles, mammals, and amphibians

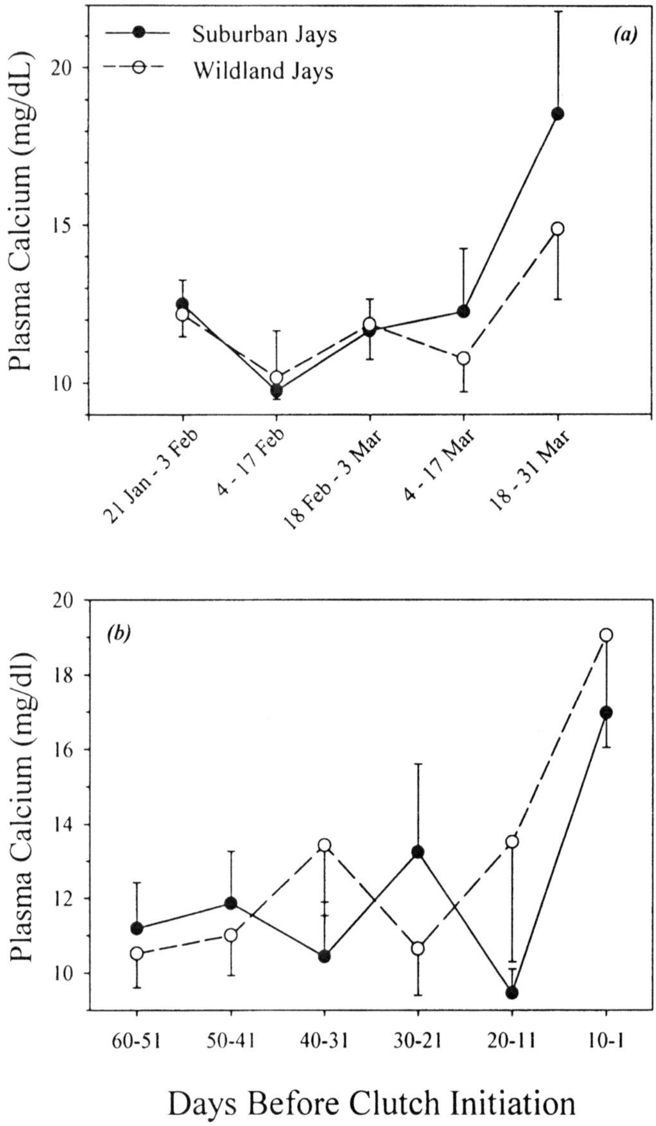

Figure 14.4. A comparison of mean (± SE) plasma calcium levels in female breeders in two populations of Florida Scrub-Jays, relative to calendar date (a) and relative to time of sampling prior to laying date (b).

(Woolfenden and Fitzpatrick 1984, 1990, personal observation). However, the relatively acidic environment of oak scrub may decrease the availability of calcium (Klasing 1988). In the suburbs, calcium is likely more abundant, occurring in pet foods and in the calcareous shell fill upon which most roads in the suburban area are constructed; thus calcium is unlikely to be limiting.

Nevertheless, we found no population differences in plasma calcium levels. The marked increase in plasma calcium levels immediately preceding clutch initiation observed in females from both populations (Fig. 14.4b) clearly shows the movement via the blood stream of endogenously and/or exogenously derived calcium to the oviduct for egg shell formation.

We also found no population differences in total body lipids. This is somewhat surprising because Schoech (1996), who provided jays in their natural habitat with supplemental food, found that supplemented female breeders had higher levels of total body lipids than did controls, but birds were not measured during the same pre-breeding period as in this study. In both populations, total body lipids decreased as laying neared. These data suggest that the differences in timing of breeding between the two populations is not a result in differences in energy availability.

The decline in total body lipids as laying neared is interesting, but not from a specifically urban perspective. Although some large species of waterfowl rely mostly on endogenous energy (i.e., lipids) and nutrient reserves for egg production (Ankney and MacInnes 1978, Ankney and Alisauskas 1991), it is commonly assumed that females of smaller species, such as passerines, must increase their food intake for egg production (Perrins 1970, 1996). The comparison between birds that depend on endogenous stores rather than exogenous dietary intake has been termed capital versus income breeders, respectively (see Drent and Daan 1980, Meijer and Drent 1999). However, such a comparison may be overly simplified since it is likely that females use a combination of capital and income during oogenesis. For example, Williams and Ternan (1999) found that female zebra finches reduce locomotor activity rather than increase food intake during oogenesis. If female zebra finches meet part of the increased energetic or nutritional demands of oogenesis by reallocation of their daily energetic expenditures, it is likely that other species may use a similar strategy. Fleischer (2000) found that suburban scrub-jays spent less time foraging and more time perching, yet increased their food intake rate when compared with scrub-jays in wildland habitats. If true, this potential energy savings does not appear to result in improved body condition.

The spatial and temporal predictability of anthropogenic foods may enable scrub-jays to increase their intake rates, while spending less time searching for and handling these foods (Fleischer 2000). In some birds, the intake rate of certain foods, especially arthropods, appears to be a cue to initiate reproduction (Benkman 1990, Winkler and Allen 1996). In Florida Scrub-Jays, the timing of breeding is not correlated with arthropod availability, but with late winter abundance of herptiles and the size of the previous autumn's acorn crop (Curry, Woolfenden, and Fitzpatrick, unpubl. data). Presumably jays assess the abundance of these foods through intake rates; thus if peanuts, the anthropogenic food most heavily used by jays during this pre-breeding period, serve as an ecological surrogate for acorns, jays may be making breeding decisions based on their availability. Their relatively high quality also may contribute to the earlier breeding by suburban jays.

For many birds early breeding confers a reproductive advantage. Early breeding birds, including Florida Scrub-Jays in natural habitats, often lay larger clutches (Perrins 1965, Boutin 1989, Meijer et al. 1990, Svensson 1995, Schoech 1996, Woolfenden and Fitzpatrick 1984, 1990, 1996, Winkler and Allen 1996). Late-nesting pairs of Florida Scrub-Jays are less likely to produce young and these young are less likely to survive (because of increased predation as the season progresses; see Woolfenden and Fitzpatrick 1984, Schaub et al. 1992). Thus we might expect that access to anthropogenic foods which leads to earlier breeding and larger clutch sizes (Bowman et al. 1998) in suburban scrub-jays might confer a reproductive advantage to these populations. However, scrub-jays almost exclusively feed their nestlings arthropods, so most anthropogenic foods may not be suitable for nestling development. Scrub-jays in wildlands may vary their time of breeding to coincide with seasonally fluctuating arthropod populations. Bowman et al. (1998) found that suburban scrub-jay reproduction was poorest relative to scrub-jays in natural habitats when the difference in laying date between the two populations was greatest. If Florida Scrub-Jays use supplemental food as a cue to initiate breeding, but nest before seasonal resources are available, then supplemental food, rather than mitigating the decrease in availability of natural food (see Bolger 2001), may further reduce its availability.

4.1 Policy and Management Recommendations

Clearly anthropogenic sources of food have an effect on the timing of breeding and clutch size of suburban populations of Florida Scrub-Jays, but the net effect of this influence may be negative rather than positive. The bulk of this food (peanuts and bird seed) is purposely provided for the birds by the human population. Bird feeding is likely to have a variety of direct and indirect benefits to wildlife, including but not limited to increased appreciation for wildlife. Suggesting that bird feeding be prohibited is unwarranted, but perhaps the public should be more aware of the changing seasonal resource needs of birds and provide food appropriate to the life history cycle of birds. Before and during the breeding season, food with readily available proteins might be appropriate, such as several new commercially available bird foods mixed with dehydrated lepidopteran larvae.

Other sources of anthropogenic foods, such as waste, may be of relatively poor quality for birds (see Pierotti and Annett 2001) and their utilization by birds also may have deleterious consequences. In addition, the attraction of mesopredators to such waste foods may have additional negative impacts on bird communities in urban areas. Urban communities should be planned so that wastes are made less available to wild birds and other animals.

ACKNOWLEDGMENTS

We thank all at Archbold Biological Station for providing a stimulating atmosphere in which to conduct research. SJS thanks Ellen Ketterson for providing him laboratory and office space while allowing him the freedom to pursue his research interests. Noah Owen-Ashley provided excellent field assistance in the wildland site and Matt Shawkey, Michelle Dent, and Barbara Beckford Nemes did the same at our suburban site. Thanks to Glenn Walsberg for his kind loan of the "original DickyJohn Ground Meat Analyzer". SJS was supported by NSF grants IBN-9722823 and IBN-9983201.

REFERENCES

Abrahamson, W. G., A. F. Johnson, J. N. Layne, and P. A. Peroni. 1984. Vegetation of the Archbold Biological Station, Florida: an example of the southern Lake Wales Ridge. Fla. Sci. 47:209-250.

Ankney, C. D., and C. D. McInnes. 1978. Nutrient reserves and reproductive performance of female lesser snow geese. Auk 95:459-471.

Ankney, C. D. and R. T. Alisauskas. 1991. The use of nutrient reserves by breeding waterfowl, p. 2170-2176. *In* B. D. Bell, R. O. Cossee, J. E. C. Flux, B. D. Heather, R. A. Hitchmough, C. R. Robertson, and M. J. Williams [EDS.], Acta XX Congressus Internationalis Ornithologici. New Zealand Ornithological Congress Trust Board, Wellington, New Zealand.

Asch, A., and D. D. Roby. 1995. Some factors affecting precision of the total body electrical conductivity technique for measuring body composition in live birds. Wilson Bull. 107:306-316.

Benkman, C. W. 1990. Intake rates and the timing of crossbill reproduction. Auk 107:376-386.

Blair, R. B., and A. E. Launer. 1997. Butterfy diversity and human land use: species assemblages along an urban gradient. Biol. Conserv. 80:113-125.

Bolger, D. 2001. Urban birds: populations, community, and landscape approaches, p. 155-177. *In* J. M. Marzluff, R. Bowman, and R. Donnelly [EDS.], Avian ecology and conservation in an urbanizing world. Kluwer Academic, Norwell, MA.

Bolger, D. T., A. V. Suarez, K. Crooks, and S. A. Morrison. 2000. Arthropods in habitat fragments: effects of area, edge and Argentine ants. Ecol. Appl. 10:1230-1248.

Boutin, S. 1989. Food supplementation experiments with terrestrial vertebrates: patterns, problems, and the future. Can. J. Zool. 68:203-220.

Bowman, R. and G. E. Woolfenden. 2001. Nest success and the timing of nest failure of Florida Scrub-Jays in suburban and wildland habitats, p. 385-404. *In* J. M. Marzluff. R. Bowman, and R. Donnelly [EDS.], Avian ecology and conservation in an urbanizing world. Kluwer Academic, Norwell, MA.

Bowman, R., G. E. Woolfenden, and J. W. Fitzpatrick. 1998. Timing of breeding and clutch size in the Florida Scrub-Jay, *Aphelocoma coerulescens*. Ostrich 69:316.

Carey, C. 1996. Female reproductive energetic, p. 324-374 *In* C. Carey [ED.]. Avian energetics and nutritional ecology. Chapman and Hall, New York.

Deeley, R. G., R. A. Burtch-Wright, C. E. Grant, P. A. Hoodless, A. K. Ryan, and T. J. Schrader. 1993. Synthesis and deposition of egg proteins, p. 205-222 *In* R. J. Etches and A. M. Verrinder [EDS.]. Manipulation of the avian genome. CRC Press, Boca Raton.

Deny, C., and H. Schmidt. 1998. Insect communities on experimental mugwort (*artemisia vulgaris* L.) Plots along an urban gradient. Oecologia 113:269-277.

Drent, R. H., and S. Daan. 1980. The prudent parent: energetic adjustments in avian breeding. *Ardea* 68:225-252.

Fleischer, A. L. Jr. 2000. The influence of time budget and rate of food handling and consumption on the timing of breeding of female Florida Scrub-Jays (*Aphelocoma coerulescens*): A comparison between populations in natural and suburban habitats. MS thesis, University of South Florida, Tampa, FL.

Ho, S-M. 1991. Vitellogenesis, p. 91-126 *In* P. K. T. Pang and M. P. Schreibman [EDS.]. Vertebrate endocrinology: Fundamentals and biomedical implications, Vol. 4. Academic Press, Orlando.

Jackson, R. L., H. Y. Lin, J. T. S. Chan, and A. R. Means. 1977. Estrogen induction of plasma vitellogenin in the cockerel: studies with phosvitin antibody. Endocrinology 101:849-857.

Kenagy, G. J., and B. M. Barnes. 1988. Seasonal reproductive patterns in four coexisting rodent species from the Cascade Mountains, Washington. J. Mammol. 69:274-292.

Klasing, K. C. 1998. Comparative Avian Nutrition. Cab International, New York.

Meijer, T., S. Daan, and M. Hall. 1990. Family planning in the kestrel (*Falco tinnunculus*): The proximate control of covariation of laying date and clutch size. Behaviour 114:117-136.

Meijer, T. S., and R. Drent. 1999. Re-examination of the capital and income dichotomy in breeding birds. Ibis 141:399-414.

Moore, F. L., and L. J. Miller. 1984. Stress-induced inhibition of sexual behavior: corticosterone inhibits courtship behaviors of a male amphibian (*Taricha granulosa*). Horm. Behav. 18:400-410.

Moore, F. L., and R. T. Zoeller. 1985. Stress-induced inhibition of reproduction: evidence of suppressed secretion of LH-RH in an amphibian. Gen. Comp. Endocrinol. 60:252-258.

Morton, J. M., R. L. Kirkpatrick, and E. P. Smith. 1991. Comments on estimating total body lipids from measures of lean mass. Condor 93:463-465.

Mumme, R. L. 1992. Do helpers increase reproductive success? An experimental analysis in the Florida scrub jay. Behav. Ecol. Sociobiol. 31:319-328.

Perrins, C. M. 1965. Population fluctuations and clutch size in the great tit (*Parus major*). J. Anim. Ecol. 34:601-647.

Perrins, C. 1970. The timing of birds' breeding seasons. Ibis 112:242-255.

Perrins, C. M. 1996. Eggs, egg formation and the timing of breeding. *Ibis* 138:2-15.

Pierotti, R., and C. Annett. 2001. The ecology of Western Gulls in habitats varying in degree of urban influence, p. 309-331. *In* J. M. Marzluff, R. Bowman, and R. Donnelly [EDS.], Avian ecology and conservation in an urbanizing world. Kluwer Academic, Norwell, MA.

Roby, D. 1991. A comparison of two noninvasive techniques to measure total body lipid in live birds. Auk 108:509-518.

Sapolsky, R. M. 1987. Stress, social status, and reproductive physiology in free-living baboons, p. 291-321 *In* D. Crews [ED.], Psychobiology of reproductive behavior: An evolutionary perspective. Prentice-Hall, New Jersey.

Schaub, R., R. L. Mumme, and G. E. Woolfenden. 1992. Predation on the eggs and nestlings of Florida scrub jays. Auk 109:585-593.

Schoech, S. J. 1996. The effect of supplemental food on body condition and the timing of reproduction in a cooperative breeder, the Florida scrub-jay (*Aphelocoma coerulescens*). Condor 98:234-244.

Schoech, S.J., R. L. Mumme, and M.C. Moore. 1991. Reproductive endocrinology and mechanisms of breeding inhibition in cooperatively breeding Florida scrub jays (*Aphelocoma c. coerulescens*). Condor 93:354-364.

Schoech, S. J., R. L. Mumme, and J. C. Wingfield. 1996. Delayed breeding in the cooperatively breeding Florida scrub-jay (*Aphelocoma coerulescens)*: inhibition or the absence of stimulation. Behav. Ecol. Sociobiol. 39:77-90.

Siegel, H. S. 1980. Physiological stress in birds. Bioscience 30:529-534.

Svensson, E. 1995. Avian reproductive timing: when should parents be prudent? Anim. Behav. 49:1569-1575.

Wasser, S. K., K. Bevis, G. King, and E. Hanson. 1997. Noninvasive physiological measures of disturbance in the northern spotted owl. Cons. Biol. 11:1019-1022.

Walsberg, G. E. 1988. Evaluation of a nondestructive method for determining fat stores in small birds and mammals. Physiol. Zool. 61:153-159.

White, H. B., III. 1991. Maternal diet, maternal proteins and egg quality. p. 1-15 *In* D. C. Deeming and M. W. J. Ferguson [EDS.], Egg incubation: Its effects on embryonic development in birds and reptiles. Cambridge University Press, Cambridge.

Williams, T. D., and S. P. Ternan. 1999. Food intake, locomotor activity, and egg laying in Zebra finches: Contribution to reproductive energy demand? Physiol. Biochem. Zool. 72:19-27.

Wilson, F. E., and B. K. Follett. 1975. Corticosterone-induced gonadosuppression in photostimulated tree sparrows. Life Sci. 17:1451-1456.

Wingfield, J. C. 1988. Changes in reproductive function of free-living birds in direct response to environmental perturbations, p. 121-148 *In* M. H. Stetson [ED.], Processing of environmental information in vertebrates, Springer-Verlag, Berlin.

Wingfield, J. C. 1994. Modulation of the adrenocortical response to stress in birds, p. 520-528 *In* K. G. Davey, R. E. Peter, and S. S. Tobe [EDS.], Perspectives in comparative endocrinology. Natl. Res. Council of Canada.

Winkler, D. W., and P. E. Allen. 1996. The seasonal decline in tree swallow clutch size: physiological constraint or strategic adjustment? Ecology 77:922-932.

Woolfenden, G. E., and J. W. Fitzpatrick. 1984. The Florida Scrub Jay: Demography of a Cooperative-breeding Bird. Princeton University Press, New Jersey.

Woolfenden, G. E., and J. W. Fitzpatrick. 1990. Florida scrub jays: a synopsis after 18 years of study, p. 241-266 *In* P. B. Stacy and W. D. Koenig [EDS.], Cooperative breeding in birds: Long-term studies of ecology and behavior. Cambridge University Press, Cambridge.

Woolfenden, G. E., and J. W. Fitzpatrick. 1996. Florida Scrub-Jay (*Aphelocoma coerulescens*), p. 1-28 *In* A. Poole and F. Gill [EDS.], The birds of North America, No. 228. The Academy of Natural Sciences, Philadelphia, and The American Ornithologists' Union, Washington, D. C.

Chapter 15

The ecology of Western Gulls in habitats varying in degree of urban influence

Raymond Pierotti[1] and Cynthia Annett[2]

[1]*Division of Biological Sciences, University of Kansas, Lawrence, Kansas 66045 USA, e-mail: pierotti@falcon.cc.ukans.edu*

[2]*Haskell Environmental Research Center, Haskell Indian Nations University, 155 Indian Ave, Box 5001, Lawrence, KS 66044 USA*

Key words: Breeding performance, California, diet choice, garbage dump, habitat selection, human influence, island, *Larus occidentalis*, settlement, Western Gull

Abstract: We studied reproductive performance, diet choice, and habitat selection in Western Gulls, *Larus occidentalis*, over a 22-year period in five colonies along the coast of California. These included colonies with heavy (Alcatraz, Anacapa), moderate (Moss Landing), and relatively little (Southeast Farallon, Santa Barbara Island) urban influence. General aspects of habitat selection were similar on all colonies, with birds seeking nest sites sheltered from the prevailing wind, and generally avoiding grassy areas. Important differences in diet choice and habitat selection were found between urban and non-urban sites, with urban gulls relying more heavily on human refuse as food, and often locating nests adjacent to human debris and artifacts. In all colonies, Western Gulls relied heavily on fish as a primary diet item, especially when rearing offspring, and birds that took predominantly fish and other pelagic prey showed superior breeding performance. Diets of less successful birds varied among sites, with urban birds dependent on human refuse, and non-urban birds being more dependent on other seabirds and scavenging. During major El Nino/Southern Oscillation events, less successful birds on non-urban sites (i.e., Southeast Farallon) flew long distances to obtain human refuse. A captive rearing experiment in which chicks were fed diets comprised of either fish or refuse, revealed that refuse was nutritionally inadequate for normal chick development. In all colonies, only a small percentage of pairs were successful as first-time breeders, and these pairs continued to be successful over many years. In contrast, most pairs were unsuccessful in their first year, only breed for 1-2 years, and showed poor performance, on both an annual and long-term basis.

1. INTRODUCTION

Gulls of the genus *Larus* are probably the seabirds most familiar to humans dwelling in urbanized regions along major bodies of water in the Northern Hemisphere. It is generally assumed that gulls have undergone massive population explosions during the 20th century, and that this has been largely the result of urbanization and the effects of increasing human populations (Kadlec and Drury 1968, Graham 1975). This assumption is based on the idea that gulls depend heavily on human refuse as a food source during the breeding season, and that the increase in refuse accompanying human population growth provided an abundant new source of food which fueled the gull population explosion (Kadlec and Drury 1968, Hunt 1972, Pierotti and Good 1994).

Recent work on the relationship between diet and reproduction (Pierotti and Annett 1987, 1990, 1991, Annett and Pierotti 1989, 1999) has brought this assumption into question and, as a result, the mechanism underlying the supposed high rate of growth in gull populations needs re-examination. A large number of factors may have had both positive and negative impacts on gull population growth. For example, as human populations have increased along coastlines over the past few hundred years, there has been extensive loss of nesting habitat for gulls, mainly because of the destruction of wetlands and increasing human use of near-shore islands. Nesting habitat has also been degraded because of the many shipping lanes adjacent to offshore nesting islands; these lanes pose a constant threat of both major oil spills and chronic low level contamination. Few well documented records exist of the size of gull populations prior to the 19th century. Thus, we are unsure if gull populations at present are unusually high, or simply have recovered to ancestral population sizes after centuries of persecution by humans (Graham 1975, Pierotti and Good 1994).

The existing paradigm is that gulls depend heavily on human refuse dumps and fish processing plants for their sustenance (Kadlec and Drury 1968, Hunt 1972). However, gulls existed for 40-50 million years before humans (Feduccia 1980), and the ability of gull populations to increase may be better related to their natural history than to any dependence on humans. Gulls are generalized foragers, capable of exploiting a wider range of food types than other aquatic birds (Ashmole 1971, Pierotti and Annett 1995). For example, gulls are better at long distance flight and aerial maneuvering than alcids or cormorants, and gulls are better plunge divers than frigatebirds and skuas (pers. obs.). In fact, with the exception of underwater pursuit, gulls are known to use every foraging mode found in other seabirds (Ashmole 1971).

In this chapter we discuss results of 22 years (1973-1994) of research on the Western Gull, *Larus occidentalis*, in a variety of nesting colonies and habitats. The historical center of distribution of Western Gulls appears to have been around San Francisco Bay. In addition to what is now the largest colony, Southeast Farallon Island, offshore of San Francisco, several islands within San Francisco Bay were probably major colonies before they were extensively developed by humans. For example, Alcatraz and Yerba Buena Islands were known historically for their large concentrations of seabirds when the Spanish first settled the region (Schoenherr et al. 1999). Tiburon Island may also have harbored significant numbers of gulls. It is possible that, prior to the arrival of Europeans, the indigenous peoples of California took seabirds and their eggs for food on these islands. However, we are aware of no data that could support or refute this point. Regardless of the possibility of such disturbance, it is possible that prior to the arrival of Europeans, these islands may have rivaled Southeast Farallon in size and importance as seabird breeding colonies.

We summarize results from our studies of the reproductive performance, diet, and habitat selection in five major colonies along coastal California, all of which have undergone urbanization to some degree. We conclude with a comparison of diets and reproductive success on Alcatraz and Southeast Farallon Islands, which are less than 50 km apart, but differ greatly in overall ecology and their degree of urban impact.

2. METHODS

2.1 Study Sites

Alcatraz Island (37° 49' 34" N, 122° 25' 20" W; Fig. 15.1A), located in San Francisco Bay, within 2 km of the city of San Francisco, had the greatest amount of urban influence. Alcatraz appears to have been a major seabird colony prior to the arrival of Europeans (Schoenherr et al. 1999). A military fort was established on the island after European settlement in the 16th century, and when the U.S. conquered California in the 1850's, Alcatraz was converted into a federal military prison. Early in the 20th century, Alcatraz was converted into one of the most notorious federal penitentiaries, operating until 1964 when it was abandoned by the federal government. From 1964 until November 1969, the island was unoccupied and Western Gulls probably nested without major human disturbance for the first time in

A

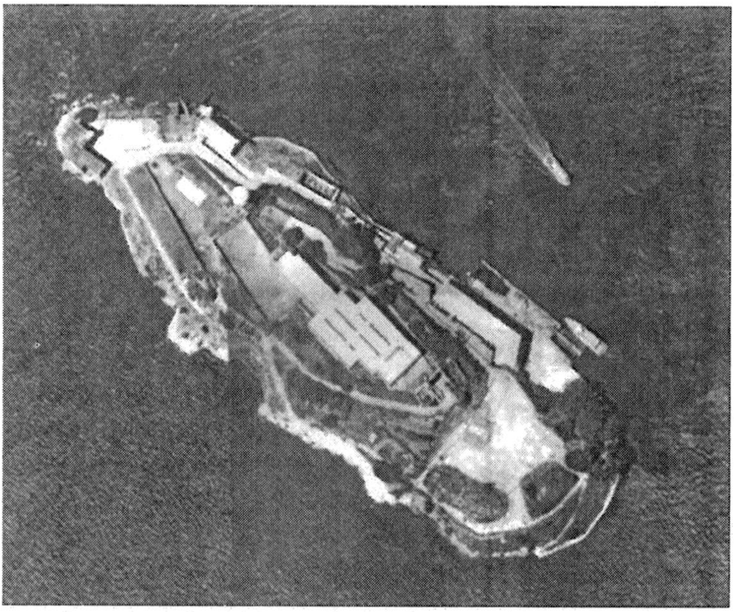

B

Figure 15.1. Aerial photographs of (A) the most urbanized study site (Alcatraz Island) and (B) the second least urbanized study site (Southeast Farallon Island). Note that in (B), development is limited to two houses, indicated by the arrow, and a winch station.

over 400 years. On 9 November 1969, Native American activists occupied Alcatraz and remained in occupation until June 1971 (Fortunate Eagle 1992). During the Native American Occupation, gull eggs were used to supplement limited food supplies, and only a few gulls were able to nest successfully (A. Fortunate Eagle, pers. comm.).

After the removal of the Native American occupants, seabirds on Alcatraz have been largely undisturbed by human activities. As a result, Western Gulls nested over much of the island and their population has continued to expand despite the hundreds of thousands of visitors that come to Alcatraz annually to tour the prison under the auspices of the National Park Service (NPS). In the 1980's, Pigeon Guillemots (*Cepphus columba*), and Pelagic Cormorants (*Phalacrocorax pelagicus*) began to breed in small numbers on Alcatraz, and in the 1990's, Brandt's and Double-crested Cormorants (*P. penicillatus* and *P. carbo*) also established small nesting colonies (Pierotti, unpubl. data).

We consider East Anacapa Island (34° 00' 41" N, 119° 25' 26" W), approximately 20 km offshore from Ventura, California, to be the next most urban influenced of our five study colonies. East Anacapa was occupied by a small ranch, as well as a United States Coast Guard (USCG) light station which was staffed until the 1960's (Schoenherr et al. 1999). Currently part of Channel Islands National Monument, East Anacapa is the most heavily visited of the Channel Islands, and is the only Western Gull nesting colony where visitors are permitted to wander through areas where gulls nest (pers. obs.). In addition, because of the island's proximity to the coast, gulls travel to areas of intense human activity and dense human populations to forage.

Moss Landing (36° 49' 04" N, 121° 46' 30" W), located midway between Santa Cruz and Monterey along the curve of Monterey Bay, had a Western Gull nesting colony along dikes in a commercial brine shrimp enterprise adjacent to California Highway 1. Although near human habitations and located on human-created habitat, this colony appeared to have less urban influence than Anacapa for two reasons. First, the brine shrimp operation was closed to the public and rarely visited by humans, and during our study in 1981-1983, even the brine shrimp operation had shut down. This area was disturbed so little by humans that Caspian and Forster's Terns (*Sterna caspia* and *S. forsteri*), which are highly susceptible to disturbance, nested successfully in the ponds during our study. Second, the salt ponds were located adjacent to Elkhorn Slough, historically the mouth of the ancient Salinas River, and currently the head of the Monterey submarine canyon. This is an area of major upwelling that results in high productivity, providing a rich supply of natural food for the colony (Bellrose 1983).

Southeast Farallon Island (SEFI) (37° 41' 53" N, 123° 00' 05" W; Fig. 15.1B), located in San Francisco County, California, but approximately 60 km west of the Golden Gate, has historically experienced considerable human impact. In the 18th and 19th centuries the major pinniped breeding populations on SEFI were nearly extirpated by hunters that also caused massive disruptions of the seabird colonies adjacent to the pinniped breeding beaches (Ainley and Boekelheide 1990). Between 1850 and 1890, seabird colonies were directly impacted by egg collecting for commercial markets in San Francisco, when hundreds of thousands of seabird eggs were taken annually (Ainley and Boekelheide 1990, Schoenherr et al. 1999). However, during the past hundred years SEFI has suffered less urban impact than the three study colonies discussed above. The USCG Light Service maintained 15-20 people on the island from 1855 until 1965 (Ainley and Boekelheide 1990), and extensive trails were constructed throughout SEFI to aid USCG personnel in moving equipment around the island, including a cement path for a cart that ran on rails. In addition, from the early 1900's until the 1940's, oil tankers routinely flushed their tanks near SEFI before entering San Francisco Bay, and several major oil spills occurred in the area as a result of oil tankers running aground (Ainley and Boekelheide 1990). Seabird populations have been recovering on SEFI since the 1960's. Point Reyes Bird Observatory (PRBO) has maintained a research crew on SEFI since the late 1960's, but the Western Gull population appears well acclimated to most activities by these individuals (Ainley and Boekelheide 1990, R. Pierotti, unpubl. data). Aside from PRBO and USCG personnel, humans have been kept off SEFI, and the island is sufficiently isolated from human populations that human refuse is generally rare in the diets of SEFI Western Gulls (Pierotti 1981, Pierotti and Annett 1995, Ainley and Boekelheide 1990).

Of our study colonies, Santa Barbara Island (SBI) (33° 28' 37" N, 119° 02' 03" W), located in the Channel Islands off Southern California approximately 85 km from Port Hueneme, appears to have experienced the least urban impact in recent times. Although SBI was a sheep ranch in the late 19th and early 20th centuries, the ranching operation did not require much of a human presence on the island because the island lacked predators on sheep (Schoenherr et al. 1999). SBI is now open to the public, but being farther offshore, it is much less frequently visited than East Anacapa. The only major trail on the island does not encroach upon Western Gull nesting areas. Like SEFI, SBI is sufficiently far offshore that Western Gulls have little human refuse in their diet; however, unlike SEFI, constant research presence does not occur on SBI (Hunt and Hunt 1975, Pierotti 1981, Pierotti and Annett 1995).

With the exception of Alcatraz, all of these areas would be considered wildlands since there are few buildings and less than 10 human residents ha^{-1} (Marzluff et al. 2001). Given the extensive development (large areas of the island are covered with concrete and buildings) and large number of daily visitors, Alcatraz should probably be classed as urban habitat. However, this is complicated `by the fact that Alcatraz has no permanent human residents, and from 1800 to 0900 (15 hr day^{-1}) there is only a single watchman present.

2.2 Field Methods

In all colonies the research protocol was similar. Active nests were marked and numbered the first time an egg was found with flagging tape, numbered stakes, or spray paint on rocks next to nests. We quantified nest-site characteristics (such as dominant vegetation and substratum), nest composition, presence of natural screens adjacent to nests, and nearest neighbor distance (except on Alcatraz and Anacapa, see below).

Temporary observation blinds were constructed on study colonies for focal animal observations of a subset of marked nests. We collected data on time budgets of breeding pairs, frequency and contents of mate feedings by males, frequency and contents of chick feedings by both parents, territory attendance by both parents, territorial interactions with neighbors, patterns of brooding, and responsiveness to disturbances (see Pierotti 1981, 1987 for more detailed descriptions of methods).

Reproductive success was assessed using several variables (see Pierotti and Annett 1991, Annett and Pierotti 1999 for details). Clutch size was determined during regular nest checks initiated prior to egg laying and continued throughout the laying period. During initial and subsequent nest-checks, we marked and numbered eggs in order of laying with a marker pen. We measured egg length and width to the nearest mm and weight to the nearest g. Hatching success was estimated by determining the number of eggs observed to hatch at each nest during nest-checks. Chicks were banded with both U.S. Fish and Wildlife Service aluminum bands and, in some cases, color bands for individual recognition. Chick growth rates were monitored by weighing chicks to the nearest 1-2 g. Fledging success at marked nests was determined by assessing the proportion of chicks which fledged (reached 500 g) at each nest.

Data on both diets and reproductive success represent the results of several years' study except on Anacapa, where data were collected only during a single year (1987) (Table 15.1). Diet data (Figs. 15.2A-C) were from representative years (except Anacapa), and no results from unusual years (e.g., El Nino/Southern Oscillation [ENSO] events) were included in the figures, or were used except for purposes of comparison among years on the same colony.

2.3 Chick-rearing Experiments

To explore factors leading to diet choice by breeding adults, we reared 28 Western Gull chicks from eggs. The eggs were removed from the Parade Grounds on Alcatraz in 1986 as part of a project to minimize breeding by gulls in this area (see habitat manipulations in Results below). These eggs were hatched in an incubator in the Department of Zoology on the University of California, Berkeley campus. Two groups of chicks, each consisting of 14 chicks, were reared on controlled diets. One group was reared on fresh anchovies (fish), the primary food type fed to young gull chicks by their parents. The other group was raised on a diet of commercial chicken (including pieces of bone), the primary component of human refuse consumed by adult Western Gulls on Alcatraz (Annett and Pierotti 1989, 1999). Chicks were kept in large outdoor enclosures at the University of California, Berkeley Animal Behavior Station and fed three times daily until satiated.

3. RESULTS

3.1 Patterns of Habitat Use

Several obvious patterns in territory and nest site choice emerged when these five colonies were compared. On islands with little urban influence, Western Gulls chose sites in habitat with little or no vegetation. On Southeast Farallon Island (SEFI), gulls nested in rocky areas dominated by Farallon weed (*Lasthenia minor maritima*), or areas of exposed rock and

Table 15.1. Reproductive performance of Western Gull pairs on colonies varying in urban influence (major ENSO events are not included). Data are annual means.

Colony[a]	Clutch size	Hatch success	Fledge success
1. Santa Barbara Island (1972 & 1975)	2.7-2.7	2.5-2.7	2.2-2.3
2. Southeast Farallon Island (CS 1971-1983 & HS/FS 1971-1974)	2.5-2.8	2.2-2.5	1.9-2.1
3. Moss Landing (1981-1983)	2.7-2.9	0.0-2.1	0.0-1.3
4. Anacapa Island (1987)	2.6	2.1	1.2
5. Alcatraz Island (1984-1988)	2.7-2.8	2.1-2.4	1.9-2.0

[a]Ranked by urbanization intensity (1 = least urban, 5 = most urban).

sand. Even though nesting sites were in short supply on this colony (15-16,000 pairs of Western Gulls on less than 40 ha), in the early 1970's gulls did not nest in areas covered by brome grass (*Bromus* spp.). As a result of the high density of nests, the average size of breeding territories was only 10 m^2 (SD = 3.9 m^2), with a range of territory sizes from 4-20 m^2 (Pierotti 1981). On Santa Barbara Island (SBI), Western Gulls nested in areas dominated by iceplant (*Mesembryanthemum* spp.) or on bare ground, and nesting territories were much larger than on SEFI (mean = 150 m^2; range 24-803 m^2), with inter-nest distances ranging from 1.8 to 21.0 m (mean = 10.9 m, SD = 4.9 m; Hunt and Hunt 1975). SBI had a much smaller gull population on a much larger island (~1500 pairs on a 260 ha island; Pierotti 1981), which may have contributed to the relatively large territory sizes. As on SEFI, gulls avoided nesting in areas dominated by brome grasses, and instead nested in areas covered by low growing species such as introduced iceplant.

On the small Moss Landing gull colony (50-60 pairs), some pairs nested on man-made dikes that separated the ponds used by a commercial brine shrimp facility, either adjacent to grasses (*Pennisetum* and *Spergularia*) and pickleweed (*Salicornia* spp.) or on unvegetated dikes. A few pairs nested on dead stumps in the middle of salt ponds. Inter-nest distance was large in this colony, because the arrangement of nests was linear, with territory length essentially equal to territory size. On dikes lacking vegetation, mean inter-nest distance (1981 mean = 29.5 m, SD = 22.9 m; 1982 mean = 27.3 m, SD = 24.0 m) was significantly longer than on dikes with vegetation that provided structure and shelter around nests (1981 mean = 13.6 m, SD = 9.4 m, 1982 mean = 11.7 m, SD = 8.7 m; Bellrose 1983).

Inter-nest distances were not measured on the most urban-influenced islands (Anacapa and Alcatraz) because nest placement was heavily influenced both by human activities and man-made structures. Gulls on Anacapa typically nested in areas dominated by low growing plants (e.g., iceplant, and in areas where landslides had occurred, creating large areas of exposed soil). In contrast, Alcatraz has extensive areas covered by concrete and buildings, much of which are in a poor state of repair, which may constrain the placement of gull nests.

On Alcatraz, gulls nested on any area that was exposed, had little or no vegetation, contained obstructions which could serve as windbreaks, and which was not regularly (i.e., several times per day) disturbed by human visitors. For example, gulls nested on the roof of the federal prison building, in the sheltered corners of the roof and next to drain structures or skylights, which anchored and sheltered the nests. Reproductive success on the prison roof was typically poor because of a combination of disturbance by workmen, and a lack of shelter for chicks once they left the nest, which left chicks exposed to heat reflected from the roof. Gulls nested next to structure on the roofs of other buildings as well; nests on the Prison Laundry and the Model Industries Building were less disturbed and better ventilated than those on the prison roof, resulting in significantly higher reproductive success.

Gulls nested in high abundance on the Parade Grounds, which had originally been the paved quadrangle between buildings that housed prison guards and their families. These buildings also housed Indians during the occupation, and after termination of the Indian occupation these buildings were bulldozed by the federal government. As a result, the Parade Grounds are currently surrounded by a ring of large rubble piles filled with broken glass, fiberglass insulation, pieces of twisted metal, and broken concrete. Most nests on the Parade Grounds were built in cracks in the concrete surface of the paved quadrangle, especially near patches of vegetation (0.5-1.0 m in height). Nests in this habitat typically had good hatching success but low fledging success, a pattern resulting from the combination of a general absence of shelter for chicks, and disturbance due to regular patrols by NPS personnel, which resulted in many chicks being attacked and killed by neighboring adults. Nesting was also disrupted whenever the Parade Grounds were used as a landing area for helicopters during emergencies.

The most natural-appearing nesting habitats on Alcatraz were in the cistern area west of the prison exercise yard, and the rocky marine terraces that made up the northern and western periphery of the island. In these areas, gulls nested on natural rock and next to vegetation on exposed soil and rock. On Alcatraz, gulls did nest in, and adjacent to, brome grasses, possibly because in many areas this was the only available material providing windbreaks for nests. However, even in these areas gulls nested around and under man-made objects, including under collapsed guard towers. The cistern was disturbed only during periodic visits by park rangers and researchers, resulting in low disturbance that contributed to high nesting success. The cistern was the main study area used in our long-term research program (Annett and Pierotti 1989, 1999).

3.1.1 Habitat Alterations

When we initiated our study on Alcatraz in 1983, many parts of the island were not used by gulls for nesting because human alterations or structures made the habitat inappropriate for their use (e.g., areas surrounded by high fences topped with barbed wire or an area of concrete over the cistern that provided no relief from wind). In contrast, areas such as the Parade Grounds had relatively high nest density but conflicted with activities of the NPS. As a result, we were asked to develop methods to alter the location of gull nesting. Specifically, the NPS was interested in moving gull nesting activity away from areas of heavy human use while inducing them to nest in areas where such conflict would be minimal.

With the help of NPS personnel and the California Conservation Corps, we developed a program of habitat alteration. One of the key manipulations consisted of removing chain link fences and barbed wire from the southern and northwestern periphery of the island. Additionally, in 1985 we instituted a program of vegetation removal from the Parade Grounds, which combined with patching the cracks in the concrete surface, created a smooth, homogeneous habitat without structure that could act as windbreaks. The results of these habitat alterations were dramatic. The number of pairs nesting on the Parade Grounds decreased from 42 in 1983 to 10 in 1986, reaching an eventual low of four nests by 1988. Similarly the number of pairs nesting in areas previously surrounded by chain-link fencing increased from seven in 1983 to 58 by 1990. However, the NPS stopped cutting the vegetation on the Parade Grounds in the late 1980's and by 1992 the number of pairs nesting on the Parade Grounds had increased back to 52. All nesting occurred adjacent to newly grown patches of vegetation.

To enhance nesting habitat on the cistern, we constructed six rings of broken concrete on the smooth concrete surface. Four of the rings were occupied during the first season, and all six were occupied by nesting gulls within two seasons after they were constructed. Four pairs used concrete rings for their nest sites for several years, and only one ring has gone unused for more than a year since they were constructed.

These findings support other experiments on Western and Herring Gulls (*Larus argentatus*) that demonstrate that cover or relief adjacent to a nest is an essential feature of successful nesting habitat (Pierotti 1976, 1982, Good 1998). Most places where gulls nested successfully on Alcatraz, (the Cistern and Periphery), were structurally similar to nest sites on SEFI; sites were in open, non-vegetated habitat with obstructions 0.5 to 1.0 m in height next to nests that provide shelter from the wind. Experimental removal of obstructions next to nests resulted in these nests blowing away (Pierotti 1981, 1982, this study). Wind speeds at ground level in nests were measured, and found to be significantly lower than wind speeds 0.5 m above nests (E. Martins, pers. comm.).

3.2 Patterns of Diet Choice

Both natural prey and human refuse are consistently available to birds on all colonies (Pierotti and Annett 1987, 1990, 1991, Annett and Pierotti 1999). Natural foods, such as pelagic fishes and invertebrates, and intertidal prey, are regularly available on a long term basis (days, weeks), but unpredictable in the short term (hours), because of the ephemeral nature of prey aggregations at the surface and fluctuations in tidal rhythms (Pierotti and Annett 1987, 1991, Annett and Pierotti 1989, 1999). In contrast, human refuse, especially at garbage dumps, is consistently available on both a long and a short-term basis but requires travel to mainland areas for foraging (Pierotti and Annett 1987, 1991, Annett and Pierotti 1989, 1999). As a result, the major difference between colonies with regard to foraging activities is in travel times to mainland areas where human-generated food (e.g., refuse) can be obtained.

Because of the extensive travel time associated with foraging on human refuse, gulls on the more remote, less urban-influenced islands such as SEFI and SBI, took primarily natural prey. On SEFI, Western Gull diets consisted almost exclusively of fish and pelagic marine invertebrates, including euphausiids and squid (Pierotti 1981, Ainley et al. 1990). However, during ENSO years, when natural prey were less available, SEFI gulls took a substantial amount of human refuse (Ainley et al. 1990). On SBI, which is located in the less productive Southern California Bight, male gulls took mostly larger fish, and females took small fish and kelp bed invertebrates (Hunt and Hunt 1976, Pierotti 1981). Human refuse was rarely observed in gull diets on SBI, and was probably taken only during periods of low marine productivity.

At Moss Landing, the diet of Western Gulls consisted almost entirely of natural prey. Prey consisted of fish, bivalves, and crabs early in the breeding season, but mainly of fish and pelagic invertebrates (squid, euphausiids) later in the breeding season (Fig. 15.2A) (Bellrose 1983). As on SEFI and SBI, human refuse was rarely observed as food on this colony. When refuse did occur it was early in the breeding season (May), prior to chick hatching, or late (July) after pairs had lost chicks. There was, however, some terrestrial influence on the diet of Moss Landing gulls. Bird eggs and small mammals (e.g., *Microtus*) were taken as prey early in the breeding season (Fig. 15.2A). Also, phalaropes (*Phalaropus* spp.), which foraged in the Moss Landing ponds during spring migration, were occasionally taken by gulls.

In contrast, gulls in the heavily urban-influenced colonies on Anacapa and Alcatraz had diets that were strongly impacted by nearby human populations. On Anacapa, the primary food items in the diet in terms of overall abundance were human refuse, pelagic red crabs (*Pleuroncodes* spp.), and fish (mostly anchovies, *Engraulis* spp.) (Fig. 15.2B). These items varied in abundance in gull diets over time, with refuse predominating during the pre-hatching period (late April through May). After chicks hatched (late May through late June), fish were much more abundant in the diet and remained a major food item throughout the remainder of the breeding season (Fig. 15.2B). Red crabs were found consistently throughout

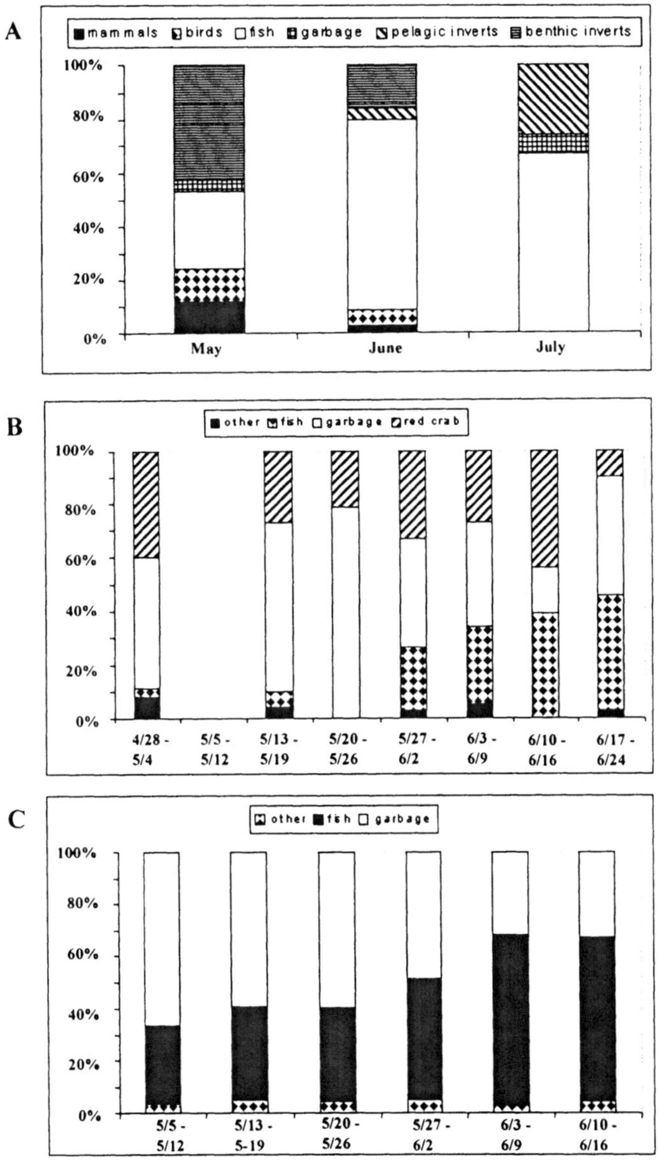

Figure 15.2. Food items found in the diets of Western Gulls in relation to date. Percentage of all items delivered by adults to chicks during the breeding season are presented for representative years (see methods) at (A) Moss Landing colony during the 1982 breeding season, (B) Anacapa colony during the 1987 breeding season, and (C) Alcatraz colony during the 1988 breeding season.

the breeding season, only declining in late June when chicks were well grown.

On Alcatraz, a similar pattern prevailed, with human refuse being the primary food taken throughout most of the early part of the breeding season, especially the incubation period. Refuse declined sharply in abundance when chicks hatched, but became more abundant in diets again when chicks were well grown in late June (Fig. 15.2C). Fish, which consisted primarily of anchovies (*Engraulis* spp.), white croaker (*Gyneonemus lineatus*), plainfin midshipman (*Poricthys notatus*), spotted cusk eel (*Chilara taylori*) and jacksmelt (*Atherinops affinis*), showed a pattern the opposite of that observed for refuse. Larger fish were relatively abundant in the diet during the period of courtship and egg laying (early May), but declined sharply once eggs were laid (mid-late May). However, once chicks hatched a sharp increase in fish occurred, especially smaller species (Fig. 15.2C) (see also Annett and Pierotti 1989). As further indication of the urban influence on the Alcatraz colony, the third most common food category observed after refuse and fish was adult Rock Doves (*Columba livia*), which were taken along the San Francisco shoreline (e.g., Fisherman's Wharf and the Marina). Certain male gulls forced adult Rock Doves into the water where they were killed and eaten.

3.3 Reproductive Performance

Mean clutch size was similar on all study colonies during most years; however, hatching success varied, with more urban-influenced colonies showing reduced hatching success (Table 15.1). Over a 13-year period, mean clutch size of pairs on SEFI ranged from 2.5 to 2.8 (Table 15.1: Pierotti 1981, Penniman et al. 1990). Over a 5-year period, hatching success ranged from 2.2 to 2.5 (Pierotti 1981). On SBI during 1972 and 1975, mean clutch sizes were comparable to those observed on SEFI; however, SBI gulls had the highest hatching success of any study colony (Table 15.1). Moss Landing gulls also produced clutch sizes comparable to both SEFI and SBI during 1981-1983 (and the largest eggs of any colony). However, Moss Landing gulls had very low mean hatching success as a result of egg predators that took from 33-100% of the eggs laid on this mainland colony over our study period (Pierotti and Bellrose 1986). Depredation occurred at night, implicating striped skunks (*Mephitis mephitis*) and raccoons (*Procyon lotor*) from a relatively distant nature reserve. On more urban influenced colonies, mean clutch size was comparable to that of less urban-influenced colonies; however, hatching success was slightly lower on Alcatraz than on SEFI and SBI, and much lower on Anacapa (Table 15.1).

Fledging success, which is a better measure of fitness than clutch size or hatching success (Annett and Pierotti 1999), revealed important differences between colonies, and some marked inter-year effects were found that are not apparent from the data on clutch size and hatching success. Gulls on SBI had the highest mean fledging success; however, SEFI and Alcatraz were comparable to SBI with the exception of ENSO years when no data were collected on SBI (Table 15.1). During ENSO years of high water temperature and low productivity, Alcatraz birds fledged only 1.6 chicks $nest^{-1}$ (1983). SEFI birds fledged 1.5 chicks $nest^{-1}$ during the 1973 ENSO, and only 0.65 chicks $nest^{-1}$ during the 1983 ENSO (Penniman et al. 1990). During the 1983 ENSO event, Alcatraz gulls did better than SEFI gulls, which indicates that more urban-influenced colonies can do better under certain conditions (i.e., when natural foods are reduced). However, gulls at Anacapa had very poor overall reproductive performance, and had lower fledging success during a typical year than did Alcatraz or SEFI during some ENSO years. Despite production of large eggs and clutches and an excellent food supply, Moss Landing had the overall poorest fledging performance because of depredation on eggs and chicks by terrestrial mammalian carnivores (Table 15.1; see also Bellrose 1983).

Overall, it seems likely that the most important influence on variation in reproductive performance is variation in quality of food used by breeders rather than direct interference by humans. Gulls on urban-influenced colonies such as Anacapa showed reduced hatching and fledging success because they took lower quality food such as human refuse. Similarly reduced hatching success on Alcatraz can also be attributed to the higher frequency of human refuse in the diet, especially since when Alcatraz gulls switched to feeding their chicks on fish their fledging success was comparable to that of less urban-influenced colonies (Table 15.1).

In a similar vein, it appears that the major impact of ENSO years is through its effect on diet quality which effects fledging success, not clutch size or hatching success. There were no major differences in clutch size nor hatching success between ENSO and other years on the colonies for which we had data (SEFI, Alcatraz). In contrast, fledging success was much lower in ENSO years than during more normal years for marine productivity, probably because adult birds were unable to provision their chicks with adequate amounts of food of appropriate nutritional value to allow normal chick survival.

3.4 Effects of Urbanization on Reproduction and Diet

Our results indicate that perhaps the most important factor in reproductive success in Western Gulls is diet, with the exception of mainland colonies that are accessible to terrestrial predators. In all colonies, and during all years, birds that took natural food showed superior breeding performance, whereas birds that took human refuse as a major component of their diets showed much lower fledging success (Pierotti and Annett 1987, 1990, 1991, Annett and Pierotti 1999). However, reproductive success varied among individuals within colonies. Levels of human disturbance were probably similar for birds taking natural and human-generated food, and thus disturbance cannot account for differences in individual breeding performance.

However, regardless of study colony we found that diets taken by less successful birds varied to a much greater degree between urban-influenced and less urban-influenced sites than did diets of successful pairs. In urban-influenced colonies birds that had low breeding success depended heavily on human refuse as their primary food. In contrast, birds in colonies less influenced by urban centers depended more heavily on other seabirds and scavenging as their primary food. This pattern was most clear during the major ENSO events of 1978 and 1983, when gulls on SEFI were much more dependent on human refuse. During those two years, 30-40% of the diet of gulls rearing chicks on SEFI consisted of human refuse, and these were the two poorest years for reproductive success (mean fledging success = 1.0 in 1978 and 0.7 in 1983; Penniman et al. 1990). This probably resulted because SEFI gulls foraging on refuse were making a daily round-trip of well over 100 km simply to obtain food.

We compared our data from over 100 individually-known pairs of gulls followed on Alcatraz (Annett and Pierotti 1999) to results published by PRBO on known-age individuals on SEFI (Pyle et al. 1991), and found somewhat similar patterns. On Alcatraz, a small percentage of pairs are successful as first-time breeders and continue to be successful over many years. These individuals take a high percentage of fish in their diets on a consistent basis. However, most breeding pairs were not successful during their first year and were unsuccessful over the long term, because most pairs breed for only 1-2 years before disappearing, never to be seen again. Less successful pairs produced smaller clutches of smaller eggs, and hatched and fledged fewer chicks than did more successful pairs, even on an annual basis (Annett and Pierotti 1999). Western Gulls on SEFI showed a similar pattern, with a large percentage of individuals having less than two years of breeding experience, with a sharp decline in the number of individuals with three or more years of breeding experience (Figs. 1-4 in Pyle et al. 1991).

We cannot examine the precise relationship between patterns of diet choice and breeding performance on SEFI more closely, since neither Pierotti (1973-1975) nor the staff of P R B O collected data on diets of individual pairs. However, on Alcatraz, where we have collected such data, the relationship between patterns of diet choice and breeding performance is very apparent. A strong positive correlation exists between overall percentage of fish in the diet of a pair and their reproductive success, both on an annual and a long-term basis (Annett and Pierotti 1999). As a result, it appears that breeding gulls which rely heavily or exclusively on human refuse as food have low reproductive success and short lifespans. As a consequence, gulls that are refuse specialists typically show very low lifetime fitness.

3.5 Chick-rearing Experiment

We investigated the relationship between breeding success and diet choice in more detail during our chick-rearing experiment. Naive, captive gull chicks fed readily upon both chicken and anchovies, and both gained mass on these diets. At the age of five days, chicks reared on the chicken diet were significantly heavier than chicks reared on the anchovy diet (mean mass for chicken = 119.4 g, SD = 17.2 g; anchovy mean = 93.5 g, SD = 13.9 g; $t_{(16)} = 3.23$; $P < 0.05$). Chicks fed chicken regurgitated bones undigested, whereas chicks reared on fish appeared to digest all fish bones.

However, by age 10-14 days, the oldest five chicks reared on the chicken diet became lethargic and would not eat. Examination of these chicks revealed that their beaks and legs had become flexible and rubber-like, and three of these chicks died. Necropsy and skeletal examination revealed that these chicks had improper skeletal development, especially in the skull (which was incompletely ossified), the breastbone (which was small and flexible), and the legs (Fig. 15.3). The remaining chicks on the chicken diet were immediately switched to a diet of anchovies combined with calcium supplements, and all survived to fledging. However, three of these chicks still developed some abnormalities in their feet, although this did not appear to affect either their growth or their ability to survive to fledging. None of the chicks that were initially raised on a diet of anchovies experienced any growth or developmental problems. For these reasons, we consider a diet consisting entirely of chicken to be nutritionally inadequate for gull chicks. Because chicken may be a major component of the refuse diet at SEFI, this may, in part, explain the low rates of fledging success here.

4. DISCUSSION

Gulls appear to be pre-adapted to co-existing with humans. Gulls use similar behaviors in both urban and non-urban settings. For example, scavenging on placentae and carrion found around pinniped breeding aggregations, foraging on invertebrates in the intertidal, and scavenging on refuse dumps, fish processing operations, or around picnic areas all require basically the same behavioral repertoire. Similarly, taking fish brought to the surface by human fishing activities is not much different from taking fish driven to the surface by foraging marine mammals or diving birds (Hoffman et al. 1981, Pierotti 1988). It is interesting to note that, like marine mammals and diving birds, human fishermen also use gulls as indicators of the location of large schools of fish (Pierotti 1988).

Western Gulls appear to choose nesting habitats that are structurally similar, regardless of nesting colony. The primary consideration in location of nests appears to be shelter from the prevailing wind (see also Pierotti 1981, 1982). However, this shelter can consist of rocks (SEFI), vegetation (SBI, Anacapa, Alcatraz, Moss Landing), mounds of soil (Anacapa, Moss Landing), or human debris and structures (Alcatraz).

With regard to reproductive performances, clutch sizes varied less among colonies than did hatching and fledging success (Table 15.1), even during major ENSO events. The Moss Landing colony suffered low hatching and fledging success because of terrestrial predators, despite the production of large clutches, large eggs, and the absence of the well known 'third chick disadvantage' in this colony (Pierotti and Bellrose 1986). Except for Moss Landing, the poorest overall performance appeared to be on Anacapa. Hatching success on Anacapa was the poorest of any colony, chick growth rates were slow, many chicks starved to death, and fledging success was the poorest of any colony studied with the exception of Moss Landing and SEFI

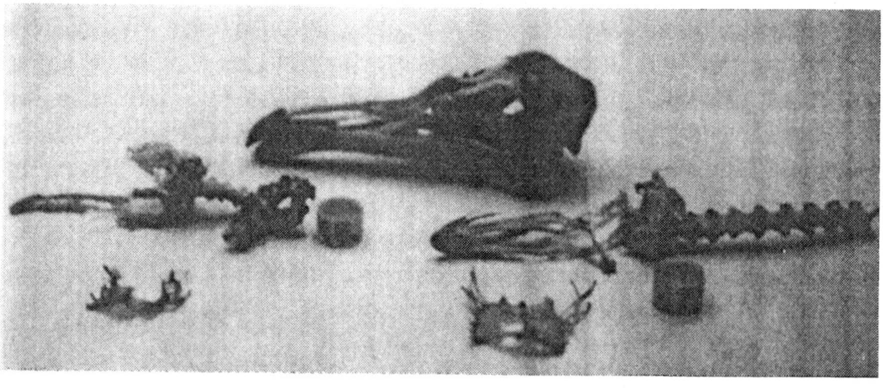

Figure 15.3. Skulls, cervical vertebrae, furculae, and breast bones of Western Gull chicks raised solely on a diet of chicken with bone (two individuals in foreground) compared with the skull of a Western Gull chick raised on a diet of fish (rear).

during the major 1983 ENSO event.

However, in terms of average population fitness it may be that gulls actually do not depend much on humans or on urban situations. Results from research summarized in this study demonstrate that gulls feeding primarily on fish or marine invertebrates have the best overall survival and breeding success, regardless of colony location. Our study of Herring Gulls (*Larus argentatus*) on the Atlantic coast showed a similar result (Pierotti and Annett 1987, 1990, 1991).

The relationship between diet and reproductive performance, as well as the large variation in individual survival and reproduction, is obscured in most studies because data are reported as averages across populations. In contrast, we have shown that relatively few pairs contribute disproportionately to future generations and that these pairs maintain a fish-dominated diet throughout their breeding years (Annett and Pierotti 1999).

It has been argued that previous breeding experience may be an important determinant of future success in individual gulls (Pyle et al. 1991), and a 'cost of reproduction' may exist because birds that breed at younger ages show reduced survival in future years (Pyle et al. 1997). However, this cost appears to be an artifact resulting from the pooling of successful and unsuccessful birds, where the latter show a major cost and the former show no cost at all (Pierotti 1991, Annett and Pierotti 1999).

5. POLICY AND MANAGEMENT IMPLICATIONS

The Western Gull is one of the rarest species of gull, with a world population of only about 40,000 breeding pairs (Sowls et al. 1980, Varoujean 1979). Approximately 40% (16,000 pairs) of the world population breeds on a single island, Southeast Farallon, which is located directly in one of the major shipping lanes leading into San Francisco Bay. At the northern end of its range, the Western Gull is undergoing extensive hybridization with Glaucous-winged Gulls (*Larus glaucescens*); intermediate phenotypes dominate the outer coast of Washington State (Hoffman et al. 1978, Bell 1997, Good 1998, Good et al. 2000). In addition, within the last 30 years, the southern subspecies of Western Gull, *L. o. wymani*, experienced a strong skew in sex ratio that altered the breeding system (Hunt et al. 1980, Pierotti 1981, Pierotti and Annett 1995, Pierotti et al. 1996; see below).

In addition to these situations that are specific to Western Gulls, there are several real potential costs to gulls that breed under highly urban-influenced conditions. First, there is increased risk of disease. Studies have shown much higher incidence of *Salmonella* in gulls breeding near large urban centers in Scotland (Monaghan et al. 1985, Girdwood et al. 1985). This increased risk of disease may be related to foraging on refuse, since gulls in Scotland that were well away from urban centers fed on natural foods, whereas urban gulls took more refuse (Monaghan et al. 1985). Western Gulls on SEFI showed an increased frequency of deaths associated with botulism during the 1983 ENSO event, when these birds took a much larger percentage of refuse than is typical for that colony.

Second, contaminants, such as pesticides, are more likely to be ingested around large urban areas. Considerable evidence exists that certain isomers of DDT act as estrogen mimics on SBI, the only island on which we studied contaminants. These estrogen mimics may have contributed to the skewed sex ratio in Western Gull populations in the Channel Islands by preventing normal development of male gonads (Hunt et al. 1980, Pierotti 1981, Pierotti and Annett 1995, Pierotti et al. 1996). Finally, there is the possibility of increased rates of mutation in gulls in urban areas (Yauk and Quinn 1996).

Our work on habitat manipulation demonstrates the importance of shelter from wind for nesting gulls, as well as the importance of understanding gull nest site preferences for managing gull colonies in a non-destructive manner. Most attempts at gull control involve culling adult birds, smashing eggs, or spraying clutches with oil (Graham 1975, Blodget 1988, Coulson 1991, Pierotti and Good 1994, Annett and Pierotti 1995). These methods are destructive, ethically questionable, and are often generally ineffective (Graham 1975, Coulson 1991, Pierotti and Good 1994, Annett and Pierotti 1995). For example, an attempt at removing gulls from a ternery in Maine, resulted in the 'managers' smashing all the tern eggs, thereby successfully destroying the tern colony (Graham 1975). A massive cull of Herring Gulls on the Isle of May (50% of 10,000 birds killed) did not significantly reduce the area used by gulls, but instead led to a doubling of inter-nest distances (Coulson 1991, Pierotti and Good 1994). In contrast, on Alcatraz we were able to change gull nesting patterns without harming a single bird by removing windbreaks in some areas while simultaneously increasing available nesting habitat in other areas. The overall effect of our habitat manipulation was to increase the number of nesting pairs of a relatively rare species of gull while still being able to resolve a conflict over land-use by humans.

We have argued that gulls can flourish in association with humans because gulls are perhaps the most generalized of marine birds in all aspects of their biology. As a consequence, gulls are capable of living close by and exploiting their relationships not only with other birds, but also with concentrations of large mammals, including humans. The generalized anatomy and behavior of gulls allow them to exploit human-derived resources in much the same way they exploit naturally occurring resources.

However, our results on reproductive performance suggest it may be best to view gulls as not necessarily doing well in urban-influenced conditions, but rather as being adaptable enough to persist under such conditions. As a consequence, gulls become familiar to humans who often regard them with contempt, rather than respecting them for their ability to survive under adverse conditions. It is probably no accident that contemporary society also treats coyotes, crows, rats, and ravens in a similar manner. Thus, contemporary western culture looks down on the most intelligent and adaptable of our fellow beings, whereas to the indigenous peoples of North America, many such animals were regarded as having special status (Pierotti and Wildcat 1997, 2000).

ACKNOWLEDGEMENTS

We thank Channel Islands National Monument, the Golden Gate National Recreation Area, the United States Fish and Wildlife Service, Point Reyes Bird Observatory, and the Monterey Bay Salt Company for access to colonies and logistical support during our work. We also thank Joelle Buffa of the San Francisco Bay National Wildlife Refuge for providing the aerial image of Southeast Farallon Island.

REFERENCES

Ainley, D. G. 1990. Introduction, p.3-22. *In* D. G. Ainley and R. J. Boekelheide [EDS.], Seabirds of the Farallon Islands: ecology, dynamics, and structure of an upwelling system community. Stanford University Press, CA.

Ainley, D. G., C. S. Strong, T. M. Penniman, and R. J. Boekelheide. 1990. The feeding ecology of Farallon seabirds, p.51-127. *In* D. G. Ainley and R. J. Boekelheide [EDS.], Seabirds of the Farallon Islands: ecology, dynamics, and structure of an upwelling system community. Stanford University Press, CA.

Annett, C. A. and R. Pierotti. 1989. Chick hatching as a trigger for dietary switches in Western Gulls. Colonial Waterbirds 12:4-11.

Annett, C. A. and Pierotti, R. 1999. Longterm reproductive output and recruitment in Western Gulls: consequences of alternate foraging tactics. Ecology 80:288-297.

Ashmole, N. P. 1971. Seabird ecology and the marine environment, p.223-286. *In* D. S. Farner, J. R. King, and K. C. Parkes [EDS.], Avian Biology, Vol. I Academic Press, New York, NY.

Bell, D. A. 1997. Hybridization and reproductive performance in gulls of the *Larus glaucescens* occidentalis complex. Condor 99: 585-594.

Bellrose, C. A. 1983. The breeding biology and ecology of a small mainland colony of Western Gulls. M.S. thesis, California State University, San Jose.

Blodget, B. G. 1988. The half-century battle for gull control. Mass. Wildlife 38:4-10.

Coulson, J. C. 1991. The population dynamics of culling Herring and Lesser Black-backed Gulls, p.479-497. *In* C. M. Perrins, J. D. Lebretton, and G. J. M. Hirons [EDS.] Bird population studies: relevance to conservation and management. Oxford Univ. Press. UK.

Feduccia, A. 1980. The origins of birds. Yale University Press, New Haven, CT.

Fortunate Eagle, A. 1992. Alcatraz: the Indian occupation of 1969-1971. Heyday Books, Berkeley, CA.

Girdwood, R. W. A., C. R. Fricker, D. Munro, C. B. Shedden, and P. Monaghan. 1985. The incidence of and significance of Salmonella carriage by Gulls. J. Hygiene 95:229-234.

Good, T. P. 1998. Breeding ecology and hybridization in the Western Gull/Glaucous-winged Gull complex. Unpublished Ph.D. thesis, University of Kansas, Lawrence, KS.

Good, T. P., J. C. Ellis, C. A. Annett, and R. Pierotti. 2000. Bounded hybrid superiority in an avian hybrid zone: effects of mate, diet, and habitat choice. Evolution 54:1774-1783.

Graham, F. 1975. Gulls: a social history. Sierra Club Books, San Francisco, CA.

Hoffman, W., D. Heinemann, and J. A. Wiens. 1981. The ecology of seabird feeding flocks in Alaska. Auk 98: 437-456.

Hoffman, W., J. A. Wiens, and J. M. Scott. 1978. Hybridization between gulls (*Larus glaucescens* and *L. occidentalis*) in the Pacific Northwest. Auk 95: 441-458.

Hunt, G. L., Jr. 1972. Influence of food distribution and human disturbance on the reproductive success of Herring Gulls. Ecology 53: 1051-1061.

Hunt, G. L., Jr., and M. W. Hunt. 1975. Reproductive ecology of the Western Gull: the importance of nest spacing. Auk: 92: 270-279.

Hunt, G. L., Jr., and M. W. Hunt. 1976. Clutch size, hatching success, and eggshell thinning in Western Gulls. Condor 75: 483-486.

Hunt, G. L., Jr., J. C. Wingfield, A. Newman, and D. S. Farner. 1980. Sex ratio of Western Gulls on Santa Barbara Island, California. Auk 97: 473-479.

Kadlec, J. A., and W. H. Drury. 1968. Structure of the New England Herring Gull population. Ecology 49: 644-676.

Marzluff, J. M., K. J. McGowan, R. Donnelly, and R. L. Knight. 2001. Causes and consequences of expanding American Crow populations, p. 333-365. *In* J. M. Marzluff, R. Bowman, and R. Donnelly [EDS.], Avian ecology and conservation in an urbanizing world. Kluwer Academic, Norwell, MA.

Monaghan, P., C. B. Shedden, K. Ensor,, C. R. Fricker, and R. W. A. Girdwood. 1985. Salmonella carriage by Herring Gulls in the Clyde area of Scotland in relation to their breeding ecology. J. Applied Ecology 22: 669-679.

Penniman, T. M., M. C. Coulter, L. B. Spear, and R. J. Boekelheide. 1990. Western Gull., p.218-244. *In* D. Ainley and R. Boekelheide [EDS.], Seabirds of the Farallon Islands: ecology, dynamics, and structure of an upwelling system community. Stanford University Press, Stanford, CA.

Pierotti, R. 1976. Sex roles, social structure, and the role of the environment in the Western Gull. M.S. thesis, California State University, Sacramento.

Pierotti, R. 1981. Male and female parental roles in the Western Gull under different environmental conditions. Auk 98:532-549.

Pierotti, R. 1982. Habitat selection and its effect on reproductive output in the Herring Gull in Newfoundland. Ecology 63:854-868.

Pierotti, R. 1987. Behavioral consequences of habitat selection in the Herring Gull. Studies in Avian Biology 10:119-128.

Pierotti, R. 1988. Interactions between gulls and otariid pinnipeds: competition, commensalism, and cooperation, p.213-239. *In* J. Burger [ED.], Seabirds and other Marine Vertebrates: commensalism, competition, and predation. Columbia Univ. Press, New York, NY.

Pierotti, R. 1991. Individual variation and the costs of reproduction. Condor 93:1039-1040.

Pierotti, R. and C. A. Annett. 1987. Reproductive consequences of specialization and switching in an ecological generalist, p.417-442. *In* A. C. Kamil, J. R. Krebs, and H. R. Pulliam [EDS.], Foraging behavior. Plenum Press, New York, NY.

Pierotti, R. and C. A. Annett. 1990. Diet and reproductive performance in seabirds. Bioscience 40:568-574.

Pierotti, R. and C. A. Annett. 1991. Diet choice in the Herring Gull: effects of constraints imposed by reproduction and ecology. Ecology 72:319-328.

Pierotti, R. and C. A. Annett. 1995. Western Gull (*Larus occidentalis*). *In* A. Poole and F. Gill [EDS.], The birds of North America. Academy of Natural Sciences, Philadelphia, PA.

Pierotti, R., C. A. Annett, and J. L. Hand. 1997. Male and female perceptions of pair-bond dynamics: monogamy in Western Gulls, *Larus occidentalis*, p.261-275. *In* P. A. Gowaty [ED.], Feminism and evolutionary biology: boundaries, intersections, and frontiers. Chapman and Hall, NY.

Pierotti, R. and C. A. Bellrose. 1986. Proximate causation and the third chick disadvantage in gulls. Auk 103:401-407.

Pierotti, R. and T. P. Good. 1994. The Herring Gull. (*Larus argentatus*). *In* A. Poole and F. Gill [EDS.] The birds of North America. Academy of Natural Sciences, Philadelphia, PA.

Pierotti, R., and D. Wildcat. 1997. Evolution, creation and Native American traditions. Journal of American Indian Science and Engineering Society 12: 70-73.

Pierotti, R., and D. Wildcat. 2000. Traditional ecological knowledge: The third alternative (commentary). Ecol. Appl. 10:1333-1340.

Pyle, P., L. B. Spear, W. J. Sydeman, and D. G. Ainley. 1991. The effects of experience and age on the breeding performance of Western Gulls. Auk 108: 25-33.

Pyle, P., N. Nur, W. J. Sydeman, and S. D. Emslie. 1997. Cost of reproduction and the evolution of deferred breeding in the Western Gull. Behav. Ecol. 8: 140-147.

Schoenherr, A. A., C. R. Feldmeth, and M. J. Emerson. 1999. Natural history of the islands of California. University of California Press, Berkeley.

Sowls, A. L., A. R. DeGange, J. W. Nelson, and G. S. Lester. 1980. Catalog of California seabird colonies. U.S. Department of Interior, Fish and Wildlife Service FWS/OBS-80/37.

Varoujean, D. H. 1979. Seabird colony catalog: Washington, Oregon, and California. U.S. Department of Interior, Fish and Wildlife Service, Portland, OR.

Yauk, C. L. and J. S. Quinn. 1996. Multilocus DNA fingerprinting reveals high rate of heritable genetic mutation in Herring Gulls nesting in an industrialized urban site. Proc. Natl. Acad. Sci. 93:1029-1033.

Chapter 16

Causes and consequences of expanding American Crow populations

John M. Marzluff[1], Kevin J. McGowan[2], Roarke Donnelly[1], and Richard L. Knight[3]

[1]*College of Forest Resources, University of Washington, Box 352100, Seattle, WA 98195 USA, email: corvid@u.washington.edu*

[2]*Department of Ecology and Evolutionary Biology, Cornell University, Corson Hall, Ithaca, NY 14853 USA*

[3]*Department of Fishery and Wildlife Biology, Colorado State University, Fort* Collins, CO 80523 *USA*

Key words: American crow, diet, dispersal, metapopulation, nest predation, population growth, source-sponge, population dynamics, urban footprint, urban sprawl, urbanization

Abstract: Corvid populations are increasing worldwide in response to urbanization. We investigated the response of American Crows (*Corvus brachyrhynchos*) to urbanization by (1) comparing rates of winter population change between urban and nonurban locations (using standard Christmas Bird Counts); (2) quantifying population size along a gradient of urbanization in western Washington; and (3) pooling studies from eastern (New York), midwestern (Wisconsin), and western North America (Washington and California) relating survivorship, reproduction, and space use to urbanization. American Crow populations tend to be densest and increasing most rapidly in urban areas of North America. This appears to be facilitated by small space needs of crows in urban relative to suburban, rural, and exurban areas. Crow survivorship is high across the urban gradient, but reproduction and hence population growth, peaks in suburban and rural settings. Local demographic considerations appear unable to account for changing winter crow populations. Rather, we hypothesize that urban crow populations may be increasing primarily as surplus crows from suburban and rural areas disperse into the city where anthropogenic food sources are easily located, rich, and concentrated. This hypothesis likely is affected by local crow sociality. In the western United States, where pre-breeders often form flocks able to exploit urban riches, our dispersal hypothesis may be accurate. But, in midwestern and eastern areas, where crows migrate south for winter or remain on territories to help rather than float as pre-breeders, dispersal may not be adequate to fuel urban population growth. Refuse, invertebrates, and small vertebrates appeared to be more common food items than the nest contents of other birds. This, and the typically diverse suite of nest predators in any area, may explain why the rate of predation on artificial nests we placed throughout the urban gradient was not highly correlated with the abundance of crows. We encourage researchers to study how urbanization affects important mechanisms (like nest predators and predation) so environmental policy will benefit from a detailed, scientific understanding of how avian communities are structured.

1. INTRODUCTION

As a rule, bird community diversity in urban landscapes is markedly lower than in comparable rural/exurban communities (Marzluff 2001). Diversity is reduced because (1) some species' habitat requirements are no longer met and (2) other species able to exploit humans ('human commensals') increase in abundance and dominate resources such as food and nesting space, or reduce the productivity of other species by preying on their nest contents. Many magpies, jays, and crows are human commensals that have increased in urban areas worldwide (Konstantinov et al. 1982, Eden 1985, Fraissinet 1989, Konstantinov 1996, Hogrefe et al. 1998; Jerzak 2001). Their abilities to exploit anthropogenic food resources and their habit of preying on other species' eggs and young suggest that their success may be partly responsible for many other species' failure to thrive in urban areas.

American Crows (*Corvus brachyrhynchos*) are human commensals throughout the coterminous United States. Standardized surveys and historical records suggest that American Crows increased in response to European colonization of North America by spreading from east to west as the American frontier was opened and developed (Marzluff et al. 1994). Crows were uncommon throughout the Great Basin and Pacific Coast through the early 1900s, except along riparian areas (Monson 1946, Richards 1971). Populations increased substantially in Oregon, Washington, and the central valleys and northern portions of California from the late 1800s to mid 1900s (Brooks 1925, Robertson 1931, Emlen 1940). As agriculture and urbanization continued to spread, so did crows (Pitelka 1942, Richards 1971). Accounts from localities such as Las Vegas Hot Springs, New Mexico (where crows were rare in 1882, but one of the commonest species in 1959; Rickard 1959) are typical of the crow population explosion. The explosion has continued throughout the latter half of the 20th century, especially in areas of urban growth and sprawl (Marzluff et al. 1994, Hogrefe et al. 1998).

Although many researchers have documented the increase in crow population size and speculated on its effects, there are no published data on the attributes of crow populations that may explain the increase. Therefore, we pooled data from eastern, mid-western, and western crow populations to determine: (1) the relationship between the degree of urbanization and crow abundance, area use, and foraging behavior; (2) if crow demography varied with degree of urbanization; and (3) if a positive relationship existed between the rate of predation on other songbird nests and abundance of crows. We use this information to develop a general model of crow population dynamics that is amenable to empirical testing.

2. METHODS

2.1 Study Areas and Research Approaches

We studied crows in six locations (Table 16.1) along a gradient (defined by Marzluff et al. 2001) from urban (Seattle, WA; Los Angeles, CA; portions of Madison, WI) through suburban (Madison; Ithaca, NY; outlying areas near Seattle), exurban (Cascade and Olympic Mountains, WA), and rural (Tompkins County, NY) areas to wildland (Olympic Peninsula, WA). We monitored nests and individually marked birds in neighborhoods, campgrounds, golf courses, parks, cemeteries, campuses, and wildland reserves (Caffrey 1992, 1999, 2000; McGowan 2001).

2.1.1 Western Washington

The western Washington study areas span the entire gradient of urbanization from downtown Seattle to the heart of Olympic National Park (47°,40' N; 122°,20' W; Fig. 16.1). Our sites were predominantly coniferous forests (Douglas-fir [*Pseudotsuga menziesii*], western hemlock *[Tsuga heterophylla*], and western redcedar [*Thuja plicata*]) with 5 – 50% deciduous overstory (big leaf maple [*Acer macrophyllum*], and red alder [*Alnus rubra*]). Overstory trees ranged from 50 to >200 years old. Canopy cover was 50 – 90%. Understories were well developed and ranged from native shrubs (mainly *Gaultheria shallon, Vaccinium parvifolium, Sambucus racemosa, Oemleria cerasiformis, Rubus spectabilis, Acer circinatum, Mahonia nervosa, Rubus parviflorus, and Symphoricarpos albus*) in exurban and wildland sites, to complex mixes of native and exotic shrubs in urban and suburban sites. Timber harvest and recreation were the dominant land-uses in the wildland and exurban sites. There, landscapes ranged from contiguous mature forest to highly fragmented, managed timberlands (Fig. 16.1). Suburban sites usually contained >50% greenspace in the form of scattered parks and riparian corridors. Urban sites had <20% greenspace, predominantly in the form of parks that ranged in size from 1 – 600 ha.

The focus of our study in western Washington was to understand the population structure and dynamics of crows in relation to urbanization and to relate crow populations to the avian community structure and the productivity of other sympatric songbirds. To accomplish this we captured and tagged adult crows, located and monitored nests, banded and observed

Table 16.1. Attributes of study areas and aspects of crow biology recorded.

Location	Years studied	1990 human population	Points studied on gradient	Aspects of biology studied
Madison WI	1982 – 1985	191, 000	Urban neighborhoods, Suburban neighborhoods, woodlots, golf courses, parks, university campus, cemetery	Fecundity
Ithaca NY	1989 – 1999	29, 000	Suburban neighborhoods, woodlots, golf courses, parks, university campus, cemetery	Fecundity, survivorship, space use, social behavior
Tompkins County NY	1989 – 1999	67,000	rural deciduous woodlots, agriculture	Fecundity, survivorship, space use, social behavior
Los Angeles CA	1985 – 1990	11 million	Urban golf course	Fecundity, survivorship, space use, social behavior
Seattle and Cascade Mountains, WA	1998 - 2000	3 million	Urban and suburban neighborhoods, parks and university campus, exurban woodlands	Fecundity, survivorship, space use, foraging, nest predation
Olympic Peninsula, WA	1995 – 2000	0 – 2,500	Suburban and exurban neighborhoods, wildland campgrounds and reserves	Fecundity, survivorship, space use, foraging, nest predation

young, and radio-tagged a sample of birds each year to determine survivorship, foraging behavior, and patterns of space use (Table 16.1). We typically banded 100 adults, monitored 30 nests, banded 90 nestlings, and radio-tagged 30 crows each year. In addition, we monitored the reproductive success and abundance of other songbirds in our crow study areas.

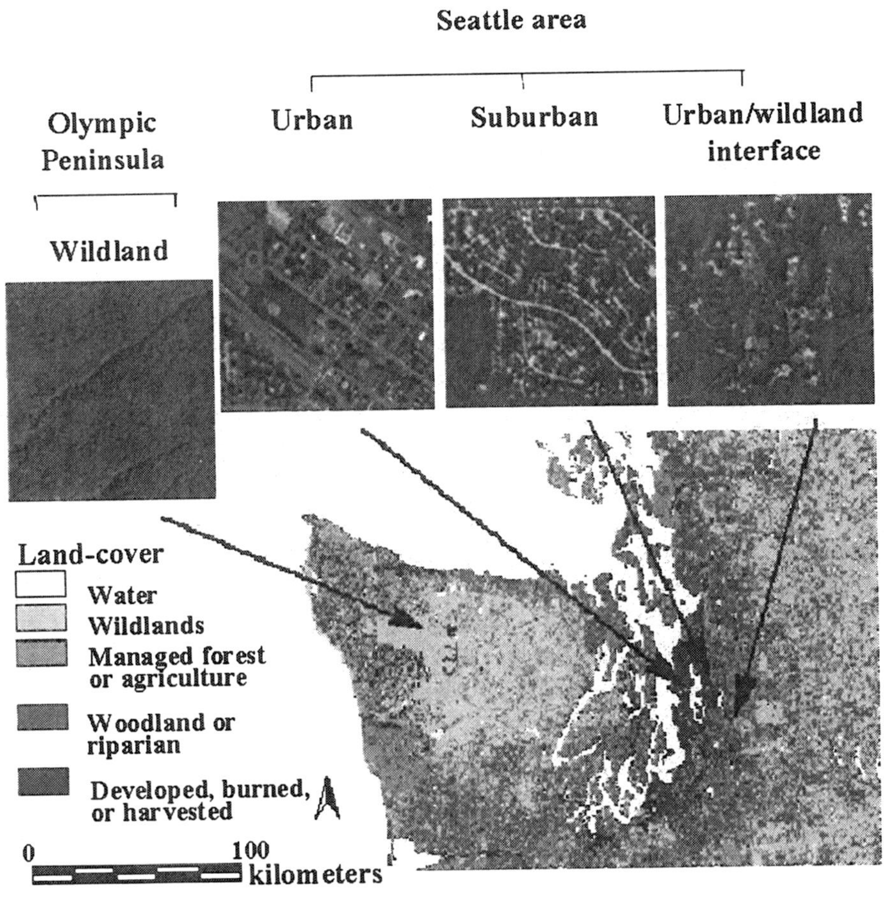

Figure 16.1. Land-cover on western Washington study areas with inset orthophotos. The Olympic Peninsula forms the left side of the figure and the Cascade Mountains form the right side. Seattle is the highly developed area in the center of the classified image.

2.1.2 Los Angeles

Crows were studied on two contiguous golf courses that were part of the Sepulveda Dam Recreation Area (34°,30' N; 118°,30' W; elevation = 2100 m; Fig. 16.2a). In this area, tree cover was approximately 15-20% and understory was lacking. The park itself had little development and a mostly continuous ground cover of exotic, mowed grasses. It included ball fields, picnic areas and the golf courses. It was surrounded by dense housing (1405 dwellings per mile^2), a large human population (3300 per mile^2), and associated services of Encino, California.

The primary goal of the Los Angeles study was to understand social organization and the correlates of breeding success in this cooperatively breeding population (Caffrey 1992, 1999, 2000).

2.1.3 Madison

Crows were studied in a variety of settings (neighborhoods of various age, parks, woodlots, cemeteries, and university campuses) in the town of Madison, Wisconsin. Madison (43°,05' N; 89°,22' W; elevation = 270 m; Fig. 16.2b) is built on an isthmus between two lakes, and includes 21 golf courses and the most restaurants per capita of any U.S. city. The population of Madison at the time of this study (1980) was 170,616 and housing and population densities around all study sites varied from urban to suburban. The study sites varied from largely concrete and exotic plant species (business district) to largely native vegetation and native species (woodlots). The sites were ranked in order of increasing domination by humans as woodlots, parks, cemeteries, university campus, neighborhoods, and business district. Impervious surfaces covered approximately 80% of the business district, but less than 1% of the woodlots. Understory vegetation was non-existent in the business district, but covered about 50% of the woodlots.

The goals of the Madison study were to understand crow demography and population dynamics in a variety of suburban settings.

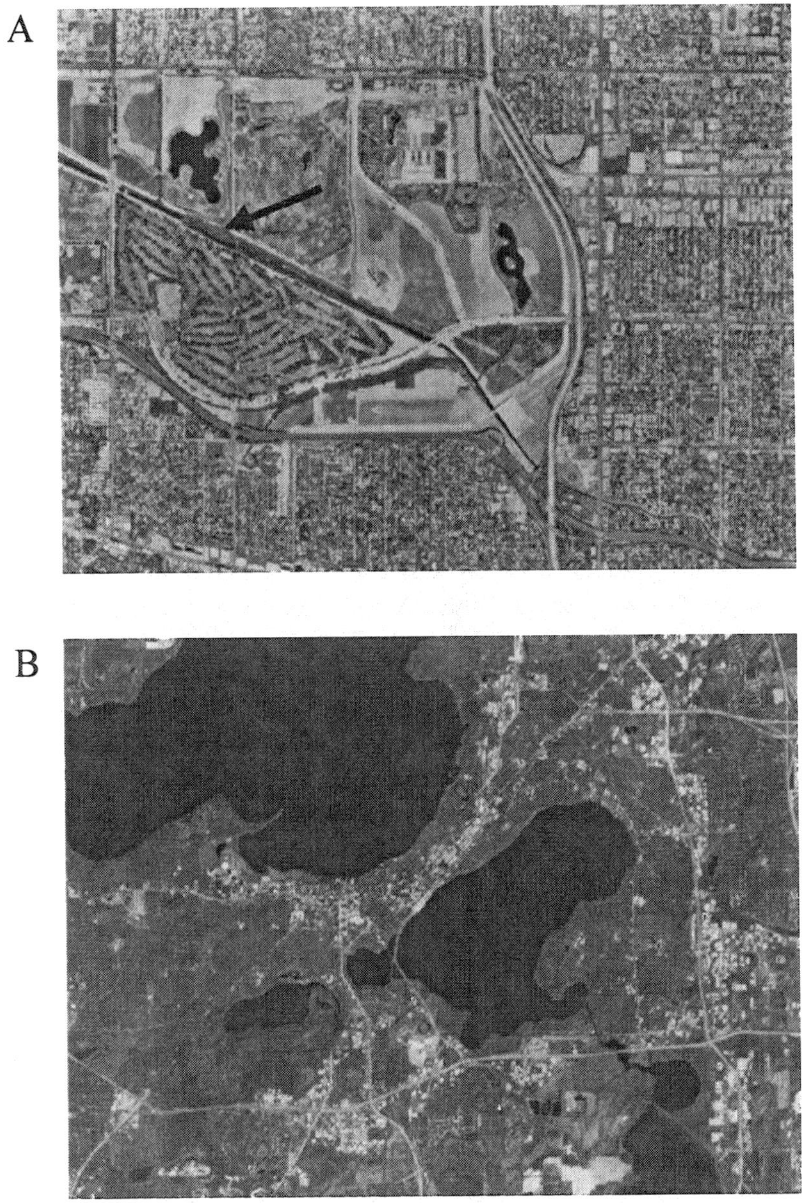

Figure 16.2. Study sites (A) in Encino, California near Los Angeles and (B) in Madison, Wisconsin. The focal golf course in Los Angeles is indicated by an arrow. The Los Angeles image is an orthophoto taken in 1998 covering ~16,000 ha. The Madison image is a Landsat image taken in 1999 covering ~7,000 ha.

2.1.4 Ithaca and Tompkins County

Crows throughout Ithaca, New York and surrounding Tompkins Country have been studied extensively by McGowan. The study areas, methods, and goals of this research are detailed in McGowan (2001, Fig. 17.1).

2.2 Abundance

2.2.1 Christmas Bird Counts

To determine if the recent growth of crow populations across the U.S. was related to the degree of urbanization, we selected 70 Christmas Bird Count (CBC; Audubon Society 2000) locations that could be classified as urban (>20% of the circle was developed in 1970) or exurban/rural (<5% of the circle was developed in 1970 and the 1996 human population size was <10,000). We used only sites that had at least 20 years of count data from 1960 – 1996 (Appendix 16.1). No single state was represented by more than two urban or two exurban/rural sites. We did not use Breeding Bird Surveys because few are conducted in urban areas (Droege 1990).

We related the number of crows (standardized per party hour) to year using standard linear regression. All crow species and subspecies were lumped (Fish Crows were rare in our sample; <0.07% of the total). We used the slope of the regression as an index of crow population growth in each circle. Slopes were compared between urban and exurban/rural sites with a Mann-Whitney *U*-test.

2.2.2 Surveys in Western Washington

We conducted fixed area (50m radius) point counts in all study sites in western Washington to determine if crow abundance (1995-2000) differed along the gradient of urbanization. We surveyed crows during the breeding season (April - August) using from 1 – 20 points spaced 250 m apart in each of our forested stands. At each point we recorded all crows heard and seen within 50 m during a 10-min count. At exurban and wildland sites we used crow calls at 2 points per stand to increase the chance that rare birds were detected. We censused each stand 2 – 5 times per season and used the maximum count (standardized as crows per point) per season to characterize abundance in the stand. Details of our survey procedure are in Luginbuhl et al. (in press).

2.3 Monitoring Crows in the Wild

We captured and marked crows in all our study sites as nestlings and adults using mist nets, noose carpets, walk-in traps, net guns, and leghold traps. We weighed each individual and measured its culmen length and wing chord. All individuals were uniquely banded with colored leg bands (Washington, Wisconsin) or patagial tags (California, New York). Field methods for New York and California are detailed elsewhere (Caffrey 1992, 1999, 2000; McGowan 2001).

A sample of individuals in the Washington study areas was instrumented with backpack-mounted radio transmitters. Transmitter harnesses were made from Teflon© ribbon following Buehler et al. (1995). We attempted to observe each radio-tagged bird at least once per week to confirm their survivorship. Tagging may influence demography, but without radio-tagging, survival could not be estimated across our large, remote study area. Any biases are likely to be slight (Vekasy et al. 1996, Withey et al. in press) and should be equal across treatments so that *relative* comparisons of demography between areas are unaffected.

In Washington, radio-tagged birds were the subjects of 1-2 hr focal observations approximately once each week throughout the spring, summer, and fall to quantify their foraging behavior and patterns of space use. We recorded the type of food, how it was obtained *(grab, pick, probe, hawk, kleptoparastize, glean*, etc.), and where it was obtained (*tree species, height and DBH*; *habitat* = cover classes defined on habitat map and structural complexity used to define our stands; *substrate* = branch, ground, various manmade structures, etc.; *microsite* = moss, lawn, asphalt, etc; distance to forest edge). We quantified the rate and success of foraging, when possible, by counting the total prey capture attempts and the percentage that were successful (indicated most reliably by swallowing) per unit time.

Locations of radio-tagged crows were plotted on maps/photos of the Washington study sites (a global positioning system was used in remote areas) and used to calculate home range statistics with a computer program, Ranges V (Kenward and Hodder 1995). During each 1-2 hr focal observation period, we plotted the entire area used by a bird, then recorded 2 - 3 locations (extreme and mid points of area used) for subsequent definition of the home range. We occasionally recorded single locations of animals at their roosts and during our travels throughout the study area. We purposely recorded few locations per day on each animal to minimize dependence among relocations. We computed maximum use areas (100% convex polygons) and core areas within them (cluster techniques; Kenward 1987). We computed minimum convex polygon home ranges to facilitate comparison with other studies and to avoid assumptions about the statistical distribution of animal locations. We defined an animal as being 'adequately sampled' if >20 locations were obtained. Investigating the increase in range size with sampling effort (incremental analysis) suggests that an average of 60-75% of the entire range is defined by 20-25 locations; 95% confidence intervals around these means include 100% definition of the area used. Our definition of adequate sampling is supported by the lack of a strong positive correlation between the number of location estimates and the home range size of adequately sampled animals (the correlation was in fact *negative*; $r = -0.30$, $n = 37$, $P = 0.07$). In the Los Angeles and Ithaca study areas, core areas and defended territories were determined by spot mapping locations of color-marked individuals (e.g., McGowan 2001).

2.4 Calculation of λ

We pooled our demographic data across study areas to estimate age-specific fecundity and survivorship in urban, suburban, rural, and wildland settings. We assumed females in all areas first bred when four years old (modal age of reproduction in Ithaca; McGowan, unpublished data) and that productivity was constant until death. We calculated the annual production of fledglings separately for each area using data from (1) urban greenspaces in Wisconsin, Los Angeles, and Seattle, (2) suburban developments and parks in Wisconsin, Seattle, and Ithaca, (3) rural Tompkins County, New York, and (4) wildland forests on the Olympic Peninsula. We assumed all crows died by age 21 years. We assumed survivorship of 1-year old crows was equal across the urbanization gradient at 48% (McGowan 2001), while annual survivorship of 2- and 3-year-olds was 72% (all estimates based on results from Ithaca; McGowan, unpubl. Data and 2001). We calculated the annual survivorship of breeders separately for each area using data from (1) urban greenspaces in Los Angeles and Seattle, (2) suburban developments and parks in Seattle and Ithaca, (3) rural Tompkins County, New York, and (4) wildland forests on the Olympic Peninsula. Survivorship was calculated using color-marked birds in Los Angeles (Caffrey 1999), Ithaca, and Tompkins County (McGowan 2001) and radio-tagged birds in Washington. We used these data to create a transition matrix for each degree of urbanization and solved the matrix for λ by finding the largest positive eigenvalue using RAMAS (Akcakaya1999). Variation in λ resulting from demographic stochasticity also was estimated using RAMAS. To do this, we assumed the following standard deviations for demographic parameters for all populations (Fecundity: 1-year-old = 0.02, 2-year-old = 0.04, 3-year-old = 0.13, >3-year-old = 0.17; Survivorship: fledglings = 0.25, 1-year-old = 0.19, 2-year-old = 0.19, >2-year-old = 0.04).

2.5 Artificial Nest Experiments

We used a variety of artificial nests to quantify the risk of nest predation. In urban and suburban Seattle we placed handmade nests in (1) native ground cover (typically sword fern, *Polystichum munitum*) to simulate those of ground nesters such as Song Sparrows (*Melospiza melodia*), Spotted Towhees (*Pipilo maculatus*), and Wilson's Warblers (*Wilsonia pusilla*), and (2) in shrubs to simulate those of American Robins (*Turdus migratorius*) and Swainson's Thrushes (*Catharus guttatus*). Both types of nests were filled with two, small (12x18 mm), blue plastic, wax-coated eggs. The wax coating allowed us to infer the type of predator from the marks on the egg (identification was calibrated using cameras to record predators at an independent subset of nests). Details of these experiments can be found in Donnelly and Marzluff (in press). On the Olympic Peninsula, we simulated nest sites of Marbled Murrelets (*Brachyramphus marmota*) by placing an artificial egg or taxidermy-mounted chicken chick in a typical murrelet canopy nesting location. Details of these experiments can be found in Marzluff et al. (2000). Ground and shrub nests remained in place until preyed on or until 27 days had elapsed. Canopy nests remained in place until preyed on or for 30 days. We determined the rate of predation within a study plot by averaging the time until predation (all nests were checked every three days) for all nests within a given stratum.

3. RESULTS

3.1 Crow Abundance and Urbanization

In western Washington, the relative abundance of crows increased with increasing degree of urbanization. Crows were rarely found during surveys in wildlands far from campgrounds (Fig. 16.3). Exurban areas also had few crows in contrast to suburban and urban areas. Crow abundance in urban areas was approximately 30 times greater than in wildlands (Fig. 16.3).

Across the U.S., crow populations tended to increase in urban areas from 1960 – 1996, but they remained relatively constant in exurban and wild areas. The annual rate of increase in crow populations was marginally greater in urban areas (mean = 27.2%, SE = 13.5% , n = 35) than in exurban, rural, and wildland areas (mean = -5.1%, SE = 22.5%, n = 35). However, because locations varied tremendously in the rate at which crow populations changed (Appendix 16.1), this difference was only marginally significant (U = 516, $P_{(1\text{-tailed})} = 0.13$). Only one rural area had a substantial increase in the crow population (Rupert, ID; 372% annual increase). In contrast, five urban areas had substantial increases (Albuquerque, NM: 425%; Hartford, CN: 187%; Sacramento, CA: 122%; Seattle, WA: 57%; and Portland, OR: 51%).

3.2 Use of Space and Foraging Behavior

Breeding adult crows occupied areas during the spring and summer that varied from slightly over 1 ha to over 3500 ha. Across areas, this variation was positively correlated with the degree of urbanization (Fig. 16.4). In the extremely dense breeding population of Los Angeles, core area size was less than 2 ha, but where crows were rare (remote portions of the Olympic Peninsula) home range size averaged 2100 ha.

Certainly, part of this difference is due to our use of (1) a variety of measures of area use (defended territories in New York, core use areas in Los Angeles, and entire traveled areas in Washington) and (2) a variety of techniques (visual observations Los Angeles and New York, radio-telemetry in Washington). However, when we examine these patterns within two of

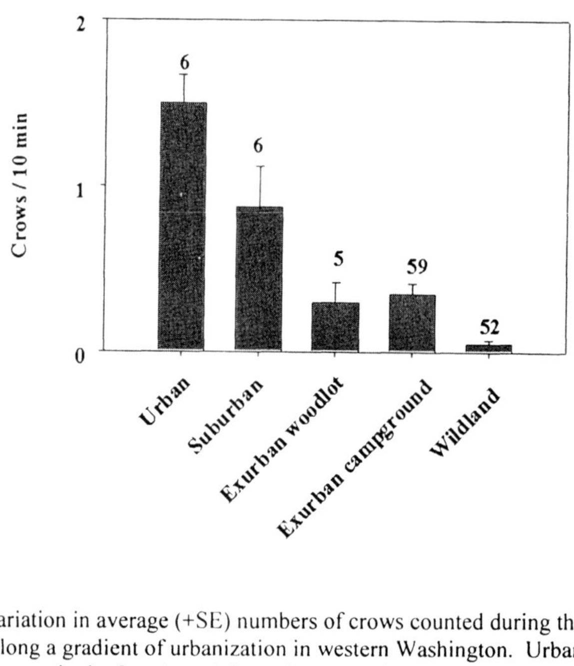

Figure 16.3. Variation in average (+SE) numbers of crows counted during the summer at various points along a gradient of urbanization in western Washington. Urban, suburban, and exurban woodlots are in the Seattle and Cascade Mountain study areas. Exurban campgrounds and wildland sites are on the Olympic Peninsula. Numbers above bars are sample sizes of independent study sites.

the study areas we still detect similar increases in area use as human settlement decreases. In upstate New York, breeding crows in the town of Ithaca used areas approximately four times smaller (mean = 8.7 ha, n = 18) than crows breeding in adjacent, rural Tompkins County (mean = 37.3 ha, n = 59; McGowan 2001; Fig. 16.4). In western Washington, crows breeding on the Olympic Peninsula had home ranges 7 times larger than those in nearby urban Seattle and suburban Snohomish (Fig. 16.4). Furthermore, on the Olympic Peninsula home range size was three times smaller for crows nesting in areas with human presence (rural and exurban developments and recreation areas; mean = 308 ha, n = 22, SE = 61.8) relative to crows nesting >5 km from such areas (mean = 2124 ha, n = 8, SE = 706; $F_{1,29}$ = 18.2, P <0.001). Variation in space use within areas of similar human settlement may be related to regional differences in social organization (crows in Los Angeles were colonial, but crows in our other urban study areas were territorial during the breeding season), breeding success (crows increased

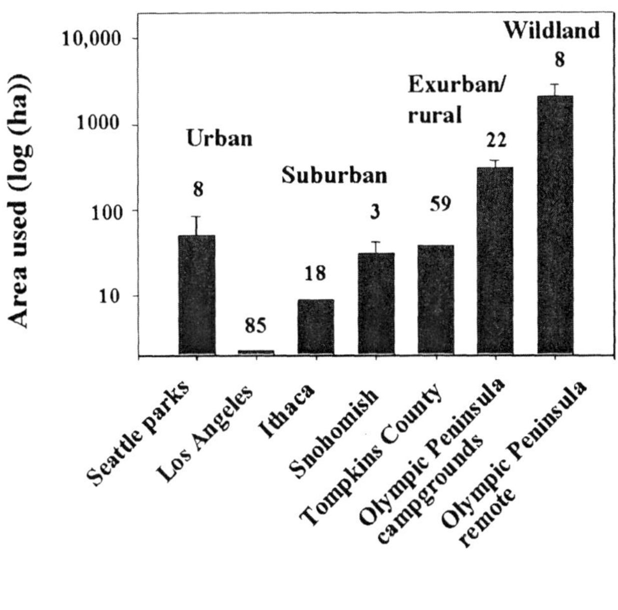

Figure 16.4. Average (+SE) space use by breeding crows in the United States. Crows in Seattle, Snohomish, and all sites on the Olympic Peninsula were monitored with telemetry using similar methodology. Likewise, crow in Ithaca and Tompkins County, New York were monitored as part of a single study with consistent methodology.

their range size in Seattle when their nests failed despite where they occurred on the urban gradient), and other aspects of locally important resources.

Our observations in western Washington suggest that differences in space use by crows along the urban gradient may be due to differences in diet. Crow diets varied from urban to wildland settings in Washington (Fig. 16.5). Human refuse (prepared meats, bread products, and vegetables) was the dominant food item eaten by crows in Seattle (65% of observed food consumption; Fig. 16.5A). Road-killed mammals, live rodents and reptiles, and invertebrates composed <25% of Seattle crow diets. In contrast, crows on the Olympic Peninsula consumed roughly equal proportions of invertebrates (38%) and human refuse (35%) regardless of their proximity to small settlements and recreation areas (Fig. 16.4B). Wildland crows often flew tens of kilometers to utilize anthropogenic food sources (Marzluff and Neatherlin, unpubl. data).

We followed radio-tagged crows in Washington to obtain unbiased estimates of their reliance on nest contents for food. We rarely saw crows prey on other species of songbirds' nests (n = 2 times) during the five years of intensive observation (>1500 hrs yr^{-1}) of 54 radio-tagged birds on the Olympic Peninsula. Nest predation was not observed during two years of observation (>800 hrs yr^{-1}) of 14 radio-tagged crows in Seattle. However, we were occasionally (n = 8 times) attracted to songbirds mobbing marauding crows. The consciousness of this mobbing may lead to inflated estimates of nest predation by casual observers of crows.

3.3 Demography and Population Dynamics

Pairs breeding in urban areas produced fewer fledglings each year than pairs breeding in suburban, rural, or wildland areas (Table 16.2). Urban breeders averaged only 60% of the number of fledglings per pair (mean = 1.1, SE = 0.17, n = 3 studies) that were produced by suburban (mean = 1.90, SE = 0.15, 3 studies) breeders. Rural (1.6, n = 1 study), and wildland breeders (1.5, n = 1 study) produced intermediate numbers of fledglings. Lower productivity by urban crows resulted primarily from an overall decrease in nest success, rather than a decrease in the number of fledglings per successful nest (Table 16.2). However, in Ithaca suburban crows had significantly higher overall nest success but fledged significantly fewer nestlings per successful nest than did rural crows (McGowan 2001). Suburban crows consistently averaged around 60% nest success.

Annual survivorship of breeding crows was high in all study areas. It appeared to be lowest in our wildland and Seattle urban study areas (Table 16.3). There, nearly 20% of crows died each year, but sample sizes are small. All observed mortality in our wildland study areas occurred on crows

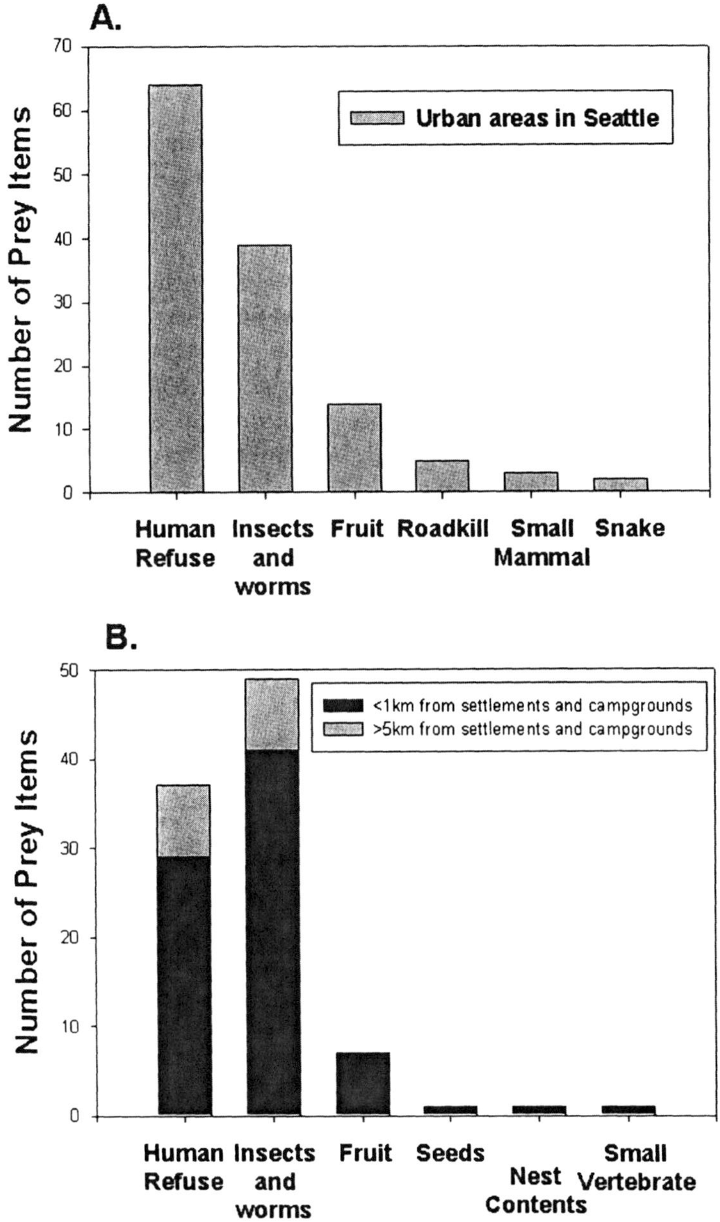

Figure 16.5. Diets of breeding crows during the spring and summer in western Washington. Numbers of each prey type that we observed crows eating are plotted separately for 12 crows in urban and suburban Seattle/Snohomish (A) and 37 exurban and wildland crows on the Olympic Peninsula (B).

nesting >5km from human settlement or recreation sites (3 of 9 such crows died). Wildland crows that bred adjacent to small settlements and recreation sites appeared long-lived (only 1 of 20 died during five years of observation).

Using the variation in annual breeder survivorship and reproduction in conjunction with our assumptions about age of first breeding, pre-breeder survival, and longevity (see Methods), we estimated λ (geometric rate of increase) for urban, suburban, rural, and wildland crow populations. This demographic analysis suggested that urban populations had the capacity for slight growth ($\lambda = 1.008$), suburban populations should grow 15.8% per year ($\lambda = 1.158$), rural populations should grow by 7.3% per year ($\lambda = 1.073$), and wildland populations should grow 3.9% per year ($\lambda = 1.039$). Incorporating observed annual variation in reproduction and survival into a model of population growth (as environmental variability) and incorporating demographic stochasticity (Akcakaya 1991) reveals that, *at best*, urban crow populations should only increase 6.3% per year. Growth of all crow populations was most sensitive to adult survivorship (elasticities: urban adult survival = 0.54; suburban = 0.36; rural = 0.47; wildland = 0.44). The predicted change in a crow population of 50 individuals living in each of these environments for 100 years is striking; suburban populations are

Table 16.2. Annual reproductive success of American Crows in areas of varying urbanization.

Site	No. breeding attempts	Years of study	% success	Mean no. fledged successful nest^{-1} yr^{-1}	Mean no. fledglings pair^{-1} yr^{-1}
Urban					
Seattle parks, campus, neighborhoods	80	1998 - 2000	53.3	2.0	1.1
Los Angeles golf course	147	1985-1990	43.0	1.93	0.83
Madison parks, campus, neighborhoods, golf course	123	1982 - 1985	44.7	3.1	1.4
Suburban					
Snohomish, WA	19	1998-2000	63.2	3.1	2.1
Madison, WI	56	1982-1985	68.1	2.9	2.0
Ithaca, NY	202	1989-1994	57.3	3.1	1.6
Rural					
Tompkins County, NY	45	1989-1994	48.0	3.6	1.6
Wildland					
Olympic Peninsula	54	1995-2000	78.0	2.0	1.5

predicted to be 4-6 orders of magnitude larger than urban, rural, or wildland populations (Fig. 16.6).

3.4 Effects on Other Birds

Although our data do not preclude the possibility that past contact with crows shaped nesting preferences of other songbirds or that crows may depredate the nests of susceptible species within a very narrow window following colonization, it seems that crows rarely prey on nests (Fig. 16.5). In Washington, we found nonsignificant, negative relationships between crow abundance and rate of predation in experiments using artificial ground nests, shrub nests, and canopy nests (Fig. 16.7). Crows have been photographed depredating artificial nests, but these events are rare (n = 2 of many hundred) and crows constitute only one of nearly 20 recorded nest predators (small and large mammals, several corvids, owls, and raptors; Luginbuhl et al., in press) in the Washington study area. Observations at over 900 artificial canopy nests designed to simulate Marbled Murrelet nests implicated crows or ravens as nest predators in only 6.2% of 451 eggs and 2.0% of 454 chicks (Luginbuhl et al., in press).

Table 16.3. Annual survivorship of breeding American Crows in areas of varying urbanization.

Area	No. crows	No. breeder years	% annual survivorship
Urban			
Seattle*	16	16	81.3
Los Angeles	23 (Female)	67 (Female)	94.3 (Female)
	27 (Male)	105 (Male)	95.1 (Male)
Suburban			
Ithaca	35	87	92.0
Snohomish*	10	24	93.3
Rural			
Tompkins County	11	20	90.0
Wildland			
Olympic Peninsula*	29	29	86.0

*Survivorship determined using radio telemetry.

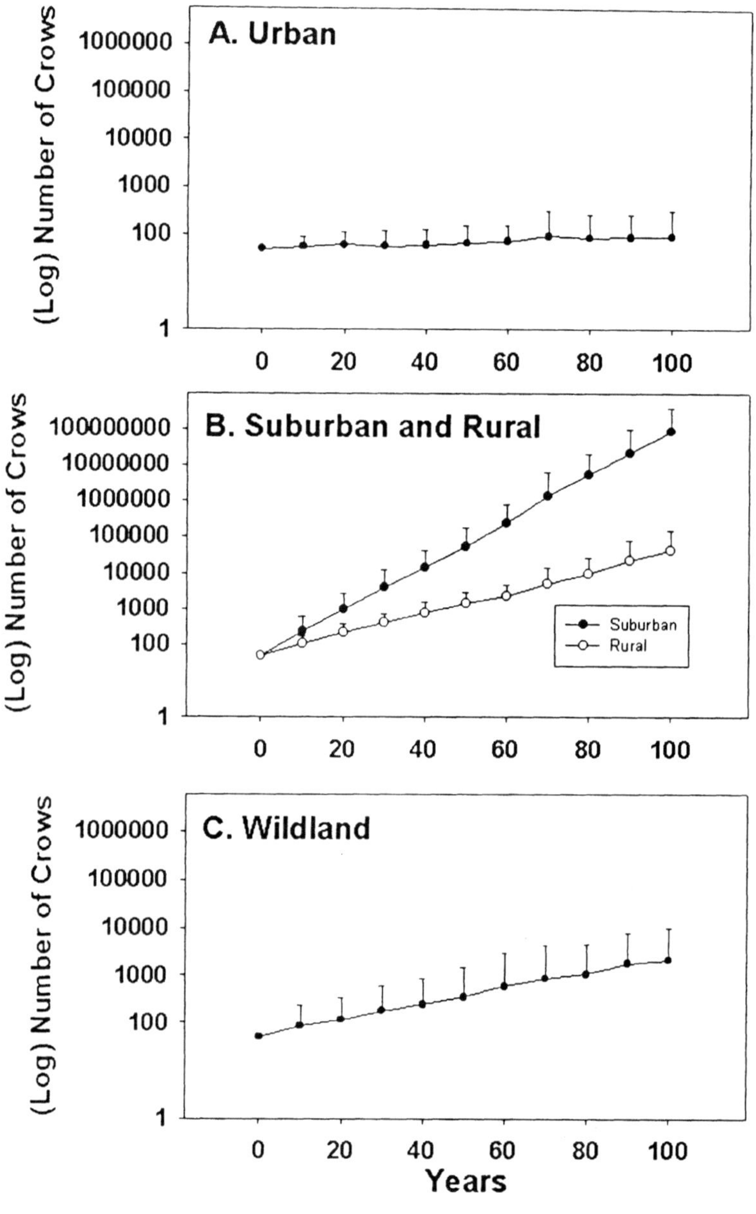

Figure 16.6. Projected change in simulated crow populations with demographic characteristics of crows in urban (A), suburban (B), rural (B), and wildland (C) areas. Average population size is plotted on a log scale (note range in B is greater than A and C). Error bars (1 SD) represent variation induced by demographic stochasticity. All populations were initiated at $n = 50$ birds.

4. DISCUSSION

4.1 Urban Life

American Crows have become human commensals or, as defined by Johnston (2001), they are synanthropic. They rarely are found breeding far (>5km) from human activity. Their abundance peaks and may be increasing most rapidly in urban and suburban areas (Fig. 16.3). Our results suggest that high abundance is possible because breeding pairs in urban areas use very little space (Fig. 16.4). This fact may be the result of food supplementation provided directly by human refuse, and indirectly by land-cover modification that replaces native forests and understories with forest-field ecotones and lawns rich in earthworms and other suitable invertebrates (Fig. 16.5; but see McGowan 2001). Without access to human settlement, crows require large areas for foraging (Fig. 16.4) which may limit population size in remote areas such as Washington's Olympic Peninsula.

There is a paradox in the commensalism between crows and humans. Despite being abundant and increasing in some urban areas (e.g., Albuquerque, Portland, and Seattle), the demographic patterns of urban crow populations are not sufficient to explain their explosive growth. Our estimates of population growth rates suggest that urban populations should grow relatively slowly, at rates less than 5% per year. (Tables 16.2, 16.3; Fig. 16.5). In contrast, the demographic patterns of suburban crow populations may allow rapid population increases. This pattern also is true of rural and wildland crow populations, albeit slower than suburban populations. Suburban and rural populations appear to be growing, as predicted by their demographic data, but exurban and wildland populations appear to be small and relatively stable, despite their capacity for growth. The difference between expected and observed population changes in urban areas is clearly seen in Seattle. There, winter counts suggest an exponentially increasing population, but demographic rates predict little population growth (Fig. 16.8).

Is this paradox real or is it simply a result of insufficient data and comparisons between breeding performance and winter populations? We think it is real. Breeder survivorship and reproduction are relatively constant between years and very similar across our different studies. Incorporating the observed variation in survivorship and fecundity into our demographic projections did not affect our conclusions (note error bars in Figs. 16.6 and 16.8). Thus it seems likely that they are accurate. A more likely source of inaccuracy involves the pre-breeding cohort. Our understanding of this age group is entirely dependent on McGowan's (2001) studies in New York.

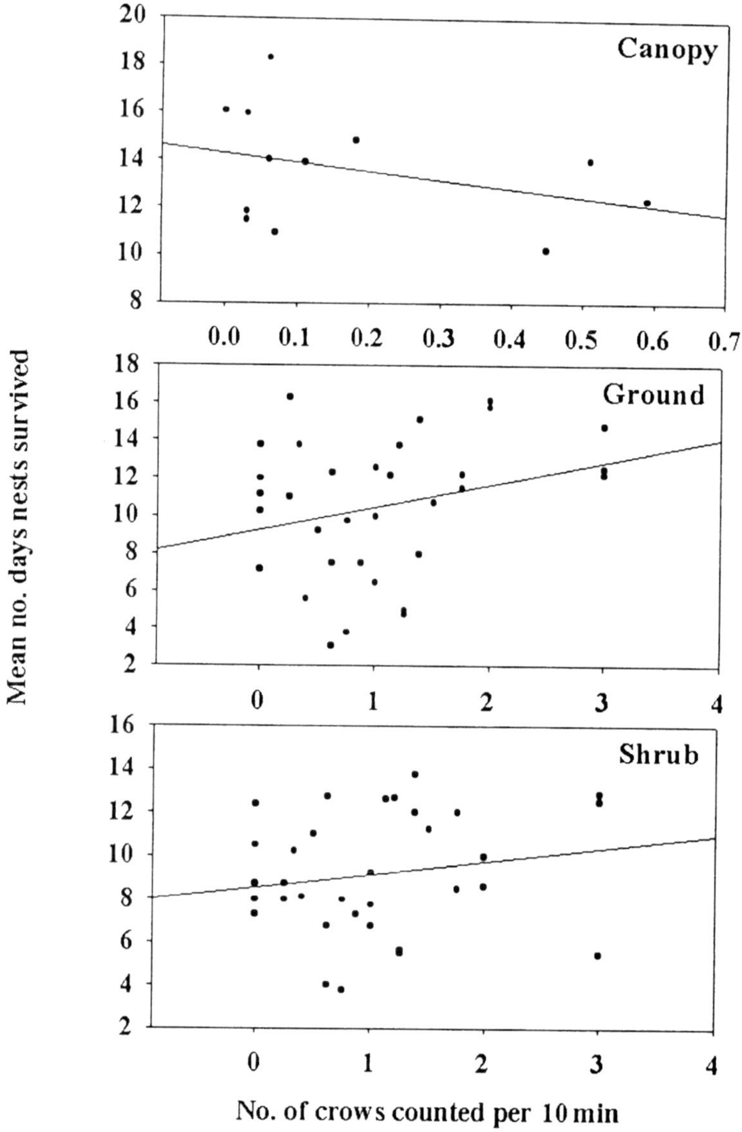

Figure 16.7. Relationships between rates of predation on artificial nests and relative abundance of crows in western Washington. Canopy nests simulated those of Marbled Murrelets on the Olympic Peninsula. Shrub and ground nests simulated thrushes and sparrows, respectively, in the Seattle and Cascade Mountain study areas. Average rates of predation within study landscapes (n = 3 - 5 study plots with 6 nests per plot in each landscape) are plotted for canopy nests. Average rates of predation in each study plot (n = 8 - 36 nests per plot) are illustrated for shrub and ground nests. Least-squares regression lines are provided for each nest type, although the relationships are not significant.

Perhaps urban crows breed at younger ages or have better survivorship in the pre-breeding period, than do other crows. However, this seems unlikely. In Seattle, breeding space appears to be saturated so early breeding would seem difficult. Young crows are also frequently killed by automobiles, suggesting low survival in urban areas, although McGowan (2001) found first year survival was higher in suburban than rural areas (Table 17.4). If the paradox is not real, our population estimates, which come from mid-winter counts (standardized CBC's; Butcher 1990), may not accurately reflect reproduction and recruitment from local surroundings. Rather they might simply indicate seasonal influxes (migration) of breeders into urban areas from the surrounding region, or that the CBC's rely heavily on counts at urban roosts, which are known to include rural crows (Caccamise et al. 1997; McGowan 2001).

Why is reproduction and survivorship lowest in urban areas? Productivity in urban areas may suffer because of a combination of (1) poor food quality (if it is abundant, it is likely of low quality relative to invertebrate rich suburban and wildland areas; McGowan 2001; Pierotti and Annett 2001), (2) high intraspecific interference during breeding, and (3) large populations of protected or subsidized nest predators (raccoons [*Procyon lotor*], eastern gray squirrels [*Sciurus carolinensis*], and Red-tailed Hawks [*Buteo jamaciensis*]). Survivorship in urban areas may be relatively low because (1) unnatural hazards are common (e.g., windows and golf balls), (2) automobiles collisions kill many fledglings, (3) predators of adults may thrive (Red-tailed Hawks and Great-horned Owls, *Bubo virginianus*, are important in this respect), and (4) toxins are common and may be concentrated in prey items. Urban environments are rapidly changing, challenging places for any animal to inhabit. The adaptability of crows is impressive, but appears to be just adequate to keep pace with the urban fast lane.

4.2 Causes of Population Growth

If the local demography of urban populations is stable, but not growing, then what is causing the urban crow population explosion? In mobile species such as crows, immigration and emigration can decouple local population trajectories from local demography. However, the usual process is for dispersal to "rescue" declining populations ($\lambda < 1$; Brown and Brown 1977, Stacey and Taper 1992). We are unaware of examples where dispersal from a productive region or population simultaneously limits growth at the source, while it fuels a population explosion at a distant site where the local breeding population appears to be at its carrying capacity (territory size is extremely small and obvious breeding habitat appears saturated). That is,

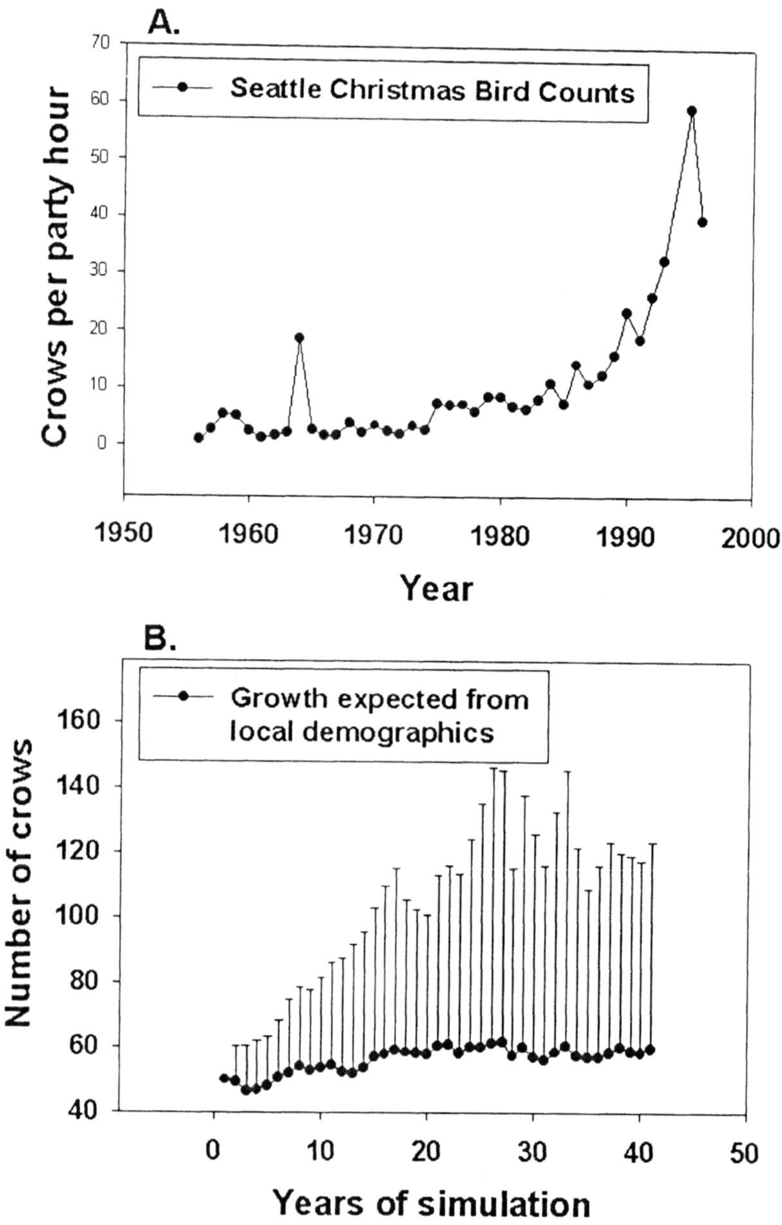

Figure 16.8. Growth in urban American Crow populations observed (A) and expected (+1 SD) from local demographic potential (B). In contrast to expectation, local demographics predict stasis, but observed growth is exponential.

nonetheless, what seems to be happening with urban crows. It is theoretically possible for dispersal from sources ($\lambda > 1$) to increase populations in sinks ($\lambda < 1$; Pulliam 1988), but in such settings sink populations typically barely persist rather than thrive (Robinson et al. 1995).

Although territoriality limits the urban breeding population of crows, crow populations may be able to grow even where natality and mortality are balanced. It is possible that unlimited urban food may enable exponential increase, fueled by immigration, in the urban non-breeding population. Additionally, suburban sprawl continually creates new breeding habitat that can be colonized by crows unable to obtain breeding space in the city core. Sprawl opens forests and juxtaposes newly created foraging habitat (lawns, roads, and dumpsters) with discrete, easily defended breeding sites (patches of trees). Refuse in the urban core is renewed daily and allows large aggregations of crows to gather at scattered, highly visible, reliable, and rich feeding areas where foraging occurs with minimal aggression because resources are not economically defensible (Brown 1969). This also occurs in many rural areas, so is not the only driver of population size.

We hypothesize that dispersal from breeding populations in suburbs and/or rural and wildlands cause urban populations to increase. Young crows produced by successful suburban, exurban, and rural breeders presumably should find it difficult to obtain breeding space in those habitats. Dispersal to the city, where food and social partners are abundant, and the shooting of crows is prohibited, would seem to be a successful strategy for a pre-breeding crow. The ability of crows to easily traverse large areas, share rich food resources, and quickly colonize rapidly expanding breeding opportunities made suitable by sprawling cities enables crow populations to increase rapidly and escape local demographic constraints on population growth.

If pre-breeding crows disperse from suburban, exurban, and rural areas into urban areas, then we should observe pre-breeding flocks in urban areas. We have observed them in Seattle and Los Angeles, but not in midwestern (Oklahoma; Caffrey, unpublished data) or eastern (New York) areas. Conspicuous differences in crow social systems may partly explain this. Crows from the northern Great Plains and in the most northern reaches of the range migrate south during winter rather than congregate in local urban areas (Aldous 1944, Ignatiuk and Clark 1991, Root 1988). Additionally, suburban and rural pre-breeding crows in New York typically remain on their territories as helpers rather than floaters (McGowan 2001). Helping occurs in western crows, but less frequently (Verbeek and Butler, 1981; Caffrey 1999). Clearly we need a better understanding of the options pursued by pre-breeding crows across the urban gradient in a variety of social systems. However, rapid crow population growth in some western cities indicated in our survey of CBC data confirms our suspicion that social systems of western crows may be especially conducive to fueling urban growth.

Suburban crow populations have the potential to bolster urban populations. For illustration, we started two hypothetical populations that were proportional in size to the populations we surveyed along the urban gradient in western Washington. This gave us initial populations of 82 suburban crows and 114 urban crows. We grew these populations for 50 years according to our calculated geometric rate of increase. This gave us nearly 145,000 suburban, but virtually unchanged numbers of urban crows. If we allow carrying capacity to be proportional to the space requirements of urban breeders (Seattle park data; Figure 16.4), then a 500 ha^2 area could support 2500 pairs in the suburbs and >5,000 pairs in the urban area. Our model suggests that the urban population would only attain its carrying capacity if crows from suburbs dispersed into the urban area at a rate proportional to the degree to which their populations were above their respective carrying capacities. During our simulated growth, suburban populations produced a sufficient number of dispersers to increase urban populations beyond their expected carrying capacity in only 35 years.

If our model of crow population dynamics is correct, then urban populations function more like a *sponge* than either a sink or a source. They soak up the surplus from nearby source populations and grow by augmenting their pre-breeding segment (Fig. 16.9). Breeder numbers likely remain constant once carrying capacity is reached because of extremely dense packing of small urban territories. Pre-breeders may wander within towns to track spatio-temporally variable food sources in the urban environment and eventually disperse back to suburbs or wildlands to breed, or quickly colonize new settlements on the urban fringe. Indeed, the sponge may function primarily in the winter, when subordinate and inferior juvenile foragers are most stressed for food resources.

The importance of urban sprawl is obvious in our model. As food resources grow in new settlements, new breeding space is created as the former wildlands or suburbs becomes new suburban or urban areas where, respectively, breeders need only 10% of their former space. The subsidized surplus of pre-breeders in nearby urban areas can quickly fill these openings. As new suburbs grow, an ever increasing number of young birds are produced to fuel growth in the original and expanding urban area. This is an interesting case of the urban 'footprint' (Wackernagel and Rees 1996) affecting the urban core. The far reaching effects of urban growth fuel changes in animal populations residing in the core. Put in an ecological context, it is an example of population regulation occurring at multiple scales as the features of the surrounding landscape (large scale) affect local populations (small scale).

4.3 Consequences of Population Growth

American Crows are nest predators that may be particularly adept at finding and destroying shrub and canopy nests (Marzluff and Balda 1992). As urbanization clearly increases crow populations, could this change result in low nesting success of other birds in urban areas? Our observations in western Washington suggest that this idea is too simplistic. Crow population increases do not necessarily correlate with increasing rates of nest predation (Fig. 16.6). Poor congruence between crow population size and nesting success of other birds is likely the result of a complex suite of nest predators whose sum, rather than individual, effect determines the rate of nest predation in any local area (Marzluff and Restani 1999, Luginbuhl et al., in press). Where crows are rare, mice, squirrels, or jays may be abundant and therefore nest predation may remain high. The singular importance of crows as limiting factors on other species also is cast into doubt by management actions that removed crows but did not document increases in nesting

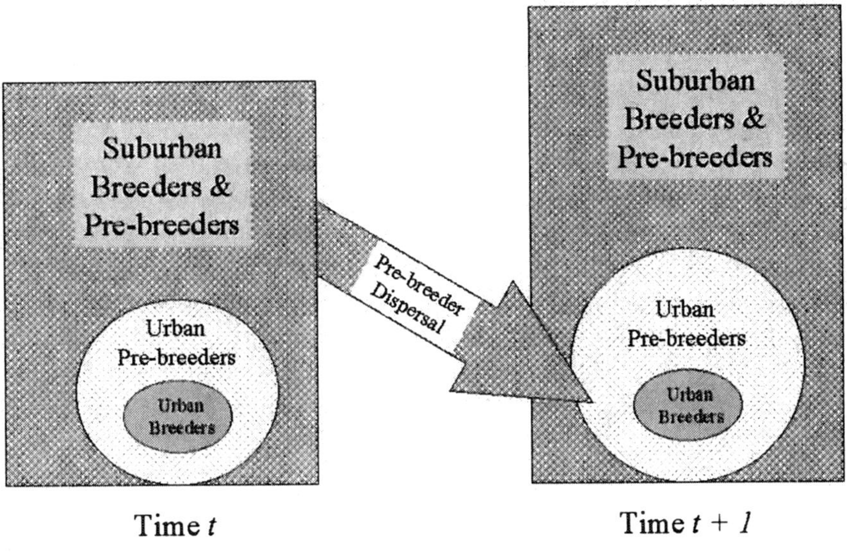

Figure 16.9. Source-sponge hypothesis of urban crow population growth. Growth is not expected in the saturated breeding segment of the urban population. Rather, the pre-breeder population swells as dispersers from surburban source populations disperse into the urban core to exploit abundant anthropogenic foods.

success of other species (Chesness et al. 1968, Parker 1984, Parr 1993, Broyer et al. 1995, Clark et al. 1995).

The suite of nest predators along the urban gradient and their resulting effect on nests is determined by a complex interaction of landscape- and local-scale factors. Where the landscape is extremely urbanized, crows, introduced tree squirrels, and rats are likely to be the dominant nest predators. They may destroy many nests if local conditions are suitable. Important local conditions might include grass cover for crows, ground cover for rats, feeders for squirrels and crows, and nesting cover for songbirds. However, in extremely manicured or altered settings nest predation may be low despite abundant crows because rats and squirrels are rare. That is, increased crow abundance may not compensate for reduced mammalian predators, in terms of the rate of nest depredation. Indeed, predators that are adapted to fragmented and urban landscapes in the western U.S. may not compensate for the loss of native predators, in general (Tewksbury et al. 1998; Haskell et al. 2001).

In the suburban landscape jays, crows, native mice, chipmunks, and squirrels are abundant. Local conditions (such as landowner practices, recreation, forest composition, forest remnant size and shape) may determine the exact mix and abundance of nest predators and resulting nest predation in a study plot. In this setting, nest predator diversity may be dictated by a few keystone predators like Red-tailed Hawks, coyote (*Canis latrans*), and short-tailed weasels (*Mustella ermina*). Regardless of the cause, the diverse predator community in this setting will likely obscure a direct relationship between crow abundance and rate of nest predation. In the wildland landscape, a similarly diverse nest predator community is determined by forest cover and patch structure, topography, presence of water, degree of recreation, and landowner practices. Nest predator communities should be extremely diverse and result in moderately high rates of nest predation that are independent of crow abundance.

4.4 Remaining Questions

Our description of crow population structure along the urban gradient is based on circumstantial evidence that needs to be tested. The generality of suburban sources needs to be validated across years and in a variety of areas. The prediction that urban breeding populations are relatively stable through time, while their non-breeding populations increase, also needs testing. In particular, estimates of adult survivorship deserve more attention, due to their relatively large influence on landscape-specific population growth. Three predictions arise from our conceptual model of urban crow populations. First, urban populations will have a larger non-breeder to breeder ratio than will suburban or wildland crows. The critical hypothesis of dispersal from wildland and suburban areas to urban areas presently has only little data (McGowan 2001). However, this hypothesis is tractable in crows with radio telemetry or observations of color-marked birds. We expect young produced in wildland, rural, and suburban areas to wander widely and disperse to urban sponges. In contrast, young produced in urban areas should be incorporated into nearby urban pre-breeding groups (if they exist) or remain as helpers in their natal territories. The second prediction is that young produced in urban settings should help parents at the nest (i.e., not disperse) more than young born in source populations further out on the urban gradient, unless urban food resources are extremely patchy (Jerzak 2001). McGowan (2001) found that although the percentage of nests with helpers did not differ between suburban and rural areas, urban families were significantly larger. The third prediction is that in rapidly urbanizing areas, like the American west, newly settled areas should be colonized by pre-breeders residing in existing and nearby urban core locations.

Our suggestion that increasing crow populations do not necessarily mean increasing rates of nest predation also should serve as a hypothesis rather than as a statement of fact. It is an appropriate null hypothesis that should be tested at various points on the urban gradient. Important contributions can be made by observing the rate of predation on real nests in areas with vastly different abundances of crows. However, ornithologists need to monitor a sizeable fraction of the suite of nest predators across this gradient to fully understand the role of crows (Marzluff and Restani 1999; Haskell et al. 2001). We predict that the rate and magnitude of predation will vary directly with crow abundance only in landscapes depauperate of other nest predators.

Population ecologists also will find urban environments to be fertile research areas. Our work on crows suggests the structuring of populations in an area of limited spatial extent with vastly different local conditions is complex. It is likely that replicates of population sources, sinks, and sponges exist within a study area of only a few hundred square kilometers. This situation would provide a fertile laboratory to quantify, perturb, and understand the viability of the metapopulation paradigm.

5. MANAGEMENT AND POLICY IMPLICATIONS

The urban gradient is an excellent experimental stage where avian communities assembled through individual- and population-level processes, can be observed before, during, and after major habitat changes. Documenting local extinctions and colonizations in rapidly developing areas, as well as in urban reserves surrounded by human-dominated 'seas', will perhaps allow urban planning to proceed in a more ecologically sustainable way. By studying successful species such as crows, we may be able to determine how they come to survive in such a radically and rapidly changing environment. What novel adjustments do they make to city life?

Urban ornithologists can make an important contribution to conservation by better elucidating how landscape-level changes interact with local conditions to determine avian community structure. The urban environment provides varied local conditions in proximity to one another within similar landscapes. For example, within the urban landscape, areas exist varying in the pattern of development, the degree to which vegetation is native or exotic, the percentage of the land-cover that is impermeable, and the degree to which homeowners feed birds or allow their cats to roam. Do these local conditions determine bird communities, or is the general urban landscape more important? We hope future studies will address these and related questions. We suggest the answers will require community surveys and detailed studies of population ecology. Studies of potential keystone species, like predators, parasites, and exotic invaders, may be especially important.

Studies of how urbanization affects mechanisms (like nest predation) that may structure avian communities can inform policy makers with reliable science. Such studies are more likely to reveal causal relationships that policy makers can use to construct sound and legally defensible decisions. Our studies of crows suggest that uncontrolled sprawl will greatly accelerate crow populations. This may not be desired. However, our results and those of others suggest that increasing crow populations are not causally linked to reduced songbird productivity. Removing crows is not a reliable way to increase productivity of songbirds. Rather, controlling sprawl itself is more likely to keep crow populations in check and provide the suitable habitat required by many species of declining songbirds.

ACKNOWLEDGEMENTS

Research on crows in western Washington was supported by grants from the National Science Foundation (DEB-9875041), the Washington Department of Fish and Wildlife, the University of Washington's Royalty Research Foundation, the U.S. Forest Service, the U. S. Fish and Wildlife Service, Rayonier, and the Washington State Department of Natural Resources. We are especially indebted to Lenny Young, Dan Varland, Marty Raphael, Scott Horton, and John Pierce for their support. Roarke Donnelly, Karla Kaczoroski, John Luginbuhl, Kim McKillip, Erik Neatherlin, and John Withey provided unpublished results and field assistance. Research was done under the auspices of UW ACUC permit #3077-03. Financial support for the Ithaca crow research was provided by U.S. Dept. Agriculture, Hatch Project Grant NYC-183429. RLK appreciates the insights and mentoring from Stanley A. Temple of the University of Wisconsin's Department of Wildlife Ecology, who also provided financial assistance to support Knight's crow research in Madison.

REFERENCES

Akcakaya, H. R. 1999. RAMAS GIS. Linking landscape data with population viability analysis. Applied Biomathematics. Setauket, NY.

Audubon Society. 2000. Audubon Christmas Bird Counts. http://www.mp1-pwrc.usgs.gov/birds/cbc.html.

Brooks, A. 1925. Anent the crow. Condor 27: 83-84.

Brown, J. H., and A. Kodric-Brown. 1977. Turnover rates in insular biogeography: Effect of immigration on extinction. Ecology 58:445-449.

Brown, J. L. 1969. Territorial behavior and population regulation in birds. Wilson Bull. 81:293-329.

Broyer, J., Fournier, J. Y., and P. Varagnat. 1995. Effect of carrion crow (*Corvus corone*) reduction on predation on artificial anatid (Anatidae) nests. Gibier Faune Sauvage 12:95-107.

Buehler, D., J. D. Fraser, M. R. Fuller, L. S. McAllister, and J. K. D. Seegar. 1995. Captive and field-tested radio transmitter attachments for bald eagles. J. Field Ornith. 66:173-180.

Butcher, G. S. 1990. Audubon Christmas bird counts. USDI FWS Biological Report 90:5-13.

Caccamise, D. F., L. M. Reed, J. Romanowski, and P. C. Stouffer. 1997. Roosting behavior and group territoriality in American crows. Auk 114:628-637.Caffrey, C. 1992. Female-biased delayed dispersal and helping in American Crows. Auk 109:609-619.

Caffrey, C. 1999. Feeding rates and individual contributions to feeding at nests of cooperatively breeding western American crows. Auk 116:836-841.

Caffrey, C. 2000. Correlates of reproductive success in cooperatively breeding western American crows: If helpers help, it's not by much. Condor 102:333-341.

Chesness, R. A., Nelson, M. M., and W. H. Longley. 1968. The effect of predator removal on pheasant reproductive success. J. Wildl. Manage. 32:683-697.

Clark, R. G., Meger, D. E., and J. B. Ignatiuk. 1995. Removing American Crows and duck nesting success. Can. J. Zool. 73:518-522.

Donnelly, R. E., and J. M. Marzluff. In press. Designing research to advance the management of birds in urbanizing areas. *In* W. Shaw and L. Harris [EDS.], Proceedings of the 4th International Urban Wildlife Conservation Symposium. Arizona State University, Tempe.

Droege, S. 1990. The North American breeding bird survey, p 1-4. *In* J. R. Sauer and S. Droege [EDS.], Survey designs and statistical methods for the estimation of avian population trends. U.S. Dept. Inter., Washington, DC.

Eden, S. F. 1985. The comparative breeding biology of magpies Pica pica in an urban and a rural habitat. J. Zool., Lond. (A) 205:325-334.

Emlen, J. T., Jr. 1940. The midwinter distribution of the crow in California. Condor 42:287-294.

Fraissinet, M. 1989. Espansione della taccola, Corvus monedula, nei capoluoghi italiani. Riv. Ital. Orn. 59: 33-42.

Haskell, D., A. M. Knupp, and M. C. Schneider. 2001. Nest predator abundance and urbanization, p. 245-260. *In* J. M. Marzluff, R. Bowman, and R. Donnelly [EDS.], Avian ecology and conservation in an urbanizing world. Kluwer Academic, Norwell, MA.

Hogrefe, T. C., and R. H. Yahner. 1998. Depredation of artificial ground nests in a suburban versus a rural landscape. J. Penn. Acad. of Science 72:3-6.

Ignatiuk, J. B. and R. G. Clark. 1991. Breeding biology of American crows in Saskatchewan parkland habitat. Canadian J. Zool. 69:168-175.

Jerzak, L. 2001. Synurbanization of the Magpie in the Palearctic, p. 405-427. *In* J. M. Marzluff, R. Bowman, and R. Donnelly [EDS.], Avian ecology and conservation in an urbanizing world. Kluwer Academic, Norwell, MA.

Johnston, R. F. 2001. The synanthropic birds of North America, p. 49-67. *In* J. M. Marzluff, R. Bowman, and R. Donnelly [EDS.], Avian ecology and conservation in an urbanizing world. Kluwer Academic, Norwell, MA.

Kenward, R. E. 1987. Wildlife radio tagging. Academic Press, Sand Diego, CA. 222p.

Kenward, R. E. and K. H. Hodder. 1995. Ranges V: software for analysing animal location data. Institute of Terrestrial Ecology, Wareham, UK.

Konstantinov 1996. Antropogenic transformations of bird communities in the forest zone of the Russian Plain. Acta Ornithologica 31: 53-65.

Konstantinov, V. M., and V. G. Babenko. 1982. Numbers and some ecological features of synanthropic populations of the corvidae under the conditions of intensive urbanization. 12:1837-1845.

Luginbuhl, J. M., J. M. Marzluff, J. E. Bradley, M. G. Raphael, and D. E. Varland. In press. Corvid survey techniques and the relationship between corvid relative abundance and nest predation. J. Field Ornithol.

Marzluff, J. M. 2001. Worldwide increase in urbanization and its effects on birds, p. 19-47. *In* J. M. Marzluff, R. Bowman, and R. Donnelly [EDS.], Avian ecology and conservation in an urbanizing world. Kluwer Academic, Norwell, MA.

Marzluff, J. M., and R. P. Balda. 1992. The Pinyon Jay. T & A D Poyser, London.

Marzluff, J. M., R. B. Boone, and G. W. Cox. 1994. Native pest bird species in the West: why have they succeeded where so many have failed? Studies Avian Biol. 15:202-220.

Marzluff, J. M., R. Bowman, and R. Donnelly. 2001. A historical perspective on urban bird research: trends, terms and approaches, p. 1-17. *In* J. M. Marzluff, R. Bowman, and R. Donnelly [EDS.], Avian ecology and conservation in an urbanizing world. Kluwer Academic, Norwell, MA.

Marzluff, J. M., Raphael, M. G., and R. Sallabanks. 2000. Understanding why forest management may affect avian species: mechanisms, experimental approaches, prioritization, and funding strategies. Wildl. Soc. Bull. 28:1132-1143.

Marzluff, J. M., and M. Restani. 1999. The effects of forest fragmentation on avian nest predation. Pp: 155-169. In: Forest Fragmentation: Wildlife and Management Implications (J. A. Rochelle, L. A. Lehmann, and J. Wisniewski eds.). Brill Academic Publishing, Leiden, The Neatherlands.

McGowan, K. 2001. Demographic and behavioral comparisons of suburban and rural American Crows, p. 367-383. *In* J. M. Marzluff, R. Bowman, and R. Donnelly [EDS.], Avian ecology and conservation in an urbanizing world. Kluwer Academic, Norwell, MA.

Monson, G. 1946. Notes on the avifauna of the Rio Grande Valley, New Mexico. Condor 48:238-241.

Parker, H. 1984. Effect of corvid removal on reproduction of willow ptarmigan and black grouse. J. Wildl. Manage. 48:1197-1205.

Parr, R. 1993. Nest predation and numbers of Golden Plovers *Pluvialis apricaria* and other moorland waders. Bird Study 40: 223-231.

Pierotti, R., and C. Annett. 2001. The ecology of Western Gulls in habitats varying in degree of urban influence, p. 309-331. *In* J. M. Marzluff, R. Bowman, and R. Donnelly [EDS.], Avian ecology and conservation in an urbanizing world. Kluwer Academic, Norwell, MA.

Pitelka, F. A. 1942. High population of breeding birds within an artificial habitat. Condor 44: 172-174.

Pulliam, H. R. 1988. Sources, sinks, and population regulation. Am. Nat. 132: 652-661.

Richards, G. L. (1971). The Common Crow, *Corvus brachyrhynchos*, in the Great Basin. Condor 73:116-118.

Rickard, W. H. 1959. Changes in winter bird species of two habitats in San Miguel County, New Mexico, after three-fourths of a century. Condor 61:438.

Robertson, J. M. 1931. Some changes in the bird life of western Orange County, California. Condor 33:204-205.

Robinson, S. K., Thompson, F. R. III, Donovan, T. M., Whitehead, D. R., and J. Faaborg. 1995. Regional forest fragmentation and the nesting success of migratory birds. Science 267:1987-1990.

Stacey, P. B. and M. Taper. 1992. Environmental variation and the persistence of small populations. Ecol. App. 2:18-29.

Tewksbury, J. J., S. J. Hejl, and T. E. Martin. 1998. Breeding productivity does not decline with increasing fragmentation in a western landscape. Ecology 79:2890-2903.

Verbeek, N. A. M., and R. W. Butler. 1981. Cooperatively breeding of the Northwestern Crow Corvus caurinus in British Columbia. Ibis 123:183-188.

Vekasy, M. S., J. M. Marzluff, M. N. Kochert, R. N. Lehman, and K. Steenhof. 1996. Influence of radio transmitters on Prairie Falcons. J. Field. Ornithol. 67:680-690.

Wackernagel, M., and W. E. Rees. 1996. Our ecological footprint : reducing human impact on the earth.: New Society Publishers, Philadelphia, PA.

Withey, J. C., Bloxton, T. D., and J. M. Marzluff. In press. Effects of tagging animals and estimating locations on inferences derived from wildlife radiotelemetry studies, p. 43-75. *In* J. J. Millspaugh and J. M. Marzluff [EDS.], Radiotelemetry and animal populations. Academic Press, San Diego, CA.

Appendix 16.1. Christmas Bird Count description and data summary.

CBC circle	Circle class[a]	Count years included	Average no. crows per party hr	% Fish Crows	Regression coefficient
Afton, MN	R	1960-1989	10.7	0.00	0.215
Aiken, SC	R	1961-1989	2.0	0.00	0.023
Albuquerque, NM	U	1962-1989	115.5	0.00	4.246
Arnett, OK	R	1967-1989	3.5	0.00	-0.064
Baker,OR	R	1960-1989	0.7	0.00	0.09
Baldwin, KS	R	1960-1989	4.6	0.00	0.089
Bismark, ND	U	1960-1989	0.0	0.00	0.001
Black Forest, CO	R	1960-1985	0.2	0.00	0.009
Boise, ID	U	1967-1989	0.3	0.00	0.055

CBC circle	Circle class[a]	Count years included	Average no. crows per party hr	% Fish Crows	Regression coefficient
Cedar, MN	R	1960-1989	5.5	0.00	0.323
Central Berkshire, MA	U	1968-1989	3.5	2.43	0.256
Chapel Hill, NC	U	1960-1989	2.7	0.00	0.016
Chicago, IL	U	1967-1989	5.5	0.00	0.296
Cincinnati, OH	U	1960-1989	8.2	0.00	-0.164
Colorado Springs, CO	U	1960-1989	4.3	0.00	0.19
Columbia, MO	R	1961-1989	2.4	0.00	0.037
Concord, NH	U	1965-1989	2.8	0.00	0.04
Dallas, TX	U	1960-1989	3.5	0.00	-0.062
Decorah, IA	R	1960-1989	5.7	0.00	0.296
Denver, CO	U	1960-1989	2.0	0.00	-0.068
Des Lacs, ND	R	1960-1989	0.0	0.00	0.001
Dubois, WY	R	1966-1989	0.0	0.00	0
Elverson, PA	R	1960-1989	10.9	0.19	-0.156
Espanola, NM	R	1960-1989	5.1	0.00	-0.064
Eugene, OR	U	1960-1989	9.1	0.00	0.018
Fargo, ND	U	1960-1989	1.2	0.00	0.109
Ferris, VT	R	1962-1989	42.2	0.00	0.084
Green Bay, WI	U	1964-1989	2.6	0.00	0.137
Gunnison, CO	R	1960-1989	1.0	0.00	0.019
Hackensack-Ridgewood, NJ	U	1960-1989	6.0	1.34	0.425
Hartford, CT	U	1960-1989	25.6	0.02	1.872
Houston, TX	U	1960-1989	0.9	0.00	0.015
Huntington, WV	U	1961-1987	4.7	0.00	-0.37
Indian, OH	R	1960-1988	0.4	0.00	0.03
Indianapolis, IN	U	1960-1989	8.9	0.00	0.121
Los Angeles, CA	U	1960-1989	0.5	0.00	0.034
Madison, SD	R	1961-1989	1.8	0.00	0.031
Maryville, MO	R	1966-1989	1.3	0.00	-0.053
Memphis, TN	U	1960-1989	0.8	0.00	0.024
Moscow, ID	R	1960-1989	0.2	0.00	0.014
Niles, MI	R	1962-1989	0.9	0.00	-0.053
Oklahoma City, OK	U	1960-1989	1.4	0.00	0.012
Omaha, NE	R	1962-1989	1.8	0.00	0.026
Orange Co (NE), CA	R	1960-1989	2.7	0.00	-0.152
Peshtigo, WI	R	1965-1989	5.3	0.00	0.441
Portland, OR	U	1960-1989	11.0	0.00	0.514
Point Reyes, CA	R	1971-1988	2.7	0.00	0.072

CBC circle	Circle class[a]	Count years included	Average no. crows per party hr	% Fish Crows	Regression coefficient
Putnam Co., NY	U	1960-1989	2.4	0.02	0.034
Rapid City, SD	U	1960-1989	0.8	0.00	0.052
Rock Run, MD	R	1964-1989	7.0	2.43	-0.216
Rupert, ID	R	1968-1989	4.8	0.00	3.722
Sacramento, CA	U	1960-1989	17.8	0.00	1.219
Scranton, PA	U	1960-1989	1.9	0.00	-0.035
Seattle, WA	U	1956-1997	10.2	0.00	0.573
Sioux Falls, SD	U	1960-1989	5.6	0.00	-0.026
Salt Lake City, UT	U	1960-1989	0.0	0.00	0
Spokane, WA	U	1960-1989	0.1	0.00	0.003
St. Paul, MN	U	1960-1989	3.2	0.00	0.215
Terre Haute, IN	R	1961-1989	2.7	0.00	-0.061
Topeka, KS	U	1960-1989	1.8	0.00	0.04
Traverse City, MI	U	1960-1989	1.1	0.00	-0.065
Trinidad, TX	R	1963-1989	4.8	0.00	0.025
Truckee, NM	U	1960-1989	0.0	0.00	-0.172
Tyler, TX	R	1960-1989	5.1	0.00	0.16
Walnut Valley, NJ	R	1968-1989	7.2	0.10	-0.302
Washita NWR, OK	R	1964-1989	1.8	0.00	-0.003
Watertown, NY	R	1960-1989	2.6	0.00	0.325
Waukegan, IL	R	1960-1989	44.6	0.00	-6.738
Wenatchee, WA	R	1961-1989	0.0	0.00	0.002
Wilmot, SD	R	1967-1987	0.5	0.00	0.04

[a]U=urban, R=rural or exurban.

Chapter 17

Demographic and behavioral comparisons of suburban and rural American Crows

Kevin J. McGowan
Department of Ecology and Evolutionary Biology, Corson Hall, Cornell University, Ithaca, NY 14853 USA, e-mail: kjm2@cornell.edu

Key words: American Crow, *Corvus brachyrhynchos*, cooperative breeding, demography, urban adaptation

Abstract: American Crows (*Corvus brachyrhynchos*) have become common suburban and urban birds in the last 50 years, using these environments for roosting and breeding. I studied crows in upstate New York from 1989 to 1999, and compared reproductive success, survival, and social behavior of crows in suburban and rural areas. Although both suburban and rural crows maintained permanent territories, suburban American Crows had smaller territories and nested at higher densities. Both suburban and rural crows periodically left their territories in fall and winter to join foraging aggregations and to roost at night. The proportion of nests attended by auxiliaries did not differ, but suburban nests were attended by larger families. Overall nest success was higher in suburban areas. Success rates during the egg stage of the nest cycle did not differ between habitat, but suburban nests had higher success during the nestling stage of the nest cycle. Survival of crows post fledging was better in suburban areas, but after one year annual survival rates did not differ between the habitats. Birds raised in one habitat tended to breed in the same habitat. Although clutch sizes and egg hatchability did not differ between the habitats, rural nests produced nearly one more young per successful nest and significantly larger young than suburban crows, suggesting that food resources were better in rural areas. Data from a drought year, a time of apparent food shortage, suggested that suburban food resources also were less dependable than those in rural areas. The increase in American Crows over the last 25 years in eastern North America may be due in part to the successful exploitation of suburban nesting habitats.

1. INTRODUCTION

Since the mid-1950's, American Crows (*Corvus brachyrhynchos*) have become common urban and suburban birds in North America, roosting and breeding in cities and towns (Jollie 1976, Houston 1977, Knight et al. 1987). Earlier, crows were more wary and avoided urban areas. Changing use of a landscape can affect such things as nesting structures employed, food used, exposure to predators, use of space (e.g., territory size and permanence, density), social structure (interaction with conspecifics, aggregations, family structure and cooperative breeding), and basic demographic factors. Some habitats may be attractive to immigrant birds, but if birds cannot reproduce successfully in these habitats, they may be an ecological trap (Gates and Gysel 1978). In Switzerland, Richner (1989) suggested that cities may be ecological traps for the Carrion Crow (*Corvus corone*). Crows may be attracted to breed in cities, but they cannot produce offspring that can effectively compete for breeding opportunities.

I examined whether American Crows nesting in a suburban environment differed in nest success, survival of the young, or in social behavior from crows in a rural environment.

2. METHODS

I studied American Crows in the city of Ithaca and the village of Cayuga Heights (Fig. 17.1a), and the surrounding rural areas of Tompkins County (Fig. 17.1b), New York from 1989 through 1999. Ithaca is a small city in the central Finger Lakes region of New York State, containing approximately 29,000 people in its 14.2 km^2 (2,062 people per km^2). Contiguous with Ithaca, Cayuga Heights is a residential area with a lower human density of 728 people per km^2. These areas are best classified as "suburban" (Marzluff et al. 2001). Just under 100,000 people live in Tompkins County (1995 data; Tompkins County Chamber of Commerce), with a density of 52 human per km^2 outside of the cities and villages. Rural areas surrounding Ithaca consist of agricultural fields, woodlots, extensive forests, cemeteries, and scattered residences. Approximately 50% of the county is wooded or reverting to forest (Smith et al. 1993), and 35% is in agriculture (1995 data; Tompkins County Chamber of Commerce).

The suburban areas used by crows consisted of residential and commercial districts, university and related properties, parks, and cemeteries. Much of the residential area contained many large mature trees. Rural areas varied from low density residential areas adjacent to agricultural and wildlands, to mixed agricultural fields and woodlots, to forest. For comparisons in this study, "suburban" crows were those located within Ithaca or Cayuga Heights with territories containing greater than 75% residential or commercial use (Fig. 17.1a). "Rural" territories fell outside of Ithaca and Cayuga Heights and contained greater than 50% forest, secondary scrubby growth, or agricultural fields and pastures. Rural territories could include residential area and some commercial use, but only if they composed less than 50% of the territory. Rural housing typically consisted of one or several well spaced homes along road frontage, bounded by extensive agricultural or wild lands (Fig. 17.1b). Territories with between 50 and 75% residential or other development were not used in analyses.

I measured territory size in the spring of 1994 and 1995 by mapping observations of tagged individuals engaging in territory defense. Exact territory size was determined for only a small percentage of families. For families where spot maps did not provide a complete territory map, I used the length of the longest axis across the territory to determine a maximum territory diameter as a measure of territory size. Because most territories

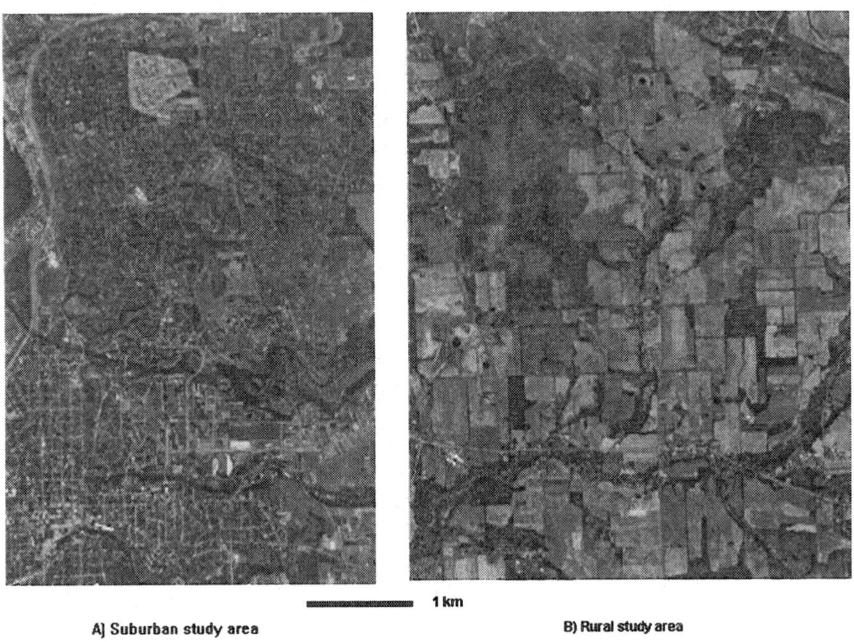

Figure 17.1. Aerial images of a) suburban study area in the city of Ithaca and the village of Cayuga Heights and b) rural study area, in the unicorporated areas of Tompkins County, east of Ithaca, NY. Photographs taken 27 March 1995.

that I accurately mapped were roughly circular, I estimated the area of all territories by assuming the longest axis represented the diameter of a circle and calculated the area of that circle. Each family of crows was used only once in the calculations; yearly fluctuations in territory size were not included. Instead, a best estimate from one year only was used for each family.

I located and monitored nests in each habitat in all years. I monitored nests from the ground for activity, and periodically climbed to the nests to examine nest contents. I marked nestlings at approximately one week before fledging (between day 24 and day 30 after hatch) with individually unique colored leg bands and patagial tags with unique combinations of letters or letters and numbers. All young were weighed and measured at that time.

Because nests were discovered at different stages in the nesting cycle, I calculated Mayfield daily survival probabilities (Mayfield 1961, 1975) to compare nest success at the various stages. Differences in nest success between the two habitats were tested using the *Z*-test developed by Hensler and Nichols (1981). A total of 215 suburban nests and 47 rural nests were used in the analyses. Incubation period was assumed to be 19 days, and nestling period 35 days.

No organized census of all marked individuals was conducted; sightings of individuals were recorded as they occurred throughout the seasons. I looked for crows at their home territories at irregular intervals throughout the year, and also searched foraging aggregations, and, less frequently, roosts for tagged individuals. An attempt was made to locate all banded individuals at the beginning of each breeding season. In addition, crows throughout the area were constantly checked for bands and tags. An extensive system of volunteer observers, recruited through public announcements and a high degree of bird awareness in the local area, contributed many important sightings of dispersing crows.

I used resightings of tagged individuals as capture/recapture data to estimate annual survival and recapture probabilities using the program MARK (White and Burnham 1999). All sightings and reports of individuals were tallied into 3 month blocks, roughly covering nestling and fledgling phases (May, June, and July); late summer and fall (Aug, September, and October); early winter (November, December, and January); and late winter and early spring, the pre- and early-nesting season (February, March, and April). All individuals entered the analyses at fledging (in the May-July block of their natal year). A total of 591 suburban crows and 162 rural ones were used in the analyses. I estimated survival assuming a constant yearly survival, but with age effects for the first and second years. Using MARK, I constructed a model to fit the data using age and environment as parameters for both survival and recaptures, then constructed another model combining environments in the survival component. I tested for a significant effect of environment by comparing the models using the Likelihood Ratio Test (Cooch and White 2000).

American Crows are known to be cooperative breeders, with extra individuals (auxiliaries) at nests in many parts of their range (Kilham 1984, 1989; Chamberlain-Auger et al. 1990; Caffrey 1992, 1999), including New York (McGowan 1995, 1996). Presence of auxiliaries and breeding group sizes were determined through nest watches and observations made during the fledgling feeding period. In suburban areas (but not rural areas; see Knight et al. 1987), crows typically mobbed me when I climbed to their nests. Although up to 75 crows from a neighborhood might join in a mobbing event, the core group of most persistent mobbers was a fairly accurate measurement of family size. In most families, not all members were individually marked; thus, I could not always determine the exact family size of groups with auxiliaries.

Values reported below are means ± SD, unless otherwise noted. Probabilities given are for two-tailed tests ($P < 0.05$), using Student's t tests, ANOVA, and log likelihood G-tests.

3. RESULTS

3.1 Territory Size and Nest Placement

Crows defended territories in both suburban and rural habitats. Crows in both habitats were present on the breeding territories throughout the year, and active territorial defence was observed in every month. However, outside of the breeding season, individuals did not restrict their activities solely to these territories, but left them periodically to forage during the day and to roost at night (see Stouffer and Caccamise 1991, and Caccamise et al. 1997 for a description of this behavior). Territory size (diameter) differed significantly between the two habitats (suburban = 0.31 km ± 0.12, n = 59; rural = 0.672 km ± 0.17, n = 18; t_{75} = 9.956, P < 0.001). Assuming territories were circular, on average, crows defended 8.7 ha in suburban areas (range 0.8 to 28.3) and 37.7 ha in rural (range 12.6 to 95.0). Within both study areas, all space was occupied, but the density of nesting crows was higher in suburban (11.5 territories per km^2) than in rural areas (2.6 territories per km^2).

Nest placement varied between suburban and rural crows. Suburban nests were significantly higher than rural nests (suburban = 18.7 m ± 5.92, n = 329, range = 4.75 to 34.2 m; rural = 14.9 m ± 4.26, n = 85, range = 4.5 to 24.0 m; t_{412} = 5.587, P <0.001), and were in taller trees (suburban = 21.2 ±

6.18, n = 327; rural = 18.0 ± 4.82, n = 85, t_{410} = 4.384, P < 0.001). Suburban nests also were placed higher in the nest tree (suburban = 87.8% of nest tree height; rural = 82.5%; t-test on arcsine transformed data, t_{410} = 5.14, P < 0.001). Suburban nests were more frequently located in coniferous trees than rural nests (96.5% conifers suburban, n = 386; 83% conifers rural, n = 100; G_1 = 19.338, P < 0.001), and crows placed nests lower in deciduous trees than in conifers (deciduous mean = 69.8% of nest tree height, n = 38; conifer = 87.5%, n = 429; t-test on arcsine transformed data, t_{465} = 11.087, P < 0.001). Although nest position in a tree was affected significantly by both habitat ($F_{1,407}$ = 7.56, P <0.01) and tree type ($F_{1,407}$ = 53.30, P < 0.001), nest height was affected significantly only by habitat ($F_{1,409}$ = 11.72, P = 0.001) and not tree type ($F_{1,409}$ = 0.70, P = >0.4). In neither analysis did an interaction between habitat and tree type exist.

3.2 Nest Success and Survival

Overall nest success was greater in suburban nests than in rural nests (Table 17.1). Survival of nests with eggs did not differ between habitats, but nests with nestlings had higher survival in the suburban habitat than in the rural habitat. Clutch size was identical in the two habitats (4.7; $n_{suburban}$ = 74, n_{rural} = 22), and the number of eggs that hatched per nest did not differ (suburban = 3.9 ± 1.02; rural = 4.1 ± 1.50; t_{47} = 0.43; P > 0.5). A total of 830 nestlings was marked and measured in 271 nests. Number of fledglings produced per nest in the populations did not differ between the habitats, but suburban nests produced fewer fledglings in each successful nest than rural nests (Table 17.2). Suburban nests experienced greater rates of brood reduction, decreasing the number of fledglings produced per nest.

Predation was the primary source of nest failure, with only a few nests lost to weather (high winds, lightning) or abandonment. Perhaps the most important nest predator at all stages of the nest cycle was the raccoon (*Procyon lotor*). Evidence of raccoon predation were feces in the nest tree and in the nest, destruction of the nest, teeth marks on plastic leg bands, and raccoons sleeping in the depredated nest. Eastern gray squirrels (*Sciurus carolinensis*) also were potentially significant egg predators. Although no direct evidence of squirrel predation was obtained, squirrels were actively chased from the vicinity of the nest tree by crows. Raccoons and Great Horned Owls (*Bubo virginianus*) were known to kill nestlings. Evidence of owl predation included decapitated nestlings, and owl feathers in the nests. Incubating or brooding adult crows (female breeders) also were killed while on the nest on several occasions, presumably by Great Horned Owls. No evidence of crows depredating other crow nests was observed.

Known predators on crow fledglings included domestic cats (*Felis catus*), raccoons, Cooper's Hawks (*Accipiter cooperii*), Red-tailed Hawks (*Buteo jamaicensis*), and Great Horned Owls. Evidence included visual observations of cats with fledglings in their mouths, Red-tailed Hawks with banded crows in their talons, remains of banded crows below Cooper's Hawk nests, and bands found in Great Horned Owl pellets. Other observed sources of mortality for fledglings included collision with automobiles, becoming trapped in vegetation, electrocution on an electrical transformer, and being killed or removed by people.

Rural crow nestlings measured at banding were heavier and larger than suburban crows in all linear body measures, except feathers (Table 17.3). By banding age, tarsus length of nestlings had attained 100% of adult size.

Post-fledging survival was significantly higher for suburban-raised than rural-raised crows ($\chi^2_3 = 11.517$, $P < 0.01$), with the greatest difference in the first year of life (Table 17.4). Annual survival after the first year was high (>70%) in both areas (Table 17.4), but the recapture rate was significantly lower in rural areas ($\chi^2_3 = 29.038$, $P < 0.01$). Although the three-age-class model (first, second, and subsequent years) provided the best fit to the data (highest AICc weight, Cooch and White 2000), it did not differ

Table 17.1. Mayfield probability of American Crow nest survival in suburban Ithaca and the surrounding rural Tompkins County, New York.

	Pdaily (SD)	Ptotal (SD)	# nests	*Z*	*P*
Eggs (19 days)					
Suburban	0.998 (0.002)	0.794 (0.008)	164		
Rural	0.989 (0.004)	0.815 (0.016)	39	1.2	n.s.
Nestlings (35 days)					
Suburban	0.991 (0.001)	0.721 (0.006)	155		
Rural	0.985 (0.004)	0.589 (0.015)	35	-8.2	<0.001
Overall (54 days)					
Suburban		0.573 (0.008)	202		
Rural		0.480 (0.015)	45	-5.5	<0.01

Table 17.2. Fledgling production by American Crows in suburban Ithaca and the surrounding Tompkins County, New York.

	SUBURBAN Mean ± SD (*n*)	RURAL Mean ± SD (*n*)	*t* (df)	*P*
Fledglings per Nest	1.62 ± 1.80 (287)	1.64 ± 2.00 (72)	0.120 (357)	0.905
Fledglings per Successful nest	3.06 ± 1.30 (151)	3.58 ± 1.46 (33)	3.339 (182)	0.044

significantly from a two-age-class model (χ^2_4 = 8.814, P = 0.07), which included only the first and subsequent years.

3.3 Dispersal

Fifty-seven individual crows marked as nestlings were later found breeding. Crows appeared to settle in areas similar to where they were raised. Of 49 crows raised in suburban territories, only two (one male and one female) bred on rural territories, and none of the eight rural-raised crows bred on suburban territories (Fisher exact test $P < 0.001$). Dispersal distance from natal site to first breeding ranged from 0 to 60 km and did not differ between the two habitats (suburban 3.2 km ± 8.80, n = 49; rural 3.3 ± 3.44, n = 8; $t_{25.7}$ = 0.06, $P > 0.5$). Three suburban-raised females traveled to other cities to breed, and one rural female bred on a rural territory on the opposite side of Ithaca from where it was raised.

Table 17.3. Size of American Crow nestlings at day 28 in suburban Ithaca and the surrounding rural Tompkins County, New York (years 1989-1999).

	SUBURBAN, n = 143	RURAL, n = 38		
	Mean ± SD	Mean ± SD	$t_{(179)}$	P
Bill (nares-tip) (mm)	22.9 ± 1.57	24.0 ± 1.46	2.743	0.007
Tarsus (mm)	58.8 ± 3.24	60.6 ± 2.18	3.157	0.002
Tail (mm)	65.6 ± 14.98	71.6 ± 15.31	2.186	0.30
Weight (g)	367 ± 54.1	411 ± 39.5	4.647	<0.001

Table 17.4. Survival and resighting (recapture) probabilities for American Crows in suburban Ithaca and the surrounding rural Tompkins County, New York.

	Age interval	Survival rate ± SE	95% CI	Resighting rate ± SE	95% CI
Suburban					
	0-1	0.52 ± 0.02	0.48 - 0.56	0.90 ± 0.02	0.85 - 0.94
	1-2	0.81 ± 0.03	0.74 - 0.86	0.90 ± 0.03	0.82 - 0.94
	Adult	0.76 ± 0.02	0.71 - 0.80	0.81 ± 0.03	0.75 - 0.86
Rural					
	0-1	0.35 ± 0.05	0.26 - 0.44	0.80 ± 0.08	0.60 - 0.91
	1-2	0.73 ± 0.14	0.41 - 0.92	0.48 ± 0.11	0.28 - 0.69
	Adult	0.72 ± 0.07	0.56 - 0.84	0.49 ± 0.09	0.32 - 0.67

3.4 Social Behavior

The occurrence of auxiliaries did not differ between suburban and rural nests. Auxiliaries were present at 238 of 299 (79.6%) suburban nests and at 37 of 62 (77.4%) rural nests (G_1 = 0.144, $P < 0.50$). However, the mean number of crows in family groups was larger in suburban families (suburban 3.7 ± 1.56, n = 286; rural 3.2 ± 0.99, n = 50; t_{334} = 2.001; $P < 0.05$). Few rural crow families exceeded 5 members (Fig. 17.2).

Both suburban and rural crows left their territories to enter winter roosts. Suburban and rural crows were found together in roosts and could be present in roosts in either habitat. Both suburban and rural crows also were observed in the same foraging flocks. Tagged crows were observed in foraging flocks and at their territory during the same day.

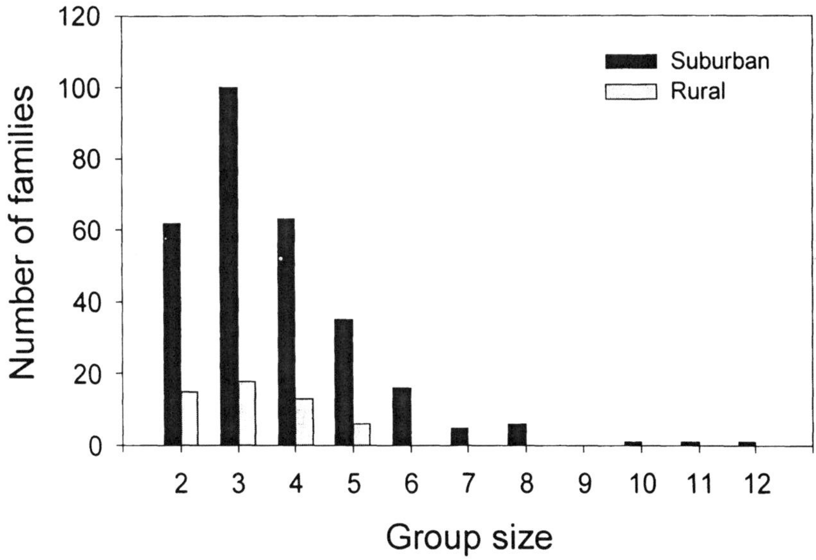

Figure 17.2. Distribution of American Crow family group sizes in suburban Ithaca (n = 286) and the surrounding rural Tompkins County, New York (n = 50), 1990-1998. Note that all groups larger than 5 were suburban groups.

4. DISCUSSION

American Crows exploit urban and suburban habitats and crow populations have been expanding slightly but significantly across the continent (Robbins et al. 1986, Marzluff et al. 1994, Sauer et al. 1999, Marzluff et al. 2001). However, in many areas, recent changes reflect a shift in the distribution of resident crows rather than changes in population size. As crows move into urban areas, both to nest and to roost in the winter, humans perceive an increase in population size. However, in New York, only a slight increase in overall crow populations has occurred (1.5%/year; BBS data, Sauer et al. 1999), but more crows appear to be using urban and suburban habitats. Recently, between 25,000 and 75,000 crows roosted in downtown Auburn, Cayuga County, New York (personal observation). Emlen (1938) reported a roost of approximately 25,000 crows in a rural area three miles south of Auburn during the winter of 1932-33. The town of Auburn has not increased substantially during these years, suggesting that crows here have moved from rural to suburban roosts. Long-time residents of Ithaca have reported that crows were not present in the city prior to the 1950's, and that their presence during the breeding season has occurred only recently.

Why have American Crows moved into urban and suburban areas to breed and roost? Houston (1977) and Knight et al. (1987) suggested that urban crows were taking advantage of the prohibition of shooting in urban areas. In addition, the 1972 extension of the Federal Migratory Bird Treaty Act of 1918 included crows for the first time, regulating and presumably decreasing hunting pressure on crows. Reduced persecution may have decreased human avoidance by crows, which then would enable them to exploit the trees, food, and relative warmth of cities

Suburban crows nested in taller trees and more frequently in conifers than did rural crows. Conifers are more abundant in suburban areas, but suburban crows selected conifers preferentially relative to their availability in Ithaca (unpubl. data). Whether rural and suburban crows have real differences in nest site selection awaits further study. Tall trees may be one of the attractions for crows to nest in urban and suburban areas, especially where available nest sites outside of towns are scarce, such as the intensively farmed Midwest and the arid West.

With few exceptions, suburban-raised crows bred in suburban areas, and rural crows bred in rural areas. All crows had opportunities to visit both habitats, at least during the fall and winter when crows made extensive movements off their territories to forage and roost. The settlement pattern of crows likely was affected to some degree by strategies used to acquire territories. Many crows settled close to their natal territories. Males in particular frequently inherited their natal territories or budded a breeding territory from their natal territory (unpubl. data). However, some crows traveled long distances to find breeding opportunities, suggesting that crows have the opportunity to sample a large number of possible breeding sites.

At least in Ithaca, suburban areas appear to be productive breeding areas for American Crows, not ecological traps. Suburban crows had higher nest success and first-year survival than did rural crows. Higher survival of suburban nests and juveniles may be a result of fewer predators in these habitats. The primary predators of crows are Great Horned Owls (Caffrey 1999, pers. obs), and perhaps several large hawks (*Accipiter gentilis*, *A. cooperii*, *Buteo jamaicensis*, *B. lineatus*) (Bent 1946). Important nest predators are the same raptor species (Chamberlain-Auger et al. 1990, Caffrey 1999, this study), as well as raccoons (Kilham 1989, this study). Populations of these predators might be lower in cities than in rural areas. Nesting density of Red-tailed Hawks in urban/suburban Syracuse, New York, and perhaps Great Horned Owls as well, was approximately one third that of rural areas (Minor et al. 1993). Although no data exist for local predator populations, all of the predators mentioned above were present in the suburban study area, and raccoons, Great Horned Owls, Red-tailed Hawks and Cooper's Hawks all are becoming common urban residents throughout North America (Minor et al. 1993, Rosenfield et al. 1996, Boal and Mannan 1998, Riley et al. 1998). Cooper's Hawk densities in urban areas may reach the maximum known for the species (Rosenfield et al. 1996, Boal and Mannan 1998), and urban raccoon populations can reach very high levels (Riley et al. 1998). The survival data in this study suggest that egg predators (e.g., raccoon) might be present in equal densities in both habitats, but that predators of nestlings and/or fledglings (e.g. Great Horned Owl) may be present in lower densities in town.

Studies in Switzerland suggested that cities may be ecological traps for Carrion Crows (Richner 1989, 1990, 1992). Richner (1989) showed that Carrion Crows breeding in cities had slightly higher overall nest success than those breeding in agricultural areas, but that they produced fewer young per successful nest, and those fledglings were of inferior quality. The offspring produced were significantly smaller than rural young and size was related to eventual breeding success. Offspring with small tarsi rarely became breeders and more urban-raised crows than rural-raised crows (79% vs. 29%) had tarsi below a threshold that predicted future breeding. Richner (1990) also found auxiliaries at urban nests, something previously unknown for the species. Crows that could not successfully breed on their own stayed and helped their parents (Richner 1990). When the urban crows were provided with supplemental food, their nest success and the size of their young increased (Richner 1992).

I observed a similar pattern of higher nest success, but fewer and smaller young in successful suburban nests. However, size at fledgling did not appear to affect the probability of attaining breeding status in American Crows. Because breeding is delayed in American Crows for up to six years (unpubl. data), relatively few data were available. Of 624 suburban crows banded, 49 (8%) are known to have become breeders, compared to only 8 of 171 (5%) rural crows ($G_1 = 2.20$, $P > 0.10$). However, suburban families had more auxiliaries than rural families (Fig. 17.2), a pattern which Richner (1990) suggested meant that urban Carrion Crows were less competitive and more likely to remain as auxiliaries. The apparent difference in group size could be the result of sampling bias. Rural crows tend to be less aggressive to people visiting their nests (Knight et al. 1987) and consequently provide fewer opportunities to accurately assess their group sizes.

The consistent pattern of fewer and smaller young from urban nests observed in two crow species (Richner 1992, this study), suggests that food may be limiting in urban populations. Although urban areas provide anthropogenic sources of food for crows, availability of natural food sources may be lower or the overall quality of food may be lower than in rural areas (Pierotti and Annett 2001).

During a severe drought in 1995, (unpubl. data, Northeast Regional Climate Center, Cornell University), I noticed several indications of hardship amongst the crows. Drought may affect crows by reducing their primary food source during nesting (earthworms and other terrestrial invertebrates; personal observation). During this year, nestlings appeared limp and flaccid instead of plump and turgid. During 1995, crows abandoned more nests in response to my visits than in any other year. During six previous years, I observed only one suspected abandonment as a result of my activities; in 1995 six abandonments occurred. One family abandoned a two-week-old nestling after I visited the nest.

Because suburban birds had access to food unlikely to have been affected by the drought, I predicted the impact of the drought would be smaller in the suburban area than in the rural area. In fact, I observed the opposite pattern. Crows in both habitats produced fewer fledglings in 1995 than in other years, although the difference was not significant (Table 17.5). Suburban crows produced fewer fledglings per nest and successful nest than rural crows, but again the differences were not significant. However, the relative differences between suburban and rural were greater in 1995 than in previous or subsequent years, suggesting the impact of the drought was greater on the suburban birds. As in previous years, suburban young were smaller, but the differences in 1995 were more pronounced (Table 17.6), and even tail differed significantly ($t_{38} = 2.470$, $P < 0.02$). The drought did not appear to affect the size of rural-raised young, but suburban young were smaller in 1995 than in other years. Whether the drought reduced the availability of natural foods for suburban crows or forced them to switch to anthropogenic sources of food of lower quality is unknown, but it suggests that food resources for suburban crows may be less predictable between years.

Why do suburban crows have smaller territories than rural crows if food is more limited? Most foraging to feed nestlings and fledglings is conducted on the territory, thus it seems logical that a larger territory would provide more food. It is possible that crowding may force smaller territory size in urban areas. Alternatively, the cues used by crows to determine the appropriate territory size may be inadequate or inappropriate in urban or suburban environments.

Table 17.5. Fledgling production by American Crows in 1995 in suburban Ithaca and the surrounding Tompkins County, New York.

	SUBURBAN Mean ± SD (*n*)	RURAL Mean ± SD (*n*)	*t* (df)	*P*
Fledglings per Nest	1.59 ± 2.89 (32)	1.79 ± 3.86 (18)	0.242 (48)	0.810
Fledglings per Successful nest	2.68 ± 1.89 (19)	3.44 ± 1.53 (9)	1.408 (26)	0.17

Table 17.6. Size of American Crow nestlings at day 26 in suburban Ithaca and surrounding rural Tompkins County, New York, during a drought year (1995) and other years.

Measurement (mm)	1989-1994, 1996-1999 Mean ± SD (*n*)	1995 only Mean ± SD (*n*)	*t* (df)	*P*
Bill (nares-tip)				
Suburban	22.7 ± 1.46 (83)	20.8 ± 1.72 (11)	3.929 (92)	<0.001
Rural	23.9 ± 1.60 (29)	22.8 ± 0.74 (11)	2.156 (38)	0.037
Tarsus				
Suburban	59.3 ± 3.71 (83)	56.0 ± 4.18 (11)	2.735 (92)	0.007
Rural	61.2 ± 1.82 (29)	62.0 ± 2.50 (11)	1.198 (38)	0.238
Tail				
Suburban	55.7 ± 11.81 (80)	48.4 ± 14.82 (11)	1.865 (89)	0.065
Rural	60.1 ± 12.79 (29)	55.5 ± 11.51 (11)	1.033 (38)	0.308
Weight (g)				
Suburban	378 ± 53.7 (82)	328 ± 57.0 (11)	2.884 (91)	0.005
Rural	398 ± 47.4 (29)	421 ± 25.3 (11)	1.488 (38)	0.145

In Ithaca and elsewhere (Stouffer and Caccamise 1991, Caccamise et al. 1997), crows defended territories, but left those areas for foraging and roosting, especially in winter. Kilham (1985) noted year-round territoriality in crows in New Hampshire, and Chamberlain-Auger et al. (1990) reported the same in Massachusetts. In Ithaca, territories appeared to be important in winter; the breeding pair and varying numbers of offspring were present every day and actively defended the territory by calls, displays, and fights. During the breeding season, crows foraged exclusively within their territory. Attendance at foraging flocks dropped off sharply when breeding began (unpubl. data).

Territory size and nesting densities found in this study are within the range reported in other studies, although these estimates vary widely. In Saskatchewan, Ignatiuk and Clark (1991) found a significant difference in the density of nesting crows between two different study sites; one in relatively open parkland, the other more highly developed for agriculture (0.59 - 0.80 vs. 0.36 - 0.40 nests/km^2, respectively). They suggested that these differences might be due to differences in production rates of young, nest site availability, or food between habitats. Sullivan and Dinsmore (1992) calculated average breeding season home ranges of crows in Manitoba of 2.6 km^2 (0.38 pairs/km^2). Chamberlain-Auger et al. (1990) reported a mean territory size of 42 ha (2.4 pairs/km^2) for crows in Massachusetts. They indicated that territory size differed significantly between an urban study area and a barrier beach, but presented no data nor indicated which was larger. Kilham (1989) reported an average territory size of 90 ha for two rural territories of resident Florida crows, although one territory eventually split into two of approximately 40 ha each. Caffrey (1992) found that crows on an urban golf course in California did not defend territories, but did maintain "core areas" on which they nested and spent most of their time. She reported a mean nesting density of 0.8 pairs/ha = 80 nests/km^2, far higher than the density found in suburban areas in this study. Emlen (1942), reported a "colony" of crows breeding in a rural walnut orchard in California, with approximately 60 pairs in the 108 acres (= 140 pairs/km^2). Still, it is not clear why territories tend to be small in urban areas. Food as well as social organization may play a role in the variability in home range size and density observed among studies and between rural and urban habitats.

Crows frequently depredate the nests of other birds (Sodersrom et al. 1998, Buler and Hamilton 2000). The movement of crows into urban and suburban areas potentially could affect songbird populations there. In Ithaca, two species frequently cited by the public as being negatively affected by crows are American Robin (*Turdus migratorius*) and Northern Cardinal (*Cardinalis cardinalis*), both common nesters in suburban Ithaca.

However, cardinals have not changed in abundance survey-wide from 1966 to 1996, and robins have increased at nearly the same rate as crows (BBS data; Sauer et al. 1999). In New York, cardinals have increased and robins have not changed in abundance (BBS data; Sauer et al. 1999), suggesting urban crow populations are not adversely affecting these two species. It is possible that urban crow populations could impact bird species with habitats that make them vulnerable to crow predation. Common Nighthawks (*Chordeiles minor*) have declined significantly over North America, and especially in New York (BBS data; Sauer et al. 1999). Urban nighthawks nest on gravel roofs of buildings. Although I have no evidence to suspect crows contributing to the decline of nighthawks, it is conceivable that urban crows could discover and then specialise on such nests. Any species with such a specialised nest site, such as a gravel or beach nest that is vulnerable once a search image has been formed, might well be adversely affected by increasing crow populations.

In conclusion, American Crows are now nesting in cities and towns in large numbers. Suburban crows have higher nest success, but produce fewer and smaller young. However, their subsequent survival is higher than that of rural crows and they appear to be equally able to recruit into the breeding population. Therefore I conclude that cities are breeding opportunities and not ecological traps for American Crows. This conclusion is supported by evidence from other regions in the U.S. (Marzluff et al. 2001). A small, steady and significant increase in American Crow populations, especially in the eastern USA (BBS data; Robbins et al. 1986, Sauer et al. 1999) is perhaps the result of exploitation of the "new" urban habitat.

ACKNOWLEDGMENTS

Financial support for part of the crow research was provided by U.S. Dept. Agriculture, Hatch Project Grant NYC-183429. I thank a large number of diligent field assistants, paid and volunteer, who have helped me find and monitor crow nests over the last decade, and, most importantly, held securely on to the other end of my rope. Thanks to Frank Joyce, David Winkler, and Mark Stanback, for advice early in the study on how to climb trees, and especially to Rob Roy Ramey II for one vital technique and the imperative to do it all safely. Thanks to Carolee Caffrey for boundless crow enthusiasm and for teaching me her tagging technique, and to Kimberly Kline for tireless construction of the tags. The manuscript was significantly improved by the comments of Jon Greenlaw, Reed Bowman, and an anonymous reviewer. Special thanks must be extended to the extensive bird-aware community of people living in and around Ithaca, who provided me with numerous crow sighting reports.

REFERENCES

Bent, A. C. 1946. Life histories of North American jays, crows, and titmice. U. S. Nat. Mus. Bull. 191:1-495.

Boal, C. W., and R. W. Mannan. 1998. Nest-site selection by Cooper's Hawks in an urban environment. J. Wildl. Manage. 62: 864-871.

Buler, J. J., and R. B. Hamilton. 2000. Predation of natural and artificial nests in a southern pine forest. Auk 117:738-747.

Caccamise, D. F., L. M. Reed, J. Romanowski, and P. C. Stouffer. 1997. Roosting behavior and group territoriality in American Crows. Auk 114:628-637.

Caffrey, C. 1992. Female-biased delayed dispersal and helping in American Crows. Auk 109:609-619.

Caffrey, C. 1999. Feeding rates and individual contributions to feeding at nests in cooperatively breeding Western American Crows. Auk 116: 836-841.

Chamberlain-Auger, J. A., P. J. Auger, and E. G. Strauss. 1990. Breeding biology of American Crows. Wilson Bull. 102:615-622.

Cooch, E., and G. White. 2000. Program Mark: a gentle introduction. http://canuck.dnr.cornell.edu/misc/cmr/mark/docs/.

Emlen, J. T., Jr. 1938. Midwinter distribution of the American Crow in New York state. Ecology 19:264-275.

Emlen, J. T., Jr. 1942. Notes on a nesting colony of Western Crows. Bird-Banding 13:143-154.

Gates, J. E., and L. W. Gysel. 1978. Avian nest dispersion and fledging success in field-forest ecotones. Ecology 59: 871-883.

Hensler, G. L., and J. D. Nichols. 1981. The Mayfield method of estimating nest success: a model, estimators, and simulation results. Wilson Bull. 93:42-53.

Houston, C. S. 1977. Changing patterns of Corvidae on the prairies. Blue Jay 35:149-155.

Ignatiuk, J. B., and R. G. Clark. 1991. Breeding biology of American Crows in a Saskatchewan parkland habitat . Can. J. Zool. 69:168-175.

Jollie, M. 1976. Species interrelationships of three corvids. The Biologist 58: 89-111.

Kilham, L. 1984. Cooperative breeding of American Crows. J. Field Ornithol. 55:349-356.

Kilham, L. 1985. Territorial behavior of American Crows. Wilson Bull. 97:389-390.

Kilham, L. 1989. The American Crow and the Common Raven. Texas A & M University Press, College Station, Texas.

Knight, R. L., D. J. Grout, and S. A. Temple. 1987. Nest-defence behavior of the American Crow in urban and rural areas. Condor 89: 175-177.

Marzluff, J. M., R. B. Boone, and G. W. Cox. 1994. Historical changes in populations and perceptions of native pest bird species in the West. Stud. Avian Biol. 15:202-220.

Marzluff, J. M., K. J. McGowan, R. Donnelly, and R. L. Knight. 2001. Causes and consequences of expanding American Crow populations, p. 333-365. *In* J. M. Marzluff, R. Bowman, and R. Donnelly [EDS.], Avian ecology and conservation in an urbanizing world. Kluwer Academic, Norwell, MA.

Mayfield, H. F. 1961. Nesting success calculated from exposure. Wilson Bull. 73:255-261.

Mayfield, H. F. 1975. Suggestions for calculating nest success. Wilson Bull. 87:456-466.

McGowan, K. J. 1995. A test of whether economy or nutrition determines fecal sac ingestion in nesting corvids. Condor 97:50-56.

McGowan, K. J. 1996. Family lives of the uncommon American Crow. Cornell Plantations 51:1-4.

Minor, W. F., M. Minor, and M. F. Ingraldi. 1993. Nesting of Red-tailed Hawks and Great Horned Owls in a central New York urban/suburban area. J. Field Ornithol. 64:433-439.

Pierotti, R., and C. Annett. 2001. The ecology of Western Gulls in habitats varying in degree of urban influence, p. 309-331. *In* J. M. Marzluff, R. Bowman, and R. Donnelly [EDS.], Avian ecology and conservation in an urbanizing world. Kluwer Academic, Norwell, MA.

Richner, H. 1989. Habitat-specific growth and fitness in Carrion Crows (*Corvus corone corone*). J. Anim. Ecol. 58:427-440.

Richner, H. 1990. Helpers-at-the-nest in Carrion Crows *Corvus corone corone*. Ibis 132:105-108.

Richner, H. 1992. The effect of extra food on fitness in breeding carrion crows. Ecology. 73:330-335.

Riley, S. P. D., J. Hadidian, and D. A. Manski. 1998. Population density, survival, and rabies in raccoons in an urban national park. Can. J. Zool. 76:1153-1164.

Robbins, C. S., D. Bystrak, and P. H. Geissler. 1986. The breeding bird survey: its first fifteen years, 1965-1978. U. S. D. I., F. W. S., Res. Publ. 157.

Rosenfield, R. N., J. Bielefeldt, J. L. Affeldt, and D. J. Beckmann. 1996. Urban nesting biology of Cooper's Hawks in Wisconsin, p. 41-44. *In* D. Bird, D. Varland, J. Negro [EDS.], Raptors in human landscapes. Academic Press, New York.

Sauer, J. R., J. E. Hines, I. Thomas, J. Fallon, and G. Gough. 1999. The North American Breeding Bird Survey, Results and Analysis 1966 - 1998. Version 98.1, USGS Patuxent Wildl. Res. Center, Laurel, MD.

Smith, B. E., P. L. Marks, and S. Gardescu. 1993. Two hundred years of forest cover in Tompkins County, New York. Bull. Torrey Botanical Club 120:229-247.

Sodersrom, B., T. Part, and J. Ryden. 1998. Different nest predator faunas and nest predation risk on ground and shrub nests at forest ecotones: An experiment and a review. Oecologia 117:108-118.

Stouffer, P. C., and D. F. Caccamise. 1991. Roosting and diurnal movements of radio-tagged American Crows. Wilson Bull. 103:387-400.

Sullivan, B. D., and J. J. Dinsmore. 1992. Home range and foraging habitat of American crows *Corvus brachyrhynchos* in a waterfowl breeding area in Manitoba. Can. Field-Nat. 106:181-184.

White, G.C., and K. P. Burnham. 1999. Program MARK: Survival estimation from populations of marked animals. Bird Study 46 Suppl. 120-138.

Chapter 18

Nest success and the timing of nest failure of Florida Scrub-Jays in suburban and wildland habitats.

Reed Bowman and Glen E. Woolfenden
Archbold Biological Station, P.O. Box 2047, Lake Placid, FL USA 33862 e-mail: rbowman@archbold-station.org

Key words: Florida Scrub-Jay, food, nest success, predation, urbanization

Abstract: Urbanization has broad and complex effects on ecological systems. Many of these effects are likely to influence nest success of birds, some positively and others negatively, thus complex interactions among effects are likely to exist that influence patterns of success. Nest success of Florida Scrub-Jays did not differ between a suburban and wildland population or across a human density gradient within the suburban population. Nesting success between egg and nestling stages did not differ in the wildland population, but nest success during the egg stage was high and success during the nestling stage low in the suburban population. As a consequence, nest contents survived longer during the nesting cycle in the suburban population than in the wildland population, even though overall rates of nest success did not differ. In the suburbs, nests with eggs were less likely to be deserted and nests with nestlings were more likely to be depredated or more likely for the nestling to starve than in the wildlands. In both habitats, nest success declined throughout the nesting season. Because suburban birds begin nesting earlier than wildland birds, nests are at a more advanced stage when nest failure rates increase. However, temporal differences in nesting are absent for second and third nest attempts, yet suburban nests still fail later in the nesting cycle. This suggests that site-specific differences exist in the predator community or the behavior of birds increasing their vulnerability to predators that affects timing of nest failure.

1. INTRODUCTION

Growth of urban areas has been widespread not only in North America, but worldwide and is expected to continue or accelerate in the future

(Douglas 1992). The process of urbanization and concomitant human modification of the natural landscape is likely to have profound effects on ecological processes occurring at a variety of scales, from landscapes to individual organisms. Because of the ubiquitous effects of urbanization, complex interactions may exist among anthropogenic and natural variables at several different levels of biological organization (McDonnell and Pickett 1990). Our work seeks to disentangle some of the interacting effects of urbanization on nest success in one species of bird.

Changes in the structure and composition of avian communities with urbanization are well documented (Graber and Graber 1963, Erz 1966, Tomialojc 1970, Batten 1972, Walcott 1974, DeGraaf and Wentworth 1981, Bezzel 1985, Natahura and Imai 1996, Rottenborn 1999). With urbanization, the density of birds tends to increase, but species diversity and evenness tends to decrease (Emlen 1974, Aldrich and Coffin 1980, Beissinger and Osborne 1982). However, avian community structure still tends to be influenced by habitat structure (Lancaster and Rees 1979, Beissinger and Osborne 1982, Mills et al. 1989), thus the aforementioned patterns do not necessarily increase or decrease linearly across urban gradients (Blair 1996). Different patterns of habitat structural changes, such as an increase in exotic vegetation cover, may favor different bird species, thus altering the composition of the bird community (Mills at al. 1989). However, other potential mechanisms underlying these community changes may include decreases in natural food availability (Burke and Nol 1998 in agricultural setting) and increases in anthropogenic sources of food or water (Emlen 1974, Lancaster and Rees 1979), increases (Crooks and Soulé 1999, Wilcove 1999, Haskell et al. 2001) or decreases (Mackin-Rogalska et al. 1988, Langen et al. 1991, Bolger et al. 1997) in the abundance of potential nest predators, or interference competition (Engels and Sexton 1994). Each of these mechanisms is likely to have a predictable effect on the demography of a species, depending on whether it tends to increase or decrease in abundance with urbanization.

The most common demographic attribute measured in urban studies of birds has been nest success. However, many of these studies have used artificial nests (see Haskell et al. 2001 for a review). These studies examine rates of predation on nests with eggs rather than a variety of effects on overall nest success and therefore may not be particularly useful in predicting the interactive effects of urbanization on avian demography. Those few studies that have examined comparative success of real nests (Howard 1974, Eden 1985, Boal and Mannon 1999, Milsap and Bear 2000) have shown varying responses to urbanization, but few examined the causes of differences in success among habitats.

Species-specific differences in the effect of urbanization on nest success may result from the many sources of variation in nest success and the complex interactions with anthropogenic variables. For example, availability of supplemental food, such as anthropogenic sources common in urban areas, may advance laying date (Boutin 1989). Changes in laying date may influence nest success because many birds experience a seasonal decline in nest success that may, in part, reflect seasonal activity patterns of different predators. Thus anthropogenic alterations in both food and predator abundance in urban habitats are likely to have different, but interacting, effects on nest success.

The Florida Scrub-Jay (*Aphelocoma coerulescens*) is a cooperative breeder endemic to Florida and declining range-wide. Nearly 30% of the extant population occurs in habitat patches surrounded by human residential development and populations in urbanizing areas appear to be declining (Breininger 1999). Predation is the primary cause of nest failure (Schaub et al. 1992, Woolfenden and Fitzpatrick 1996). In addition, supplemental food available to the jays in suburban habitats appears to advance their laying date (Bowman et al. 1999). The objectives of this study were to compare nest success between a suburban and wildland population and to gain insights on the interactive effects of timing of breeding and habitat in explaining patterns of nest failure.

2. METHODS

Between 1993 and 1998, we studied the nesting success of Florida Scrub-Jays near Lake Placid, Highlands County, Florida, at Placid Lake Estates, a 1,500 ha residential housing subdivision, and at Archbold Biological Station, a 2,000 ha natural preserve (Fig. 18.1). The two sites, 8 km apart, are connected by contiguous scrub habitat. Development began in Placid Lakes, hereafter referred to as the suburban site, in the mid-1960's and has extended south so that a gradient in housing density exists, from a maximum density of 180 houses/40 ha to fewer than 20 houses/40 ha. Scrub-jays occur in scrub patches throughout the density gradient, but most (~80%) occur at human densities less than 40 houses/40 ha.

Within the suburban site, we mapped each house and used the procedure KRIG (Mitchell 1999) to construct a contour map of housing density based on isocline intervals of 20 houses/40 ha. Because only a relatively small proportion of our suburban study site had housing densities exceeding 40 houses/40 ha, we pooled all density classes above 40 houses/40ha into a single high density class. Thus, for analytical purposes, we defined three different housing density classes: low density (<20 houses/40 ha), medium density (20-39 houses/40 ha), and high density (≥40 houses/40 ha). Every year, each scrub-jay territory was assigned a housing density class based on the spatial location of the territory centroid relative to the housing density isoclines. New houses were mapped and added to the GIS point coverage. The KRIG procedure was performed once annually, just prior to the breeding season, and new isoclines were constructed.

At both study sites, all scrub-jays were uniquely banded and periodically censused. During the breeding season, we attempted to find every nest in every territory. Most nests were found during building or egg-laying. Once found, nests were checked every 3-4 days to determine completed clutch size, the number of eggs that hatched, and the number of young that fledged. Nest visits were kept as brief as possible, usually less than 20-30 sec., just time enough to accurately record nest contents. We banded and measured all nestlings on day 11 post-hatching. During subsequent checks, nests were viewed with binoculars from 5-10 m. Nests were considered successful if one or more nestlings fledged. We assigned a cause for all nest failures. Deserted nests were untended by adults for at least one day prior to failure. We assigned hatch failure at nests where incubation continued past the anticipated hatch date based on an 18 day incubation period (Woolfenden and Fitzpatrick 1984). Depredated nests often showed physical disturbance, but we assigned predation for most losses that could not be otherwise explained. Starvation of young, wind damage, and human causes (typically destruction of habitat) of nest loss were rare enough that they were pooled for contingency table analysis; however, we report actual proportions for each cause. We quantified causes of nest failure as the proportion of all nests that failed because of a specific cause.

We calculated nest success during the egg and nestling stages and for both stages combined. The egg stage occurred from the laying of the first egg to the hatching of the first egg. The nestling stage was from the hatching of the first egg to fledging. Overall nesting success was the proportion of nests with eggs that fledged at least one young. Egg stage success was defined as the proportion of nests with eggs in which at least one egg hatched. Nestling stage success was defined as the proportion of nests with nestlings that fledged at least one young. We chose this traditional method of reporting nest success, rather than the commonly used Mayfield method (Mayfield 1961, 1975), because we found more than 95% of all nests before or during egg laying. The Mayfield method corrects a bias that exists when a large proportion of nests is found later in the nesting cycle.

Florida Scrub-Jays often attempt multiple nests in a single year and all nest attempts were included in our database. To avoid lack of independence, we calculated the mean annual nest success in each territory during each year. These data were not normally distributed; therefore, we used the non-

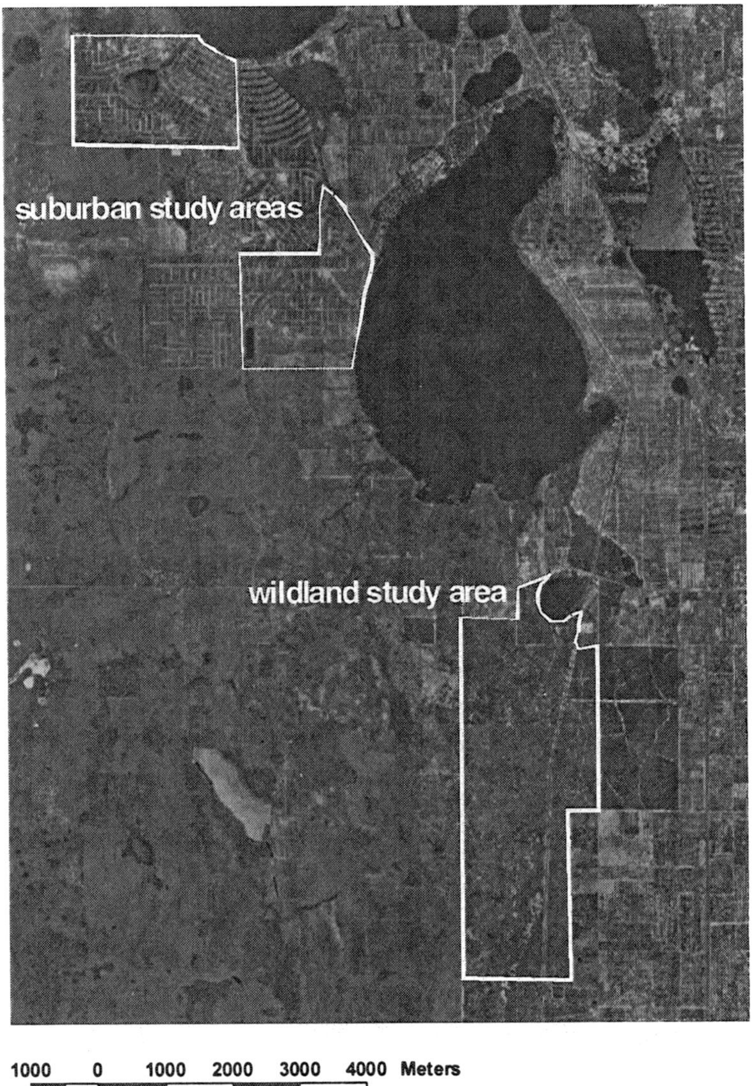

Figure 18.1. Aerial photograph of suburban and wildland study areas near the southern terminus of the Lake Wales Ridge, Lake Placid, Highlands County, Florida.

parametric Kruskal-Wallis test to examine the effect of year or habitat (suburban vs. wildland) or housing density class on mean nest success. However, parametric analysis of variance tends to be robust to deviations from normality and homogeneity of variances, thus we used two-way analysis of variance to test the interaction between these variables. Where interactions were significant and Kruskal-Wallis and ANOVA results did not differ, we report results of both tests. For pair-wise comparisons, we used Dunnett's C test, which does not assume equal variances (SPSS 1999). To test the independent effects of time of laying and habitat on nest success, we used logistic regression using first-nest attempts only. We used Chi-square tests to examine variation in causes of nest failure between habitats and between housing density classes.

For each failed nest, we determined the time of nest failure relative to laying of the first egg. The failure date was estimated as the midpoint between the date the nest was last active and the date it was first determined that the nest had failed, which usually was only a few days. For analyses of the temporal patterns of nest survival, the data set used (1993 - 1998) for the suburban site comprised 765 nests and for the wildland site 507 nests. Survival data were analyzed using the Kaplan-Meier procedure (SPSS 1999). Statistical comparisons of cumulative survival curves were made using the Breslow or generalized Wilcoxon test (SPSS 1999).

Scrub-jays in the suburban site annually begin nesting earlier than do scrub-jays in the wildland site (Bowman et al. 1999), thus we used ANCOVA to control for the effects of laying date, when examining site-specific differences in the time of nest failure. To avoid non-independence of data, we analyzed first, second, and third nest attempts separately.

3. RESULTS

3.1 Patterns of Nest Success

For all years combined, nest success of scrub-jays did not differ between the suburban site (0.43) and the wildland site (0.49) (Kruskal-Wallis: χ^2 = 2.40, P = 0.122); however, in 1996, nest success was significantly higher in the wildland site than in the suburban site (Table 18.1). Nest success did vary between years (χ^2 = 20.72, P = 0.001), but no interaction existed in the effect of year and habitat on nest success. In the suburban site, success during the egg stage (0.74) was higher than during the nestling stage (0.57). In the wildland site, success was similar during both stages (0.66 and 0.69, respectively). For all years combined, nest success during the egg stage was significantly higher in the suburban site (0.74) than in the wildland site

(0.66) (χ^2 = 5.09, P = 0.024); however, although the pattern was similar within years, the differences were significant only during 1998. Nest success during the nestling stage was significantly lower in the suburban site (0.57) than in the wildland site (0.69) (χ^2 = 7.21 P = 0.013), but within years, the differences were significant only during 1996.

Within the suburban site, the overall success of scrub-jay nests did not differ among housing density classes (χ^2 = 1.83, P = 0.399). Nest success did vary between years (χ^2 = 12.72, P = 0.026) (Table 18.2), but no interaction existed in the effect of year and human density on nest success. During 1994, overall nest success was highest in the low housing density class, but in 1995, overall nest success was highest in the high housing density class. In all three suburban density classes, success during the egg stage was higher than during the nestling stage. No other significant patterns within years existed.

In both habitats, overall nest success declined significantly ($F_{1,10}$ = 6.97, P < 0.001) through the nesting season. The general pattern of decline was similar in both habitats (Fig. 18.2a). However, in the suburban site, nests initiated very early in the nesting season had lower nest success than

Table 18.1. Mean annual success [± 1 SE (n)] of all nest attempts per scrub-jay territory during the egg and nestling stages and for the entire nesting cycle in a suburban and wildland population. Samples sizes indicate the number of breeding groups with nests.

Site/Year	Eggs	Nestlings	Overall
Suburbs			
1993	0.74 ± 0.04 (65)	0.73 ± 0.05 (60)	0.55 ± 0.05 (65)
1994	0.82 ± 0.03 (63)	0.57 ± 0.05 (62)	0.47 ± 0.05 (63)
1995	0.69 ± 0.04 (81)	0.52 ± 0.05 (70)	0.36 ± 0.04 (81)
1996	0.77 ± 0.04 (70)	0.56 ± 0.05 (65)	0.45 ± 0.05 (70)
1997	0.71 ± 0.05 (64)	0.48 ± 0.06 (55)	0.35 ± 0.05 (64)
1998	0.73 ± 0.05 (54)	0.57 ± 0.06 (47)	0.42 ± 0.06 (54)
Cumulative	0.74 ± 0.02 (397)	0.57 ± 0.03 (359)	0.43 ± 0.02 (397)
Wildlands			
1993	0.62 ± 0.05 (46)	0.76 ± 0.06 (40)	0.47 ± 0.05 (46)
1994	0.78 ± 0.05 (48)	0.70 ± 0.06 (44)	0.59 ± 0.06 (48)
1995	0.65 ± 0.06 (46)	0.62 ± 0.08 (34)	0.42 ± 0.07 (46)
1996	0.77 ± 0.05 (63)	0.77 ± 0.05 (57)	0.64 ± 0.06 (64)
1997	0.59 ± 0.05 (65)	0.65 ± 0.07)48)	0.42 ± 0.06 (65)
1998	0.56 ± 0.06 (46)	0.58 ± 0.09 (33)	0.37 ± 0.07 (46)
Cumulative	0.66 ± 0.03 (315)	0.69 ± 0.04 (256)	0.49 ± 0.03 (315)

predicted by a fit regression line. A similar pattern of seasonally declining nest success was evident at all housing density classes within the suburban site (Fig 18.2b). The general pattern of seasonal decline was similar at all density classes ($F_{2,10} = 2.37$, $P = 0.10$). Date of laying the first egg of the first clutch, rather than habitat, was a significant predictor of nest success (logistic regression, $\beta = -0.014$, $P = 0.004$), but the model explained relatively little of the overall variation (adj. $r^2 = 0.14$).

Table 18.2. Mean annual success [± 1 SE (*n*)] of all nest attempts per scrub-jay territory during the egg and nestling stages and for the entire nesting cycle at different housing density classes (low class, <20 houses/40 ha; medium class, 20-39 houses/40 ha; high class ≥40 houses/40 ha) in a suburban population. Sample sizes indicate the number of breeding groups.

Site/Year	Eggs	Nestlings	Overall
Low Density			
1993	0.76 ± 0.06 (31)	0.78 ± 0.07 (29)	0.59 ± 0.07 (31)
1994	0.92 ± 0.03 (33)	0.65 ± 0.07 (33)	0.58 ± 0.07 (33)
1995	0.66 ± 0.06 (46)	0.46 ± 0.08 (38)	0.30 ± 0.06 (46)
1996	0.79 ± 0.04 (43)	0.56 ± 0.06 (41)	0.46 ± 0.06 (43)
1997	0.69 ± 0.06 (41)	0.54 ± 0.08 (35)	0.39 ± 0.07 (41)
1998	0.73 ± 0.06 (32)	0.55 ± 0.09 (28)	0.43 ± 0.08 (32)
Cumulative	0.75 ± 0.02 (226)	0.58 ± 0.03 (204)	0.45 ± 0.03 (226)
Medium Density			
1993	0.74 ± 0.08 (18)	0.69 ± 0.10 (16)	0.52 ± 0.10 (18)
1994	0.67 ± 0.06 (15)	0.46 ± 0.11 (15)	0.30 ± 0.08 (15)
1995	0.70 ± 0.07 (22)	0.52 ± 0.09 (20)	0.33 ± 0.06 (22)
1996	0.68 ± 0.09 (16)	0.54 ± 0.12 (14)	0.40 ± 0.10 (16)
1997	0.73 ± 0.11 (13)	0.32 ± 0.10 (11)	0.36 ± 0.09 (13)
1998	0.73 ± 0.10 (14)	0.65 ± 0.12 (12)	0.45 ± 0.10 (14)
Cumulative	0.71 ± 0.01 (98)	0.53 ± 0.04 (88)	0.38 ± 0.04 (98)
High Density			
1993	0.72 ± 0.07 (16)	0.67 ± 0.11 (15)	0.52 ± 0.10 (16)
1994	0.77 ± 0.08 (15)	0.51 ± 0.11 (14)	0.38 ± 0.09 (15)
1995	0.82 ± 0.09 (13)	0.71 ± 0.11 (12)	0.60 ± 0.11 (13)
1996	0.83 ± 0.10 (11)	0.58 ± 0.13 (10)	0.53 ± 0.13 (11)
1997	0.77 ± 0.11 (10)	0.43 ± 0.15 (9)	0.33 ± 0.13 (10)
1998	0.71 ± 0.13 (8)	0.50 ± 0.19 (7)	0.34 ± 0.16 (8)
Cumulative	0.77 ± 0.04 (73)	0.58 ± 0.05 (67)	0.46 ± 0.05 (73)

3.2 Causes and Timing of Nest Failure

During the egg stage, the causes of nest failure differed between the suburban site and the wildland site (Chi-square: $\chi^2 = 44.7$, DF = 3, $P < 0.001$). Nest desertion during the egg stage was lower in the suburban site (0.036) than in the wildland site (0.122). In both sites, most desertions occurred during laying of the first nesting attempt. Predation as a cause of nest failure during the egg stage was lower in the suburban site (0.179) than in the wildland site (0.204). Whole clutch losses to hatching failure, defined as clutches that survive the entire incubation period but fail to hatch, were higher in the suburban site (0.029) than in the wildland site (0.007). Wind accounted for more nest losses in the suburban site (0.013) than in the wildland site (0.007).

During the nestling stage, the categorical causes of nest failure did not differ between the two habitats (Chi-square: $\chi^2 = 2.8$, DF = 2, $P = 0.09$). Predation as a cause of nest failure during the nestling stage was higher in the suburban site (0.400) than in the wildland site (0.307). In the suburban site, starvation of the entire brood affected 1.7% of all nests and losses caused by desertion, wind, and human causes affected less than 1% of all nests. In the wildland site, no starvation-related losses occurred and losses caused by desertion affected less than 1% of all nests.

Within the suburban site, the cause of nest failure during the egg stage varied significantly among housing density classes ($\chi^2 = 19.95$, DF = 6, $P = 0.003$). Predation as a cause of nest failure was highest at medium housing density and lowest at high housing density (range 11.8 to 23.0%). Hatching failure was highest at high housing density and lowest at medium housing density (range 1.3 to 4.7%) and desertion was highest at low housing density and lowest at medium housing density (range 2.6 to 4.1%). Wind-related losses increased with housing density from 0.2 to 1.2%. Within the suburban site, no significant differences existed in the causes of nest failure during the nestling stage. Predation was the primary cause of nest loss and ranged from 20.9 to 27.0%.

Patterns of nest success suggested that a difference in the timing of nest failure existed between the two habitats. Nests failed significantly later ($F_1 = 52.02$, $P < 0.001$) after the laying of the first egg in the suburban site (20.6 ± 0.5 days) than in the wildland site (14.7 ± 0.6 days). Cumulative nest survival curves were significantly different between the suburban site and the wildland site during both the egg stage ($\chi^2 = 27.65$, $P < 0.001$) and the nestling stage ($\chi^2 = 3.89$, $P = 0.048$) (Fig. 18.3). In the suburban site, nests failed later in the nesting cycle and breeding pairs averaged more nest attempts per season than in the wildland site (1.93 ± 0.04 vs. 1.62 ± 0.04). Therefore, breeding females in the suburban site spent more time tending

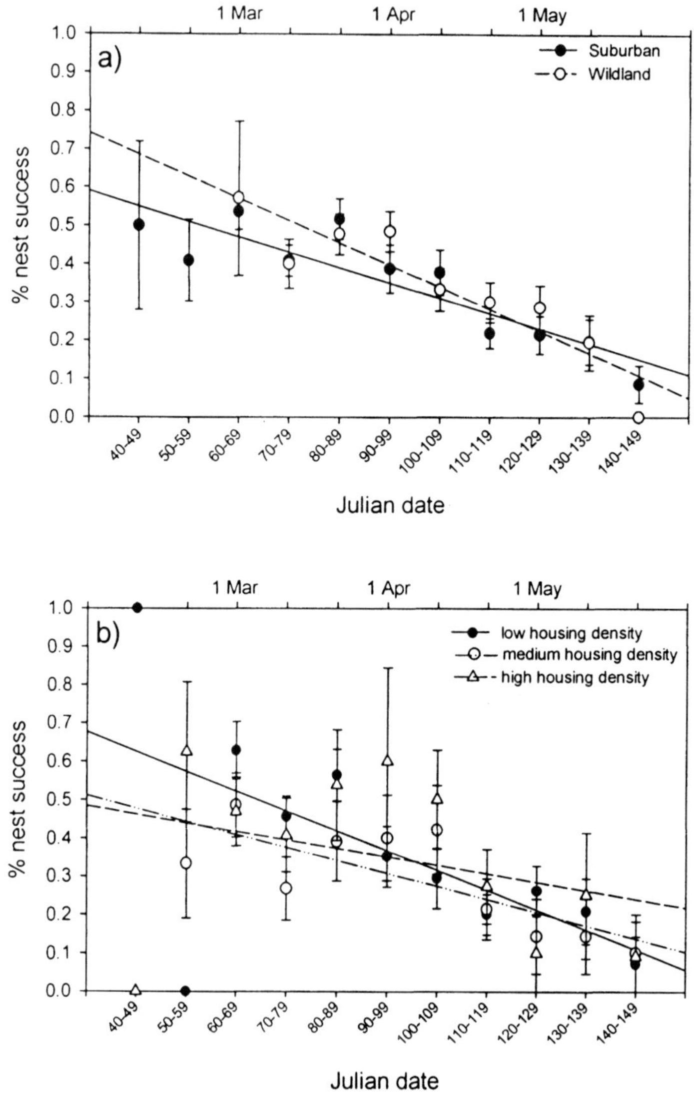

Figure 18.2. Seasonal success (± 1 SE) of Florida Scrub-Jay nests initiated during 10-day intervals during the nesting season in a) a suburban and wildland site and b) at three housing density classes (low density, <20 houses/40 ha; medium density, 20-39 houses/40 ha; high density, ≥ 40 houses/40 ha) within the suburban site, 1993 to 1998.

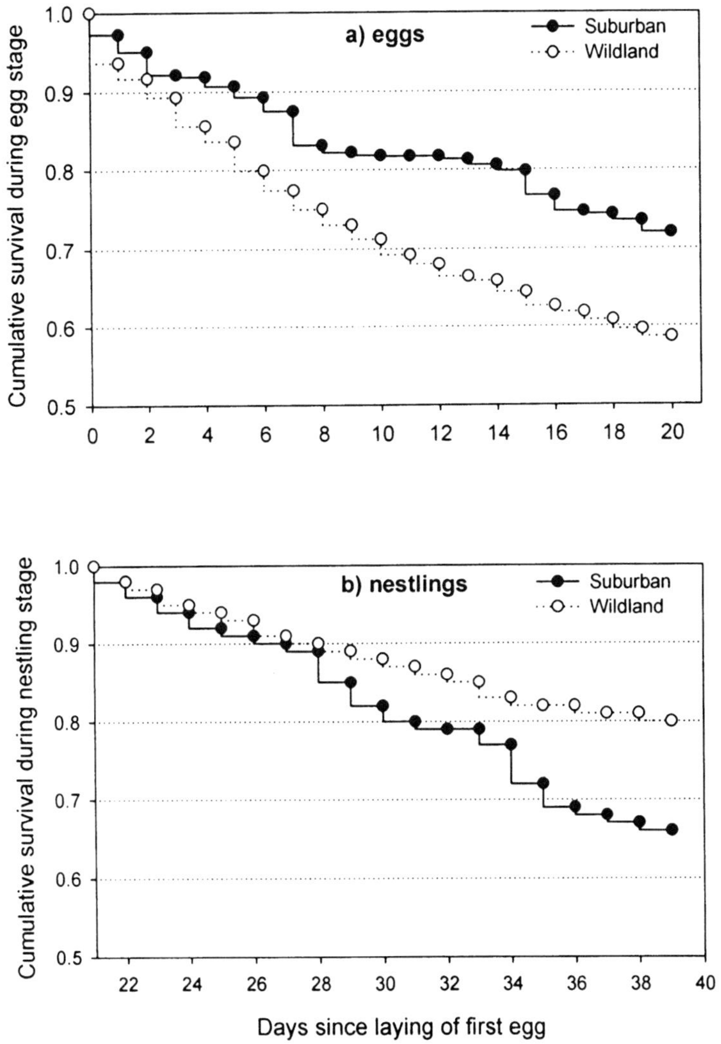

Figure 18.3. Cumulative survival of Florida Scrub-Jay nests during the a) egg and b) nestlings stages in a suburban and wildland site, 1993 to 1998. Statistical significance of pairwise comparisons determined by Breslow (generalized Wilcoxon) rank test.

nests (40.1 days) that were ultimately unsuccessful than did females in the wildland site (24.1 days).

3.3 Interaction of Laying Date, Nest Attempt, and Habitat on Time of Nest Failure

Because scrub-jays breed earlier in the season in the suburban site than in the wildland site (Bowman et al. 1999), we examined the relative effects of laying date and habitat on time of nest failure. For all nests combined, in both habitats, timing of nest failure did not vary with laying date or with nest attempt. However, for each nest attempt, the effect of habitat and laying date on time of nest failure differed.

For first-nest attempts, birds that laid eggs early in the season tended to fail late in the nesting cycle, regardless of habitat (Table 18.3). When we corrected for the difference in timing of laying, the adjusted mean time of nest failure in the suburban site was 18.1 days and 16.3 days in the wildland site. First-nest attempts were laid an average of 17 days earlier in the suburban site than in the wildland site (Fig. 18.4). However, time of nest failure influenced the laying date of subsequent nest attempts. The later a nest failed in the nesting cycle, the later the laying date for subsequent attempts; thus the difference in laying date between habitats for second-nest attempts was only five days. For second-nest attempts, laying date had less of an effect on time of nest failure, but habitat had an independent and significant effect. When we corrected for the difference in timing of laying, the adjusted mean time of nest failure was 22.4 days in the suburban site and

Table 18.3. Analysis of covariance of the effects of habitat (suburbs vs. wildland) and laying date on the timing of nest failures for first, second, and third-nest attempts of Florida Scrub-Jays, 1993 to 1998. No interaction existed between habitat and laying date for any nest attempt, thus the assumption of equality of slopes was justified for each nest attempt.

Source	DF	MS	*F*	*P*
First nest attempt (n = 392)				
Habitat	1	278.3	1.97	0.161
Laying date	1	994.9	7.06	0.008
Second nest attempt (n = 288)				
Habitat	1	3815.3	33.14	0.001
Laying date	1	418.5	3.64	0.058
Third nest attempt (n = 101)				
Habitat	1	497.7	13.97	0.002
Laying date	1	28.9	0.23	0.632

14.7 days in the wildland site. This pattern was even stronger for third-nest attempts, by which time the mean laying date in the suburban site was five days later than in the wildland site (Fig. 18.4). When we corrected for the difference in timing of laying for third nest attempts, the adjusted mean time of nest failure was 20.8 days in the suburban site and 15.9 days in the wildland site.

4. DISCUSSION

Nesting success in our two populations of Florida Scrub-Jays followed the patterns of other passerines in many respects. At both sites, nesting success declined throughout the season and nesting success varied between years. Predation was the primary cause of failure. Overall nest success was similar in our suburban (0.43) and wildland (0.49) habitats and did not differ across a range of different human densities within the suburbs. Of the relatively few studies that simultaneously compared nest success of urban and rural birds, the overall patterns seem to be ambiguous, in part because the focal species often responds differently to urbanization. Howard (1974) reported that nest success of American Robins (*Turdus migratorius*) was low in forests, high in low human density habitats (farms, cemeteries, parks), and intermediate in areas of high human density. Along an urban gradient in Ohio, robin density peaked in low human density habitats (Blair 1996), in concordance with the demographic data. In Florida, Burrowing Owls (*Athene cunicularia*) are thought to be declining with increasing urbanization, but nest success did not differ along a gradient from low to high human residential density (Milsap and Bear 2000). Conversely, Cooper's Hawks (*Accipiter cooperii*) sometimes nest at high densities in urban areas, but have lower nest success than when nesting in lower human density habitats (Boal and Mannon 1999), in part because urban nestlings have a higher incidence of *Trichomonas* infection (Boal et al. 1998). American Crows also are less successful where they are most dense (Marzluff et al. 2001). Success of breeding Black-billed Magpies (*Pica pica*) was similar in urban and rural populations, but the sources of nest failure differed (Eden 1985). Predation of urban nests was lower, but these nests suffered higher hatching failure and nestling starvation.

Although nest success of scrub-jays did not differ between habitats, differences in the timing of nest failures existed and investigating these differences give insight into the causes of nest failure in Florida Scrub-Jays. In addition, interactive effects on the timing of nest failure differed between nest attempts and investigating these differences reveals potential effects of ecological differences between the habitats.

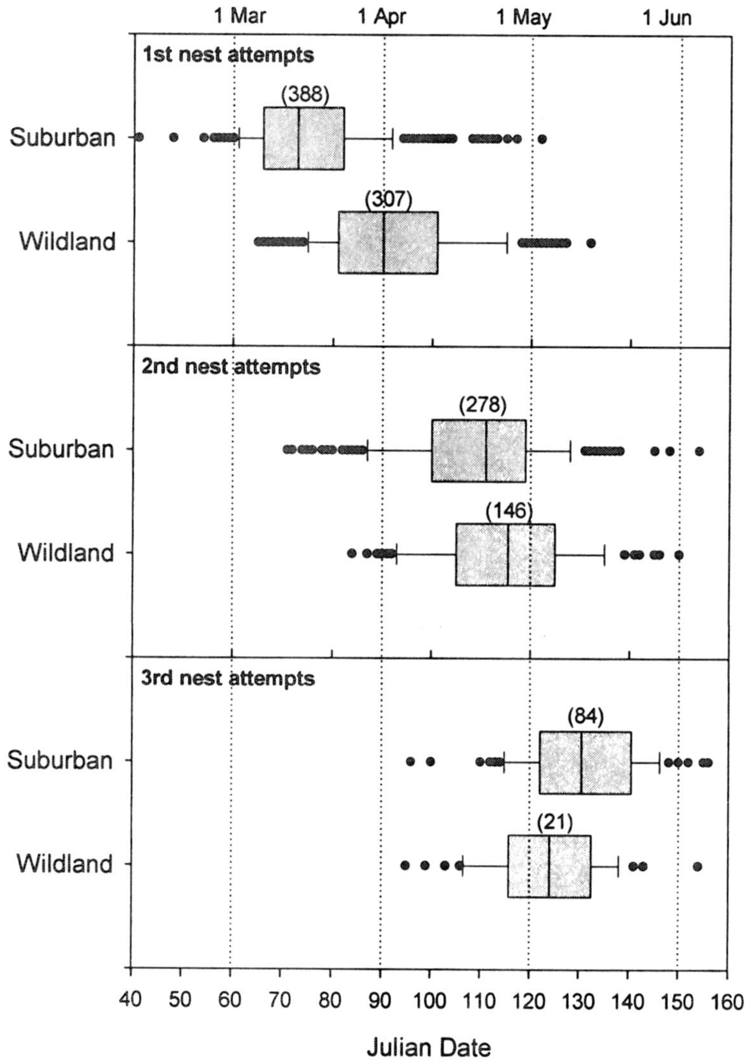

Figure 18.4. Horizontal box plot of the time of laying for first-, second-, and third-nest attempts of Florida Scrub-Jays in a suburban and wildland site, 1993 to 1998. Vertical lines within the shaded boxes represent the median laying date. The shaded box represents the timing during which the middle 50% of the sample initiated laying. Error bars represent the timing of laying for the first 10th and 90th percent of the sample. Filled circles represent outliers that initiated laying very early or very late.

One major difference between the habitats was that nesting failures in the suburban site were lower during the egg stage than during the nestling stage, whereas in the wildlands, failures were the same during these two stages. During the egg stage, nesting failures were lower in the suburbs (0.26) than in the wildlands (0.34). Although the proportion of nesting failures from hatching failure and wind was slightly higher in the suburbs than in the wildlands, the proportion of losses caused by predation and, especially, nest desertion was lower, accounting for the overall lower nesting failure rates. We suggest that food availability accounts for habitat-specific differences in desertion rates during laying. Scrub-jays often desert early nesting attempts, especially during cold weather. Cold weather likely reduces local food availability that may provide a cue that predictably abundant food is not yet available for raising nestlings. In the suburbs, anthropogenic sources of food may augment natural food sources so that suburban jays may be less likely to abandon early nesting attempts. Within the suburbs, nest desertion was highest at low human density, where a few jay territories do not have access to anthropogenic foods. If nest desertion is a response to cues about future conditions, birds that do not abandon their nests because of inappropriate cues may be more likely to attempt raising nestlings under adverse conditions. In part, this might explain why the earliest nesting jays in the suburbs had lower nest success than predicted and why date, and not site, explained the difference in the timing of nest failures for first nest attempts, but not for subsequent nest attempts.

Losses caused by hatching failure were higher in the suburban population. The suburban site has a higher proportion of inexperienced breeders than does the wildland site (24.1% vs. 13.7%, $\chi^2 = 11.8$, $P = 0.001$; see also Breininger 1999) and, in the suburbs, inexperienced pairs had higher rates of hatching failure than pairs with at least one experienced member (7.6% vs. 2.0%). Losses caused by wind were higher in the suburbs and tended to increase with human density. High nests in tall, isolated shrubs are most vulnerable to wind. Fire suppression, one consequence of urbanization, leads to taller oaks and concomitantly higher jay nests. Habitat destruction often results in isolated oaks that are used by scrub-jays for nesting in the suburban habitat.

During the nestling stage, nesting failures were higher in the suburban site (0.43) than in the wildlands (0.31). Predation accounted for most losses in both habitats, but was higher in the suburbs. The proportion of nesting failures from starvation was low in both habitats, but higher in the suburbs than in the wildlands. Despite the availability of anthropogenic foods, natural food availability is likely to change with urbanization. The relative abundance of many arthropods, including lepidopterans, decreases with increasing urbanization (Blair and Launer 1997, Denys and Schmidt 1998,

Gunnarsson and Hake 1999). Lepidopteran larvae and orthopterans are less abundant, and the mass of 11-day old nestlings is lower, in our suburban study site than in our wildland site (unpubl. data), suggesting that lower arthropod availability results in less food for nestlings. At least in the suburbs, entire broods may occasionally starve. In other bird species, begging call rates and intensity may increase with short-term food deprivation (Iacovides and Evans 1998, Leonard and Horn 1998). Increased begging has been associated with an increased predation risk (Halupka 1998, Dearborn 1999), suggesting that higher rates of nestling predation in our suburban site may be indirectly affected by differences in food availability between the sites.

Predation is the primary cause of nest failure in Florida Scrub-Jays (Schaub et al. 1992, Woolfenden and Fitzpatrick 1996) and differences in predation likely contribute to habitat-specific timing of nest failure. Schaub et al. (1992) found that the activity of diurnal snakes and birds increased through the scrub-jay breeding season and that daily nest predation rates were correlated with an index of diurnal predator activity. Because scrub-jays nest earlier in the suburbs, a greater proportion of those nests are likely to contain nestlings rather than eggs as predator activity increases. This explains the difference in the timing of nest failures between habitats for first nest attempts. However, these temporal differences in nest initiation become increasingly smaller for second and third attempts, yet nests in the suburban site continued to fail later in the cycle than nests in the wildland site. To us, this suggests that ecological differences between the sites alter the vulnerability of nests with eggs and nestlings as we have suggested above or that differences exist in the community of predators between the two habitats.

Scrub-jay nest predators are certain species of birds, mammals, and snakes. Snakes tend to be the major predators of the contents of Florida Scrub-Jay nests (Schaub et al. 1992). The eastern indigo (*Drymarchon corais*) and eastern coachwhip (*Masticophis flagellum*), both known predators of scrub-jays (Hailman et al. 1994, pers. obs.), occur in our study sites, but we have no data on their relative abundance. Few data exist regarding the effects of urbanization on snake communities; however, habitat loss, fragmentation, and alterations are likely to results in declines among many species of reptiles (Gibbons et al. 2000) and snakes in particular (Shine 1991). Even very low density human development (exurban) can seriously impact populations of large snakes such as the eastern indigo (Moler 1992). However, certain avian nest predators, such as Blue Jays (*Cyanocitta cristata*), Fish Crows (*Corvus ossifragus*), and Common Grackles (*Quiscalus quiscalus*), are more common in our suburban study area and tend to increase with urbanization (Blair 1996). Furthermore, scrub-jay densities are higher in the suburban scrub, and they rob other scrub-jay nests of eggs (Woolfenden 1974). Some predators that prey on nestlings are likely to be abundant in suburbs (foxes, Adkins and Stott 1998; raccoons, Riley et al. 1998; black rats, Bolger et al. 1997; among others), but these predators also have access to sources of anthropogenic foods and this may decrease the degree to which they prey on nests (see Doncaster and Macdonald 1997, Schmidt 1999).

Although nest success did not differ between the wildland and suburban sites, an interaction existed between laying date, nest attempt, and habitat on the timing of nest failure. These interactions may exist because of more complex interactions between changes in food availability, both anthropogenic and natural, the timing of breeding, the composition and activity patterns of the predator community, and behavioral responses of birds in response to these variables. But if nest success does not differ, then are these interactions important? Only female scrub-jays incubate and brood their young, thus their activity remains tied to their nest, potentially increasing their vulnerability to predators (Slagsvold and Dale 1996). Females in the suburbs spend, on average, 16 more days tending nests than birds in wildlands, even though they are no more successful. The potential costs of incubation and brooding are high and the fitness consequences to these additional costs may exist (Monoghan and Nager 1997). In addition, the breeding period for suburban jays is longer (Fig. 18.4) and coincides with a period when adult mortality rates increase. Annually adult mortality rates are highest when care of dependent fledglings and the onset of definitive prebasic molt overlap (Woolfenden and Fitzpatrick 1984).

4.1 Policy and Management Implications

Ecological changes associated with urbanization are complex and widespread and determining effects at multiple scales may require thoroughly testing the complex interactions that exist between anthropogenic and natural variables. In this case, we have identified that these interactions exist and influence the timing of nest failures, but we have not yet identified the specific pathways among variables. Policy and management recommendations would be premature without knowing specific causes and effects, and the long-term consequences of changes in the timing of nest failure. Future research should focus on the effects of anthropogenic food on behavior of both birds and their predators, whether the increased time suburban birds tend nests represents an increase in reproductive effort, and the potential costs and fitness consequences of this effort.

ACKNOWLEDGEMENTS

Both these long-term studies have depended upon the contributions, both intellectually and in the field, of numerous colleagues, students, field assistants, and volunteers. We greatly appreciate their participation. Archbold Biological Station has provided excellent support throughout the duration of these projects. This manuscript has benefited from discussions with our colleagues and from the careful and thoughtful reviews of R.L. Curry and J.S. Greenlaw. Our work has been supported by grants from the United States Fish and Wildlife Service, the Florida Game and Fresh Water Fish Commission, and the National Science Foundation.

REFERENCES

Adkins, C. A., and P. Stott. 1998. Home ranges, movements, and habitat associations of red foxes *Vulpes vulpes* in suburban Toronto, Ontario, Can. J. Zool. 244:335-346.

Aldrich, J. W., and R. W. Coffin. 1980. Breeding bird populations from forest to suburbia after thirty-seven years. Am. Birds 34:3-7.

Batten, L. A. 1972. Breeding bird species diversity in relation to increasing urbanisation. Bird Study 19:157-166.

Beissinger, S. R., and D. R. Osborne. 1982. Effects of urbanization on avian community organization. Condor 84:75-83.

Bezzel, E. 1985. Birdlife in intensively used rural and urban environments. Ornis Fennica 62:90-95.

Blair, R. B. 1996. Land use and avian species diversity along an urban gradient. Ecol. Appl. 6:506-519.

Blair, R. B., and A. E. Launer. 1997. Butterfly diversity and human land use: species assemblages along an urban gradient. Biol. Conserv.1:113-125.

Boal, C W., R. W. Mannan, and K. S. Hudelson. 1998. Trichomoniasis in Cooper's Hawks from Arizona. J. Wildl. Manag. 34:590-593.

Boal, C. W., and R. W. Mannan. 1999. Comparative breeding ecology of Cooper's hawk in urban and exurban areas of southwestern Arizona. J. Wildl. Manage. 63:77-84.

Bolger, D. T., A. C. Alberts, R. M. Sauvajot, P. Potenze, C. McCalvin, D. Tran, S. Mazzoni, and M. A. Soulé. 1997. Response of rodents to habitat fragmentation in coastal southern California. Ecol. Appl. 7:552-563.

Boutin, S. 1989. Food supplementation experiments with terrestrial vertebrates: patterns, problems, and the future. Can. J. Zool. 68:203-220.

Bowman, R., G. E. Woolfenden, and J. W. Fitzpatrick. 1999. Timing of breeding and clutch size in the Florida Scrub-Jay, p. 2465-2479. *In* N. J. Adams and R. H. Slotow [EDS.], Proc. 22nd Intl. Ornithol. Congr., Durban, BirdLife South Africa.

Breininger, D. R. 1999. Florida Scrub-Jay demography and dispersal in a fragmented landscape. Auk 116:520-527.

Burke, D. M., and E. Nol. 1998. Influence of food abundance, nest-site habitat, and forest fragmentation on breeding ovenbirds. Auk 115:96-104.

Crooks, K. R., and M. E. Soulé. 1999. Mesopredator release and avifaunal extinctions in a fragmented system. Nature 400:563-566.

Dearborn, D. C. 1999. Brown-headed cowbird nesting vocalizations and risk of predation. Auk 116:448-457.

DeGraaf, R. M., and J. M. Wentworth. 1981. Urban bird communities and habitats in New England. Trans. 46th. N. Amer. Wildl. Nat. Res. Conf. 1981:396-413.

Denys, C., and H. Schmidt. 1998. Insect communities on experimental mugwort (*Artemisia vulgaris* L.) Plots along an urban gradient. Oecologia 113:269-277.

Douglas, I. 1992. The case for urban ecology. Urban Nat. Mag. 1:15-17.

Doncaster, C. P., and D. W. Macdonald. 1997. Activity patterns and interactions of red foxes (*Vulpes vulpes*) in Oxford city. J. Zool. (Lond.) 241:73-87.

Eden, S. F. 1985. The comparative breeding biology of magpies *Pica pica* in an urban and rural habitat (Aves: Corvidae). J. Zool. (Lond.) 205:325-334.

Emlen, J. T. 1974. An urban bird community in Tucson, Arizona: derivation, structure, regulation. Condor 76:184-197.

Engels, T. M., and C. W. Sexton. 1994. Negative correlation of blue jays and golden-cheeked warblers near an urbanizing area. Conserv. Biol. 8:286-290.

Erz, W. 1966. Ecological principles in the urbanization of birds. Ostrich Suppl. 6:357-363.

Gibbons, J. W., D. E. Scott, T. J. Ryan, K. A. Buhlman, T. D. Tuberville, B. S. Metts, J. L. Greene, T. Mills, Y. Leiden, S. Poppy, and C. T. Winne. 2000. The global decline of reptiles, déjù amphibians. BioScience 8:653-666.

Graber, R. R., and J. W. Graber. 1963. A comparative study of bird populations in Illinois, 1906-1909 and 1956-1958. Ill. Natl. Hist. Surv. 28:383-528.

Gunnarsson, B., and M. Hake. 1999. Bird predation affects canopy-living arthropods in city parks. 1999. Can. J. Zool. 77:1419-1428.

Hailman, J. P., K. J. McGowan, and G. E. Woolfenden. 1994. Role of helpers in the sentinel behaviour of the Florida Scrub Jay (*Aphelocoma c. coerulescens*). Ethology 97:119-140.

Halupka, K. 1998. Vocal begging by nestlings and vulnerability to nest predation in meadow pipits *Anthus pratensis*: to what extent do predation costs of begging exist? Ibis 140:144-149.

Haskell, D., A. M. Knupp, and M. C. Scneider. 2001. Nest predator abundance and urbanization, p. 245-260. *In* J. M. Marzluff, R. Bowman, and R. Donnelly [EDS.], Avian ecology and conservation in an urbanizing world. Kluwer Academic, Norwell, MA.

Howard, D. V. 1974. Urban robins: a population study, p. 67-76. *In* J. H. Noyes and D. R. Progulske [EDS.], Wildlife in an urbanizing environment. Coop. Ext. Serv., Univ. Mass., Springfield, MA.

Iacovides, S., and R. M. Evans. 1998. Begging as graded signals of need for food in young ring-billed gulls. Anim. Behav. 56:79-85.

James, F.C., and C.E. McCulloch. 1985. Data analysis and the design of experiments in ornithology, p. 1-63. *In* R.F. Johnston [ED.], Current Ornithol., Vol. 2. Plenum Press, New York, NY.

Langen, T. A., D. T. Bolger, and T. J. Case. 1991. Predation on artificial bird nests in chaparral fragments. Oecologia 86:395-401.

Lancaster, R. K., and W. E. Rees. 1979. Bird communities and the structure of urban habitats. Can. J. Zool. 57:2358-2368.

Leonard, M. L., and A. G. Horn. 1998. Need and nestmates affect begging in tree swallows. Behav. Ecol. Sociobiol. 42:431-436.

Mackin-Rogalska, R., J. Pinowski, J. Solon, and Z. Wójcik. 1988. Changes in vegetation, avifauna, and small mammals in a suburban habitat. Polish Ecol. Studies 14:293-330.

Marzluff, J. M., K. J. McGowan, R. Donnelly, and R. L. Knight. 2001. Causes and consequences of expanding American Crow populations, p. 333-365. *In* J. M. Marzluff,

R. Bowman, and R. Donnelly [EDS.], Avian ecology and conservation in an urbanizing world. Kluwer Academic, Norwell, MA.
Mayfield, H. F. 1961. Nesting success calculated from exposure. Wilson Bull. 73:255-261.
Mayfield, H. F. 1975. Suggestions for calculating nest success. Wilson Bull. 87:456-466.
McDonnell, M. J., and S. T. A. Pickett. 1990. Ecosystem structures and function along urban-rural gradients: an unexploited opportunity for ecology. Ecology 71:1232-1237.
Mitchell, A. 1999. The ESRI guide to GIS analysis. Vol. 1: Geographic patterns and relationships. ESRI Press, Redlands, CA.
Mills, G. S., J. B. Dunning, Jr., and J. M. Bates. 1989. Effects of urbanization on breeding bird community structure in southwestern desert habitats. Condor 91:416-428.
Milsap, B. A., and C. Bear. 2000. Density and reproduction of Burrowing Owls along an urban development gradient. J. Wildl. Manage. 64:33-41.
Moler, P. E. 1992. Eastern indigo snake, p. 181-186. *In* P. E. Moler [ED.], Rare and endangered biota of Florida. Vol. 3, Amphibians and reptiles. Univ. Press of Florida. Gainesville, FL.
Monoghan, P., and R. G. Nader. 1997. Why don't birds lay more eggs? Trends in Ecol. Evol. 12:270-274.
Natahura, Y., and C. Imai. 1996. Spatial structure of avifauna along urban-rural gradients. Ecol. Res.11:1-9.
Riley, S. P. D., J. Hadidian, and D. A. Manski. 1998. Population density, survival and rabies in raccoons in an urban national park. Can. J. Zool. 76:1153-1164.
Rottenborn, S. C. 1999. Predicting the impacts of urbanization on riparian bird communities. Biol. Conserv. 88:289-299.
Schaub, R., and R. L. Mumme, and G. E. Woolfenden. 1992. Predation on the eggs and nestlings of Florida Scrub Jays. Auk 109:585-593.
Schmidt, K. A. 1999. Foraging theory as a conceptual framework for studying nest predation. Oikos 85:151-160.
Shine, R. 1991. Australian snakes: a natural history. Cornell University Press, Ithaca, NY.
Slagsvold, T., and S. Dale. 1996. Disappearance of female pied flycatchers in relation to breeding stage and experimentally induced molt. Ecology 77:461-471.
SPSS, Inc. 1999. Advanced Models 10.0. SPSS, Inc. Chicago, IL.
Tomialojć, L. 1970. Quantitative studies on the synanthropic avifauna of Legnica and its environs. Acta Ornithol. 9:293-302.
Walcott, C. T. 1974. Changes in bird life in Cambridge from 1980 to 1964. Auk 91:151-160.
Wilcove, D. S. 1999. The condor's shadow: the loss and recovery of wildlife in America. W. H. Freeman and Co., New York.
Woolfenden, G. E. 1974. Nesting and survival in a population of Florida Scrub Jays. Living Bird 12:25-49.
Woolfenden, G. E., and J. W. Fitzpatrick. 1984. The Florida Scrub -Jay: demography of a cooperative-breeding bird. Monogr. Pop. Biol. No. 20. Princeton Univ. Press, Princeton. 406pp.
Woolfenden, G. E., and J. W. Fitzpatrick. 1996. Florida Scrub-Jay (*Aphelocoma coerulescens*), p. 1-28. *In* A. Poole and F. Gill [EDS.], The Birds of North America, No. 228. The Acad. Natl. Sci., Philadelphia, and The American Ornithologists' Union, Washington, DC.

Chapter 19

Synurbanization of the magpie in the Palearctic

Leszek Jerzak
Pedagogical University, Al. Wojska Polskiego 69, 65-762 Zielona Gora, Poland, e-mail: ljerz@lord.wsp.zgora.pl

Key words: Anthropogenic food supplementation, human persecution, magpie, nest site, Palearctic, persecution, Poland, population growth, predation, reproductive success, urbanization

Abstract: Black-billed Magpie (*Pica pica*) populations in Palearctic towns have grown dramatically in abundance and density over the last 50 years. Human population growth, largely concentrated in urban areas, has increase anthropogenic foods and reduced potential nest predator density. To examine whether these factors facilitated growth of urban magpie populations, I surveyed breeding magpies in 12 urban and two non-urban landscapes in western Poland and studied the breeding ecology of magpies in the largest of the cities, Zielona Gora, between 1982 and 1996. Magpie abundance increased three times faster in urban than rural settings and breeding success was slightly higher for urban pairs, suggesting that urban population growth resulted from changes in urban landscapes rather than emigration. Growth of the urban magpie population was explained by an increase in suitable, natural nesting substrate and a positive net rate of reproduction. Nest predation was a primary source of nest loss and was negatively associated with nest height. The maturation of poplars planted by urban revitalization projects facilitated nesting. Urban birds also took advantage of the ameleoriated climate and anthropogenic food; they built nests and laid eggs earlier in the year and renested more frequently than rural birds, foraged in the presence of humans, and supplemented their diet with material from open trash bins. The population appeared to be regulated by density dependent juvenile mortality, but the Zielona Gora population continues to grow and maximum breeding population density has not been reported in the literature. Magpies have the ecological and behavioral flexibility to succeed in urban environments. In time, the unique selective environment of urban areas may lead to the evolution of urban-adapted traits in this species.

1. INTRODUCTION

The Black-billed Magpie (*Pica pica*) is a common bird in most parts of the Palearctic, and a portion of western North America (Bock and Lepthien 1975, Birkhead 1991; Fig. 19.1). It is native to all parts of its range except Japan and Ireland, were it was introduced nearly 400 years ago (Birkhead 1991, Takeishi 1995). North American populations are attributed to colonization from northeast Asia approximately 4,000 years ago, after the last glaciation (Birkhead 1991). Magpies are found in tundra, taiga, southern tropical forests, and the Himalayan Mountains (Dementiev 1954, Goodwin 1986, Birkhead 1991).

Throughout its range, the magpie is associated with human activity. It preferentially nests near human settlement and attains greatest density in

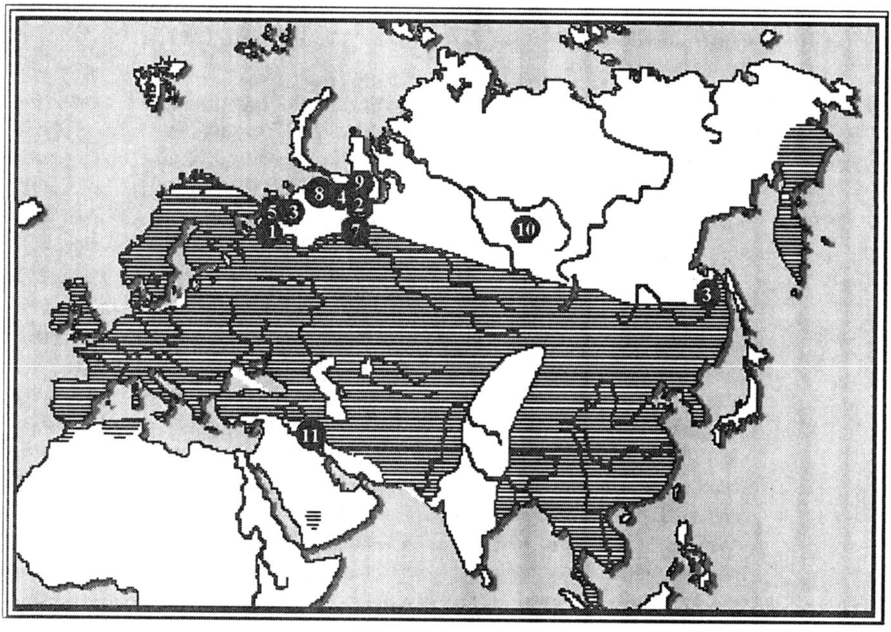

Figure 19.1. Palearctic distribution of the Black-billed Magpie. Numbers indicate cities outside the known distribution where magpie colonization and population growth have occurred: (1) coast of White Sea (Kaliakin et al. 1989), (2) Vorkhuta River Valley (Morozov 1989), (3) Amur River Valley (Babenko 1989), (4) Bolshezemelskaya Tundra (Morozov 1989), (5) coast o Barents Sea (Morozov 1989), (6) Kanin Peninsula (Morozov 1989), (7) Kolva River Valley (Morozov 1989), (8) Usinsk-Narian Mar (Morozov 1989), (9) Mezen, Pechora, Usa, and Vorkhuta Rivers Valleys (Asoskova et al. 1997), (10) Central Siberian Uplands (Rogacheva 1988), (11) Karakumy Desert (Rustamov 1958).

urban areas in Denmark (Moller 1978, Fjeldsa 1981), Eurasia (Stepanyan 1997), France (Balanca 1984a), Germany (Bahrmann 1968, Deckert 1980, Dittrich 1981), Japan (Takeishi 1995), Switzerland (Schifferli and Fuchs 1981), and western Russia (Konstantinov and Babenko 1977). This association has been attributed to lower rates of nest predation by crows (Baeyens 1981) and the presence of food subsidies (Takeishi 1995) near human activity. Over the last 50 years, magpies have colonized many cities within and outside its historical range (Fig. 19.1) and exhibit dramatic population growth in urban and suburban landscapes (Fig. 19.2, Table 19.1); this population growth has, with some exceptions, proceeded eastward across the Palearctic. Despite these trends, our understanding of why magpies are so strongly associated with humans remains rudimentary. Moreover, we do not know the upper limit of magpie population density in urban areas. Consequently, many basic and applied questions regarding the biology of magpies still exist. The goals of this paper are to (1) describe the geographic range over which urban and suburban magpies have increased and (2) investigate mechanisms potentially responsible for that growth. To accomplish these goals, I will draw heavily on the literature, including the newly accessible and translated eastern literature from groups such as the Corvidae Working Group (e.g., Konstantinov 1984). I will also draw on a recent survey of breeding magpie density in western Poland and a detailed study of magpie breeding biology in a Polish city, Zielona Gora, where the magpie population has grown appreciably in the recent past.

2. METHODS

2.1 Surveys

I estimated the abundance of breeding magpies by counting nests. This method allowed me to cover a greater area, relative to counting individual birds, without reducing the likelihood of detecting breeding activity. The method was kept constant for the duration of the study and was conducted by the same observer in each plot. Thus, the results are directly comparable between years and plots. Some magpies built multiple nests, so counting nests visited only once may overestimate the number of breeding pairs. For this reason, every nest was visited two or three times during a single season to more accurately estimate the number of breeding pairs. A breeding pair was recorded if an active nest was being constructed, if a pair was observed displaying sexual activity, or if a nest contained fresh sticks. Nests were counted for the first time ~1 May, before leaves emerged on the trees.

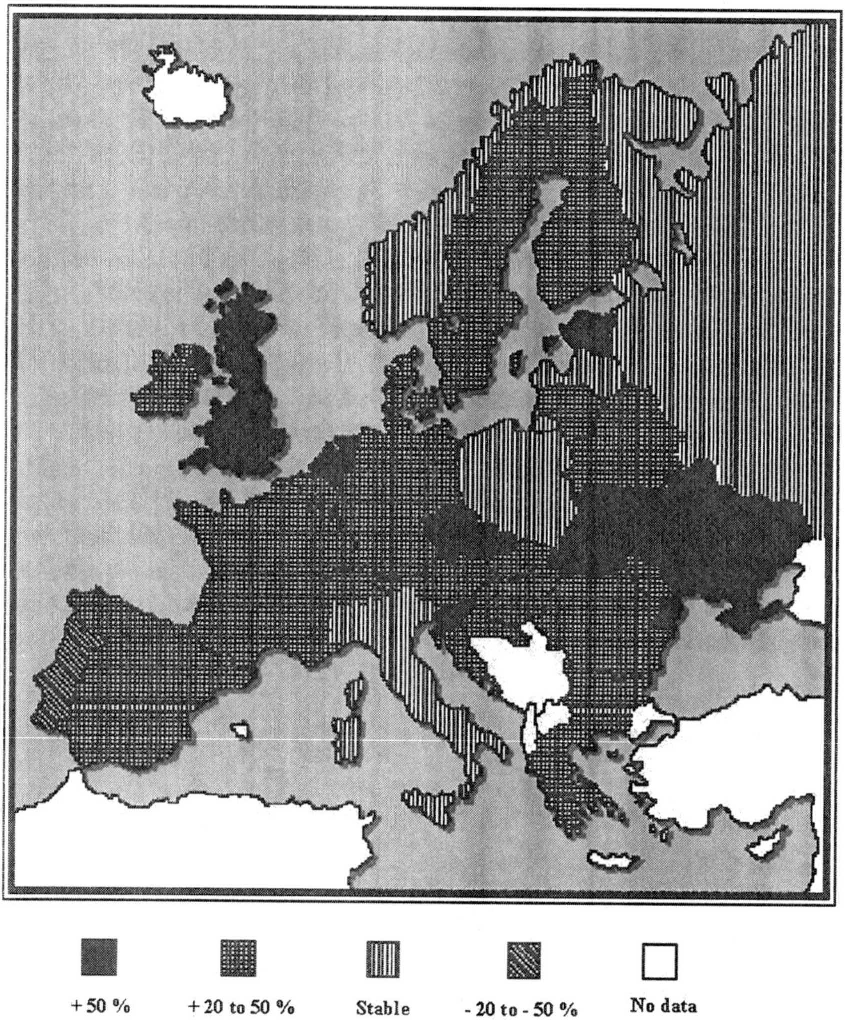

Figure 19.2. Relative magpie population increase across Europe.

Hereafter, nest density is referred to as breeding pair density in the text and abbreviated as breeding pairs per km^2 in tables.

I counted all active nests at one exurban (Brody), one rural (Gorzyca), four suburban (Babimost, Cybinka, Osno Lubuskie, and Zbaszynek), and eight urban study sites (Miedzyrzecz, Rzepin, Skwierzyna, Slubice, Sulechow, Sulecin, Swiebodzin, and Zielona Gora) in western Poland

Table 19.1. Timing of urban and suburban magpie population growth.

Years	Town	References
1950s	Poznan-POL	Klejnotowski 1972
	Sverdlovsk-RUS	Bolotnikov, Litvinov 1992
1960s	Berlin-GER	Lehman et al. 1986
	Bydgoszcz-POL	Klejnotowski 1974
	Byelorus towns-BYE	Ivaniutenko 1989
	Cracow-POL	Ferens 1957, Harmata 1985
	Dublin-IRE	Kavanagh 1987
	Komsomolsk/Amur-RUS	Babenko 1989
	Koszalin-POL	Gorski 1997
	Moscow-RUS	Blagosklonov 1984
	Olsztyn-POL	Okulewicz 1971
	Saratov-RUS	Tuchin, Varshavsky 1984
	Sofia & other-BUL	Nankinov 1981
1970s	Aerdeen-UK	Sharrock 1976
	Banska Bistrica, etc. SLO	Salaj 1981, 1988
	Eliste-RUS	Kukish 1984
	Habarovsk-RUS	Tagirova 1989
	Harkov-UKR	Krivicky et al. 1984
	Irkuck-RUS	Lipin et al. 1984
	Perm-RUS	Bolotnikov, Litvinov 1992
	Riga-LAT	Blagosklonov 1984
	Rostov-RUS	Bielik 1989
	St. Petersburg-RUS	Ilinsky, Malchevsky 1984
	Sverdlovsk-RUS	Niekrasov 1989
	Ternopol-UKR	Talposh et al. 1989
	Voronez-RUS	Sarychev, Ivanchev 1984, Vorobiev, Lihacky 1984
	Warsaw-POL	Klejnotowski 1972, Luniak 1990
1980s	Alma Aty-KAZ	Kovshar 1988
	Bukhara-UZB	Bakaiev 1984
	Chernihov-UKR	Samofolov 1984
	Donieck-UKR	Tatarchenko 1984
	Izhevsk-RUS	Braude 1984
	Kazan-RUS	Vodoladzskaya 1997
	Lipeck-RUS	Aleksandrov et al. 1989
	Samarakand-UZB	Sagitov 1984
	Volsk, Khavynsk, Balashov, Marks, Novouzensk-RUS	Tuchin, Varshavsky 1984
	Vorkuta-RUS	Morozov 1987

between 1982 and 1996 (Table 19.2). The amount of forest and farmland surrounding sites varied. Suburban sites were characterized by small buildings, farms with gardens and fruit trees, and parks with tall trees. The largest and most urbanized town, Zielona Gora (~120,000 people on 20 km^2), was surveyed annually while the other sites were surveyed biennially. Zielona Gora exemplified central European towns and, consequently, most of the urban sites; it had an old, city center surrounded by buildings dating mostly from the 19th century, parks < 10 ha, and rows of trees (typically *Populus spp., Tilia spp., Acer spp.* and *Aesculus hippocastanum*). Lawn area was minimal. Since urbanization intensity declined with distance from the urban core, some comparisons are based on sites in (urban) and far (suburban) from Zielona Gora's center.

2.2 Productivity and Foraging

I studied nest site selection, breeding phenology, and nest success in Zielona Gora in 1989, 1991, 1993, and 1995 (see Jerzak 1995). Nests were observed from the ground at 7-day intervals until fledging. Data collection began on 1 February each year. A newly completed nest, in which magpies were not observed, was considered to have been abandoned prior to laying and did not contribute to productivity calculations. A nest in which birds were seen sitting, but not feeding young, was considered to have failed at the laying or incubating stage. A nest in which incubation was seen, followed

Table 19.2. Classification of western Poland study sites and associated magpie density.

Annual % increase in breeding pairs per km^2 1982-1996	Site	Class	Matrix	Breeding pairs per km^2 in 1995
3.9	Brody	Exurban	Forest	0.2
0.5	Gorzyca	Rural	Farmland	0.5
0.5	Babimost	Suburban	Farmland	5.7
41.7	Cybinka	Suburban	Mixed	3.0
2.8	Osno Lubuskie	Suburban	Forest	2.9
39.6	Zbaszynek	Suburban	Mixed	7.0
4.4	Miedzyrzecz	Urban	Mixed	7.3
10.4	Rzepin	Urban	Forest	1.8
4.8	Skwierzyna	Urban	Mixed	2.2
2.8	Subice	Urban	Farmland	6.0
6.7	Sulechow	Urban	Farmland	9.6
4.6	Sulecin	Urban	Forest	6.2
7.7	Swiebodzin	Urban	Farmland	4.9
12.8	Zielona Gora	Urban	Forest	16.9

by arrivals of adults with food, was considered to have failed at the nestling stage if the fledging of young was not observed. If ≥1 young fledged, the nest was considered successful.

Many of the nests in towns were located near the top of tall trees, so direct observation of contents was often impossible. Therefore, I inferred events such as the onset of nestling provisioning from observation of behavior and back-calculated laying dates using published values for the nesting cycle (Gotzman and Jabonski 1972, Goodwin 1986, Cramp 1994). I estimated the number of young fledged per nest by direct observation and following family groups immediately after fledging. Renesting attempts were difficult to document, because foliage often obscured the nest. To increase the chances of detecting renesting, I surveyed territories in which first nest attempts had failed through July. I used nest building and/or breeding behavior (e.g. nestling provisioning) to determine if renesting occurred . In many instances throughout this chapter, the nesting data that I collected are supplemented with data from the Polish Nest Record Scheme.

Additionally, I quantified non-natural, anthropogenic food resources and compared the degree to which birds were habituated to humans in urban and rural sites. Anthropogenic foods were quantified by mapping the distribution of trash bins in 1988. Habituation was examined by approaching birds and recording the distance between the observer and the bird when the bird flew away or flushed. Habituation trials were conducted in an urban park in Zielona Gora and near the rural village of Ochla, outside Zielona Gora.

3. RESULTS AND DISCUSSION

3.1 Distribution and Abundance

Throughout the Palearctic, magpies were more abundant in urban and suburban landscapes than rural landscapes (Birkhead 1991, Dott 1994). Exceptions existed, but they probably resulted from different landscape classification methods (e.g., Vines 1981, Balanca 1984a, Birkhead 1991). Some urban populations were extremely dense and varied across urban gradients within cities, generally declining with distance from the city center (Vines 1981, Balanca 1984a, Birkhead 1991). Increases in urban magpie abundance and density have also been documented across the Palearctic (Table 19.3). Yet, magpie populations are stable or declining slightly in some European countries, suggesting increases in urban magpies are associated with urbanization rather than a response to something affecting most or all of the bird's range (e.g., climate change or pesticide use).

Magpie abundance increased in all towns studied in western Poland (Table 19.2). Suburban populations grew the fastest, but the highest density occurred in urban populations. Rural magpies increased, but far less than their counterparts in other landscapes. Interestingly, the matrix surrounding the study sites may have impacted the change in magpie abundance and resulting density. Of the four sites showing the largest population increase, two were surrounded by forest and two were surrounded by a mixture of farmland and forest. However, sites showing the greatest densities at the termination of this study, were distributed equally across the four matrix types. Therefore, urban cores surrounded by forest and mixtures of forest and agriculture had the greatest potential for magpie population growth.

In 1929, the entire city of Zielona Gora consisted of what is now considered only the city center; at that time magpie density was ~1.4 breeding pairs per km^2 (Gruhl 1929). By 1971, the city had expanded and magpie densities in the city center had reached 7 breeding pairs per km^2, and by 1987 the density was 15.6 breeding pairs per km^2. The percent increase in

Table 19.3. Increase in urban magpie populations. Most studies were conducted during the last 25 years

Region	Trend or % increase	Reference
Arhus-DEN	+	Henriksen 1989
Banska Bistrica-SLO	+	Salaj 1981
Berlin-GER	+	Lehman et al. 1986, Witt 1985, Witt 1997
Cracow-POL	+	Harmata 1985
Kazan-RUS	+	Vodoladzskaya 1997
Moscow-RUS	+	Blagosklonov 1984
Saga-JAP	+	Takeishi 1995
Voronezh-RUS	+	Sarychev and Ivanchev 1984
Warsaw-POL	+	Luniak et al. 1997, Lesinski 1998
West Berlin-GER	5.8[b] per year	Gorski 1997, Witt 1989
Slupsk-POL	9.8[a] per year	Gorski and Kotlarz 1997
Manchester-UK	10.3[a] over 2 years	Tatner 1982b
Poznan-POL	10.4 - 28.0[b] over 19 years	Gorski 1997 and Mizera 1988
Koszalin region-POL	10.4[a] per year	Gorski 1997
Dublin, IRE	13.0[b] over 13 years	Kavanagh 1987
European Russian towns	50.0-500.0[a] over 7 years, +	Lihackij and Vorobiev 1984, Konstantinov et al. 1996
Sheffield-UK	97.6[b] over 40 years	Eden 1985, Clarkson and Birkhead 1987
Osnabruck-GER	280[a] over 9 years	Kooiker 1995
Warsaw-POL	900[b] over 13 years	Lesinski 1998

[a]Based on abundance.

[b]Based on density.

population density over those 58 years (1014%) is staggering! Similarly, magpies increased over the entire city over the course of this study; the number of breeding pairs increased from 117 to 342. The growth trajectory for the city-wide population did not appear to be levelling off and was not parallel to that in neighboring rural and exurban populations.

Looking at the internal structure of the city more closely, it appeared that the number of birds in the city center grew more rapidly than outside the city center. Furthermore, within the city center birds increased the fastest in areas where urban revitalization or renovation occurred. The planting of lawns and poplars and use of open trash-bins may have facilitated population growth in this portion of the city.

3.2 Proximate Mechanisms Facilitating Population Growth

3.2.1 Breeding Phenology

Klejnotowski (1972a), Tatner (1982a), and Birkhead (1991) reported that magpies built nests and laid eggs earlier in urban areas than surrounding rural or exurban areas. Indeed, this appears to be a general rule (Table 19.4) (Bowman and Woolfenden 2001, Schoech and Bowman 2001). Birds in comparatively urban settings may build nests and lay eggs earlier than birds in other settings (1) due to physiological changes associated with artificial light or warmer temperatures and earlier foliage emergence, (2) because warmer temperatures (up to 10°C; Bezzel 1985, Jones 1986) allow earlier foraging on invertebrates, and earlier incubation, and brooding (Hogstedt 1981, Reese and Kadlec 1984, Eden 1985, Kavanagh 1986, Hochachka and Boag 1987), (3) to advertise territory occupation in dense magpie populations (Jerzak 1995), or (4) to increase the probability that fledged young will survive their first year (Lack 1966, Perrins 1970).

Zielona Gora birds build nests at about the same time as other urban populations. However, the lag between nest initiation and egg laying was slightly longer in Zielona Gora and the only two non-urban sites for which dates are available.

3.2.2 Nest Sites

Magpies nest on wide variety of substrates including shrubs, trees, anthropogenic substrates. Most studies indicate that shrubs and trees were the most common substrate (Klejnotoski 1974, pers. obs.) and that anthropogenic substrates were used when natural structures were not available (Payn 1991, Wallschlager 1994, Eguchi and Takeishi 1997). In western Poland, the vast majority of magpies built nests in trees, while only exurban and suburban pairs nested in shrubs. Nest-trees were usually *Betula spp.*, *Poplar spp.* (44% of all nests, Jerzak 1992), and *Tilia spp.* and their combined density was positively correlated with the number of breeding magpie pairs (Jerzak 1997). This correlation appears to hold in geographically distant sites (Tatner 1982b, Birkhead et al. 1986, Bokotey 1977), suggesting that urban magpies may be limited by habitat (e.g., suitable nest trees surrounded by lawn).

Poplars trees may have been particularly important to magpies, since they were reported as common and, occassionally preferred, nest substrates in other regions (Klejnotowski 1974, Lemke 1977, Sarychev and Ivanchev 1984, Tuchin and Varshavski 1984, Harmata 1985, Witt 1985, Doo Pyo and Tae Hoe 1986, Lehmann et al. 1986, Wink 1967, Salaj 1981). In many cases, "preference" was probably an overstatement as all trees on the landscape were not quantified. Poplars were probably a common nest substrate for the simple reason that they were extremely popular with programs designed to vegetate urban neighborhoods. Since the 1950s, poplars were frequently planted in and near urban areas throughout the former Soviet Union, eastern Europe, and Poland, (Birkhead 1991, pers. obs.). Poplars were probably not suitable for nesting until they were quite tall, ~20 years old (Tatner 1982b, Jerzak 1997), so tree planting regimes during the 1950s and 1960s may, in part, explain the boom in urban and

Table 19.4. Nest and clutch initiation timing of Black-billed Magpies across the Palearctic.

Site	Class	Earliest date	Mean date	Ref
Nests				
Zielona Gora-POL	Urban	2-16 Feb.	9-21 Mar.	This study
Zielona Gora-POL	Urban	10-16 Feb.	16-18 Mar.	Jerzak 1995
Poznan-POL	Rural	7 Mar.		Klejnotowski 1972a, 1974
Poznan, POL	Urban	25 Jan.		Klejnotowski 1974
Supsk-POL	Urban	1-10 Feb.	1-20 Mar.	Gorski and Kolarz 1997
Arhus-DEN	Suburb.		<15 Mar.	Henriksen 1989
Tomsk-RUS	Urban		16-21 Mar.	Kuranov 1984
Clutches				
Dublin-IRE	Urban	1-5 Mar.	8 Apr.	Kavanagh 1986
Zielona Gora-POL	Urban	30 Mar.-6 Apr.	16-22 Mar.	This study
Zielona Gora-POL	Urban	25 Mar.-6 Apr	10-16 Mar.	Jerzak 1995
Slupsk-POL	Urban		10-20 Apr.	Gorski and Kotlarz 1997
Sheffield-UK	Urban		10 Apr	Eden 1985
Manchester-UK	Urban		8-13 Apr.	Tatner 1982a
Sheffield (non-urban)-UK	Rural		15 Apr	Birkhead 1991
Anglesey (non-urban)-UK	Rural		12 Apr	Seel 1983

suburban magpie populations over the last two decades. Few urban nests were found in shrubs or small trees, which suggests that urban magpies may be sensitive to human disturbance or may be vulnerable to terrestrial predators. Studies have shown that magpies tended to build nests near the tops of even tall trees (Dott 1994, Takeishi and Eguchi 1994, Jerzak 1997, Lesinski 1998) and that the abundance of tall trees is better correlated with magpie density and distribution than is the overall abundance of trees (Klejnotowski 1974, Harmata 1985, Jerzak 1997).

Magpies are territorial, so potential nest trees must be somewhat distance from other breeding pairs. It follows that when all suitable trees are occupied, that the number of breeding pairs would cease to rise or that new pairs would nest on other, less productive substrates.; That is, nest site limitation would ensue and nest site abundance and distribution would help define carrying capacity. Much larger nearest nest distances were measured in non-urban landscapes than in urban landscapes (Table 19.5). Zielona Gora's mean nearest nest distance was not unusual and was probably rising, given the population's positive trajectory.

3.2.3 Food Supply

Food supply influences breeding success and, consequently, shapes life history traits (Lack 1954, Perrins 1970). The cascading effect on nesting phenology, juvenile survivorship, and annual success has already been mentioned. Given the potential impact of this resource, it deserves more critical review.

Studies of magpie diets have been conducted in rural and urban settings (Csiki 1919, Eigelis 1964, Holyoak 1968, Balanca 1984a). They were omnivores in both settings, taking both plant material (e.g., grains and fruits) and ground invertebrates (e.g., adult orthopterans, coleopterans, and hymenopterans and larval lepidopterans), but taking more plant material in the winter and more animal foods in summer (Kalmbach 1927, Owen 1956, Csiki 1919, Balanca 1984a, Eigelis 1964, Tatner 1983, Clarkson 1984, Henriksen 1991). More to the point, they were also generalists and foraged on garbage when it was available. Garbage was more abundant in urban than suburban settings, so garbage comprised a greater proportion of the diet of urban magpies (e.g., 23-40% of urban diet; Balanca 1984a, Clarkson 1984, Henriksen 1991) and urban birds were often seen foraging in and around garbage cans (Harmata 1985). Breeding adults with access to supplemental food during a period of adverse weather showed lower nestling mortality rates (Hogstedt 1981, Dhindsa and Boag 1990). However, high-garbage diets may not always be beneficial (Pierotti and Annett 2001). Kuranov

(1984) found that the inner organs of nestling magpies in the city center and suburbs developed differently, possibly causing reduced urban fledging weights.

The greater availability of garbage in urban areas may have enabled magpies to attain higher densities in urban areas, as has been suggested by previous studies of magpies (Nowakowski 1996, Vodoladzskaya 1997) and urban avifauna (Marzluff et al. 1998). In Zielona Gora, birds consumed meat scraps, bread, etc. from garbage bins (pers. obs.) and breeder density was strongly correlated with the distribution of those bins (Fig. 19.3). Garbage may have allowed pairs to provision young even before higher quality invertebrates were available. Given that nestlings required diets high in invertebrates at critical stages in development, urban territories that lack greenspace producing locally abundant invertebrates (Pisarski 1990) may have lower fledging success or juvenile survivorship, even when garbage is readily available (Bowman and Woolfenden 2001). However, territories without much greenspace may still average a net reproductive rate ≥ 1. The omnivorous and generalist foraging habits of the magpie may predisposed it to synanthropy (Johnston 2001).

Table 19.5. Distance to nearest active nest.

Class	Site	Mean ± SD	Reference
Urban	Zielona Gora-POL	177.3 ± 88.5	This study
Urban	Sheffield-UK	145 ± 46.6	Birkhead 1991
Urban	Haren-HOL	264 ± 116.4	Birkhead 1991
Urban	Berlin (plot 1)-GER	331 ± 121.4	Witt 1985
Urban	Berlin (plot 2)-GER	378 ± 138.8	Witt 1985
Urban	Osnabruck-GER	188	Kooiker 1995
Urban	Arhus (yr 1)-DEN	356	Henriksen 1989
Urban	Arhus (yr 2)-DEN	281	Henriksen 1989
Non-urban	SW GER	331 ± 121.4	Ellenberg et al 1989
Non-urban	Coto Donana-SPA	126 ± 65.9	Alvarez et al. 1974
Non-urban	Paris basin-FRA	150	Balanca 1984
Non-urban	Padana Plain-ITA	509	Fasola et al. 1988
		Range or nearest	
Urban	Saga-JAP	76.3	Takeishi 1995
Urban	Osnabruck-GER	45	Kooiker 1995
Urban	Warsaw-POL	40	Lesinski 1998
Urban	Svierdlovsk-RUS	80-100	Nekrasov 1989
Urban	Saratov-RUS	30-50	Tuchin et al. 1984
Urban	Habarovsk-RUS	30-100	Tagirova 1989

3.2.4 Habituation to Humans

In urban landscapes, birds may be more tolerant of human intrusion and allow humans to get closer to them before flushing from the ground and low perches (Graczyk 1963, Strawinski 1971, Luniak et al. 1997). Urban magpies exhibited shorter flushing distances in urban settings across their range relative to non-urban birds and the difference is almost certainly a result of persecution (Kenney and Knight 1992; Fernández-Juricic et al. 2001; Table 19.6). As expected, birds in Zielona Gora's urban park flushed at much closer distances than those outside the city (Table 19.6). However, in the suburbs people perceive magpies as a pest and cut down magpie nest trees and destroyed nests (pers. obs.). Although I did not measure the flushing distances of suburban magpies, it probably would have been intermediate between rural/exurban and urban flushing distances because disturbance at the nest may not necessarily translate to greater flushing distances. Moreover, distinguishing among the contexts in which a nest predator is encountered (e.g., at a nest vs. while foraging) is probably extremely beneficial when foraging amid humans, especially on garbage,

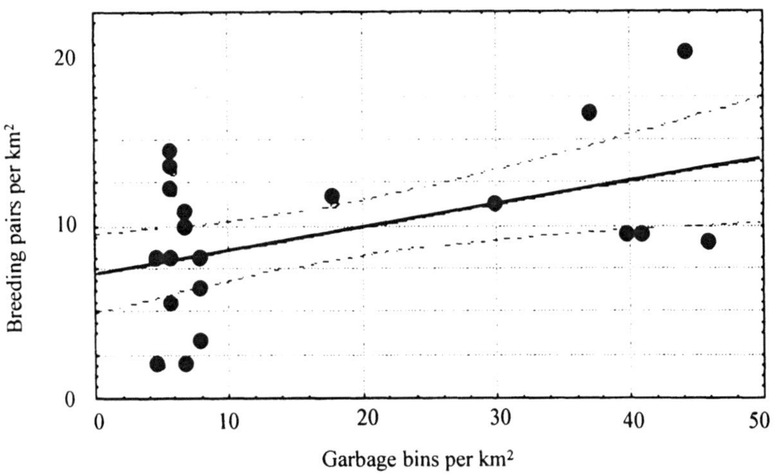

Figure 19.3. Correlation of the density of breeding pairs of Black-billed magpies and garbage bin density in Zielona Gora, Poland in 1988. Least squares regression ($r = 0.46$, $P < 0.05$) was used to generate the best-fit line and the 95% confidence interval.

and not out of the realm of possibilities for birds as intelligent as corvids (c.f., Cilento and Jones 1999).

3.3 Productivity

Urban magpies breed as well or more successfully than rural/exurban magpies (Table 19.7). Moreover, the estimated difference in productivity indices is likely to be conservative as the inclusion of renesting events by individual studies was often ambiguous. Failure to detect or consider renesting events could reduce the difference between landscapes indices because renesting attempts were more likely in the warmer urban climates and could have increased the estimated productivity for that landscape.

What factors contribute to the greater productivity in urban landscapes? Previous studies concluded that reduced crow density and the availability of anthropogenic food explained the relatively greater productivity in urban landscapes. Warmer climates and reduced persecution by humans were probably important, as well. These advantages may, however, be offset by stress and reduced clutch size from territory packing and heightened intraspecific competition and/or nest depredation by conspecifics and by increases in predators that are also supported by anthropogenic food or benefit by the loss of top- or meso-predators. However, Takeishi (1995) did not find any evidence of smaller clutches or broods in more urban settings. Rather, density in Sheffield, England seemed to reduce juvenile survivorship.

Table 19.6. Distance at which human presence caused magpies to flush from the ground or low perches.

Site	Class	Mean ± SD (m)	Range	Ref
Zielona Gora-POL	Urban	6.4 ± 2.6	2-14	This study
Poznan-POL	Urban		2-30	Klejnotowski 1974
Fort Collins-US	Urban	9.4		Kenney and Knight 1992
Cracow-POL	Urban		5-10	Harmata 1985
Ochla-POL	Rural	36.3 ± 10.5	15-72	This study
NE Colorado-US	Rural, Persecuted	64.7		Kenney and Knight 1992
NE Colorado-US	Rural, non-persecuted	28.9		Kenney and Knight 1992

Table 19.7. Productivity of Black-billed Magpie's across the Palearctic.

Mean no. fledglings per pair	% pairs successful[a]	Site	Class	Reference
1.13	50[b]	Zielona Gora-POL	Urban	This study
1.22	53	Manchester-UK	Urban	Tatner 1982
1.3	57	Slupsk-POL	Urban	Gorski and Kotlarz 1997
1.38	43.9	Ulm-GER	Urban	Mack and Jurgens 1999
1.39		Bayeruth-GER	Urban	Sachteleban et al. 1992
1.48	55	Dublin-IRE	Urban	Kavanagh 1986
1.91		Saarbrucken-GER	Urban	Gast 1984
1.97	65.7	Sheffield-UK	Urban	Eden 1985
1.2-1.95	50.6-69.7	Tomsk-RUS	Urban	Kuranov 1984
1.48	**54.96**	**Grand mean[c]**	**Urban**	
0.4	17	JAP	Non-urban	Eguchi 1995
0.86		GER	Non-urban	Sachteleban et al. 1992
0.98	38.4	HOL	Non-urban	Baeyens 1981
1.1	35.3	FRA	Non-urban	Balanca 1984b
1.18	36.8	GER	Non-urban	Mack 1998
1.3	43.7	SWE	Non-urban	Hogstedt 1981
1.44	31.7	DEN	Non-urban	Moller 1991
1.44	23.2-26.0	POL	Non-urban	Klejnotowski 1972a
1.45	38.7	SPA	Non-urban	Arias de Reyna et al. 1984
1.58	40.4	Spain	Non-urban	Redondo 1991
1.81	57.4	UK	Non-urban	Birkhead 1991
2.1	66	UK	Non-urban	Love and Summers 1973
2.6		Zeysko-Burenska Plains, Russia	Non-urban	Dugincov 1989
0.69-0.94	42.4	UK	Non-urban	Vines 1981
1.37	**36.34**	**Grand mean[c]**	**Non-urban**	

[a]Published values are unlikely to include any second attempts, because they are difficult to detect. Therefore, this is considered as a conservative estimate of annual % pair success.
[b]Based upon first nests only.
[c]Calculated with midpoints of given ranges.

Studies examining causes of nest failure concluded that lower and non-urban nests were more likely to fail and that failure was often due to predation. The predators invariably included crows (*Corvus corones* and *C. macrorhynchos*), humans and, in some cases, conspecifics, snakes, domestic cats, weasels (*Mustella mustella*) and martens (*Martes foina*) (Klejnotowski 1972b, Baeyens 1981, Vines 1981, Tatner 1982b, Balanca 1984b, Tuchin and Varshavskij 1984, Eguchi 1995, Jerzak 1995, Gorski and Kotlarz 1997). Crows and humans were likely to be the primary drivers of decreased nest success in less urban landscapes as crows were not abundant in heavily urbanized enivronments (pers. obs.) and there were social and legal constraints on what measures humans could take to eradicate urban magpies (Takeishi 1995). Humans may be the lesser threat in rural environments, since Balanca (1984b) found greater success and lower rates of depredation by crows in nests near houses in the French countryside.

In Zielona Gora, I observed 139 first nest attempts. Of these, 27 (19.4%) were abandoned prior to laying. Of the nests with clutches, 24 (21.4%) and 32 (28.6%) failed during the egg and nestling stages, respectively, and 56 (50.0%) successfully fledged at least one young. The number of fledglings per nest varied according to nest class (2.27 per successful nest, 1.13 per nest with eggs, 0.91 per all nest attempts). Twenty-seven renests followed 56 first attempts with eggs laid (48.2%). Eleven renesting attempts (40.7%) were successful. Assuming these data are more conservative than most of those reported in Table 19.7, this population's productivity probably falls at the low end of the range for urban populations.

I was able to identify the cause of failure at 41 magpie nests in Zielona Gora. Of those nests, 17 (41%) were depredated by Hooded Crows and eight (20%) were depredated by humans. Hooded crows bred near suburban magpie nests and were observed occasionally making longer flights, away from the nest, into the urban center.

4. CONCLUSIONS

Magpies have successfully coexisted with humans across a wide range of urban intensity in the Palearctic for some time. Recent and rapid growth in urban magpie populations has occurred at multiple scales (Zielona Gora, western Poland, and the Palearctic) and appears to be unrelated to population trends in other landscapes. The difference between landscape growth rates estimated by this study and reported in the literature was striking considering inconsistent landscape classification and the potential for underestimating annual success in urban populations. Limited growth of magpie populations in exurban and rural western Poland supports the hypothesis that the dramatic growth in urban magpie populations was largely a response to changes in the urban environment, rather than changes throughout the bird's range. The elements of the urban environment that appear responsible for these differential population growth rates include reduced crow density and thus, reduced nest predation rates, an abundance of anthropogenic food, increased availability of nesting sites because of urban program planting poplar trees, and habituation to humans. The matrix surrounding urban and suburban areas may also have had an effect, but it is not clear if that was a result of interactions between the matrix and magpie natural history or an interaction between the matrix and the character of the urban/suburban study site.

The ecological and behavioral adaptations that enable magpies to take advantage of heavily urbanized settings are not new. Rather, they have been within the magpie's repertoire for some time and corroborate Gliwicz et al.'s (1994) assertion that species amenable to synurbanization have broad habitat requirements, omnivorous and generalist feeding habits, and show behavioral plasticity. This is not to say that the urban selection regime (e.g., greater juvenile survival for early breeding) will not lead to new adaptations, demographic and otherwise, in the future (Johnston 2001).

ACKNOWLEDGEMENTS

I am very grateful to Prof. Chuck Trost (Idaho State University), Dr. Brendan Kavanagh (Royal College of Surgeons in Ireland), Roarke Donnelly (University of Washington) and Dr. John Marzluff (University of Washington) for their role in improving the clarity and grammar of this chapter's text. Special thanks to Prof. Maciej Luniak (Museum & Institute of Zoology PAS) for comments and Mike Blair with BTO for help in using data from The EBCC Atlas of European Breeding Birds.

REFERENCES

Aleksandrov, V. N. and A. I. Zemlianin. 1989. Distribution and density of Corvidae in town Lipeck, p.120-121. *In* V. M. Konstantinov and S. M. Klimov [EDS.], Corvidae in natural and anthropogenic landscapes. Lipeckij Gosud. Ped. Institut, Lipeck 2.

Arias de Reyna, L. M., P. Recuerda, M. Corvillo, and A. Cruz. 1984. Reproduccion de la urraca (*Pica pica*) en Sierra Morena (Andalucia). Donana, Acta Vertebrata 11:79-92.

Babenko, V. G. 1989. Corvidae in the Amur Lowlands and their connections with anthoropogenic landscapes, p.47-49. *In* V. M.. Konstantinov and S. M. Klimov [EDS], Corvidae in natural and anthropogenic landscapes. Lipeckij Gosud. Ped. Institut, Lipeck 2.

Bahrmann, U. 1968. Die Elster. Die Neue Brehm - Bucherei. A. Ziemsen Verlag. Wittenberg Lutherstadt.

Bakaiev, S. B. 1984. Corvidae in anthropogenic landscapes in the arid zone of Uzbekstan. Ekol., biocenotich. i hoziast. znachenie vronovyh ptic. Nauka, Moskva: 111-114.

Balanca, G. 1984a. Le regime alimentaire d'une population de pies bavardes (*Pica pica*). Gibier Faune Sauvage 3:37-61.

Balanca, G. 1984b. Le determinisme du succes de la reproduction chez une population de pies bavardes (*Pica pica*). Gibier Faune Sauvage 4:5-27.

Bezzel, E. 1985. Birdlife in intensively used rural and urban environments. Ornis Fennica 62:90-95.

Bielik, V. P. 1989. Distribution and density of Corvidae in Rostov Province, p.41-43. *In* V. M.. Konstantinov and S. M. Klimov [EDS.], Corvidae in natural and anthropogenic landscapes. Lipeckij Gosud. Ped. Institut, Lipeck 2.

Birkhead T. R. 1991. The magpies. T. & A.D. Poyser, London, UK.

Birkhead, T. R., and K. Clarkson. 1985. The magpie as an aid to teaching behaviour and ecology. J. of Biol. Educ. 19:163-168.

Birkhead, T. R., S. F. Eden, K. Clarkson, S. F. Goodburn, and J. Pellatt. 1986. Social organisation of a population of magpies *Pica pica*. Ardea 74:59-68.

Blagosklonov, K. N. 1984. Corvidae in towns of Eastern Europe. Ekol., biocenotich. i hozyayst. znachenie vranovih ptic. Nauka, Moskva: 64-67.

Bock, C. E., and L. W. Lepthien. 1975. Distribution and abudance of the black-billed magpie (*Pica pica*) in North America. Great Basin Nat. 35:269-272.

Bokotey, A. A. 1997. Numbers and distribution of the Magpie *Pica pica* in Lvov (Ukraine). Acta Ornithol. 32:5-7.

Bolotnikov, A. M., and Litvinov 1992. Evolutionary strategy of common Corvidae species, p.7-9. *In* V. M. Konstantinov, A. N. Hohlov [ED.], Ekologicheskiye problemy vranovyh ptic. Materialy III sovieshchenya, Stavropol.

Bowman, R., and G. Woolfenden. 2001. Nest success and the timing of nest failure of Florida Scrub-Jays in suburban and wildlands, p. 385-404. *In* J. M. Marzluff, R. Bowman, and R. Donnelly [EDS.], Avian ecology and conservation in an urbanizing world. Kluwer Academic, Norwell, MA.

Braude, M. I. 1984. The characteristics of the bird fauna in the town of Izhevsk and analysis of mortality factors. Ptici i urbabisirovannyi landshaft. Kaunus 30-32.

Cilento, N. J., and D. Jones. 1999. Aggression by Australian Magpies *Gymnorhina tibicen* toward human intruders. Emu 95:85-90.

Clarkson, K. 1984. The breeding and feeding ecology of the Magpie *Pica pica*. Ph.D. dissertation, University of Sheffield, UK.

Clarkson, K., and T. R. Birkhead. 1987. Magpie in Sheffield - a recipe for success. BTO News 151:8-9.

Cramp, S. [ED.]. 1994. The birds of the Western Palearctic. Vol. 6: 54-75.

Csiki, E. 1919. Positive daten ueber die Nahrung unserer Voegel. Aquila 26:76-107.

Deckert, G. 1980. Siedlungsdichte und Nahrungssuche bei Elster, *Pica pica pica* (L.) und Nebelkrache, *Corvus corone cornix* (L.). Beitr. Vogelkd. 26:305-334.

Dementiev, G. P. 1954. Birds of the Soviet Union. Moscow, Sov. Nauka. 5.

Dhindsa, M., D. A. Boag. 1990. The effect of food supplementation on the reproductive success of Black-billed Magpies *Pica pica*. Ibis 132:595-602.

Dittrich, W. 1981. Siedlungsdichte und Habitatwahl der Elster (*Pica pica*) in Nord Bayern. J. fur Ornithologie 122:181-185.

Dott, H. E. M. 1994. Density of breeding Magpies and Carrion Crows in southeast Scotland in 1992-93. Scottish Birds 17:205-211.

Doo Pyo, L., and K. Tae Hoe. 1986. A comparative study of the breeding density of magpies *Pica pica sericea* Gould, between urban and rural areas. Bull. Inst. Ornithol., Kyung Hee University 1:39-51.

Dugincov, V. A. 1989. Breeding biology of the Magpie in Zeisko-bureiskoi Plain, Far East, p.22-23. *In* V. M. Konstantinov and S. M. Klimov [EDS.], Corvidae in natural and anthropogenic landscapes. Lipeckij Gosud. Ped. Institut, Lipeck 3.

Eden, S. F. 1985. The comparative breeding biology of Magpies *Pica pica* in an urban and a rural habitat (Aves: *Corvidae*). J. of Zool. (London) 205:325-334.

Eguchi, K. 1995. Seasonal change in breeding success of the Black-billed Magpie *Pica pica sericea*. Japanese J. of Ornith. 44:73-80.

Eguchi, K., and M. Takeishi. 1997. The ecology of the Black-billed Magpie *Pica pica sericea* in Japan. Acta Ornithol. 32:33-37.

Eigelis, Y. K. 1964. Feeding habitats and economic importance of the magpie (*Pica pica* L.) in deciduous and pine stands of the steppe and forest-steppe of the European part of USSR. Zoologicheskiy Zhurnal 43:1517-1529.

Ferens, B. 1957. Birds of Cracow, its protection and restoration. Ochr. Przyr. 24:279-336.

Fernández-Juricic, E., M. D. Jiminez, and E. Lucas. 2001. Bird tolerance to human disturbance in urban parks of Madrid (Spain): Management implications, p. 261-275. *In* J. M. Marzluff, R. Bowman, and R. Donnelly [EDS.], Avian ecology and conservation in an urbanizing world. Kluwer Academic, Norwell, MA.

Fjeldsa, J. 1981. Bestand og habitat hos Husskade *Pica pica* pa Midtsjaelland. Dansk Orn. Foren. Tidsskr. 75:31-39.

Gast, F. 1984. Die Elster (*Pica pica*) als Biomonitoring fur die Belastung von Nahrungsnetzen durch Umweltchemikalien. Ph.D. dissertation, University of Saarbrucken.

Goodwin, D. 1986. Crows of the world. British Museum of Natural History, London, UK.

Gorski, W. 1997. Urban and rural populations of Magpie *Pica pica* in the Koszalin Region, NW Poland. Acta Ornith. 32, 1:51-59.

Gorski, W., and B. Kotlarz. 1997. Changes and breeding ecology in an urban population of the magpie *Pica pica* in Slupsk, NW Poland. Acta Ornith. 32:61-67.

Gotzman, J., and B. Jablonski. 1972. Nests of our birds. PZWS, Warszawa.

Graczyk, R. 1963. Experimantal research on the ethology of the thrushes, *Turdidae*. Roczniki WSR 17:21-71.

Gruhl, K. 1929. Tier- und Pflanzenwelt des Kreises Grunberg und seiner naheren Umgebung. Levisohn, Grunberg.

Harmata, W. 1985. Magpie *Pica pica* in Cracow - its importance and role in the urban environment. Chrońmy Przyr. Ojcz. 4:24-31.

Henriksen, K. 1989. Breeding density and nesting habitats of Magpies *Pica pica* and Hooded Crow *Corvus corone cornix* in a suburban area in Denmark. Dansk. Orn. Foren. Tidsskr. 83:55-59.

Henriksen, K. 1991. Winter diet of Magpies *Pica pica* near Arhus, eastern Jutland. Dansk. Orn. Foren. Tidsskr. 85:169-172.

Hochachka, W. M., and D. Boag. 1987. Food shortage for the Black-billed Magpie (*Pica pica*): an experiment using supplemental food. Can. J. Zool. 65:1270-1274.

Hogstedt, G. 1981. Effect of additional food on reproductive success in the magpie (*Pica pica*). J. of Anim. Ecol. 50:219-229.

Holyoak, D. 1968. A comparative study of the food of some British Corvidae. Bird Study 15:147-153.

Ivaniutenko, A. N. 1989. Synanthropic Corvidae in White Russia, p.15-17. *In* V. M. Konstantinov and S. M. Klimov [EDS.], Corvidae in natural and anthropogenic landscapes. Lipeckij Gosud. Ped. Institut, Lipeck 2.

Jerzak, L. 1992. Breeding density and nest site characteristics of the magpie (*Pica pica* L.) in Lubuska land towns in 1985 and 1987. Przyroda Œrodk. Nadodrza 2, WSP Zielona Gora 65-77.

Jerzak, L. 1995. Breeding ecology of an urban magpie *Pica pica* population in Zielona Gora (SW Poland). Acta Ornith. 29:123-133.

Jerzak, L. 1997. Magpie *Pica pica* nest sites in urban habitats in Poland. Acta Ornith. 32:69 - 76.

Johnston, R. F. 2001. The synanthropic birds of North America, p. 49-67. *In* J. M. Marzluff, R. Bowman, and R. Donnelly [EDS.], Avian ecology and conservation in an urbanizing world. Kluwer Academic, Norwell, MA.

Jones, L. R. 1986. The effect of photoperiod and temperature on testicular growth in captive Black-billed Magpies. Condor 88:91-93.

Kalmbach, E. R. 1927. The magpie in relation to agriculture. USDA GTR 24:1-29.

Kavanagh, B. P. 1986. Population dynamics of the magpie (*Pica pica*) in an urban environment. Ph.D. dissertation, Dublin University.

Kavanagh, B. P. 1987. The breeding density of the magpie in Dublin city. Irish Birds 3:387-394.

Kavanagh, B. P., L. Jerzak, and W. Gorski. 1991. Factors affecting the breeding performance of the Magpie (*Pica pica*) in three European cities: Dublin, Slupsk, Zielona Gora, p.71-81. *In* J. Pinowski, B. P. Kavanagh, and W. Gorski [EDS.], Nesting mortality of Granivorous birds due to microorganisms and toxic substances. PWN, Warszawa.

Kenney, S. A., and R. L. Knight. 1992. Flight distances of Black-billed Magpies in different regimes of human density and persecution. Condor 94:545-547.

Klejnotowski, Z. 1972a. Biology of magpie - *Pica pica*. Roczniki WSR w Poznaniu. Ornitologia Stosowana LVI 4:69-98.

Klejnotowski, Z. 1972b. Ecology of magpie - *Pica pica*. Roczniki WSR w Poznaniu. Ornitologia Stosowana LVI 4:45-67.

Klejnotowski, Z. 1974. Urbanization of magpie *Pica pica* (L.) in Poland. Roczniki AR w Poznaniu. Ornitologia Stosowana LXX 7:77-87.

Konstantinov, V. M. 1984. Ecology, biotic and agricultural importance of Corvidae. Izdatelstvo "Nauka", Moskva.

Konstantinov, V. M., and V. G. Babenko. 1977. Patterns of annual density dynamics of sinanthropic Corvidae in cultivated landscape in middle stripe of the European part of the USSR. Ornitologia 13:100-109.

Konstantinov, V. M., W. Nowicki, and A. G. Pichurin. 1996. Recent changes in the avifauna of cities in European Russia and Eastern Poland - results of a questionnaire. Acta Ornith. 31:59-66.

Kooiker, G. 1995. Siedlungsokologische Untersuchungen an einer urbanen Elsterpopulation (*Pica pica*) einer nordwestdeutschen GroBstadt. Okol. Vogel 18:85-106.

Kovshar, V. A. 1988. Colonisation of the magpie in the town of Alma-Aty, Kazakhstan. Vtoroye Vsesoyuz. Sovieshch. Vranovyie ptici v iestestviennyh i antropogennyh landshaftah, 1-4 Noyabria, Lipeck.

Krivickij, I. A., A. V. Kovalev, and N. G. Sadovskaja. 1984. About new behavioural reactions in some bird species in conditions of urban landscape. Ptici i urbanizirovannyi landshaft. Kaunus 79-80.

Kukish, A. J. 1984. O gniezdovyh poseleniah vranovyh ptic v gorode Eliste i okrestnostiah.

Kuranov, B. D. 1984. Biologicheskiye osobiennosti urbanizirovannoi populacyi soroki.

Lack, D. 1954. The Natural Regulation of Animal Numbers. Oxford University Press, UK.

Lack, D. 1966. Population studies of birds. Clarendon Press, Oxford, UK.

Lehmann, R., G. Degen, and G. Jaeschke. 1986. Bestandsuntersuchungen an der Elster, *Pica pica* (L.) in drei Berliner Innerstadtbezirken. Pica 11:2-14.

Lemke, W. 1977. Zum Brutbestand der Elster (*Pica pica*) in Stadtgebiet von Cuxhaven 1969-1976. Vogelk. Berichte aus Niedersachs 2:41-47.

Lesinski, G. 1998. Population increase and adaptations of magpie *Pica pica* in outskirts and suburban built-up area of Warsaw in 1983-1997. Kulon 3:185-193.

Lipin, S. I., V. D. Sonin, and Y. A. Durnev. 1984. Ways of avifauna formation in urban lanscape of the town of Irkutsk. Ptici i urbanizirovannyj landshaft. Kaunus 87-88.

Love, J. A., and R. W. Summers. 1973. Breeding biology of magpies in Aberdeenshire. Scottish Birds 7:399-403.

Luniak, M. 1990. Avifauna of cities in Central and Eastern Europe – results of the international inquiry, p.131-149. *In* M. Luniak [ED.], Urban ecological studies in Central and Eastern Europe. Ossolineum, Wroclaw.

Luniak, M., P. Kozowski, and W. Nowicki. 1997. Magpie *Pica pica* in Warsaw - abundance, distribution and changes in its population. Acta Ornith. 32:77-86.

Mack, U. 1998. Populationsbiologie und Raumnutzung der Elster (*Pica pica* L. 1758) in einem urbanen Okosystem - Untersuchungen im Grossraum Ulm. Okolgie der Vogel 20.

Mack, U., and M. E. Jurgens. 1999. Aaskrahe, Elster und Eichelhaher in Deutschland. Bundesamt fur Naturschutz, Bonn.

Marzluff, J. M., F. R. Gehlbach, and D. A. Manuwal. 1998. Urban environments: Influences on avifauna and challenges for the avian conservationist, p.288. *In* J. M. Marzluff and R. Sallabanks [EDS.], Avian conservation: research and management. Island Press, Washington, DC.

Mizera, T. 1988 An ecological study of the synanthropic avifauna of the Solacz District of Poznañ in 1975-1984. Acta Zoologica Cracoviensia 31:3-64.

Moller, A. P. 1978. Population density, nest placing, and population changes of the Magpie *Pica pica* in Denmark. Dansk orn. Foren. Tidsskr. 72:197-215.

Morozov, V. V. 1987. New data on fauna and distribution of birds for the Eastern part of Bolshozemelskaya (Great Land) tundra. Ornitologia 22:134-147.

Nankinov, D. 1981. State of urbanisation and synanthropisation of Bulgarian birds]. Ornitol. Info. Biuletin 10:9-29.

Nekrasov, E. S. 1989. About the origin of urban Corvidae populations in Svierdlovsk, p.124-126. *In* V. M. Konstantinov and S. M. Klimov [EDS.], Corvidae in natural and anthropogenic landscapes. Lipeckij Gosud. Ped. Institut, Lipeck 2.

Nowakowski, J. J. 1996. Changes in the breeding avifauna of Olsztyn (NE Poland) in the years 1968-1993. Acta Ornith. 31:39-44.

Okulewicz, J. 1971. Birds of Olsztyn and its victinity. Acta Ornith. 13:127-172.

Owen, D. 1956. Food of nestling Jays and Magpies. Bird Study 3:257-265.

Parslow, J. L. F. 1973. Breeding birds of Britain and Ireland. Berkhamstead, T. & A. D. Poyser, London, UK.

Payn, W. H. 1991. Magpies nesting on television aerials. Brit. Birds 84:513.

Perrins, C. M. 1970. The timing of birds' breeding seasons. Ibis 112:242-255.

Pierotti, R., and C. Annett. 2001. The ecology of Western Gulls in habitats varying in degree of urban influence, p. 309-331. *In* J. M. Marzluff, R. Bowman, and R. Donnelly [EDS.], Avian ecology and conservation in an urbanizing world. Kluwer Academic, Norwell, MA.

Pisarski, B. 1990. The invertebrate fauna of urbanized areas in Warsaw, p. 98-111. *In* M. Luniak [ED.], Urban ecological studies. Proceedings of the International Symposium Warszawa- Jablonna, 24-25 september 1986.

Reese, K. P., and J. A. Kadlec. 1984. Supplemental feeding: possible negative effects on Black-billed Magpies. J. Wildl. Manage. 48:608-610.

Sachteleben J., T. Blick, A. Geyer, T. Krober, and S. Ponisch. 1992. Bruterfolg, Siedlungsdichte und Raumnutzung der Elster (*Pica pica*) in unterschiedlichen Habitaten. J. fur Ornith. 133:389-402.

Sagitov, A. K. 1984. Changes in avifauna of the town of Samarakanda. Ptici i urbanizirovannyj landshaft. Kaunus 118-119.

Salaj, J. 1981. Manifestations of synanthropization of the magpie (*Pica pica* L.) in Slovakia. Biologia (Bratislava) 36:669-671.

Salaj, J. 1988. Effects of bird penetration into inhabited areas on anthropophobia and anthropophilia of some species of the families *Turdide* and *Corvidae*. Biologia (Bratislava) 43:153-158.

Samofolov, M. F. 1984. About avifauna of the town of Chernihov. Ptici i urbanizirovannyj landshaft. Kaunus 120-122.

Sarychev, V. S., and V. P. Ivanchev. 1984. Urbanization of magpie in the region of the Upper Don River area. Ptici i urbanisirovannyj landshaft. Kaunus 126-127.

Schifferli L., and E. Fuchs. 1981. Brutbestandesaufnahme von Rabenkraehe *Corvus corone corone* und Elster *Pica pica* im aargauischen Reusstal. Ornithol. Beobachter 78:233-243.

Schoech, S., and R. Bowman. 2001. Variation in the timing of breeding in two Florida Scrub-Jay populations: Do physiologic measures reflect different environments?, p. 291-308 *In* J. M. Marzluff, R. Bowman, and R. Donnelly [EDS.], Avian ecology and conservation in an urbanizing world. Kluwer Academic, Norwell, MA.

Seel, D. C. 1983. Breeding of the magpie on Anglesey, North Wales. (Bangor Occasional Paper Nr 15) Bangor, Institute of Terrestrial Ecology.

Sharrock, J. T. R. 1976. The atlas of breeding birds in Britain and Ireland. Berkhamstead, T. & A. D. Poyser, London, UK.

Stepanyan, L. 1997. Some geographical variations of magpie behavior between European Russia, Middle and Central Russia and Korea. Acta Ornith. 32:132.

Sterbetz, I. 1964. Beitraege zur Erforschung der wirtschaftlichen Bedeutungen der Elster (*Pica pica*) in Ungarn. Angew. Ornithol. 2:30-36.

Strawiński, S. 1971. About birds, people and towns. Wiedza Powszechna, Warszawa.

Tagirova, V. T. 1989. Synantropisation of magpie in the town of Habarovsk, p.135-136. *In* V. M. Konstantinov and S. M. Klimov [EDS.], Corvidae in natural and anthropogenic landscapes. Lipeckij Gosud. Ped. Institut, Lipeck 2.

Takeishi, M. 1995. A study on the ecological actors in the population dynamics and distribution of the Black-billed Magpie *Pica pica sericea* in Japan. Bull. Kitakyushu Mus. Nat. Hist. 14:55-97.

Takeishi, M. and K. Eguchi. 1994. Nest-site Characteristics in the Black-billed Magpies *Pica pica sericea*. Jap. J. of Ornith. 42:53-59.

Talposh, V. S., Y. M. Antoniuk, and M. I. Maihruk. 1989. Corvidae of the town of Tarnopol, p.102-103. *In* V. M. Konstantinov and S. M. Klimov [EDS.], Corvidae in natural and anthropogenic landscapes. Lipeckij Gosud. Ped. Institut, Lipeck 2.

Tarachenko, A. A. 1984. Characteristic of avifauna of the town of Zhitomir. Ptici i urbanizirovannyj landshaft. Kaunus 133-134.

Tatner, P. 1982a. The breeding biology of magpies *Pica pica* in an urban environment. J. Zool., London 197:559-581.

Tatner, P. 1982b. Factors influencing the distribution of magpies *Pica pica* in an urban environment. Bird Study 29:227-234.

Tatner, P. 1983. The diet of urban magpies. Ibis 125:90-107.

Tuchin, A. V., and S. N. Varshavskij. 1984. The distribution and number of synantropic magpie populations in the region of Lower Volga River. Ekologia, biocenoticheskoe i hozyaistviennoe znachenie vranovyh ptic. Izdatielstvo "Nauka", Moskva 128-131.

Vines, A. V. 1981. A socio-biology of Magpies *Pica pica*. Ibis 123:190-202.

Vodolazhkaya, T. I., and E. F. Naumova. 1989. Distribution and density of Corvidae in the town of Kazan, p.158-159. *In* V. M. Konstantinov and S. M. Klimov [EDS.], Corvidae in natural and anthropogenic landscapes. Lipeckij Gosud. Ped. Institut, Lipeck 2.

Vodoladzskaya, T. 1997. Magpie ecology in urban landscape of Kazan, Russia. Acta Ornith. 32:132.

Wallschlager, D. 1994. Ungewohnlicher Brutzplatz der Elster (*Pica pica*). Berliner Ornith. Bericht 4:23-25.

Wink, M. 1967. Zur Verbreitung der Elster (*Pica pica*) in Bonn. Charadrius 3:192-194.

Witt, K. 1985. Bestande von Elster (*Pica pica*) und Nebelkrahe (*Corvus corone cornix*) auf Berliner Probeflachen 1984. Orn. Ber. F. Berlin (West) 10:154-175.

Witt, K. 1989. Haben Elstern (*Pica pica*) einen Einfluss auf die Kleinvogelwelt einer Grossstadt? Die Vogelwelt Jg. 110:142-150.

Witt, K. 1997. On the abundance of magpie *Pica pica* during the breeding and wintering seasons in Berlin. Acta Ornith. 32:121-127.

Chapter 20

Macaw abundance in relation to human population density in the western Amazon basin

Daniel M. Brooks[1] and Alfredo J. Begazo[2]
[1]*Houston Museum of Natural Science, Department of Vertebrate Zoology, One Hermann Circle Dr., Houston, TX 77030-1799 USA,* e-mail: *dbrooks@hmns.org*

[2]*University of Florida, Department of Wildlife Ecology and Conservation, P.O. Box 141932, Gainesville, FL 32614 USA*

Key words: Agriculture, *Ara*, Colombia, human settlement, macaw, Peru, pet trade, sustainable use, urbanization

Abstract: We assess whether macaw population size varies with human population density in the western Amazon basin. Several factors, including collecting baby macaws for the pet trade, increase dramatically in areas with higher human population densities in the Amazon basin. We use a 'quasi-experimental' design to determine population size variation of four *Ara* macaw species in two regions of the Peruvian and Colombian Amazon, each having a low human population density site (~7 - 12 villages, 100 – 150 people per village) and high human population density site (30 - 50 villages each) for comparison. In region one, the abundances of all four species of *Ara* macaws were negatively related to human density. However, this trend held for only two of the four *Ara* species in region two. We discuss the relative importance of various factors (seasonal fruit abundance, overharvest of macaws for protein or the pet trade, and palm tree harvest) to the patterns observed, as well as the policy and management implications of these patterns.

1. INTRODUCTION

As human populations grow, they affect bird populations. The quality and magnitude of these impacts are contemporary issues in avian conservation. One dimension of this topic concerns the variation in bird abundance with human population density. Studies of birds in sparsely inhabited Neotropical areas can provide interesting insight into how urbanization affects birds. The western Amazon basin proves an ideal testing ground for this topic because there have been colonies and villages of people in some regions (e.g., Francisco de Orellana, Departamento Loreto, Peru) for ~450 yr.

The human population in rural Amazonia is composed of mestizo (mixed ethnic background), indigenous tribes, and the detribalized descendants of the indigenous people that are known in Peru as ribereños or river people (Egoavil 1992). The origin of the ribereños can be traced to the exploitation of the Peruvian Amazon during the 19th century. The transformation of Amazonian people from tribal groups to ribereños began with the earliest European immigration and continued with the detribalization process imposed by missionaries, expansion of the slave trade, and influx of immigrants during the rubber boom. Because of the need for labor and the promise of prosperity, rubber exploitation attracted people from the Peruvian coast, the Andean region, Brazil, Europe, North America, Asia, and detribalized Amerindians (Dourojeanni 1990). Hence, the non-native population increased from 18,000 in 1876 to 120,000 in 1920 (San Roman 1975). After the abrupt ending of the rubber exploitation, many immigrants returned to their place of origin, but others established themselves in the Peruvian Amazon where they turned their attention to other extractive activities (San Roman 1975).

Currently, ribereños greatly exceed indigenous people. Data from a 1981 census revealed that 280,000 nontribal people live in the rural sector of Departamento Loreto, comprising 85% of the entire rural population. The remaining 15% are mainly indigenous people with a population of 50,000 (Egoavil 1992). Ribereño villages are located at the margins of major rivers and some are more than a century old (FPCN 1994). Indigenous groups and ribereños practice agriculture, fishing, hunting, small-scale lumber extraction, and collection of minor forest products. However, ribereños are more involved in market economies, at both the regional and international level, than are indigenous groups (Padoch 1988). Psittacids, especially *Ara* macaws (Munn 1992), are highly threatened in several areas of their range due to over-collecting for the pet trade, among other factors (Collar and Juniper 1992). Such over-collecting is one of several factors exacerbated by an ever-expanding human population (Begazo 1999). Our objectives herein are to determine whether macaw population size varies with human population density in the western Amazon basin.

2. METHODS

2.1 Habitat and History of Human Use

Regional habitats have been overviewed elsewhere (e.g., Brooks 1997), but will be described here briefly. The habitats sampled are primarily situated along water, within forest, or a transition/edge situation between these habitats. Regions along the immediate edges of water (e.g., marshes, lakes, streams and rivers) are often dominated by sawgrass (*Spartina spp.)* or cane, followed by thick undergrowth further inland from the water. Island edge vegetation is similar, unless the island is young, in which case it is monocultured with rapidly colonizing plant species that are primary successional island specialists (e.g., *Cecropia* spp., *Gynerium* spp., and *Heliconia* spp.). Aquatic edge vegetation is separated from interior rainforest by dense forest with thick undergrowth, with canopy height often not exceeding 10 - 15 m. The exception is floodplain, which is characterized by continuous short stems and grasses (e.g., *Tridescantia* spp.) and tall, leafy trees with trunks spaced 5 - 20 m apart. Primary, interior rainforest is high in plant diversity, is characterized by a dark understory due to few penetrable light gaps, and contains tall trees that form part of the canopy, buttresses, or canopy emergents (e.g., *Cedrela* spp., *Ceiba pentandra*, *Ficus insipida* and *Inga* spp.), often exceeding 35 m. Other tall trees in the region include palms (e.g., *Euterpes prectoria*, *Mauritia flexuosa*, *Scheelea* spp., *Socratea* spp.), often occupying a gradient of habitats from river edge to interior forest.

While the sites chosen may have been heavily influenced by humans, it is unlikely that this influence was significant in the 1600's, let alone more recently. While extraction of rubber and animal pelts may have influenced human habitation throughout Amazonian macaw habitat, the forest was still intact at the more pristine sites. The main forms of threats to macaw populations are more apparent at sites with more permanent civilization and higher human population densities. These threats increased after rubber extraction was no longer profitable, with people in higher populated areas turning to extraction of live animals for biomedical research and pets. The increase in human population density coupled with better transportation, availability of firearms and ammunition, and the destruction of Palm trees (e.g., *Mouritia flexuosa* and *Euterpe* spp.) are factors that are likely to decrease macaw abundance.

2.2 Study Sites

We use a 'quasi-experimental' design to estimate the impact of human density on macaw abundance in two remote, continuously forested regions of the Peruvian and Colombian Amazon (for more detailed description see Brooks 1997, 1998; Brooks et al. 1999; Begazo 1997, 1999). Within each region, we selected a site with high human density (~30 - 50 villages each, with approximately 100 - 150 people per village) and a site with low human density (7 - 12 villages) yielding a total of four study sites. Regions 1 and 2 are located north and south of Iquitos, Peru, respectively. We refer to sites by their human population density (high or low) and the region in which they occurred (1 or 2). High 1 is where the Napo River drains into the Amazon River in Peru (~2°45'S; 72°55'W; Fig. 20.1A). Low 1 is along the main river stems of the eastern Peruvian and Colombian Amazon (~2°45'S; 72°55'W - 2°15'S; 70°00'W; Fig. 20.1B). High 2 (~4°22'S; 73°08'W) and Low 2 (~4°31'S; 72°24'W) are located near Tamshiyacu, and Javari-Mirim, respectively. Note that differences in human population density and, potentially, corresponding habitat quality are not apparent from satellite imagery (Miller et al. 2001). Mean annual temperature in the western Amazon basin is 26° C and annual rainfall exceeds 2500 mm (Gorchov et al. 1995).

2.3 Sampling

Three one-week repeated surveys took place at High 1 in May 1995, May 1996 and May 1998. Low 1 was surveyed for a week in June 1996. High 2 was sampled for 13 days in February 1997, and Low 2 was sampled for two weeks in June 1997. We used strip transect sampling methods at each site, recording birds that could be accurately detected visually using unlimited distance contacts (Ralph 1981). We covered 89, 75, 62, and 69 km week^{-1} by boat and foot at High 1, Low 1, High 2, and Low 2; respectively. All abundance estimates were weighted by converting the total number of individuals observed into individuals per week.

3. RESULTS AND DISCUSSION

3.1 Human Population Density and Macaw Abundance

During three weeks of surveys at High 1, we observed two *Ara arauna* (mean = 0.6 birds week^{-1}), zero *A. macao* (0.0), four *A. severa* (1.3), and seven *A. manilata* (2.3). During one week of surveys at Low 1, we observed

A

B

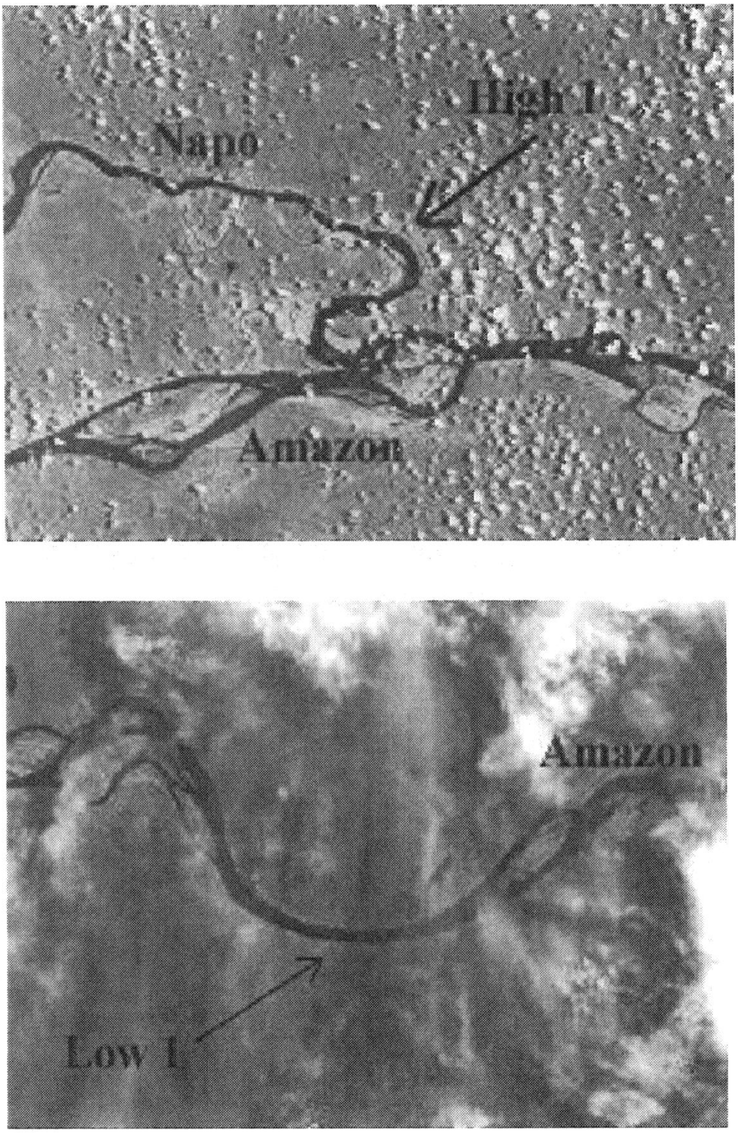

Figure 20.1. (A) Study site with high human population density (High 1) just upstream of the confluence of the Napo and Amazon Rivers. (B) Study site with low human population density (Low 1) on the Amazon River. Both images have some cloud cover (spotty in A, wispy in B), were taken by Landsat in 1999, and cover ~4800 ha. Variation in human population density is not evident in these images.

nine *A. arauna*, two *A. macao*, six *A. severa*, and four *A. manilata*. During 13 days of surveys at High 2, we observed 16 *A. arauna* (8.6), seven *A. macao* (3.7), nine *A. severa* (4.8), and zero *A. manilata* (0.0). During two weeks of surveys at Low 2, we observed 32 *A. arauna* (mean = 16 birds week^{-1}), six *A. macao* (3), five *A. severa* (1.5), and three *A. manilata* (2.5).

The differences between the average density of all four species is strong between High 1 and Low 1 (Fig. 20.2). Thus, *Ara* macaw abundance appears to be negatively correlated with human population densities at these sites. In Region 2, a similar trend holds for *A. arauna* and *A. manilata* between the High and Low sites; however, the opposite pattern occurs for *A. macao* and *A. severa*; the former is slightly higher and the latter dramatically higher at the high human density site. (Fig. 20.2).

These results must be interpreted cautiously considering their preliminary nature. While more intensive sampling is desired, this is often difficult, because macaws travel great distances on a daily basis (e.g., Terborgh 1986, Munn 1992). In light of this, we consider the relative affects of seasonal and annual fluctuations in fruit abundance (c.f. Terborgh 1986) as well as harvesting rates (c.f. Begazo 1997) as additional factors explaining the

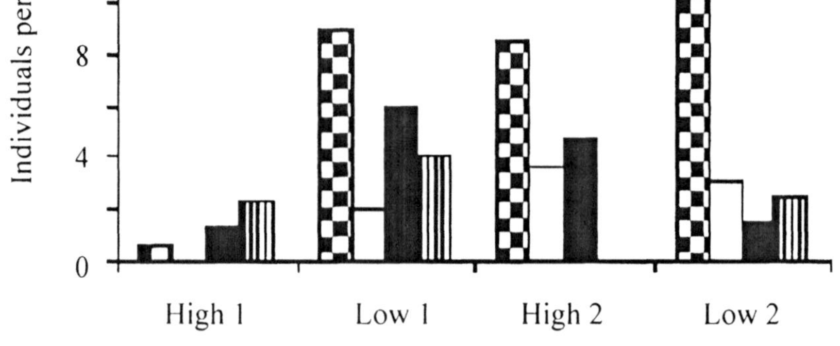

Figure 20.2. Relative abundance of macaws at sites with high and low human population density.

differences in the density of macaws between sites of high and low human population density.

3.2 Variation in Macaw Abundance

Seasonal fluctuations in fruit abundance may partly explain the reversed pattern for *A. severa* (and *A. macao*) between High 2 and Low 2; these two sites were sampled during high water (February) and low water (June) seasons, respectively. Alternatively, annual fluctuations in fruit crop abundance (c.f., Terborgh 1986) may influence population abundance, because *Ara* macaws are moderate to strong specialists on specific fruits (e.g., Roth 1984). However, seasonal or annual fluctuations in fruit crop are weak explanations, considering the strong pattern shown for the other species (*A. arauna* and *A. manilata*) and other cases (i.e., High 1 vs. Low 1) when comparing sites of high versus low human population density. Moreover, large parrots typically do not fluctuate in seasonal abundance (e.g., Galetti 1997, Brooks 1998).

Rather, the main factor dampening macaw population levels in regions of higher human population density is likely overharvest, perhaps due to the pet trade rather than for a protein source. Multiple visits to the local markets by both authors in Iquitos (the large market town and central city in Amazonian Peru) failed to reveal macaw carcasses for sale. In contrast, baby macaws are often available locally as pets. Additionally, data from a number of Amazonian sites (Table 20.1) suggest that less than eight large macaws (i.e., *A. arauna* and *A. macao*) are harvested per site for meat each year on average. Moreover, macaws ranked in the lower 50% of all species taken per site (mean = 20th prey item consumed out of 38 species) and of only bird species taken for food (mean = 7th prey item consumed out of 15 species) (Table 20.1). For most rural Amazonians, macaw trapping appears to be opportunistic, but is likely an important activity during parts of the year in at least some Amazonian villages (Gonzalez 1999). Macaw trapping is likely to be more pervasive in villages located near large expanses of *M. flexuosa,* that are nesting areas mainly for *A. arauna*. This may partly explain the strong difference between populations of *A. arauna,* a species popular with the pet trade (Thomsen and Brautigam 1991), at High 1 and Low 1 ($X^2_{(1)}$ = 7.35, $P < 0.01$). This corroborates the data of Gonzalez (1999) from Region 2, where 1718 psittacids (*Amazona* and *Ara*) were harvested from nests over a 4 year period (range = 166 – 680 birds year^{-1}) from 3890 ha of *M. flexuosa* palm swamps.

4. POLICY AND MANAGEMENT IMPLICATIONS

Although laws may be established to prevent wildlife trade, the same laws, especially when not properly enforced, may actually exacerbate the effects of wildlife harvesting (Bodmer et al. 1997). For example, Thomsen and Mulliken (1992) indicated that Peru established laws restricting export of certain species, requiring a trapping license, and establishing a quota system in 1983, with trade being restricted to Amazonian species (i.e., those east of the Andes). Despite this, trade levels increased dramatically from <20,000 birds exported in 1983 to >50,000 exported in 1984 (the year after the law was passed; Thomsen and Brautigam 1991), and trade levels did not return to <20,000 birds exported yr^{-1} until 1986. Indeed, Peru was listed as one of the top five exporters of psittacids (Thomsen and Brautigam 1991).

The greatest impact of human activity on *Ara* macaws may be the harvest of palm trees and the resulting loss of nesting sites. At Manu in the southern Peruvian Amazon, the availability of nest sites appears to limit macaws (AJB, unpubl. data). The few natural nests available were fiercely defended by pairs that guarded them well before the breeding season began. When artificial nestboxes were erected in *Mauritia flexuosa* stands and in other places, these boxes were sometimes occupied by macaws within a few hours, and fights over the newly available nests were common. Additionally, it was common to see pairs patrolling the occupied cavities, apparently in search of an opportunity to seize a nest.

Palm trees are often cut to harvest and sell the fruit (Bodmer and Brooks 1997, Galetti and Aleixo 1998). Thousands of 25 – 30 year old palm trees are cut every year to sell the fruit for a value of $10 - 20 per tree. Consequently, both macaw food, in the form of palm fruits, and nesting sites

Table 20.1. Macaw harvest at various sites in the Amazon.

References	Location	# Macaws harvested yr^{-1}	Rank[a] / all sp.	Rank[a] / all birds
Hames 1979	Venezuela	6.4 *macao*	25 / 64	9 / 25
Vickers 1991	N. Ecuador	4.8 *macao*	19 / 48	7 / 20
Vickers 1991	N. Ecuador	0.8 *arauna*	38 / 48	12 / 20
Yost and Kelly 1983	E. Ecuador	16.0 *macao*	24 / 36	9 / 14
Begazo unpubl.	C. Peru	9.3[b]	-	-
Ayres 1991	S. Brazil (1978)	14.0[b]	6 / 26	3 / 9
Ayres 1991	S. Brazil (1980)	1.0[b]	7.5 / 8	1 / 2
	Mean	7.47[b]	20 / 38	7 / 15

[a]These data refer to the number in a ranked list where macaws were consumed (average) divided by the number of all vertebrates (all species) or birds (last column) consumed. For example 1 = most frequently consumed, maximum number = least frequently consumed.
[b]Data for *A. arauna* and *A. macao* combined.

are diminished. Additionally, any tree with a cavity containing baby macaws, parrots or toucans is also cut down to obtain the chicks.

Abating the impact of human activities near human settlements is an extremely difficult task. The conservation of Amazonian diversity will more likely be successful if strategies operate within the traditional exploitation system of rural Amazonians. Hunting patterns of ribereño people include frequent searches for wildlife in the areas surrounding the human settlements (short trips) to procure meat for the day's meal, and less frequent searches farther away from the human settlements (long trips) to procure larger wildlife. Given the socio-economic realities of rural Amazonians, it is unlikely that they will refrain from shooting large birds that are easy to procure, or refrain from cutting palm trees near settlements.

A strategy that considers both socio-economic and biological factors may help conserve Amazonian birds. This strategy includes the concept of a functional refuge (McCullough 1987, Joshi and Gadgil 1991). A functional refuge is the area or circumstance that protects resident animal species from being hunted to extinction. Functional refugia take advantage of spatial characteristics, such as being distant from human settlements, or comprised of habitats that humans find difficult to access, or they depend on human characteristics such as being located near areas where humans lack motivation to access those areas (McCullough 1987). Behavioral and biological attributes of Amazonian birds suggest that a functional refuge for tinamous can be realized in proximity to human settlement, while functional refugia for large cracids, trumpeters, and macaws may need to be located very far from human settlements (Begazo 1999, DMB pers. obs.). Hunters in the study site for which the macaw data were obtained took game within a 25 km radius of their settlement (Begazo 1999). Therefore, functional refuges that abate population decline and local extinction of macaws needs to exclude settlement within 25 km of their borders. Such distant areas are not subject to hunting or removal of chicks due to the impracticality of such activities.

In conclusion, our data suggest that overall, populations of *Ara* macaws are higher in regions with lower human population densities. While we hypothesize various cause and affect relationships, future studies of longer duration are needed to test our hypotheses. The studies of Gonzalez (1999) are more quantitative, and more adequately assess how chick harvest affects macaw abundance. Studies correlating variation in macaw populations with palm harvest and natural variation in resource abundance, especially fruit, would further our knowledge of how humans impact macaws.

ACKNOWLEDGEMENTS

We are indebted to CONEPAC and INRENA for permitting us to work in the region and to The Houston Museum of Natural Science for providing computer time. Roarke Donnelly and John Marzluff expertly provided editorial comments on the manuscript. Finally, thanks to Roarke Donnelly (Univ. Washington), Doug Muchoney, Marc Steininger and Thomas Brooks (Conservation International), and Boston University for providing the satellite imagery.

REFERENCES

Ayres, J. M., D. Magalhães-L, E. Souza-M. and J. L. K. Barreiros. 1991. On the track of the road: changes in subsistence hunting in a Brazilian Amazonian village, p.82-92. *In* J.G. Robinson and K.H. Redford [EDS.], Neotropical Wildlife Use and Conservation. University of Chicago Press, IL.

Begazo, A. J. 1997. Use and conservation of the Cracidae in the Peruvian Amazon, p.449-459. *In* S.D. Strahl, S. Beaujon, D.M. Brooks, A.J. Begazo, G. Sedaghatkish and F. Olmos [EDS.], The Cracidae: their biology and conservation. Hancock House Publ., Blaine, WA.

Begazo, A. J. 1999. Hunting of Birds in the Peruvian Amazon. Ph.D. dissertation, University of Florida, Gainesville.

Bodmer, R. E., R. Aquino and P. E. Puertas. 1997. Alternativas de manejo para la Reserva Nacional Pacaya-Samiria: un analisis sobre el uso sostenible de la caza, p.65-74. *In* T. G. Fang, R. E Bodmer, R. Aquino y M. Valqui [EDS.], Manejo de Fauna Silvestre en la Amazonia. La Paz, Bolivia.

Bodmer, R. E., and D. M. Brooks. 1997. Status and action plan of the lowland tapir (*Tapirus terrestris*), p.46-56. *In* D. M. Brooks, R. E. Bodmer and S. Matola [EDS.], Tapirs: status survey and conservation action plan. IUCN, Gland, Switzerland.

Brooks, D. M. 1997. Son la competencia, el tamano y la superposicion de dietas pronosticadores de la composición de Ramphastidae?, p.283-288. *In* T.G. Fang, R.E Bodmer, R. Aquino y M. Valqui [EDS.], Manejo de Fauna Silvestre en la Amazonia. La Paz, Bolivia.

Brooks, D. M. 1998. Competition and coexistence in Neotropical birds: a latitudinal comparison. Ph.D. dissertation, Texas A&M University, College Station.

Brooks, D. M., L. Pando-V. and A. Ocmin-P. 1999. Comparative behavioral ecology of Cotingas in the northern Peruvian Amazon. Orn. Neotrop. 10: 193-206.

Collar, N. J., and A. T. Juniper. 1992. Dimensions and causes of the parrot conservation crisis, p.1-24. *In* S. R. Beissinger and N. F. R. Snyder [EDS.], New World parrots in crisis: solutions from conservation biology. Smithsononian Institution Press, Washington, DC.

Dourojeanni, M. 1990. Amazonia, Que hacer? CETA (Centr. Estud. Teol. Amaz.). Iquitos, Peru.

Egoavil, E. O. 1992. Perfil demografico de la region Loreto. IIAP (Instit. Invest. Amaz. Per.). Iquitos, Peru.

FPCN (Fundación Peruana para la Conservacion de la Naturaleza). 1994. Estudio socio-economico de las poblaciones vecinas a la Reserva Nacional Pacaya-Samiria. Inform. Fase I. FPCN, Iquitos, Peru.

Galetti, M. 1997. Seasonal abundance and feeding ecology of parrots and parakeets in a lowland Atlantic forest of Brazil. Ararajuba 5:115-126.

Galetti, M., and A. Aleixo. 1998. Effects of palm heart harvesting on avian frugivores in the Atlantic rain forest of Brazil. J. Appl. Ecol. 35:286-293.

Gonzalez, J. A. 1999. Use and conservation of parrots and macaw in *Mauritia flexuosa* palm swamps in the northeastern Peruvian Amazon. Proc. VI Neotrop. Ornithol. Congr.: 193.

Gorchov, D. L., F. Cornejo, C. F. Ascorra and M. Jaramillo. 1995. Dietary overlap between frugivorous birds and bats in the Peruvian Amazon. Oikos 74:235-250.

Hames, R. B. 1979. A comparison of the efficiencies of the shotgun and the bow in Neotropical forest hunting. Hum. Ecol. 7:219-251.

Joshi, N., and M. Gadgil. 1991. On the role of refugia in promoting prudent use of biological resources. Theor. Pop. 40:211-229.

McCullough, D. R. 1987. The theory and management of *Odocoileus* populations. *In* C. Wemmer [ED.], Biology and management of the Cervidae. Smithsonian Institution Press, Washington, DC.

Miller, J., J. Fraterrigo, J. Wiens, and T. Hobbs. 2001. Urbanization, avian communities, and landscape ecology, p. 117-137. *In* J. M. Marzluff, R. Bowman, and R. Donnelly [EDS.], Avian ecology and conservation in an urbanizing world. Kluwer Academic, Norwell, MA.

Munn, C. A. 1992. Macaw biology and ecotourism, or "when a bird in the bush is worth two in the hand", p.47-72. *In* S. R. Beissinger and N. F. R. Snyder [EDS.], New World parrots in crisis: solutions from conservation biology. Smithsonian Instution Press, Washington, DC.

Padoch, C. 1988. People of the floodplain forest, p.127-141. *In* J. S. Denslow and C. Padoch [EDS.], People of the tropical forest. University of California. Press, Berkeley.

Ralph, C. J. 1981. Terminology used in estimating numbers of birds. Estimating numbers of terrestrial birds. Stud. Avian Biol. 6:577-578.

Roth, P. 1984. Repartição do habitat entre Psitacídeos simpátricos na Amazonia. Acta Amaz. 14:175-221.

San Roman, J. V. 1975. Perfiles Historicos de la Amazonia Peruana. Edic. Paulinas, Lima, Peru.

Terborgh, J. 1986. Population densitites of Amazonian birds: implications for conservation. Ibis 128:165-166.

Thomsen, J. B., and A. Brautigam. 1991. Substainable use of Neotropical parrots, p.359-379. *In* S. R. Beissinger and N. F. R. Snyder [EDS.], New World parrots in crisis: solutions from conservation biology. Smithsonian Institution Press, Washington, DC.

Thomsen, J. B., and T. A. Mulliken. 1992. Trade in Neotropical psittacines and its conservation implications, p.221-240. *In* S. R. Beissinger and N. F. R. Snyder [EDS.], New World parrots in crisis: solutions from conservation biology. Smithsonian Instution Press, Washington, DC.

Vickers, W. T. 1991. Hunting yields and game composition over ten years in an Amazon Indian territory, p.53-81. *In* J. G. Robinson and K. H. Redford [EDS.], Neotropical wildlife use and conservation. University of Chicago Press, IL.

Yost, J. A. and P. M. Kelley. 1983. Shotguns, blowguns, and spears: the analysis of technological efficiency, p.189-224. *In* R. B. Hames and W. T. Vickers [EDS.], Adaptative responses of native Amazonians. Academic Press, New York, NY.

Chapter 21

Waterbird production in an urban center in Alaska

Michael R. North
Minnesota Department of Natural Resources, 1601 Minnesota Drive, Brainerd, Minnesota 56401 USA, e-mail: michael.north@dnr.state.mn.us

Key words: Alaska, Clean Water Act, Cook Inlet, grebe, loon, reproduction, restoration, urbanization, waterfowl, wetland

Abstract: I conducted waterbird brood surveys near Anchorage, Alaska in 1991. I found 519 waterfowl, grebe, and loon broods containing 2840 young representing 17 species in the Anchorage Bowl. Brood density in Anchorage exceeds that of major waterbird rearing areas in the wildlands of Alaska. Forty-six of 51 (90%) waterbodies, including all man-made waterbodies, supported broods. Man-made waterbodies (n = 22) supported 82% of all broods and 85% of all young. All species used man-made waterbodies to a greater extent than natural waterbodies except Pacific Loons (*Gavia pacifica*) and Bonaparte's Gulls (*Larus philadelphia*). Man-made waterbodies and urbanization were particularly important for Red-necked Grebes (*Podiceps grisgena*), Canada Geese (*Branta canadensis*), Mallards (*Anas platyrhynchos*), American Wigeon (*Anas americana*), Northern Pintails (*Anas acuta*), and Greater Scaup (*Aythya marila*). Man-made waterbodies may have allowed Gadwalls (*Anas strepara*), Canvasbacks (*Aythya valisineria*), and Lesser Scaup (*Aythya affinis*) to expand their range into Cook Inlet. The primary waterbird species that appear to be adversely affected by urbanization are Common Loons (*Gavia immer*), Pacific Loons, Horned Grebes (*Podiceps auritus*), Ring-necked Ducks (*Aythya collaris*), goldeneyes (*Bucephala* spp.), and Sandhill Cranes (*Grus canadensis*). Wetland habitats in Anchorage need to be protected by strict adherence to the Section 404(b)(1) guidelines and standard mitigation sequencing principles of the Clean Water Act, including mandatory compensatory replacement.

1. INTRODUCTION

I conducted waterbird brood counts in and around Anchorage, Alaska, in 1991-1992 to collect data that might be used to protect waterbird habitats, to estimate brood production, and to quantify brood habitat use and brood distribution. Anchorage is the only large urban center in Alaska and the only one in North America north of 55°N. Its population fluctuates around 240,000 residents. Anchorage is also arguably unique because it is a major urban area located amidst vast wilderness.

Although Anchorage supports large populations of nesting waterbirds such as loons, grebes, Canada Geese (*Branta canadensis*), ducks, gulls, and Arctic Terns (*Sterna paradisaea*), no comprehensive surveys have been conducted. Previous survey work (e.g., Rothe et al. 1983, Barnes and Trapp 1985, Nysewander and Trapp 1985) has been either area-specific or species-specific. Waterbird populations and their habitats have recently undergone extensive changes in Anchorage (Tande 1988, Anonymous 1993a). As late as 1954, waterfowl were considered unimportant resources in Anchorage because of their limited numbers (Anonymous 1954); the primary species were Mallards (*Anas platyrhynchos*) and American Wigeon (*A. americana*). Wetlands determined to have low value through a comprehensive functional assessment methodology are authorized by the U.S. Army Corps of Engineers to be filled or drained under a General Permit administered by the Municipality of Anchorage. The General Permit does not require an environmental impact assessment nor mitigation. Higher value freshwater wetlands and waters, and all intertidal and subtidal wetlands require a Section 404 individual permit before they may be filled or drained, but such permits are regularly granted without mitigation. Compensatory mitigation is not required because developers and politicians have argued successfully that mitigation opportunities are generally not available. Without mechanisms for requiring avoidance, minimization, or compensatory mitigation, wetland conservation relies on effective, persuasive arguments about the functions and values of urban wetlands. Developers routinely espouse a counter argument that urban wetlands are low value for wildlife compared to the rest of Alaska simply because they occur in an urban setting.

Based on methodical surveys and casual observations I made of waterbird reproduction in Anchorage and numerous other areas of Alaska from 1983-1990, I developed two hypotheses: 1) that on a per area basis waterbird production in Anchorage exceeded production in many major waterbird rearing areas in Alaska, including the Cook Inlet area, and 2) that waterbird species richness was no lower in Anchorage than in the remainder of the Cook Inlet area.

More than 70 thousand swans, 1 million geese, 12 million ducks (Lensink and Derksen 1990), and millions of loons, grebes, cranes, gulls, and terns breed annually in Alaska. Major waterbird rearing areas are surveyed annually for brood production by the U.S. Fish and Wildlife Service (USFWS) and include coastal areas (Yukon-Kuskokwim Delta, Bristol Bay lowlands, Kenai Peninsula-Susitna River Valley [upper Cook Inlet], and Kotzebue Sound), and interior river valleys (Yukon Flats, Innoko River, Koyukuk River, Tanana-Kuskokwim Rivers, Copper River/Gulkana) (Hodges 1991). Waterbird breeding populations also are surveyed annually on the North Slope (Brackney and King 1992). Other important waterbird rearing areas receive periodic production surveys, including the Copper River Delta and Seward Peninsula (Hodges 1991). Anchorage lies within the Kenai Peninsula-Susitna River Valley region, thus this region provided the most relevant data for direct comparison with Anchorage.

2. METHODS

2.1 Study Area

Anchorage is near the north end of Cook Inlet in southcentral Alaska. The wildlands of the greater Cook Inlet run from the Matanuska-Susitna Valley in the north to Kachemak Bay in the south, and include lowland areas on the Kenai Peninsula west of the Kenai Mountains and lowland areas on the Alaska Peninsula east of the Alaska Range. The Municipality of Anchorage is rural, when taken as a whole, and covers 5071 km^2, of which 10% is inhabited, 6% is on military bases, and 84% is in state park, national forest, or other wilderness areas (Municipality of Anchorage 1997). The Anchorage Bowl (Fig. 21.1) includes urban and suburban landscapes, covers 326 km^2, and constitutes 62% of the inhabited area. My survey area covered 298 km^2 of the Anchorage Bowl, from Potter's Marsh in the south to Ship Creek in the north, and from Cook Inlet in the west to treeline in the Chugach Mountains in the east. The survey area contains numerous natural bogs, lakes and ponds. Nine major creek systems run through the city. Man-made waterbodies include sediment basins, water-filled borrow pits, and four lakes created by damming creek outlets. I classified Lake Hood/Spenard as man-made, because these two formerly natural lakes have been joined and their shorelines substantially altered by dredging to create the largest floatplane base in the United States (Anonymous 1990). The dredging began in the 1940's, but most occurred in the 1970s and substantially increased the amount of shoreline edge. The hydrology of many natural wetlands has been altered by development. Prolonged melting from

A

B

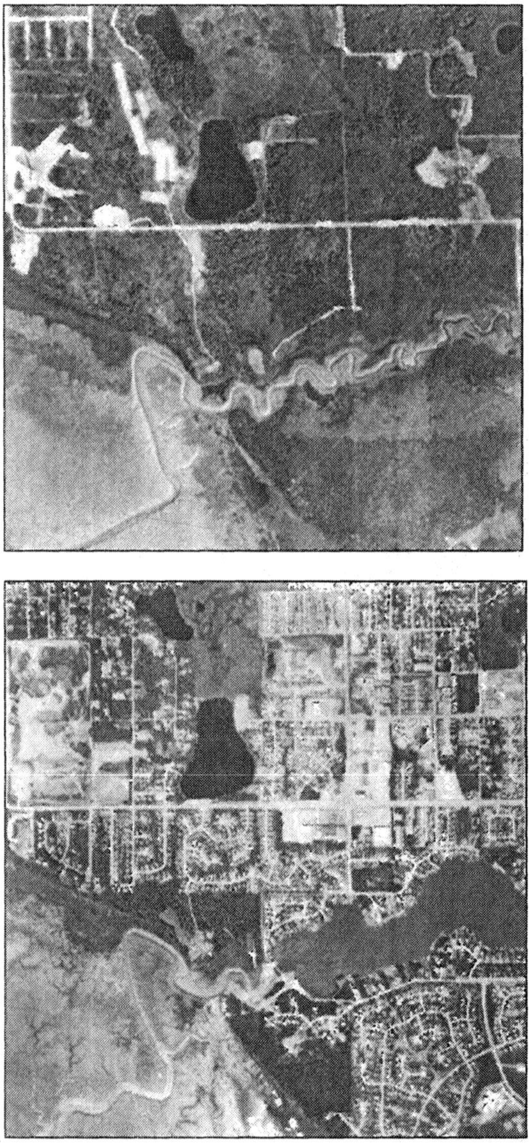

Figure 21.1. Aerial photograph showing development conditions within western Anchorage Bowl, Alaska (A) circa the 1950's and (B) June 1990. Sundi, Jewel, and Campbell Lakes are located near the upper left, center, and lower right of the image, respectively. Each image is ~0.8 km on a side.

municipal snowdumps throughout the spring and early summer has also altered the hydrology of surrounding wetlands and created ponds where none had previously existed. Many waterbodies have been lost entirely or greatly reduced in size from filling, including Delaney Lake, Hillstrand Pond, and four unnamed lakes surrounding Connor's Lake.

Anchorage has grown in boom and bust cycles. In 1991, Anchorage had a population of 237,907, of which 194,293 resided in the Anchorage Bowl (population density 5.94 people ha^{-1}). Peak population in the Anchorage Bowl occurred in 1985 when 203,441 people resided there (6.22 people ha^{-1}). From 1980-1984, the population of Anchorage increased by about 69,000 people; from 1986-1988 it decreased by about 27,000 people. From 1988-1994 it again increased by about 31,600 people. From 1990-1994, Anchorage had a population growth rate of 12%, compared to the national average of 4%. By 2025, Anchorage is estimated to have a population of 366,000 (range 271,000 - 511,000). An increasing proportion of population growth is occurring within the Anchorage Bowl. The proportion of Anchorage residents living in the Anchorage Bowl was 77% in 1960, 81% in 1970, 83% in 1980, and 84% in 1996. The population of Anchorage has a high turnover rate. In 1996, 40% of the current population had moved to Anchorage between 1990-1996; only 29% lived there prior to 1980. Thus, the majority of the Anchorage population is likely unaware of the changes in wetland and lake habitats and waterbird populations that have occurred due to development. In 1990, 52% of Anchorage Bowl residents lived in multi-family housing, 40% lived in single-family homes, and 8% lived in mobile homes. New housing construction peaked in 1982-1984, with nearly 19,000 new units built. From 1987-1992, only 2300 new housing units were constructed. In 1991, the value of new residential construction was $97 million; the value of new commercial construction was $42 million. Anchorage has a highly mobile population. In 1991, there were 189,808 registered passenger vehicles, pick-up trucks, commercial trucks, and buses in Anchorage. In 1990, only 6% of Anchorage households did not have a vehicle, which was below the national average of 10%. In 1990, only 2.5 % of the population commuted to work using public transportation, compared to 6.5% in other U.S. metropolitan areas. Annual operations at the Lake Hood floatplane base and adjacent onshore small plane landing strip ranged from 82,000 to 89,000 trips from 1987-1995, down from a peak of 100,000 in 1985.

2.2 Surveys

I conducted surveys for waterbird broods on all known lakes (n = 21), natural ponds and bogs (n = 16), and man-made ponds and ditches (n = 14) in the Anchorage area between 21 May and 25 July 1991. For the purposes of this study, I considered the following taxonomic groups to be waterbirds: Gaviidae, Podicipedidae, Anatidae, Gruidae, Laridae, and Phalaropidinae. I limited my surveys to waterbodies that were lacustrine; palustrine and seasonally, semipermanently, or permanently flooded; and riverine systems. All waterbodies except five were surveyed once during the first survey period (21 May - 30 June) to count broods of early nesting species (e.g., Canada Geese, Mallards, Northern Pintails [*Anas acuta*]). Because of time constraints, only the most productive and/or readily accessible waterbodies were surveyed during the second survey period (1-25 July) for broods of later nesting species (e.g., American Wigeon and Greater Scaup [*Aythya marila*]). Four waterbodies were surveyed only in July and Campbell Lake was surveyed only after July 25. Some waterbodies were surveyed numerous times. Undoubtedly, some small ponds were never found. Small waterbodies were scanned from shore with binoculars or spotting scopes. Large waterbodies were surveyed either from shore or by canoe. Only readily accessible reaches of rivers were surveyed. For each waterbody I recorded the number of nests or broods observed for each species, number of eggs or young in each clutch or brood, and the estimated age of each brood. I also estimated the percent of the waterbody I surveyed and its origin (i.e., natural or man-made). Many of the natural lakes in town are bordered by residential lots. Campbell Lake, a large man-made reservoir, has no public access and, during 1991, I was able to survey only 25% of it. However, Campbell Lake was resurveyed once in 1992 after access was gained. Data from waterbodies that were not surveyed in their entirety are considered minimum estimates and were not extrapolated to provide an estimate for the entire lake. Extrapolation would introduce new biases (e.g., differing visibility of species that use near-shore habitats versus open-water habitats). Numerous small ponds dot a coastal wetland that extends for about 10 km along the coast of Turnagain Arm, but they were not surveyed because they were inaccessible. Numerous large lakes exist on the Elmendorf and Fort Richardson military bases north of Ship Creek, but these were not surveyed because of time constraints and access restrictions. All young observed were assumed to have survived for the purposes of comparison with statewide production estimates (Hodges 1991). This is clearly not the case, thus the data should be viewed as an estimate of potential production. Each brood observed was identified to species, and the number of young were counted and placed into 5-day (for waterfowl) or 10-day (for other waterbirds) age categories to facilitate not recounting the brood on subsequent surveys. I was able to adequately and consistently age young based on their size, color, and patterns of feather emergence. For broods I believed I observed multiple times, I used the highest count of young. Several unattended ducklings were observed and tallied. I was able to identify many unattended young to species by their color and time of year (i.e., most were yellow Mallard ducklings observed before any other waterfowl except Northern Pintails had broods, and pintail young were dark gray or black). In some instances, nests were found during surveys and tallied in lieu of broods, but I do not report nests where the young may have been also tallied as broods. Clutches of unknown size were assumed to contain the mean number of eggs found in other clutches. Broody hens were assumed to have the average number of young found in other broods, following methods employed by Hodges (1991). The number of broods represented in scaup gang broods was assumed to be equivalent to the number of adult females present, or calculated by dividing the gang brood size by the average size of other individual broods, following methods used by Hodges (1991). Both species of scaup were combined because I was not always able to identify individuals to species, although most broods were certainly Greater Scaup. Average brood size was not calculated for Canada Goose gang broods. Surveys for Mew Gulls (*Larus canus*) and possibly Herring Gulls (*Larus argentatus*) were incomplete, as parts of two primary breeding areas (Potter's Marsh and Lake Hood/Spenard) were inaccessible for searching on foot, and chicks were capable of hiding in tall vegetation.

I compared results of my surveys to data from USFWS surveys throughout the state conducted in 1991 (Hodges 1991). These surveys provide the only comparative data. In the USFWS survey, lakes (in square plots) were surveyed once from the ground by foot and/or canoe, and the survey was timed to occur midway between the peak occurrence of dabbler and diver broods. Observations were recorded as brood of known size, broody hen, partial brood, gang brood, or orphan brood (Hodges 1991).

2.3 Statistical Analyses

To examine whether urbanization affected species richness, I compiled lists of species known to breed formerly and presently in the greater Anchorage area, and in the upper Cook Inlet area I tested for differences in the presence or absence of species using the sign test. (Steele and Torrie 1960). Comparisons were made between species present in the current Anchorage study area and the former/greater Anchorage area; and between the current Anchorage study area and upper Cook Inlet. Ties were excluded from the analysis. To examine if man-made or natural waterbodies were used more frequently, I conducted chi-square tests for independence that incorporated Yates' correction for continuity (Dixon and Massey 1951, Sokal and Rohlf 1973) on the number of waterbodies used by each species relative to the number available. To determine if man-made or natural waterbodies supported different numbers of broods, I conducted chi-square tests for goodness of fit (Sokal and Rohlf 1973) on the total number of broods found on each type of waterbody for each species. For the latter analyses, I based expected use on the assumption that use would be proportional to availability. For all analyses, $\alpha=0.05$.

3. RESULTS

3.1 Species Richness

I found 17 species of waterbirds breeding in Anchorage in 1991; additionally, Common Goldeneye (*Bucephala clangula*) was found breeding in Anchorage in 1992 and Common Loon (*Gavia immer*) in 1993 (Table 21.1). Potter's Marsh and Lake Hood/Spenard had the greatest number of breeding (or probable breeding) species, with 14 and 12 species, respectively. Breeding species composition was not significantly different between the current Anchorage study area and the former/greater Anchorage area ($\chi^2_{(1)} = 2.77$, $P < 0.10$). In total, 25 waterbird species were known to have previously bred in the greater Anchorage area, including Red-throated Loon (*G. stellata*), Trumpeter Swan (*Cygnus buccinator*), Ring-necked Duck (*Aythya collaris*), Harlequin Duck (*Histrionicus histrionicus*), Oldsquaw (*Clangula hyemalis*), Barrow's Goldeneye (*Bucephala islandica*), Sandhill Crane (*Grus canadensis*), Glaucous-winged Gull (*Larus glaucescens*), and Red-necked Phalarope (*Phalaropus lobatus*) (Anchorage Audubon Society 1983, Rothe et al. 1983).

In the upper Cook Inlet area, exclusive of my Anchorage study area, 32 species are known to have bred. Breeding species composition was significantly different between my Anchorage study area and the Cook Inlet area ($\chi^2_{(1)}=8.45$, $P < 0.005$). Although the greater Anchorage area and the upper Cook Inlet area supported a greater number of breeding species than the current study site, three species were found breeding exclusively in the latter, more urbanized area: Gadwall (*Anas strepera*), Canvasback (*Aythya valisineria*), and Lesser Scaup (*Aythya affinis*).

3.2 Habitat Use and Abundance

Numbers of loon, grebe, and waterfowl broods and young observed on Anchorage waterbodies are presented in Table 21.2. Forty-six of 51 (90%) waterbodies (excluding creeks) surveyed supported broods of waterfowl or other waterbirds. Only two natural lakes and three natural ponds did not contain broods or nesting pairs, but one lake supported a pair of Red-necked

Grebes (*Podiceps grisegena*) that probably nested in most years. From these surveys, a minimum of 477 waterfowl broods containing 2735 young (2596 in broods, 139 unattended) were produced in the Anchorage Bowl south of Ship Creek in 1991. These totals include a minimum of 531 Canada Goose

Table 21.1. Waterbird species that bred in the current Anchorage study area from 1987-1994, formerly and/or in the greater Anchorage area prior to 1983, and in the Cook Inlet region.

Species	Current Anchorage study area	Former / greater Anchorage area	Cook Inlet
Common Loon	x	x	x
Pacific Loon	x	x	x
Red-throated Loon		x	x
Horned Grebe	x	x	x
Red-necked Grebe	x	x	x
Trumpeter Swan		x	x
Canada Goose	x	x	x
Gr. White-fronted Goose			x
Mallard	x	x	x
Gadwall	x		
Northern Pintail	x	x	x
American Wigeon	x	x	x
Green-winged Teal	x	x	x
Northern Shoveler	x	x	x
Canvasback	x		
Ring-necked Duck		x	x
Lesser Scaup	x		
Greater Scaup	x	x	x
Oldsquaw		x	x
Harlequin Duck		x	x
Surf Scoter			x
White-winged Scoter			x
Barrow's Goldeneye		x	x
Common Goldeneye	x	x	x
Common Merganser			x
Red-breasted Merganser			x
Sandhill Crane		x	x
Red-necked Phalarope		x	x
Parasitic Jaeger			x
Arctic Tern	x	x	x
Aleutian Tern			x
Bonaparte's Gull	x	x	x
Mew Gull	x	x	x
Herring Gull (hybrids)	x	x	x
Glaucous-winged Gull		x	x
Species richness	19	25	32
Sources[a]	1,2,3	2,4,5	2,4,5,6

[a]1 = this study, 2 = pers. obs., 3 = Tankersley 1994, 4 = Rothe et al. 1983, 5 = Anchorage Audubon Society 1983, 6 = Rosenberg 1986.

young (1.8/km^2), 1711 dabbling duck young (5.7/km^2) and 493 diving duck young (1.7/km^2). An additional 38 grebe and four loon broods containing 101 and four young, respectively, were found in the Anchorage Bowl. Mallards were the most abundant and widespread species, with broods occurring on 64.7% of the waterbodies surveyed (Table 21.3). Scaup were the second most abundant species of duck, and American Wigeon were the second most abundant species of dabbling duck (Table 21.2).

Mew Gulls nested in five colonies in Anchorage: large colonies occurred at Potter's Marsh, Lake Hood/Spenard, and Business Park Wetlands; a small colony of at least 11 pairs occurred at Eastchester/Westchester Lagoon; and a colony of a few dozen pairs was discovered in the middle of a heavy industrial area at the Port of Anchorage north of Ship Creek. The colonies at Lake Hood/Spenard and Potter's Marsh were not accessible for pair counts, but at least 92 young were produced at Potter's Marsh. Fourteen other isolated pairs were distributed about Anchorage among various types of wetlands. Herring Gulls were rare breeders in Anchorage (Table 21.3). Three pairs raised at least three young on man-made waterbodies in mid-

Table 21.2. Total number and percent of broods and young on man-made (n = 22)[a] and natural (n = 31)[b] waterbodies in the Anchorage Bowl, 1991.

	Man-made waterbodies		Natural waterbodies		
Species	Broods	Young	Broods	Young	$\chi^2_{(1)}$ (P)[e]
Pacific Loon	0[c] (0)[d]	0 (0)	4 (100)	4 (100)	3.0 (<0.1)
Red-necked Grebe	30 (83)	81 (83)	6 (17)	17 (17)	25.3 (<0.01)
Horned Grebe	2 (100)	3 (100)	0 (0)	0 (0)	3.0 (<0.1)
Canada Goose	73 (77)	438 (81)	21 (23)	93 (19)	49.0 (<0.01)
Mallard	178 (80)	1003 (82)	48 (20)	242 (18)	129.9 (<0.01)
Northern Pintail	19 (95)	101 (98)	1 (5)	2 (2)	23.1 (<0.01)
American Wigeon	52 (90)	216 (87)	6 (10)	33 (13)	53.9 (<0.01)
Gadwall	2 (100)	6 (100)	0 (0)	0 (0)	3.0 (<0.1)
Green-winged Teal	6 (67)	33 (57)	3 (33)	25 (43)	3.5 (<0.1)
Northern Shoveler	5 (100)	17 (100)	0 (0)	0 (0)	6.9 (<0.01)
Anas spp.	0 (-)	30 (-)	0 (-)	3 (-)	----
Canvasback	3 (100)	22 (100)	0 (0)	0 (0)	3.9 (<0.05)
Scaup spp.	56 (93)	449 (95)	4 (7)	22 (5)	64.9 (<0.01)
Herring Gull	4 (100)	5 (100)	0 (0)	0 (0)	5.4 (<0.05)
Bonaparte's Gull	1 (50)	1 (33)	1 (50)	2 (67)	0.1 (>0.5)
Arctic Tern	6 (100)	13 (100)	0 (0)	0 (0)	8.4 (<0.01)
Totals	437 (82)	2418 (85)	94 (18)	443 (15)	

[a] Includes 1 ditch containing 1 Mallard brood (6 young).

[b] Includes 2 creeks containing 3 Mallard broods (17 young) and 1 Green-winged Teal brood (10 young).

[c] Total number.

[d] Percent.

[e] Chi-square goodness of fit test on number of broods.

Anchorage, a few pairs nested at Potter's Marsh north of the Mew Gull colony, and the species was a probable breeder at Lake Hood/Spenard. Bonaparte's Gulls (*Larus philadelphia*) nested in few areas in 1991. There was one brood each at Lake Hood/Spenard and Goose Lake; however, three to eight pairs probably nested and raised young at Sundi Lake. The only Arctic Tern reproduction I found in my survey area occurred at a colony adjacent to Lake Hood. The colony consisted of six pairs and potentially fledged 10 of 13 chicks hatched. Five or six pairs existed at Potter's Marsh, although I observed no evidence of breeding there in 1991. Isolated pairs were observed at one other man-made and three natural lakes, and two pairs nested in the Mew Gull colony at the Port of Anchorage.

A greater number of broods of virtually every species of waterbird occurred on man-made waterbodies than on natural waterbodies in Anchorage (Table 21.2). Excluding Mew Gulls, 22 man-made waterbodies supported 82.3% (n = 437) of all broods and 84.5% (n = 2418) of all young. The number of broods on man-made waterbodies was significantly greater than the number on natural waterbodies for Red-necked Grebe, Canada Goose, Mallard, American Wigeon, Northern Pintail, Northern Shoveler (*Anas clypeata*), Canvasback, scaup spp., Herring Gull, and Arctic Tern; differences also approached significance for Horned Grebe (*Podiceps*

Table 21.3. Comparative use of man-made (n = 21) and natural (n = 29) waterbodies (excluding creeks and ditches) by loon, grebe, and waterfowl broods in the Anchorage area, 1991. Actual number of lakes used and broods observed are noted in parentheses.

	Man-made waterbodies		Natural waterbodies		
Species	% with broods	Mean brood size	% with broods	Mean brood size	$\chi^2_{(1)}$ (P)[a]
Pacific Loon	0	-	13.8 (4)	1.00 (4)	1.55 (>0.1)
Red-necked Grebe	28.6 (6)	2.70 (30)	13.8 (4)	2.83 (6)	0.87 (>0.1)
Horned Grebe	9.5 (2)	1.50 (2)	0	-	0.93 (>0.1)
Canada Goose	61.9 (13)	6.00 (73)	27.6 (8)	4.43 (21)	4.56 (<0.05)
Mallard	71.4 (15)	5.63 (177)	58.6 (17)	5.00 (45)	0.40 (>0.1)
Northern Pintail	19.0 (4)	5.32 (19)	3.4 (1)	2.00 (1)	1.79 (>0.1)
American Wigeon	42.9 (9)	4.15 (52)	17.2 (5)	5.50 (6)	2.80 (<0.1)
Gadwall	9.5 (2)	3.00 (2)	0	-	0.93 (>0.1)
Green-winged Teal	14.3 (3)	5.50 (6)	6.9 (2)	7.50 (2)	0.15 (>0.5)
Northern Shoveler	4.8 (1)	3.40 (5)	0	-	0.03 (>0.5)
Canvasback	4.8 (1)	7.33 (3)	0	-	0.03 (>0.5)
Scaup spp.	23.8 (5)	8.02 (56)	10.3 (3)	5.50 (4)	0.79 (>0.1)
Herring Gull	19.0 (4)	1.25 (4)	0	-	3.70 (<0.1)
Bonaparte's Gull	4.8 (1)	1.00 (1)	3.4 (1)	2.00 (1)	0.25 (>0.1)
Arctic Tern	4.8 (1)	2.17 (6)	0	-	0.03 (>0.5)

[a]Chi-square test for independence on percent of man-made vs. natural waterbodies used by broods.

auritus), Gadwall, and Green-winged Teal (*Anas crecca*). A greater proportion of man-made waterbodies were used by all species except Pacific Loon (*Gavia pacifica*), relative to natural waterbodies; however, these differences were significant only for Canada Geese. Differential use of man-made and natural waterbodies also approached significance for American Wigeon and Horned Grebe. Pacific Loons and non-breeding Common Loons preferred natural lakes in the Anchorage area with few exceptions, and in 1991, successful reproduction occurred only on natural lakes. On four lakes where loons nested in 1991, only two other waterbird broods (one Canada Goose, one Mallard) were found.

I did not test for significant differences in mean brood sizes between man-made and natural lakes because either a large disparity in sample sizes existed, or the differences in brood sizes seemed small (e.g., for Mallards); Canada Goose brood sizes were the exception to this rule, but Canada Goose brood size is masked by gang broods.

Within Anchorage, the highest waterfowl nesting density occurs west of the Old Seward Highway; intermediate nesting density occurs north of Tudor Road between Muldoon Road and the Old Seward Highway; and the lowest nesting density occurs east of the Old Seward Highway and south of Tudor Road, and includes the area east of Muldoon. Specific waterbodies were regionally important, or important for some species. Lake Hood/Spenard supported 42.6% (203 of 477) of all waterfowl broods and 46.3% (1265 of 2735) of all waterfowl young observed in the study area, including all Northern Shoveler and Lesser Scaup broods observed in 1991. University [Behm] Lake, a water-filled gravel pit, was the most important waterbody for American Wigeon. Cheney Lake, another water-filled gravel pit, supported a large number of Mallard broods and young, and had the second highest number of Red-necked Grebe broods. Potter's Marsh was the only site in Anchorage where Canvasbacks nested, and was one of only two sites where Gadwalls and Horned Grebes nested.

3.3 Population Changes

The only breeding waterbird trend data available for Anchorage are from a series of surveys of Lake Hood/Spenard from the early 1970s through the mid-1990s (Table 21.4). Waterfowl use of Lake Hood/Spenard increased from 1971 to 1983-1985 (Barnes and Trapp 1985) and continued to 1991 (Table 21.4). Numbers of Canada Goose, Mallard, Northern Pintail, and scaup young increased. Some species, such as Green-winged Teal and Northern Shoveler showed no discernible pattern. The number of American Wigeon young dropped between the mid-1980s and the 1990s.

3.4 Relative Abundance and Density

When compared to major waterfowl rearing areas in Alaska, the Anchorage area ranks high in density of young birds. Statewide, Hodges (1991) observed a total of 2024 loon, grebe, and waterfowl broods from a sample area of 1292 km^2. Extrapolating from my survey area of 298 km^2 to a comparable size of 1292 km^2 yields 2252 loon, grebe, and waterfowl broods. This suggests that waterbird production is comparable to, and may even exceed that of major waterfowl rearing areas in the state. Compared to statewide waterfowl production numbers for 1991, Anchorage produced a higher density of dabbling ducks than all other waterfowl rearing areas except the Yukon Flats, Yukon-Kuskokwim Delta, and Seward Peninsula (Table 21.5). For production of diving ducks in 1991, Anchorage ranks intermediate among the major rearing areas surveyed in Alaska (Table 21.5). Anchorage produced greater densities of young Canada Geese and Red-necked Grebes than any other waterbird rearing areas in Alaska (Table 21.6). Production of Pacific Loons (0.013/km^2) and Horned Grebes (0.010/km^2) was about a third to a fourth of that estimated for all rearing areas in the state (0.04/km^2 for each species; from Hodges 1991).

All species that bred in Anchorage, except Horned Grebes and Green-winged Teal, attained higher densities of young in the Anchorage Bowl than they did in the Cook Inlet (Kenai-Susitna) area (Table 21.5). In addition, the densities of young Red-necked Grebes, Canada Geese, Mallards,

Table 21.4. Relationship of current waterbird production at Lake Hood/Spenard to historic data.

	No. young counted					Mean brood size			
Species	'71[a]	'83[b]	'85[a]	'91[c]	'93[c]	'71[a]	'83[a]	'91[c]	'93[c]
Horned Grebe	7	0	0	0	0	2.3	--	--	--
Canada Goose	5	16	28	193	236	5	5.3	6.2	GB[d]
Green-winged Teal	19	0	0	17	3	6.3	--	5.7	3
Mallard	56	206	230	539	44	6.2	5.6	5.3	2.6
Northern Pintail	0	13	7	66	56	--	2.6	6	3.6
Northern Shoveler	0	0	0	17	9	--	--	3.4	3
American Wigeon	67	134	200	61	59	5.6	4.2	2.7	3.3
Scaup spp.	0	244	291	372	219	--	7.6	8.5	6.3
Goldeneye spp.	0	0	5	0	0	--	--	--	--
Arctic Tern	--	0	0	13	0	--	--	2.2	--

[a]From Barnes and Trapp (1985).

[b]From 15 July 1983 survey in Barnes and Trapp (1985). Nysewander and Trapp (1985) recorded 20 breeding adult arctic terns here in May and June 1983.

[c]This study.

[d]GB=gang brood.

American Wigeon, Canvasbacks, and scaup spp. were higher in the Anchorage Bowl than statewide (Table 21.5). Red-necked Grebes were nine times more abundant, Canada Geese 15 times more abundant, and Mallards 11 times more abundant in the Anchorage Bowl than statewide .

Brood size for most species was relatively consistent between the Anchorage Bowl and statewide rearing areas (Table 21.6). Notable exceptions included Red-necked Grebe, Canada Goose, Green-winged Teal, Canvasback, and scaup spp., which had larger average brood sizes in Anchorage, and Northern Shoveler which had a smaller average brood size in Anchorage.

4. DISCUSSION

Waterbird species richness was higher in the Cook Inlet than the Anchorage Bowl and did not differ between greater/former Anchorage Bowl and my study area, causing me to reject my second hypothesis. The lack of support for my hypothesis may have resulted from the greater variety of habitats and larger number of birds counted in both cases. Regarding the comparison of past and present species richness in the Anchorage Bowl, it is important to note that (1) the larger area included breeding habitat for Barrow's Goldeneyes, Harlequin Ducks, and Oldsquaw and (2) the relatively recent displacement of sensitive species such as the Sandhill Crane. The

Table 21.5. Density of waterbird young (young per km^2) in Anchorage (this study) and major waterfowl rearing areas in Alaska in 1991 (Hodges 1991).

Survey Region	Area (km^2)	Red-necked Grebes	Canada Geese	Dabbling Ducks	Diving Ducks
Anchorage	298	0.4	1.8	5.8	1.7
Kenai-Susitna	26483	0	0	0.4	0.7
Gulkana/Copper River	14395	0	0.1	1.2	1.7
Tanana-Kuskokwim River	50008	0	0	1.5	0.9
Yukon Flats	18783	0.2	0.2	7.8	3.5
Innoko River	15537	0	0	2.1	0.1
Koyukuk River	16506	0	0	4.4	1.4
Bristol Bay Lowlands	57032	0	0	3.6	0.4
Yukon-Kuskokwim Delta	76550	0.1	0.5	8.4	2.2
Kotzebue Sound	17607	0.1	0	3.9	2.2
Seward Peninsula	870	0.3	0.7	7.5	3.2
Kodiak Island	148	0	0	2.7	0
Yakutat Forelands	39	0	0	0	1.3

lack of Ring-necked Duck broods in the study area is notable, because the species breeds on Elmendorf AFB (Rothe et al. 1983).

I could not test my hypothesis that urban waterbird production exceeded those of other prime waterbird rearing areas in Alaska. However, some data seemed to support the hypothesis. The combined density of loon, grebe, and waterfowl broods slightly exceeded that of the statewide combined estimate. The density of Red-necked Grebe and Canada Goose broods exceeded those of all other waterbird rearing areas in Alaska (Table 21.6). Only three of 12 waterbird rearing areas in Alaska produced a higher density of *Anas* spp. young than did Anchorage (Table 21.6). On a statewide basis, six of 13 species that breed in Anchorage attained their highest brood densities in Anchorage. When compared to the Cook Inlet region, 10 of 13 species attained higher brood densities in Anchorage than the surrounding USFWS survey strata.

With the exception of loons, I found a higher proportion of broods of all species on man-made waterbodies than on natural waterbodies. Although the differences were significant only for Canada Goose broods, man-made waterbodies were significantly more productive than natural waterbodies in the Anchorage Bowl (Table 21.2). Man-made waterbodies appear to be extremely important for Canada Geese, American Wigeon, and scaup. Canada Goose used a significantly greater percentage and had larger broods on man-made waterbodies than on natural waterbodies. American Wigeons showed similar, although nonsignificant patterns, but their broods were larger on natural waterbodies (Table 21.3). Over 90% of scaup broods and

Table 21.6. Comparison of waterbird young density and mean brood sizes (young per km^2 / mean brood size) in Anchorage (from this study), the greater Cook Inlet, and in Alaska in 1991 (from Hodges 1991).

Species	Anchorage	Cook Inlet	Alaska
Pacific Loon	0 / 1.0	0 / 2.0	0 / 1.2
Common Loon	0 / -	0 / 1.0	0 / 1.2
Red-necked Grebe	0.4 / 2.7	0 / -	0 / 1.4
Horned Grebe	0 / 1.5	0 / 5.0	0 / 1.7
Canada Goose	1.8 / 5.6	0 / -	0.1 / 3.7
Mallard	4.2 / 5.5	0.4 / 5.4	0.4 / 4.6
Gadwall	0 / 3.0	0 / -	0 / 3.6
American Wigeon	0.9 / 4.3	0.1 / 4.0	0.6 / 4.8
Green-winged Teal	0.2 / 6.4	0.2 / 5.0	1.2 / 3.6
Northern Pintail	0.4 / 5.2	0 / 6.0	0.7 / 4.0
Northern Shoveler	0.1 / 3.4	0 / -	0.3 / 4.6
Canvasback	0.1 / 7.3	0 / -	0 / 5.0
Scaup spp.	1.6 / 7.9	0.2 / 5.5	0.9 / 5.2

young were produced on man-made waterbodies (Table 21.2), but no difference existed in the proportions of man-made and natural waterbodies used for brood rearing (Table 21.3), perhaps because most scaup were produced on only a few waterbodies or natural waterbodies represented attractive but poor quality nesting habitat. Natural wetlands, however, appear to be important for Canada Goose nesting. Many nests were found in natural bogs dominated by sweetgale (*Myrica gale*); fewer were found on islands in reservoirs (unpubl. data). Although man-made waterbodies were important for Mallards (71% received brood use and contributed 80% of total production), 59% of natural lakes were also used by Mallards for brood-rearing.

Most Red-necked Grebe production in the Anchorage Bowl occurred on man-made waterbodies. This differs from a population in the Northwest Territories where 88% of the waterbodies used were natural; only 4% of man-made waterbodies were used (Fournier and Hines 1998). However, at this site all man-made waterbodies were borrow pit ponds <1.1 ha in size.

Only Pacific Loons used natural lakes to a greater degree than man-made lakes. On the four lakes that produced Pacific Loon broods, only two other broods were produced (one Canada Goose, one Mallard). Loons are aggressive in excluding other waterbirds from their territories (North 1994, McIntyre and Barr 1997), and this may in part explain the apparent preference of man-made waterbodies by other waterbird species in Anchorage. Also, man-made lakes are probably more productive because they are more eutrophic, especially those created on creek systems. Creeks in Anchorage carry large loads of nutrients and other pollutants downstream (Anonymous 1993b; pers. obs.). However, many man-made lakes are probably not suitable for Pacific and Common Loons because loons are visual predators and need clear water for successful foraging (McIntyre and Barr 1997).

Data from surveys of Lake Hood/Spenard indicate that populations of Canada Geese, Mallards, Northern Pintails, and scaup have likely increased substantially in Anchorage as a result of urbanization. Much of the decreases from 1991-1993, especially for scaup, can be attributed to a nest-control program initiated by Anchorage International Airport. In 1993, shrubs were removed from a large island and pigs placed on it to eat eggs. Horned Grebes appear to have declined on Lake Hood/Spenard since 1971. Horned Grebe reproduction was probably affected by habitat changes, or by nest failure because of waves generated by float planes. Causes of fluctuations for Green-winged Teal and American Wigeon are not apparent.

Man-made waterbodies may have enabled Gadwalls, Canvasbacks, and Lesser Scaup to expand their breeding range into the Cook Inlet region, and Northern Shovelers into Anchorage. Cook Inlet was not identified as a breeding locale for Canvasbacks or Lesser Scaup by Bellrose (1980). In addition to the three Canvasback broods I observed at Potter's Marsh in 1991, I observed seven other Canvasback broods there from 1987-1994. Only Greater Scaup had previously been documented breeding in Anchorage, but I positively identified several broods of Lesser Scaup. Urbanization has been described as contributing to a comparable expansion of Mallards into American Black Duck (*Anas rubripes*) range (Heusmann 1988).

Pacific and Common Loons appear to have been adversely affected by urbanization. In 1991, only 14% of the natural waterbodies supported Pacific Loon broods, and none supported Common Loon broods (although Common Loons nested on a natural lake in the Anchorage Bowl in 1993 [Tankersley 1994]). On Elmendorf AFB, Common Loons did not nest but occupied 30% (n = 5) of 17 waterbodies surveyed, while Pacific Loons nested on 6% (n = 1) of those waterbodies (Rothe et al. 1983). On the Kenai Peninsula, Rosenberg (1986) found Common Loons occurred on 72% of the lakes surveyed. In Alaska, the highest density of summering Common Loons occurs in the Cook Inlet area (Hodges et al. 1996).

Red-necked Grebes appeared to benefit from urbanization, while Horned Grebes appeared to be adversely affected. This trend for Horned Grebes is supported by brood data from Lake Hood/Spenard. Although the density of Red-necked Grebes was much higher in Anchorage (0.35/km^2) than in any other rearing areas in Alaska, overall density in Alaska is lower than in the Northwest Territories (2.2/km^2, from data in Fournier and Hines 1998).

Comparative data on brood density in Anchorage and Cook Inlet (Table 21.6), brood use of man-made and natural waterbodies (Table 21.2), and population changes at Lake Hood/Spenard (Table 21.4) provide corroborative evidence on spatial and temporal scales that the primary waterfowl species that benefit from urbanization are Canada Geese, Mallards, Northern Pintails, American Wigeon, and scaup. Statewide, densities of summering Canada Geese, American Wigeon and scaup are low in the Cook Inlet area (Hodges et al. 1996). Yet, these species are among the most abundant in the Anchorage Bowl, with brood densities higher than anywhere in the state. Impacts of urbanization that favor these species may include a reduced predator population (particularly avian predators), waterbodies that are more eutrophic, a greater diversity of waterbody types, an increase in shallow-water feeding habitats (see Adams et al. 1985), and release from competition with loons. Mallards also benefit by food handouts from humans, which allow some to remain in Anchorage year-around. Overwintering by Mallards was first documented on an Audubon Christmas Bird Count in 1947; since then the wintering population has increased slowly to over 2000 (USGS 2000). Species that may be adversely affected by urbanization are more difficult to identify. However, the primary waterbird species that appear to be adversely affected by urbanization are Pacific and Common Loons, Horned Grebes, Ring-necked Ducks, goldeneye spp., and Sandhill Cranes. Impacts of urbanization that may be detrimental to these species include greater amounts of human disturbance, waterbodies that are less oligotrophic, and habitat fragmentation. Horned Grebes are an enigma. Although this survey found Horned Grebe broods only on man-made waterbodies in 1991, two of four broods I found in other years were on natural waterbodies. It may be that the benefits of man-made wetlands for this species are offset by other consequences of urbanization including heightened competition from other waterbirds showing a positive response urbanization.

A valid criticism of my conclusions may be that Canada Geese and Mallards are undesirable in an urban environment. Conversely, the recreational value of feeding waterfowl is valued by a segment of urbanites that wildlife professionals have traditionally eschewed or not recognized. However, this study shows that Canada Geese and Mallards are not the only waterbird species that respond positively to urbanization. Moreover, Anchorage wetlands produce waterfowl that provide hunting opportunities in rural Alaska and the Pacific Flyway. Heusmann (1981) found that only 10-15% of urban Mallards remained in urban settings year-around. Band recoveries show an exchange of Canada Geese between Anchorage and the Willamette Valley in Oregon, the north end of Cook Inlet, and the Nushagak Peninsula in southwestern Alaska (Campbell 1992, Bird Banding Lab data). Band recoveries also show an exchange of Mallards and Northern Pintails between Anchorage, popular hunting areas on the west side of Cook Inlet, British Columbia, California and Washington state (unpubl. data).

5. POLICY AND MANAGEMENT IMPLICATIONS

The results of my surveys indicate that 1) waterfowl production has increased substantially in Anchorage since the early 1970's, 2) the density of waterfowl young in Anchorage is comparable to that of major waterfowl rearing areas in Alaska, 3) waterbird production occurs primarily on man-made and man-altered waterbodies, 4) all man-made waterbodies were used by waterbirds for reproduction, 5) the urban environment appears to have allowed Canada Geese, Mallards, Northern Pintails, Northern Shovelers, and scaup populations to increase substantially, 6) the urban environment may have allowed Gadwalls, Canvasbacks and Lesser Scaup to extend their breeding range into Cook Inlet, 7) waterbodies in Anchorage have wildlife values that should be protected by adhering strictly to the Section 404(b)(1) guidelines of the Clean Water Act, and 8) wildlife values lost due to unavoidable wetland impacts should be compensated for through wetland creation and restoration.

A wide variety of waterbody habitats need to be preserved in order to conserve the full suite of waterbird species that breed in Anchorage. If the Anchorage Bowl becomes more populated at projected rates, severe pressure will exist to develop more wetland areas. Conservation of wetland areas can only be accomplished by strict adherence to the 404(b)(1) guidelines and standard mitigation sequencing principles (e.g., avoidance, minimization, compensation). Based on the widespread use of all types of waterbodies in Anchorage, I believe that creation of new ponds or enlargement or enhancement of existing waterbodies will support and enhance waterbird reproduction in Anchorage.

ACKNOWLEDGEMENTS

I would like to thank the two anonymous reviewers and David McGillivary for authorizing completion of the brood surveys using U.S. Fish and Wildlife Service resources.

REFERENCES

Adams, L. W., L. E. Dove, and T. M. Franklin. 1985. Mallard pair and brood use of urban stormwater-control impoundments. Wildl. Soc. Bull. 13:46-51.

Anchorage Audubon Society. 1983. Birds of Anchorage, Alaska: a checklist.

Anonymous. 1954. A report on fish and wildlife resources in relation to the water-use developments of the Anchorage area, Territory of Alaska. Admin. rep., US Dept. Interior, Fish Wildl. Serv., Juneau, AK

Anonymous. 1990. The Alaska wilderness milepost, 5th ed. Alaska Northwest Books, Bothell, WA.

Anonymous. 1993a. Anchorage wetland trends study (1950 to 1990). Unpubl. rep., U.S. Fish Wildl. Serv., Anchorage. AK.

Anonymous. 1993b. Chester Creek drainage. Draft Water Quality Assessment Report, Alaska Dept. Environm. Conserv., Anchorage.

Barnes, L. J., and J. L. Trapp. 1985. Trends in waterbird populations and species composition at the Lake Hood seaplane base, Anchorage, Alaska: 1971-1985. Unpubl. rep., US Fish Wildl. Serv., Anchorage, AK.

Bellrose, F. C. 1980. Ducks, geese and swans of North America. Wildl. Manage. Inst. and Stackpole Books, Harrisburg, PA.

Brackney, A. W., and R. J. King. 1992. Aerial breeding pair surveys of the Arctic Coastal Plain of Alaska: abundance and distribution of waterbirds 1986-1990. Unpubl. rep., US Fish Wildl. Serv., Fairbanks, AK.

Campbell, B. H., D. Rosenberg, and T. Rothe. 1992. Alaska Department of Fish and Game Division of Wildlife Conservation waterfowl program annual report, 1 July 1991-30 August 1992. Unpubl. rep., Alaska Dept. Fish Game, Anchorage.

Dixon, W. J., and F. J. Massey, Jr. 1951. Introduction to statistical analysis. McGraw-Hill Book Co., Inc., New York, NY.

Fournier, M. A., and J. E. Hines. 1998. Breeding ecology and status of the Red-necked Grebe. *Podiceps grisgena*, in the subarctic of the Northwest Territories. Can. Field-Natur. 112:474-480.

Heusmann, H. W. 1981. Movements and survival rates of park Mallards. J. Field Ornithol. 52:214-221.

Heusmann, H. W. 1988. The role of parks in the range expansion of the Mallard in the northeast, p.405-412. *In* Weller, M. W. [ED.], Waterfowl in Winter. University of Minnesota Press, Minneapolis.

Hodges, J. I. 1991. Alaska waterfowl production surveys, 1991. Unpubl. rep., US Fish Wildl. Serv., Juneau, AK.

Hodges, J. I., J. G. King, B. Conant, and H. A. Hanson. 1996. Aerial surveys of waterbirds in Alaska 1957-94: population trends and observer variability. Natl. Biol. Serv. Information Technology. Rep. 4, Washington, DC.

Lensink, C. J., and D. V. Derksen. 1990. Evaluation of Alaskan wetlands for waterfowl, p.45-84. *In* Alaska: regional wetland functions. Proceedings of a workshop held at Anchorage, Alaska May 28-29, 1986. Publ. No. 90-1, Environ. Inst., Univ. Mass., Amherst, MA.

McIntyre, J. W., and J. F. Barr. 1997. Common Loon (*Gavia immer*). *In* A. Poole and F. Gill [EDS.], The Birds of North America. Academy of Natural Sciences, Philadelphia, PA.

Municipality of Anchorage. 1997. Anchorage indicators, 1997. Vols. 1 and 2. http://www.muni.org.

North, M. R. 1994. Yellow-billed Loon (*Gavia adamsii*). *In* A. Poole and F. Gill [EDS.], The Birds of North America. Academy of Natural Sciences, Philadelphia, PA.

Nysewander, D. R., and J. L. Trapp. 1985. Surveys of colonial waterbirds in upper Cook Inlet 1983. Unpubl. rep., US Fish Wildl. Serv., Anchorage, AK.

Rosenberg, D. H. 1986. Wetland types and bird use of Kenai Lowlands. Unpubl. rep., US Fish Wildl. Serv., Anchorage, AK.

Rothe, T. C., S. H. Lanigan, P. A. Martin, and G. F. Tande. 1983. Natural resource inventory of Elmendorf Air Force Base, Alaska. Unpubl. rep., US Fish Wildl. Serv., Anchorage, AK.

Sokal, R. R., and F. J. Rohlf. 1973. Introduction to biostatistics. W. H. Freeman Co., San Francisco, CA.

Steel, R. G. D., and J. H. Torrie. 1960. Principles and procedures of statistics, with special reference to the biological sciences. McGraw-Hill Book Co., Inc., New York, NY.

Tande, G. F. 1988. Changes in Anchorage wetlands between 1982 and 1988. Unpubl. rep., US Fish Wildl. Serv., Anchorage, AK.

Tankersley, N. 1994. June 1 letter to Loon Watch volunteers. Alaska Dept. Fish Game, Anchorage.

United States Geological Survey. 2000. Audubon Christmas bird counts. http://www.mp1-pwrc.usgs.gov/birds/cbc.html.

Chapter 22

Creating a homogeneous avifauna

Robert B. Blair
Department of Zoology, Miami University, Oxford, OH 45056 USA, e-mail: blairrb@muohio.edu

Key words: Conservation, diversity, extinction, gradient, homogenization, invasion, urbanization

Abstract: I compared birds on urban gradients in two ecoregions of the United States by censusing summer resident bird populations at six sites in central California's coastal chaparral and southwest Ohio's eastern broadleaf forest. These sites represented comparable gradients of urban land-use which ranged from relatively undisturbed to highly developed and included biological preserves, recreational areas, golf courses, residential neighborhoods, office parks (or apartment complexes), and business districts. Species richness and Shannon diversity peaked at moderately disturbed sites and were significantly correlated between comparable land-use types in the two ecoregions. Bird abundance and biomass peaked at moderately disturbed sites as well but were not significantly correlated between ecoregions. The pre-development bird species (assumed to be those found at the most undisturbed sites) dropped out gradually as the sites became more urban and the number of remaining species was significantly correlated between ecoregions with only three (CA) and one (OH) species remaining in the most urban sites. Taxonomically, the bird communities at the least urbanized sites (the preserves and recreational areas) were very different with an average Jaccard's index of species similarity of 0.065 while the most disturbed sites (the business district, apartment complexes/office park, residential areas and golf courses) had an average similarity of 0.185. The species assemblages along the gradient shifted gradually, demonstrating local extinction of and local invasion by different species as the sites become more urban.

1. INTRODUCTION

"The earth's biota is being homogenized rapidly as human activities increasingly introduce species outside their natural range." (Lodge 1993)

"Replacement of native endemic species with opportunistic and human commensal species could ultimately produce a relatively homogenized biosphere." (McKinney 1998)

"Global ecological homogenization, with commensurate loss of species, is the logical endpoint of the anthropogenic spread of exotics." (Lodge et al. 1998)

The world's biota is becoming homogenized at an unprecedented rate (Vermeij 1991, Vitousek et al. 1996, McKinney 1998) and statements such as those listed above reflect a shifting concern in the field of conservation biology. Human activities are aiding the dispersal and range expansion of tens of thousands of species (Lodge 1993), leading some scientists to wonder if we are creating a world of "cockroaches, pigeons, and starlings" (P. Ehrlich pers. comm.).

Ecologists studying the phenomenon of homogenization have focused on identifying characteristics common to invading species or invaded habitats (Reichard et al. 2001). Traits of invasive species include high dispersal rates, single-parent or vegetative reproduction, high genetic variability, phenotypic plasticity, large native range, eurytopy, polyphagy, and human commensalism (Lodge 1993). Attributes of invasible habitats include climatic similarity to the native habitat of the invading species, early successional stage, low species diversity, low predator richness, and recent disturbance (Lodge 1993). However, many exceptions to these traits exist and predicting which species successfully will invade or which habitats will be invaded is an inexact science (for brief introductions to the invasion literature, see Lodge 1993, Vitousek et al. 1996, Hobbs and Mooney 1998).

Many of the traits of invasible habitats are the result of changes in the landscape brought about by human land-use and urbanization. However, few ecologists have asked if analogous communities in different ecoregions are actually becoming more homogeneous (i.e., similar) due to human land-uses. In other words, biotic homogenization is assumed to result from biological invasion, but this assumption rarely has been tested explicitly. In this study, I examine the avifauna of two ecoregions to test whether one type of human activity—urbanization—leads to more homogeneous biotic communities.

The establishment of non-indigenous bird species is peculiarly linked with humans. During the 1800s, western European settlers introduced many species throughout the world because they wanted birds from their homelands in their new environs. The American Acclimatization Society is infamous for its persistent attempts to introduce all of the species mentioned in Shakespeare's writings to New York City's Central Park and for its most notorious introduction, the European Starling (*Sturnus vulgaris*) (Wetmore 1964, Ehrlich et al. 1988). North America's population traces from a release by the Society of 80 birds in 1890 and 40 more in 1891 (Wood 1924). A similar group on the West Coast unsuccessfully tried to establish starlings in the City Park of Portland, Oregon with releases in 1889, 1890, and 1892 (Long 1982). Currently, many non-indigenous species are becoming established through the intentional or inadvertent release of pets (Van Bael and Pruett-Jones 1996, Garrett and Smithson 1999). Less direct, but still human-induced, is the range expansion following manipulation of the landscape due to urbanization (Mills et al. 1989).

Case (1996) took a global perspective on the establishment of invasive bird species by examining introductions and extinctions on islands throughout the world. He found that the best correlate of successful invasion was the number of native species that had gone extinct on that island during the previous 3000 years. He interpreted the number of extinctions as a reflection of the direct and indirect effects of human activity, including habitat destruction, and invasion by deleterious organisms such as exotic predators, herbivores, and parasites. He also surmised that the correlation between introductions (a.k.a. invasions) and extinctions arose because native birds are usually more common to (or even restricted to) native habitats while introduced birds occupy open and disturbed habitats. In addition, Case suggested that conversion of land to urban and agricultural uses decreases habitat availability for native species and increases habitat availability for introduced species.

In this study, I examine the idea that urbanization leads to the local extinction of species and invasion by different species and, consequently, creates more homogeneous bird communities. Specifically, I extend previous research on an urban gradient in California's coastal chaparral ecoregion by comparing it to a similar gradient in Ohio's eastern broadleaf forest ecoregion (sensu Bailey et al. 1994). In California, I found that urbanization affects not only the distribution and abundance of individual species of birds, but also community measures such as species richness, Shannon diversity, bird abundance, and bird biomass (Blair 1996, Blair 1999). Here, I examine (1) if the effects of urbanization on the avian community are similar in such disparate ecoregions, and (2) if increasing urbanization leads to an increasingly homogeneous avifauna.

2. METHODS

2.1 Field Sites

The sites chosen for this study represent forms of development typical in an urban-suburban matrix and comprise the full gradient of human land-use intensity ranging from relatively undisturbed wildland to highly developed urban areas. In each ecoregion, the sites include a biological preserve, recreational area, golf course, residential neighborhood, office park (California) or apartment complex (Ohio), and business district.

In California, the six sites were located within a circle of 3 km radius centered at Stanford University near Palo Alto, California USA (37°20' N, 122° 15' W; typical elevation <100 m). These sites were judged to be as ecologically similar as possible prior to development. Assessment of historical ecological similarity was based on the presence of coast live oak (*Quercus agrifolia*), valley oak (*Q. lobata*) or blue oak (*Q. douglasii*) trees greater than 100 years of age and on accounts of predevelopment vegetation (Cooper 1926).

The California sites (Figs. 22.1 and 22.2) include:

(1) a biological preserve—The Jasper Ridge Biological Preserve is 512 ha property which has been used for biological research since the 1890s. Prior to 1974, portions of the preserve were leased for cattle grazing and public recreation. Currently, access is limited to researchers and visitors on docent-led tours. This site can be considered wildland with none of it being developed.

(2) an open-space recreational area—The Stanford University foothills preserve was heavily grazed until 1989 and currently is used by joggers, dog-walkers, hikers, and equestrians. Motorized access is limited to maintenance vehicles. This site, too, could be considered wildland because it has no development, though it has a history of human manipulation.

(3) a golf course—The Stanford University Golf Course borders the open-space recreation area and was originally developed in 1930. At the smallest scale, this site would also be considered wildland because it has few permanent structures, no permanent residents, and <4% coverage in buildings and pavement. However, at a more appropriate scale, I view it as exurban or suburban.

(4) a residential area—This area consists of single-family detached houses which were built mostly between 1930 and 1960. The larger residential zone that includes this site shares borders with the open-space recreational area and the office park. This site is distinctly suburban with ~42% of the area covered in buildings and pavement.

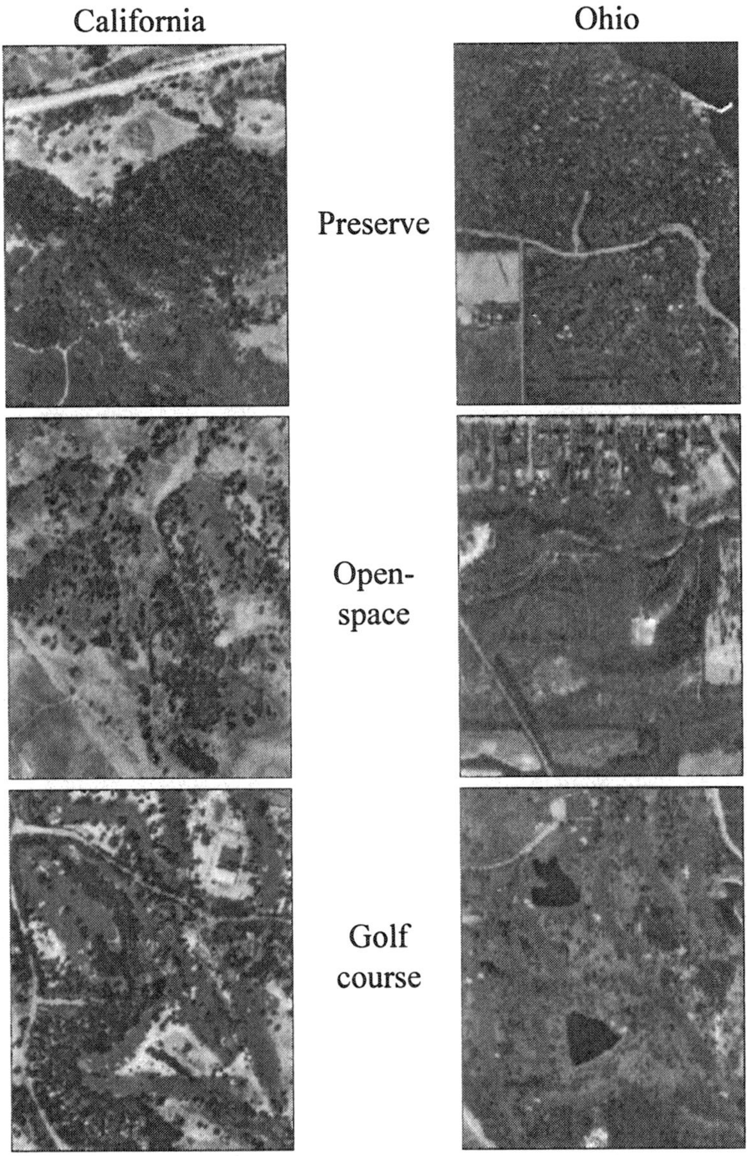

Figure 22.1. Aerial views of the preserves, open-spaces, and golf courses in California and Ohio. Images were taken in August 1998 in California, in March 1994 in Ohio, and cover ~50 ha.

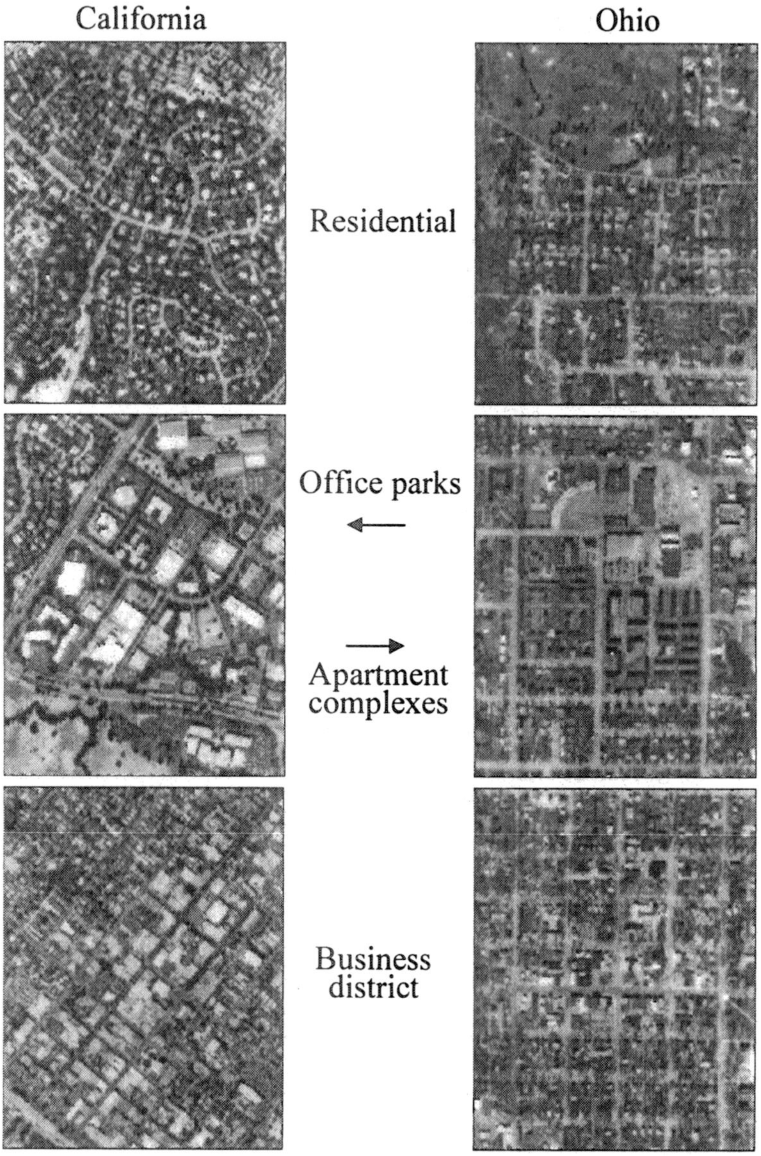

Figure 22.2. Aerial views of residential areas, apartment complexes/office parks, and business districts in California and Ohio. Images were taken in August 1998 in California, in March 1994 in Ohio, and cover ~50 ha.

(5) an office park—The Stanford Research Park was created in the mid-1950s. The development of the specific study site began in 1965 and consists primarily of low-rise office buildings, parking lots, and landscaped areas. The office park borders the residential area. This site can be considered borderline urban with approximately 66 percent of the area covered in buildings and pavement. However, it is lushly landscaped.

(6) a business district—Downtown Palo Alto is separated from Stanford University buildings by an approximately 1 km wide band of eucalyptus and oak woodland. This site is distinctly urban with 86% of the area covered in buildings and pavement.

In Ohio, the six sites were located within a 10 km radius of Miami University, Oxford, Ohio USA (39°30' N, 84° 45' W; typical elevation ~290 m). These sites were judged to be as ecologically similar as possible prior to development and originally consisted of climax beech-maple forest according to historical maps and reference materials (Gering and Blair 1999).

The sites (Figs. 22.1 and 22.2) include:

(1) a nature preserve—Hueston Woods Nature Preserve is a 67 ha woodlot comprised of mature beech-maple forest with the exception of a narrow service road, a small parking area, and moderate-use dirt footpaths. A surrounding state park (975 ha) consists of young and mature deciduous forest, old fields, infrequent stands of conifers, and a 250 ha lake (Ray and Vankat 1982). This site can be considered wildland with little development.

(2) a recreational area—Peffer Memorial Park was purchased by Miami University as two separate parcels of pasture land and agricultural field in 1955 and 1966. A large portion of the 80 ha recreational park is managed by Miami University as a multiple-use trail system for residents of the local community. Currently, the vegetation is predominantly secondary growth of low stature. This site, too, can be considered wildland with no development, though it has a history of being substantially manipulated by humans.

(3) a golf course—The 102 ha Hueston Woods Golf Course borders the nature preserve and was established in 1980. The course accommodates 10,000 players between March and November. Rough consists primarily of native trees (e.g., maple, beech) and grasses. The site has <2% of its immediate area covered with buildings or pavement, but the surrounding landscape dictates that it should be considered exurban or suburban.

(4) a residential district—This residential community contains mostly detached, single-family houses built between 1960 and 1974. Dense lawns, gardens, shade trees, and native and ornamental flora are frequently located around the houses (Beissinger and Osborne 1982). This site can be considered suburban with ~32% of its area covered with buildings and pavement.

(5) apartment developments—This area consists of multi-level apartment complexes and intervening parking lots with isolated trees and small-scale landscaping. The development of apartment complexes occurred between 1960 and 1980. We conducted our surveys in the apartment complexes rather than an office park, as used in California, because no office parks exist near Oxford, Ohio. The building height, density, layout, parking, and plantings are comparable to that of an office park. This site can be considered marginally urban with approximately 61 percent of its area covered with buildings and pavement.

(6) a business district—This is a 4 x 2 block city area consisting primarily of low-rise office buildings, parking lots, and one-half block of landscaped urban park. Initial construction of these properties began in the early to mid 1800s. This site is distinctly urban with ~81% of the area covered in buildings and pavement.

2.2 Ordering of Gradients

I used two methods to rank sites from most natural to most urban. The first method was the Delphi Technique which can be used when hard data are lacking, but expertise in a system is available (Zuboy 1981, Blair 1994, Blair 1996). In California, faculty, staff, and students (n = 14) working with the Center for Conservation Biology at Stanford University were asked to rank the sites independently and anonymously from most natural to most urban, and to give the criteria they used. The results of the ranking were then distributed to the participants and the process was repeated until the group reached consensus on the ordering.

In Ohio, faculty and graduate students (n = 12) in the Department of Zoology at Miami University ranked the sites. The initial survey did not yield a group consensus, so the results were redistributed to participants along with measurements of canopy cover (m^2 ha^{-1}) and tree basal area (m^2 ha^{-1}) from each site. The participants again ranked the sites, taking into account the site-specific data as well as the results of the initial survey.

The second method was to order the sites in accordance with percent of area covered by buildings, pavement, lawn, grassland, and trees or shrubs. This produced an identical ranking to that of the Delphi Technique. Alberti et al. (2001) discuss more quantitative techniques for measuring degree of urbanization.

2.3 Land-cover Quantification

For each study site, I visually estimated the area covered by buildings, pavement, lawn, grassland, and trees or shrubs in an approximately 50 m radius (depending on the particular scale of the map or photograph) centered at each survey point and then converted the total amount to percentage of site covered. In California, I estimated the cover of each land-use type from recent aerial photographs provided by Stanford University, the City of Palo Alto, and Jasper Ridge Biological Preserve. In Ohio, I used ortho-digitally corrected land-use maps available from the City of Oxford and aerial photographs available from the Department of Geography at Miami University.

2.4 Bird Surveys

I estimated the relative abundance of all perching or singing birds during peak breeding season using variable circular-plots at 16 survey points in a 4 x 4 matrix within each site where each point was at least 100 m from its nearest neighbor (Reynolds et. al 1980). In Hueston Woods and Peffer Memorial Park (both in Ohio), sampling points were located along foot trails instead of in a 4 x 4 matrix.

In California, points were surveyed a total of eight times in June and July, 1992, and four times in June, 1993. The open-space site was visited only six times in 1992 because it burned on 10 July. In Ohio, points were surveyed a total of eight times in June and July 1996, and four times in June 1997. Daily surveys began at dawn and continued until all 16 points at a site were covered, ~2 hrs. Each point was visited for five minutes. This method results in a relative abundance that estimates density (birds ha^{-1}) of all species within each site (see Blair 1996).

2.5 Community Measures

Species richness is the number of species seen at each site and is directly comparable among sites because sampling effort and area were equal at all sites. Species diversity was calculated with the Shannon Index (Shannon and Weaver 1949, Magurran 1988), which incorporates both species richness and evenness.

The abundance for all birds is the sum of the abundances for all species within a site. The biomass for all birds is bird abundance multiplied by average mass of each species as reported in Dunning (1993). Rock Doves (*Columba livia*) were omitted from the California biomass data because they overwhelmed all other species (Blair 1996). The assemblages of species found at the biological preserves in California and Ohio were taken as the minimally-disturbed standard (Karr and Chu 1998), or the best representatives of the bird communities that existed in the areas prior to development. Percent of woodland species in more urban sites is based on this minimally-disturbed standard.

2.6 Validation of Sites

I conducted validation surveys to assess whether the assemblages of birds at sites included in this study were typical of sites with similar land-use. In each ecoregion, I surveyed 24 additional sites that were within 20 km of the study sites. The set of validation sites included four different business districts, office parks (CA) and apartment complexes (OH), residential areas, golf courses, open-space reserves which were heavily used for recreation, and open-space reserves which were infrequently used by hikers (because no additional biological preserves were available). In California, I visited the original study sites and the 24 validation sites for one day each in July 1993 using the same bird sampling protocols as used in the main portion of this study. In Ohio, an assistant or I visited these sites in June and July of 1997. To test whether these sites were typical of other sites of land-use within each ecoregion, I performed a cluster analysis using the Goodman-Kruskal gamma correlation coefficient on the count of individuals within each species (not densities) of all birds observed. This distance metric is recommended for rank order or ordinal scales (SYSTAT 1992).

2.7 Comparisons Between Ecoregions

I compared species richness, diversity, abundance, biomass, and percent of woodland species between the two ecoregions with Pearson correlations (SYSTAT 1992). I used two methods to determine if increasing urbanization led to a more homogeneous avifauna. First, I conducted a land-use type by land-use type comparison (e.g., golf courses in both ecoregions) using Jaccard's Index of Similarity, which measures percent species overlap between two sites (Magurran 1988). Jaccard's Index may range from 0 (indicating no species overlap) to 1 (indicating complete species overlap). Second, I performed a multivariate cluster analysis using normalized percent disagreement on the average daily density of all species in all of the 12 formal study sites, a metric that is useful when there is high ß-diversity and, consequently, many empty cells (Jongman et al. 1995, SYSTAT 1992).

3. RESULTS

3.1 Ordering the Urban Gradient

The Delphi Technique led to similar orderings of the sites along the gradient in both ecoregions. In California, the ranking of the sites in order of increasing urbanization was biological preserve, open-space recreational area, golf course, residential area, office park, and business district. In Ohio, the ranking of the sites was nearly identical except that exactly half of the respondents ranked the apartment complexes and half ranked the business district as most urban. However, land-cover data (see below) suggested that the business district is the more urban (Blair 1996, Gering and Blair 1999).

The assessments of land-cover revealed unimodal distributions of percentage area covered by buildings, pavement, lawn, grassland, open water, and trees and shrubs along the urbanization gradient (Figs. 22.3 and 22.4). Cluster analysis using these measures generally grouped each land-use type across ecoregions. For example, the residential areas in California and Ohio were more similar to one another according to cover characteristics than to other sites (Fig. 22.5). The major exception was the open-space reserve in California. This location had large areas of grassland, a cover type largely unique to that site. The distinct placement of the open-space reserve in California forced the open-space reserve in Ohio to group with the preserves.

3.2 Birds Along the Gradient

The abundances of all bird species varied to some degree across the gradients (Figs. 22.6 and 22.7). Of the 40 bird species in the California survey, 31 had continuous, unimodal distributions (within the limits of 1 SE) of abundance across the urban gradient. In other words, the estimated densities of these species were highest at one end of the gradient and decreased gradually as urbanization increased or decreased. Of the 44 species encountered in Ohio, 38 had unimodal distributions within the limits of 1 SE.

3.3 Comparison of Community Measures

Species richness ($n = 6$, $r = 0.959$, $P = 0.002$), Shannon diversity ($n = 6$, $r = 0.826$, $P = 0.043$), and the percent of woodland species remaining ($n = 6$, $r = 0.913$, $P = 0.011$) were significantly correlated between similar land-use types in the two ecoregions (Fig. 22.8). Bird abundance ($n = 6$, $r = 0.488$, $P = 0.326$) was not significantly correlated between similar land-use types in

California and Ohio. However, abundances in both California and Ohio displayed similar, unimodal patterns of high levels at intermediate sites along the gradient and lower levels at the more urban or more natural sites. If sites were shifted by one degree of urbanization along the gradient (i.e., pairing Ohio business district and California office park, Ohio apartment and California residential, Ohio residential and California golf course, Ohio golf course and Calfornia open space, Ohio open space and California preserve) then the correlation was significant ($n = 5$, $r = 0.955$, $P = 0.012$). Bird

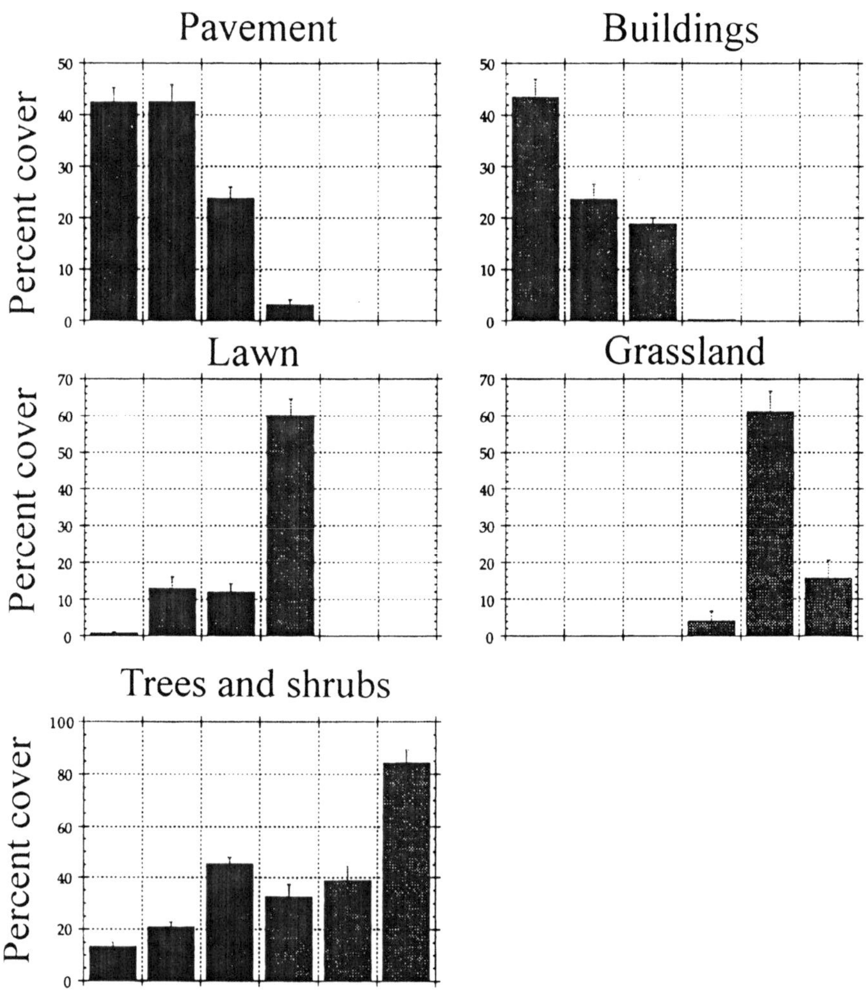

Figure 22.3. Percent area (± SE) covered by pavement, buildings, lawn, grassland, and trees and shrubs in a California preserve, open-space recreational area, golf course, residential area, office park, and business district (in order from left to right).

biomass ($n = 6$, $r = 0.717$, $P = 0.109$) was not significantly correlated between similar land-use types in California and Ohio though both states displayed similar, unimodal patterns of high levels at intermediate sites along the gradient and lower levels at the more urban or more natural sites.

3.4 Homogenization of Avifauna

Species similarities between the formal study sites in California and Ohio exhibited a distinct break midway along the urbanization gradient, more

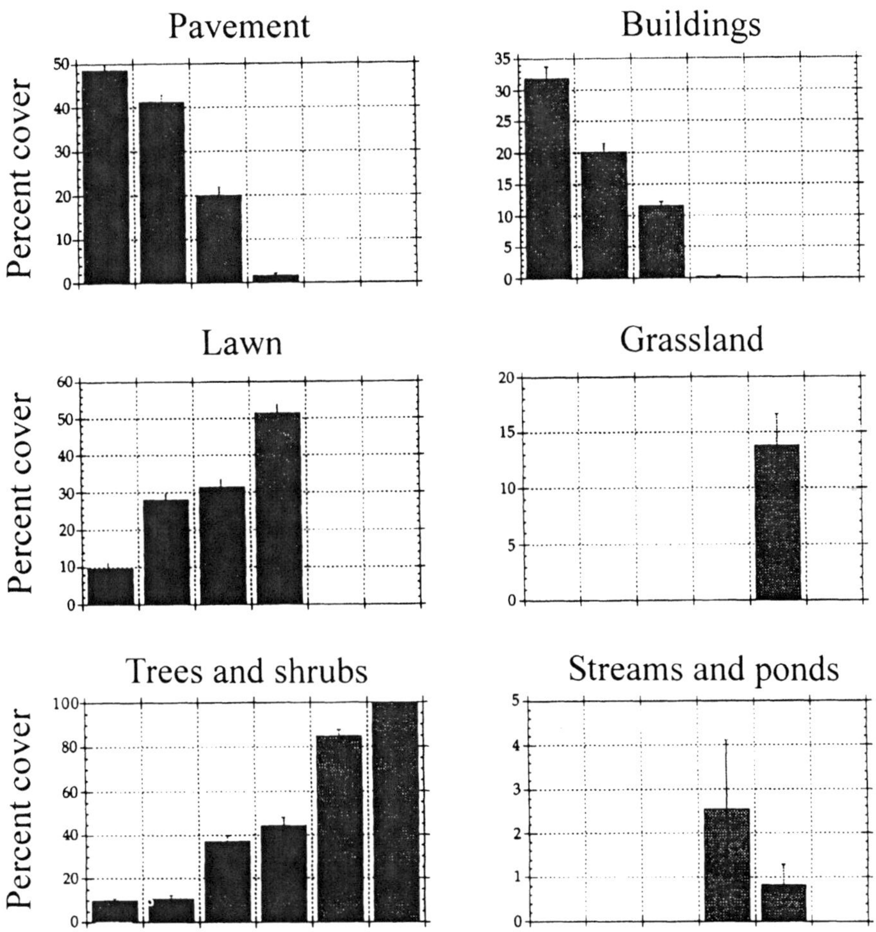

Figure 22.4. Percent area (± SE) covered by pavement, building, lawn, grassland, and trees and shrubs, and open water in an Ohio preserve, open-space recreational area, golf course, residential area, apartment complexes, and business district (in order from left to right).

specifically between golf courses and open-space (Fig. 22.9). These results suggest that the land-use types fall into two groups: relatively urbanized (business district, office park/apartment complexes, residential, and golf courses) with an average of 0.185 and relatively undeveloped (the open-space reserve and biological preserve) with an average of 0.065. Instead of increasing monotonically with urbanization, it appears that species similarity jumps once the landscape is entirely manipulated.

The cluster analysis similarly revealed this pattern; more natural sites were relatively unique while more urban sites were more similar to one another (Fig. 22.10). Based on the abundance of all species, the dendrogram shows that the biological preserves, open-space reserves, and golf courses fall out in highly differentiated groups based on their ecoregion while the residential areas, office park/apartment complexes, and business districts fall into one cluster regardless of ecoregion.

3.5 Validation of Sites

In each ecoregion, the species assemblage at the six formal study sites was representative of the species found at validation sites with similar levels of development. In California, all of the business districts, all of the office

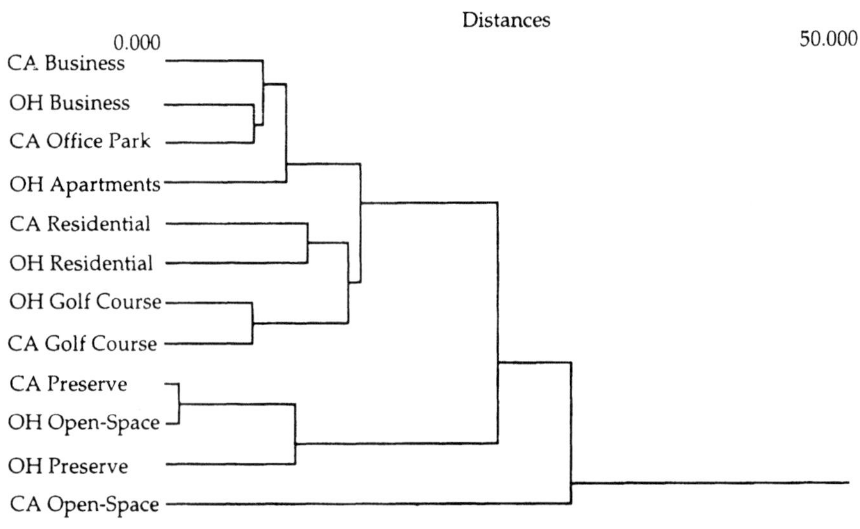

Figure 22.5. Cluster analysis of sites in California and Ohio based on percent area covered by pavement, buildings, lawn, grassland, trees and shrubs and the metric of Euclidean distance.

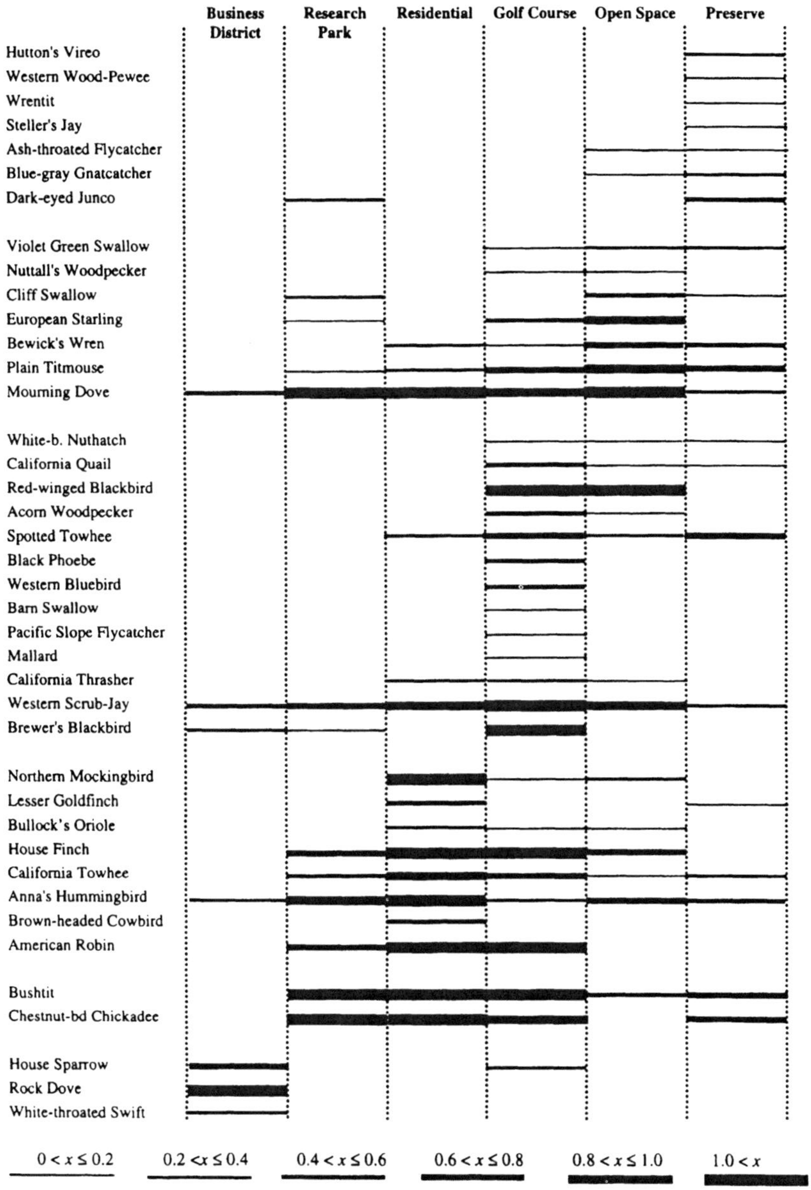

Figure 22.6. Distribution and abundance of summer resident birds in California. Line width represents birds ha^{-1}.

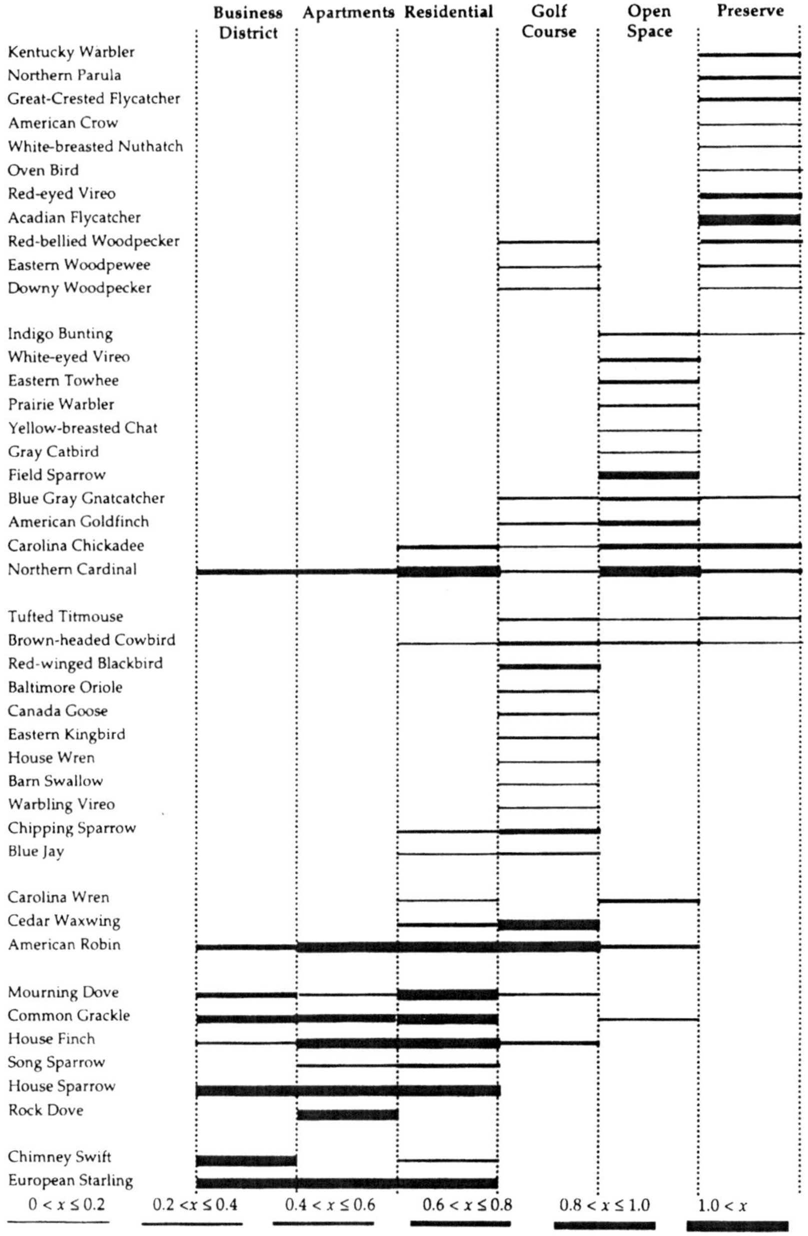

Figure 22.7. Distribution and abundance of summer resident birds in Ohio. Line width represents birds ha^{-1}.

parks, all of the golf courses, all of the preserves, four of the open- space recreational areas, and four of the residential sites were grouped by cluster analysis into distinct subsets. One of the residential sites and one of the open-space recreational areas (not the formal study sites) grouped away from other sites within its category (Fig. 22.11). In Ohio, the pattern was similar

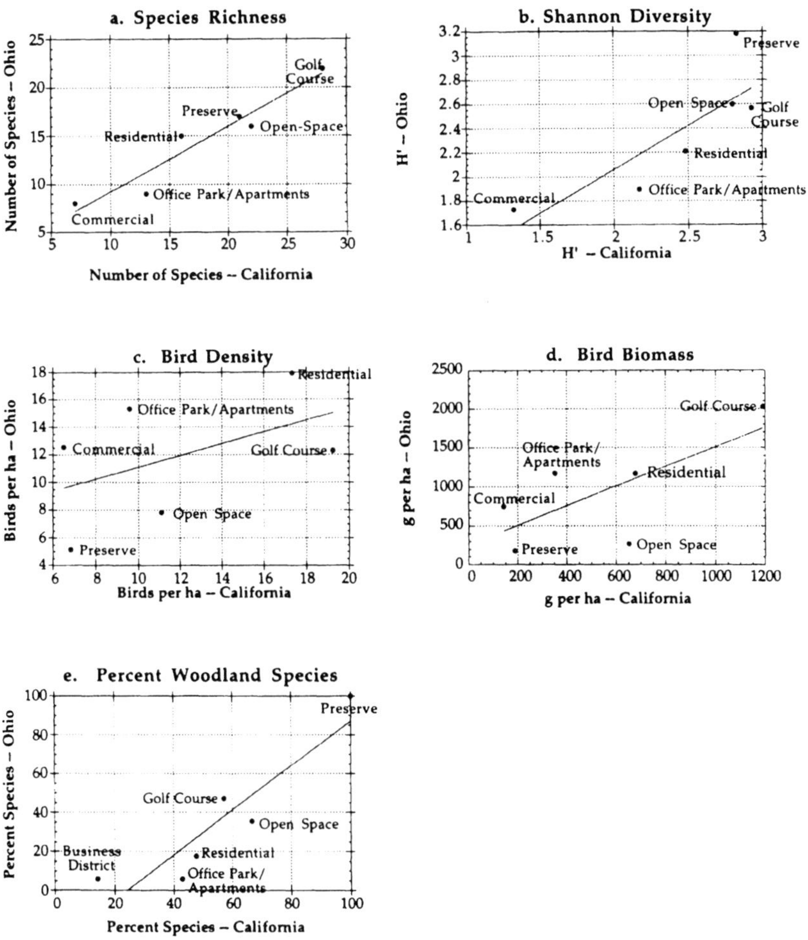

Figure 22.8. Comparison between California and Ohio of community measures (a) species richness ($n = 6$, $r = 0.959$, $P = 0.002$), (b) Shannon diversity ($n = 6$, $r = 0.826$, $P = 0.043$), (c) bird abundance (number of individuals ha^{-1}) ($n = 6$, $r = 0.488$, $P = 0.326$), (d) bird biomass (g ha^{-1}) ($n = 6$, $r = 0.717$, $P = 0.109$), and (e) percent of woodland species ($n = 6$, $r = 0.913$, $P = 0.011$). Lines were generated by least squares regressions.

but less clear because the sites display a strong trend of nestedness (i.e., species in relatively depauperate sites tended to be subsets of those in relatively rich sites; Fig. 22.12; Bolger et al. 1991, Atmar and Patterson 1993). The golf courses, preserves, and open-space reserves formed distinct, though nested, groups. The remaining three land-use types were nested as well. The apartment complexes, however, bridged the transition between residential areas and the business districts. In other words, some apartment complexes are very "residential" while others are more "commercial." Also, the Oxford business district was more similar to some of the apartment complexes than most business districts. These results suggest that, in general, the formal study sites in both California and Ohio are more similar in both species composition and individual species abundances to sites of similar land-use in their ecoregion. It also suggests that none of the formal study sites are wildly anomalous.

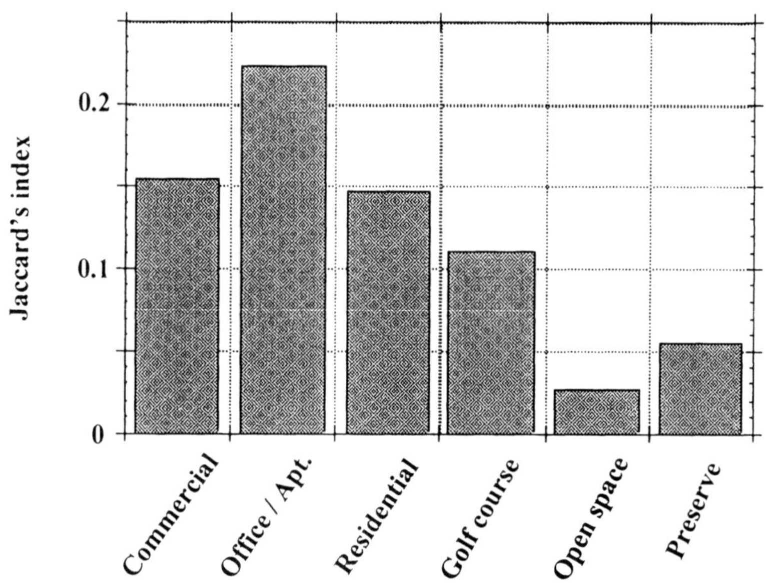

Figure 22.9. Jaccard's index of similarity suggests that species overlap for the sites in California and Ohio is, on average, three times higher in the land-use types that are entirely manipulated by humans (business district, research park/apartment complex, residential area, and gold course) than in those sites that are less manipulated (open space/reserves and biological preserves).

4. DISCUSSION

Humans have the unique ability to shape the landscape according to their vision. If it is too wet, we drain it. If it is too dry, we irrigate. If it is too hot, we plant trees for shade and construct buildings to air condition. If it is too cold, we plant trees as windbreaks and construct buildings to heat. If it doesn't look like "home", we alter the flora to make it look more familiar. If it doesn't sound like "home", we introduce songbirds with which we are familiar.

This ability and these desires have resulted in a wholesale conversion of the landscape in urban and suburban areas into a comforting and surreal homogeneity throughout the United States. One could argue that you could drop a blindfolded person outside of a fast-food restaurant in Portland, Oregon; Portland, Maine; or Portland, Ohio; remove the blindfold and they would be hard pressed to identify the state in which they had been dropped. This "sameness" is created through uniformity in building types and scale, transportation patterns, and even ornamental plantings (Clay 1994). This conversion of the landscape—urbanization— also has its effect on the fauna.

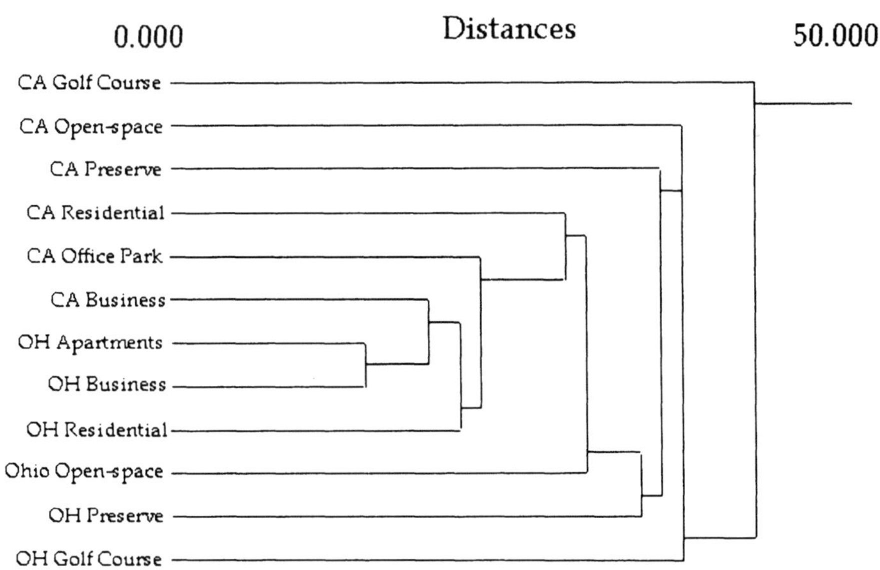

Figure 22.10. Cluster analysis of sites based on abundance of all bird species using the distance metric of normalized percent disagreement. The office park in California and the apartments in Ohio are considered equivalent land-uses in this study.

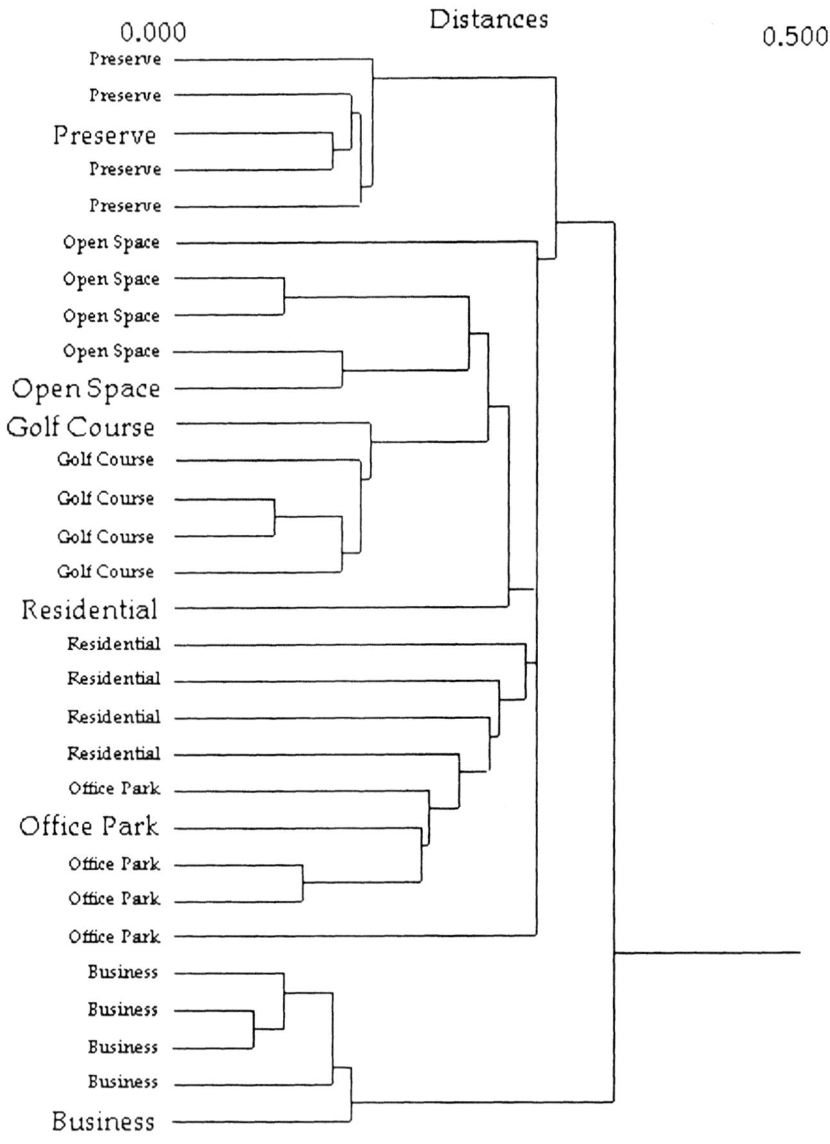

Figure 22.11. Cluster analysis comparing formal study sites in California (in larger font) and 24 other sites (four of each land-use type) based upon counts from 1 day surveys of all sites using the distance metric of Goodman-Kruskal gamma correlation coefficient.

Urbanization affects the distribution and abundance of birds (Marzluff 2001). My study demonstrates that the avifauna changes in a predictable pattern as land is converted from forested wildland to rural to suburban to urban, whether in California or Ohio. Species richness initially increases but then drops below the level of that of the wildland. Shannon diversity, bird

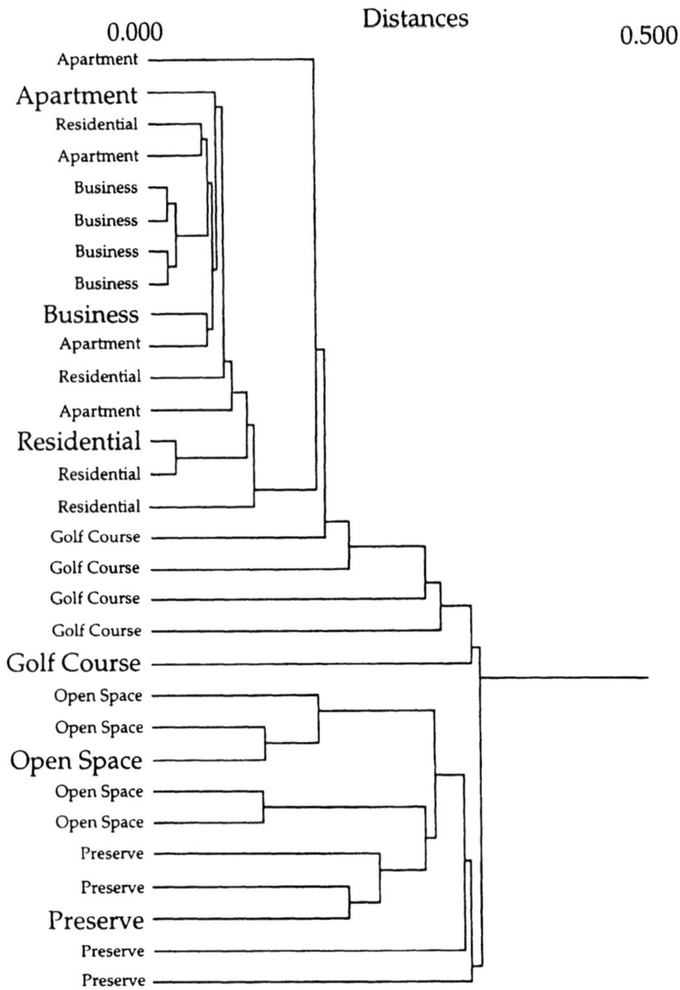

Figure 22.12. Cluster analysis comparing formal study sites in Ohio (in larger font) and 24 other sites (four of each land-use type) based upon counts from 1 day surveys of all sites using the distance metric of Goodman-Kruskal gamma çorrelation coefficient.

abundance, and total bird biomass all follow similar patterns. The species found in undisturbed woodland are gradually extirpated until virtually none of those species exist in the most urban areas. This study also demonstrates that these patterns are identical in such disparate environments as California's coastal chaparral and Ohio's eastern broadleaf forest.

This similarity in pattern is somewhat surprising because of the environmental differences, both climatic and economic, between the ecoregions. The coastal chaparral has a Mediterranean climate with mild temperatures that rarely dip below freezing or rise above 30° C. The region has a rainy season between November and April and has virtually no precipitation between May and October. Palo Alto is in the heart of the Silicon Valley, a megalopolis that extends 100 km from San Francisco to San Jose. The area has experienced an unprecedented building boom since the early 1970s.

The eastern broadleaf forest presents an interesting contrast. It experiences a different climate with the temperature frequently remaining below freezing in winter, and often rising above 30° C in summer. It has moderate precipitation throughout the year with the highest levels in May and June. Oxford is more than 50 km from any major urban area and its population has remained relatively stable at 8,000 permanent residents and 16,000 students since the early 1970s.

Clergeau et al. (1998) also conducted a comparison of avifauna along urban gradients between two cities with differing climates. They surveyed the birds of Quebec City, Canada with its cold inland temperate climate and Rennes, France with its warm Atlantic temperate climate. They found that the summer bird fauna responded to urbanization similarly in both cities, suggesting that pronounced climatic differences do not mask the effect of urbanization on the avian communities. In fact, it appears that urbanization overwhelms the environmental differences. Thus, both Clergeau et al. and this study demonstrate that avifaunas in different ecoregions respond similarly to urbanization.

It is worth noting that Clergeau et al. found that three species (House Sparrow [*Passer domesticus*], European Starling, and Rock Dove), as a group, dominated the avifauna at most sites in both cities in both summer and winter. This pattern is reflected in this study in that this same group of species dominated the two most urban sites in both California and Ohio.

The similarity of response by the avifauna in this study to increasing urbanization (e.g., species richness, Shannon diversity, abundance, deletion of woodland species) also leads to an increasingly more homogeneous avifauna. In the simplest terms, the most natural sites (the preserves and open-space reserves) in California and Ohio had an average overlap in species of ~7%. In contrast, the most developed sites (the business districts, office parks/apartments, residential areas, and golf courses) had an average overlap in species of ~19% percent—nearly three times the rate of the most natural sites.

The more nuanced, multivariate cluster analysis which compares all sites to all others reveals that the most natural sites along the gradient (the preserves, open-space reserves, and golf courses) divide along ecoregions implying that the avifauna of all three land-use types are fairly unique to Ohio or California. In contrast, those sites that are more developed (the business districts, office park/apartments, and residential areas) coalesce regardless of ecoregion. In other words, the assemblage of species is relatively unique to one or the other ecoregion in natural sites but many of the same species are found in urban sites irrespective of ecoregion.

These comparisons suggest that the more humans manipulate the environment, the more similar the bird communities become. However, it also points to regional differences in the entire fauna. For example, White-throated Swifts (*Aeronautes saxatalis*) are found in downtown Palo Alto while Chimney Swifts (*Chaetura pelagica*) are found in uptown Oxford. This highlights that the two are ecologically similar but they cannot be directly compared with indices of species overlap. In the entire study, 40 species were regularly surveyed in California and 44 species were regularly found in Ohio. Only 13 species were common to both regions and this leads to a total species overlap between California and Ohio of ~18%. Consequently, though urbanization does not lead to identical bird communities between the ecoregions, it does lead to more similar communities. This similarity, but not identity, is presumably due to a combination of environmental and dispersal limitations on many species.

Urbanization is not a single environmental parameter such as temperature, pH, or geographic orientation. Rather, urbanization is a layered, patchy network of environmental parameters that play out at many different scales (Alberti et al. 2001, Hostetler 2001). At the territory level, urbanization may affect physical environmental parameters and the dispersion of resources. At the population level, urbanization may affect the viability and longevity of isolated populations (Bolger et al. 1991). At the landscape level, urbanization involves changes in patch size, configuration, connectivity, and the amount of edge (Marzluff 2001, Miller et al. 2001). Urbanization may also affect ecological processes (Vitousek et al. 1996; Bolger 2001). For example, predation rates on artificial nests decline monotonically along the wildland to urban gradient in Ohio (Gering and Blair 1999). Here, I suggest that urbanization may affect two other ecological processes in addition to predation: local extinction and invasion.

In this study, woodland species, or those representing the minimally disturbed standard, disappeared along a gradient of increasing urbanization

(Fig. 21-8e, also see Blair 1996). The most sensitive species, or those that go locally extinct with the degree of change that occurs from the preserve to the open-space areas, include Hutton's Vireo (*Vireo huttoni*), Western Wood-Pewee (*Contopus sordidulus*), Steller's Jay (*Cyanocitta stelleri*) and Wrentit (*Chamea fasciata*) in California and Kentucky Warbler (*Opornis formosus*), Northern Parula (*Parula americana*), Great-Crested Flycatcher (*Myiarchus crinitus*), American Crow (*Corvus brachyrhynchos*), White-breasted Nuthatch (*Sitta carolinensis*), Ovenbird (*Seiurus aurocapilus*), Red-eyed Vireo (*Vireo olivaceus*), and Acadian Flycatcher (*Empidonax virescens*) in Ohio. Judging from these lists and other studies (Stenberg 1988, Germain et al. 1998), small incremental shifts along the urbanization gradient may have large effects on the avifauna.

If these species were evenly dispersed across the landscapes prior to European settlement, then this is evidence that habitat conversion associated with urbanization has caused the local extinction of these species. Conversely, species that are restricted to more urbanized sites can be assumed to be local invaders that have been able to exploit environments changed by human settlement and urbanization.

The processes of extinction and invasion happen repeatedly along the urban gradient. For example, in California, the Ash-Throated Flycatcher (*Myiarchus cinarascens*) goes extinct locally with urbanization. It is found at its highest abundance in the preserve, to a lesser degree in the open-space reserve, and not at all in more urbanized sites. In contrast, the Western Scrub-Jay (*Aphelocoma californica*) increases in abundance until the middle of the gradient (the golf course) but then declines as sites become even more urbanized. The jay can be considered both an invader (capitalizing on urbanization) but also threatened by more intense urbanization. Finally, Rock Doves are found only at the most urbanized site and are an example of a particularly successful invader aided by urbanization. Notice the shift in species abundances as your eye travels down Figs. 22.6 and 22.7. This pattern of urban avoiding, suburban adaptable, and urban exploiting species is found repeatedly in both California and Ohio.

This study suggests that human manipulation of the landscape is leading to both extinction and invasion on a very local scale. Local extinction and invasion are the patterns that Hobbs and Mooney (1998) highlight when they argue that the extinction debate should be broadened beyond the extinction of species. They argue that population extinctions and invasions (or, in their terms, deletions and additions) should be addressed as a threat to biodiversity because species extinction is simply the final endpoint of many population extinctions.

5. RESEARCH NEEDS

Future research along the lines of this study should focus on the effects of urbanization at different levels of biological organization. First, at the landscape level, it would be useful to explore how urbanization affects the heterogeneity of the landscape and how landscape heterogeneity is linked to faunal heterogeneity.

Second, at the level of this study, that of the species assemblage or community, it would be useful to develop theory and gather empirical data on the upper limits of similarity between ecoregions. Is a tripling of species overlap from 0.065 in natural areas to 0.185 in urban areas a notable increase in homogeneity?

Third, at the population level, it would be useful to explore the changes in the demographics of certain species along an urban gradient. This study assumes that a high abundance of birds is "good" and that a high density of a particular species indicates better habitat for that species. This may not be the case because reproduction and abundance may not necessarily be linked (Van Horne 1983, Vickery et al. 1992).

Finally, this study must be taken with the warning that it is correlative. It should be fairly easy to test empirically the ideas presented herein by measuring the changes in avifauna as areas continue to undergo suburban- and urbanization.

6. POLICY AND MANAGEMENT IMPLICATIONS

Determining what forces have led to the current patterns of urban development in the United States and the consequent homogenization of the landscape is far from simple. Current land-use patterns in the United States can be attributed to many factors. These range from historical, local attributes such as the proximity to transportation and potable water to current, global, economic conditions such as the effects of the North American Free Trade Agreement on the United States manufacturing base. The landscapes included in this study largely have been urbanized since 1945 and the automobile has been one of the strongest factors to influence development patterns in these areas. The outcome of these development decisions has been suburban sprawl. For example, the metropolitan area nearest to the Ohio study sites, Cincinnati, grew 12% between 1990 and 1996 while its population increased by only 2% (Sierra Club 1998).

This study has strong implications for management at two different scales: the individual preserve and the geographical region. At the preserve level, managers should note that species richness is not necessarily a good measure of the "wildness" of a site and that species composition is more important in making management decisions. In both ecoregions, the minimally disturbed standard sites (the preserves) had fewer species than slightly more disturbed sites such as the golf courses. This indicates that preserve managers should target the species found in the minimally disturbed standard (i.e., the woodland species) and not on overriding community measures such as species richness, Shannon diversity, or total bird abundance as their metrics for management "success." The second lesson that may be derived from this study is that the effects of urbanization are similar in broadly comparable ecoregions, (e.g., those with forest cover). This implies that the results of other studies, such as those that appear in this volume, may be applied across ecoregions and, with caution, to each manager's particular situations.

At the regional level, this study illustrates two valuable principles for land-use planners. First, any level of urbanization affects native biodiversity. The open-space reserves in California and Ohio do not support all of the species that are found in the preserves and, more intensely developed sites support even fewer species. This implies that urbanization should be concentrated when possible. Second, this study shows that increasingly more intense land-use leads to increasingly more similar avifauna (i.e., that current land-use patterns are creating a homogeneous avifauna). It also shows that this homogenization occurs at relatively less severe uses including golf courses and residential sites.

ACKNOWLEDGMENTS

I want to thank many people who have been involved with this project including Alistair Hobday, Charlie Quinn, Brad Purcell, and Julie Whipkey who worked with me in the field; Mike Vanni and Erica Fleishman who helped with the manuscript and cleared my muddy thinking; and John Marzluff and Reed Bowman who brought together those researchers working on birds and urbanization.

REFERENCES

Alberti, M., E. Botsford, and A. Cohen. 2001. Quantifying the urban gradient: linking urban planning and ecology, p. 89-115. *In* J. M. Marzluff, R. Bowman, and R. Donnelly [EDS.], Avian ecology and conservation in an urbanizing world. Kluwer Academic, Norwell, MA.

Atmar, W. and B. D. Patterson. 1993. The measure of order and disorder in the distribution of species in fragmented habitat. Oecologia 96:373-382.

Bailey, R. G., P. E. Avers, T. King, W. H. McNab, eds. 1994. Ecoregions and subregions of the United States (map). Washington, DC; U.S. Geological Survey. Scale 1:7,500,000; colored. Accompanied by a supplementary table of map unit descriptions compiled and edited by McNab, W. H., and R. G. Bailey, Prepared for the U.S. Department of Agriculture, Forest Service.

Beissinger, S. R. and D. R. Osborne. 1982. Effects of urbanization on avian community organization. Condor 84:75-83.

Blair, R. B. 1994. Birds, butterflies, and conservation on an urban gradient in central California. Ph.D. dissertation. Stanford University, Stanford, California.

Blair, R. B. 1996. Land-use and avian species diversity along an urban gradient. Ecol. Appl.. 6:506-519.

Blair, R. B. 1999. Birds and butterflies: surrogate taxa for assessing biodiversity? Ecol. Appl.. 9:164-170.

Bolger, D. 2001. Urban birds: population, community, and landscape approaches, p. 155-177. *In* J. M. Marzluff, R. Bowman, and R. Donnelly [EDS.], Avian ecology and conservation in an urbanizing world. Kluwer Academic, Norwell, MA.

Bolger, D. T., A. C. Alberts, and M. E. Soulé. 1991. Occurrence patterns of bird species in habitat fragments: sampling, extinction, and nested species subsets. Am. Nat. 137:155-166.

Case, T. J. 1996. Global patterns in the establishment and distribution of exotic birds. Biol. Conserv. 78:69-96.

Clay, G. 1994. Real places: an unconventional guide to America's generic landscape. The University of Chicago Press, IL.

Clergeau, P., J. L. Savard, G. Mennechez, and G. Falardeau. 1998. Bird abundance and diversity along an urban-rural gradient: A comparative study between two cities on different continents. Condor 100:413 - 425.

Cooper, W. S. 1926. Vegetational development upon alluvial fans in the vicinity of Palo Alto, California. Ecology 7:1-30.

Dunning, J. B. Jr. 1993. CRC handbook of avian body masses. CRC Press, Boca Raton, FL.

Ehrlich, P. R., D. S. Dobkin, and D. Wheye. 1988. The Birder's Handbook. Simon and Schuster, New York, NY.

Garrett, K. L. and W. S. Smithson. 1999. Urbanization and naturalized exotic bird species: a close fit. Avian ecology and conservation in an urbanizing world symposium, Annual meeting of the Cooper Ornithological Society, Portland, OR.

Gering, J. C. and R. B. Blair. 1999. Predation on artificial bird nests along an urban gradient: predatory risk or relaxation in urban environments? Ecography. 22:532-541.

Germaine S. S., S. S. Rosenstock, R. E. Schweinsburg, and W. S. Richardson. 1998. Relationships among breeding birds, habitat, and residential development in Greater Tucson, Arizona. Ecol. Appl. 8:680-691.

Hobbs, R. J. and H. A. Mooney. 1998. Broadening the extinction debate: Population deletions and additions in California and western Australia. Conserv. Biol. 12: 271-283.

Hostetler, M. 2001. The importance of multi-scale analyses in avian habitat selection studies in urban environments, p. 139-154. *In* J. M. Marzluff, R. Bowman, and R. Donnelly [EDS.], Avian ecology and conservation in an urbanizing world. Kluwer Academic, Norwell, MA.

Jongman, R. H. G., C. J. F. Ter Braak, and O. F. R. Van Tongeren. 1995. Data analysis in community and landscape ecology. Cambridge Univ. Press, United Kingdom.

Karr, J. R. and E. W. Chu. 1998. Restoring life in running: Using multimetric indexes effectively. Island Press, Covelo, CA.

Lodge, D. M. 1993. Biological invasions: Lessons for ecology. Trends Ecol. Evol. 8:133-137.

Lodge, D. M., R. A. Stein. K. M. Brown, A. P. Covich, C. Brönmark, J. E. Garvey and S. P. Losiewski. 1998. Predicting impact of freshwater exotic species on biodiversity: Challenges in spatial scaling. Aust. J. Ecol. 23:53-67.

Long, J. L. 1982. Introduced birds of the world. Universe Books, New York, NY.

Magurran, A. E. 1988. Ecological diversity and its measurement. Princeton Univ. Press, NJ.

Marzluff, J. M. 2001. Worldwide increase in urbanization and its effects on birds, p. 19-47. *In* J. M. Marzluff, R. Bowman, and R. Donnelly [EDS.], Avian ecology and conservation in an urbanizing world. Kluwer Academic, Norwell, MA.

McKinney, M. L. 1998. On predicting biotic homogenization: species-area patterns in marine biota. Global Ecol. Biogeo. Letters 7:297-301.

Miller, J., J. Fraterrigo, J. Wiens, and T. Hobbs. 2001. Urbanization, avian communities, and landscape ecology, p. 117-137. *In* J. M. Marzluff, R. Bowman, and R. Donnelly [EDS.], Avian ecology and conservation in an urbanizing world. Kluwer Academic, Norwell, MA.

Mills, G. S., J. B. Dunning Jr., and J. M. Bates. 1989. Effects of urbanization on breeding bird community structure in southwestern desert habitats. Condor 91:416-428.

Ray, M. A., and J. L. Vankat. 1982. A vegetation map of Hueston Woods State Park and Nature Preserve, p.22-29. *In* G. E. Willeke [ED.], Hueston Woods State Park and Nature Preserve: Proceedings of symposium, Miami University, Oxford, OH.

Reichard, S., L. Chalker-Scott, and S. Buchanan. 2001. Interactions among non-native plants and birds. *In* J. M. Marzluff, R. Bowman, and R. Donnelly [EDS.], Avian ecology and conservation in an urbanizing world, p. 179-223. Kluwer Academic, Norwell, MA.

Reynolds, R. T., J. M. Scott, and R. A. Nussbaum. 1980. A variable circular-plot method for estimating bird numbers. Condor 82:309-313.

Shannon, C. E. and W. Weaver. 1949. The mathematical theory of communication. Univ. of Illinois Press, Urbana.

Sierra Club. 1998. The dark side of the american dream: the costs and consequences of suburban sprawl. [Online] http://www.sierraclub.org/sprawl/report98/.

Stenberg, K. 1988. Urban macrostructure and wildlife distributions: regional planning implications. Ph.D. dissertation. Univ. of Arizona, Tucson.

SYSTAT 1992. SYSTAT: Statistics, Version 5.2. SYSTAT Inc., Evanston, IL.

Van Bael, S. and S. Pruett-Jones. Exponential population growth on Monk Parakeets in the United States. Wilson Bull. 108:584-588.

Van Horne, B. 1983. Density as a misleading indicator of habitat quality. J. Wildl. Manage. 47:893-901.

Vermeij, G. J. 1991. When biotas meet: Understanding biotic interchange. Science 253:1099-1104.

Vickery, P. D. and M. L. Hunter, J. V. Wells. 1992. Is density an indicator of breeding success? Auk 109:706-710.

Vitousek, P. M., C. M. D'Antonio, L. L. Loope, and R. Westbrooks. 1996. Biological invasions as global change. Amer. Scientist 84:468-478

Wetmore, A. 1964. Song and garden birds of North America. National Geographic Society, Washington, DC.

Wood, C. A. 1924. The Starling family at home and abroad. Condor. 26:123-136.

Zuboy, J. R. 1981. A new tool for fisheries managers: the Delphi Technique. N. Amer. J. Fish. Manage. 1:55-59.

Chapter 23

Avian community characteristics of urban greenspaces in St. Louis, Missouri

Jeffrey M. Azerrad and Charles H. Nilon
Department of Fisheries and Wildlife Sciences, 302 Anheuser-Busch Natural Resources Building, University of Missouri-Columbia, Columbia, MO 65211-7240 USA, e-mail: nilonc@missouri.edu

Key words: Avian community, cavity nesters, greenspace, Missouri, St. Louis, urbanization, vegetation

Abstract: We studied bird communities among six vegetation classes in St. Louis to determine how a range of greenspace types function as avian habitat. During 1997 and 1998, 67 bird species were recorded along strip transects among the six habitats sampled. The open woodland habitat supported a unique community, and provided habitat for cavity and middle/upper canopy nesting species. The important species that specialized in woodlands were the Red-headed Woodpecker (*Melanerpes erythrocephalus*) and House Wren (*Troglodytes aedon*). All other habitats separated out based on the interpretation of species data in an ordination analysis. We measured 13 habitat variables, and of these, canopy closure, shrub density, forb cover, and canopy height best explained bird species variation, suggesting these features are important for managing bird habitat in St. Louis.

1. INTRODUCTION

An extensive literature focuses on the ecology of birds in different urban land-use types. Such studies investigated bird communities in cemeteries (Thomas and Dixon 1974, Lussenhop 1977), parks (Gavareski 1976, Suhonen and Jokimaki 1988, Cicero 1989, Antikainen 1992), woodlots (Tilghman 1987), residential areas (Geis 1974, DeGraaf and Wentworth 1986, Germaine et al. 1998), and under different levels of urbanization (Nilon et al. 1995, Blair 1996, Natuhara and Imai 1996). These studies examined the response of bird species and communities to the habitat patch size and to the structural diversity of urban areas, and studied the abundance of both native and exotic bird species among a variety of land-uses.

Although such investigations evaluate a broad range of land-uses, little is understood about bird communities that inhabit the range of habitat types associated with each land-use. A single land-use often contains patches of vegetation that vary in origin and disturbance history, and are influenced by the types of sites that surround them (Pickett et al. 1997). Marked variation among landscapes within small geographical areas can be traced to the human history of urban areas (Flores et al. 1998). The factors creating spatially diverse areas include abandonment of developed land, soil compaction, and intensive park management. These factors produce unique and diverse areas, important in defining bird communities inhabiting urban greenspaces.

Matthews et al. (1988) identified 31 cover types that occur within 26 land-use types in seven New York metropolitan areas. Sukopp and Weiler (1988) and Roberts (1994) have described the range of remnant natural areas, ruderal areas, and planted sites that form distinct vegetation patches in urban open spaces (Zipperer et al. 1997). Therefore, identifying habitat heterogeneity at small spatial scales and studying how these areas function as breeding bird habitat is a crucial part of managing wildlife in cities.

The types of birds found among different habitat types in urban greenspace are not only of interest for ecological reasons. Greenspaces are often considered essential for the well-being of urban residents and are given priority by planners and land managers (Barker and Graf 1989, Rielly and Page 1995). The range of habitat types found in urban open spaces are the places where urban residents have day-to-day encounters with nature and birds are part of that experience (Burgess et al. 1988). The same features of a site that are bird habitat also are "recreation habitats" for people (Nilon and Huckstep 1998).

We studied breeding birds in habitat types that occur in open spaces in the North City neighborhoods of St. Louis, Missouri. As part of the North City Sustainable Communities Project, our study focused on identifying sites that might be important for conservation and, more importantly, sites that provide opportunities for residents of these inner city neighborhoods to have day-to-day contact with nature. Our objectives were to: (1) compare bird species richness and abundance among the major habitat types occurring in greenspaces in North City St. Louis; (2) identify habitat features that directly influence avian richness and abundance within the dominant habitat types; and (3) identify the habitat types that contain a unique assemblage of bird species.

2. METHODS

2.1 Study Area

St. Louis is located in the Ozark Border Natural Division of Missouri, and is situated along the transition between grasslands and deciduous forests (Thom and Wilson 1980, Schroeder 1997). The North City Neighborhoods of St. Louis, Missouri is a 3000 ha urban area adjacent to the Mississippi River (Wayman 1979). The area is characterized by a mix of industrial sites, highway and railroad right-of-ways, and residential neighborhoods (Fig. 23.1). The Missouri Resource Assessment Partnership's land-cover classification for St. Louis City classifies most of the study area as "urban impervious" and "urban vegetated" land. These terms refer to urban environments with vegetation including small parks, tree-lined streets, and yards and urban environments covered by paved or built surfaces, respectively. The study area also contains some of the largest open spaces in the city, with large cemeteries, city parks, abandoned industrial areas, and remnant floodplain forests.

2.2 Habitat Classification

We classified habitats in all open spaces in the study area using an entitation procedure described by Hunt (1986) and Rogers and Rowntree (1988). Entitation is a classification system based on plant formation (cover type) and the dominant life form of understory (< 5m) vegetation and is a modification of entitation methods described by Mueller-Dombois and Ellenberg (1974). We located all patches ≥ 1 ha in the study area on 1:2400 scale aerial photographs obtained from the St. Louis Metropolitan Sewer District and classified each patch into one of five plant formation types (Table 23.1). We visited the study area between January and March 1997 to verify the plant formation classification and to complete the description of the understory vegetation structure.

The study area contained 99 patches ranging in size from 1 to 218 ha, encompassing 701 ha of greenspace (Table 23.2). These patches were classified into five plant formations (cover types) that could be further subdivided into 10 plant formation groups (Table 23.2). We studied four plant formation groups that made up 78% of the patches and 93% of the greenspace in the study area: deciduous closed forest without evergreens (forest), deciduous woodland with evergreen (woodland), scrub thickets (thicket), and herbaceous meadows, pastures, or grasslands (meadow/lawn). During the second field season we studied two less common plant formation

Figure 23.1. USGS digital orthophoto quadrangle taken in 1996 of the study area (circumscribed by the white line) in North City Neighborhood, St. Louis, Missouri. The image covers approximately 1700 x 1300 m.

groups: herbaceous tall and perennial forbs (tall/perennial forbs) and herbaceous emergent wetlands (emergent wetlands). We studied these sites to determine if they contained additional species that were not found in other formation groups. The tall/perennial forb and emergent wetland formation groups were not included in any analysis of bird-habitat relationships.

2.3 Bird Counts

We determined that a minimum of 12 patches in each formation group would allow us to detect a 5% difference in bird abundance among the formation groups. All patches in the study area ≥ 20 ha were included in our study. We also selected all patches completely surrounded by built areas (n = 6). We then randomly selected additional patches in each formation group. Twelve patches were selected in both the forest and meadow/lawn formation groups. We were denied access to eight patches in the woodland and thicket groups. As a result, we selected all remaining patches in the woodland (n = 5) and thicket (n = 4) formation groups.

Bird counts were conducted along 100 m strip transects within each of the selected patches (Christman 1984). DeGraaf et al. (1991) recommended using strip transects to count birds in patches that are highly variable and have high amounts of edge. The number of transects/patch was assigned in proportion to the area of the patch. One transect was located in patches < 12 ha, three in patches between 12 and 20 ha, six in patches between 20 and 60 ha and 8 in patches larger than 60 ha. We completed bird counts on 58 transects in 1997 and 70 transects in 1998 when additional transects were added to patches in the woodland, and meadow/lawn. The tall/perennial forb

Table 23.1. Plant formations and life form categories for classifying greenspaces.

Category	Criteria
Plant formation (cover type)	
Closed forest	Trees ≥ 5 m tall with interlocking crowns.
Woodland	Trees ≥ 5 m tall without interlocking crowns. > 40% tree cover without interlocking crowns.
Scrub	Woody vegetation 0.5 - 5.0 m tall.
Herbaceous	Grass and grass-like plants, woody vegetation < 40%.
Scarcely vegetated	Bare mineral soil, sand, or pavement dominates.
Life forms of vegetation (<5 m tall)	
Phanerophytes	Plants that grow taller than 0.5 m and do not die back below that height.
Chamaephytes	Branches or shoots remain < 0.5 m above ground.
Hemicryptophytes	Shoots die back to ground level. Buds remain at ground level.
Cryptophytes	Buds or shoots survive below ground.

and emergent wetland formation groups were added to the study in 1998. We counted birds in all the patches of these two formation groups.

All bird counts were conducted between May and July 1997 and 1998. Counts were 8 min in length and conducted between the hours of 0530 to 0930 during calm, rain-free mornings. Eight counts were made on each transect and all birds heard or seen within 50 m of the transect line were recorded. We did not count birds flying through or over the patches. Count data were averaged for each species to derive mean number of detections per count as a measure of abundance. We used the SIMSTAT general linear model procedure to test for between-year differences ($P < 0.05$) in mean species richness per count among the forest, woodland, thicket, and meadow/lawn formation groups (Peladeau 1996).

2.4 Habitat Variables

We measured 12 habitat variables within five 0.04 ha circular plots in each strip transect using methods described by James and Shugart (1970). The plots were arranged on alternating sides of the transects at 25 m intervals and located 20 m from the transects at right angles. Within each plot we measured canopy cover, canopy height, shrub stem density (< 3 cm dbh) ha^{-1}, and ground cover (percent cover by grass, forbs, artificial surface, leaf litter, woody litter, rocks, bare soil, water, shrubs). These habitat variables were only measured in the forest, woodland, thicket, and

Table 23.2. Plant formation groups identified in the North St. Louis study area.

Formation group	No. of patches	Area (ha)
Closed forest		
Temperate evergreen	1	3.2
Deciduous without evergreens	15	66.1
Woodland		
Deciduous with evergreen	6	223.9
Deciduous without evergreen	8	15.2
Scrub		
Thicket	9	9.9
Herbaceous		
Meadow, pasture, grassland	47	348.6
Tall and perennial forb	2	10.9
Emergent wetland	2	5.6
Scarcely vegetated		
Artificial surface	3	5.8
Compacted surface	6	11.6
Total	99	700.8

meadow/lawn formation groups. For emergent wetlands and tall/perennial forb groups we were interested only in determining species composition.

Principal components analysis (PCA) was used to determine the habitat variables that described the greatest amount of variation among the strip transects. We selected variables associated with principal components that explained ≥ 50% of variation among transects and that had loadings ≥ 0.4 or ≤ -0.4. We used a Kruskal-Wallis Analysis of Variance (ANOVA) to determine if there were differences ($P < 0.05$) in these selected variables among the formation groups (Peladeau 1996). The selected variables were used as environmental variables in a canonical correspondence analysis (CCA) of the bird community data (ter Braak 1986).

2.5 Bird Communities

CCA was used to identify patterns in bird species composition and abundance and to correlate environmental variables with the bird community data along ordination axes (ter Braak 1986). Species scores and species abundance was used to determine the bird species important for characterizing the ordination axes. We selected bird species in the highest 25% and lowest 25% of CCA axis scores and the top third of bird species abundance values. We tested for year effects by comparing the ranked sum of the means for each of these species using a repeated measures ANOVA in PC-SAS Version 6.12. We used Sorensen's distance and a group average clustering procedure to determine percent dissimilarity (PD) in species composition and abundance among the six formation groups (McCune and Mefford 1997).

We classified each bird species observed into one of three nesting guilds identified and described in the North American Breeding Bird Survey: ground/lower, middle/upper, and cavity (Sauer et al. 1997). This classification, based on functional life history requirements, was an additional tool to help us understand bird use of greenspace habitats. We used one-way ANOVA to determine if there were differences ($P < 0.05$) in the mean number of detections per count of each nesting guild among the formation groups.

3. RESULTS

3.1 Richness and Relative Abundance

Sixty-seven bird species were detected during the two years of the study (Appendix 23.1). European Starling (*Sturnus vulgaris*), Red-winged

Blackbird (*Agelaius phoeniceus*), Common Grackle (*Quiscalus quiscula*), American Robin (*Turdus migratorius*), and Northern Cardinal (*Cardinalis cardinalis*) were the five most abundant species detected in the study area (Fig. 23.2). Species richness ranged from 16 to 52 species among the formation groups and was highest in the meadow/lawn formation group (Fig. 23.3). There was no difference (P = 0.68) in species richness per count between 1997 and 1998.

3.2 Community Analysis

There were differences in percent canopy cover ($P < 0.001$), canopy height ($P < 0.001$), shrub stem density ($P < 0.001$), percent ground cover by grass ($P < 0.001$), percent ground cover by leaf litter ($P < 0.001$), percent ground cover by woody litter ($P = 0.001$), percent ground cover by shrubs ($P < 0.001$), and percent ground cover by trees ($P < 0.001$) among the formation groups (Table 23.3). Two principal components (Table 23.4)

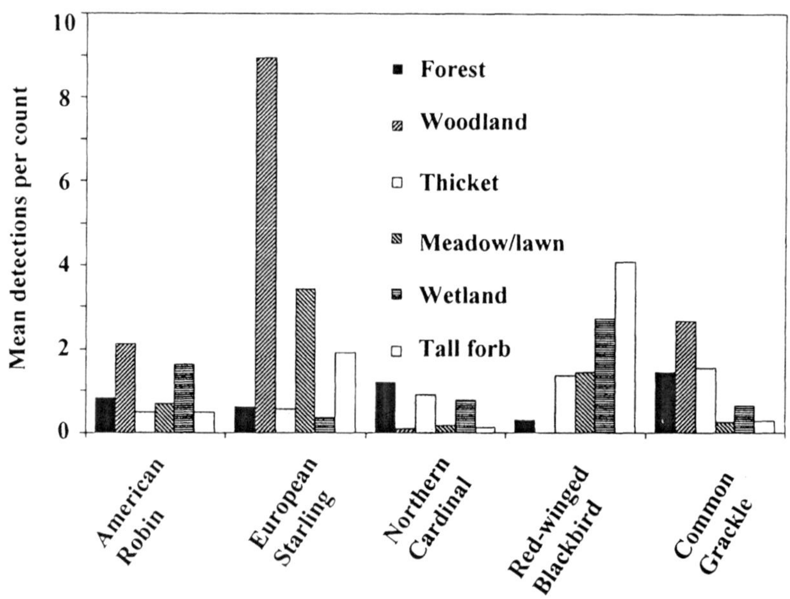

Figure 23.2. Mean detections per count for the five most abundant species in North City greenspaces, 1997 and 1998.

explained 50.3% of the variation among strip transects in habitat variables. Component one, explaining 37.1% of variation among transects, had negative loadings for canopy cover and number of shrub stems indicating a separation of sites with trees and a shrub understory. Component two, explaining an additional 13.2% of the variance was characterized by positive loadings for ground cover by forbs and negative loadings for canopy height indicating a separation between forested and grassland transects. Based on the results of the PCA, we used percent canopy (% canopy), canopy height, shrub stem density, and percent ground cover by forbs (% forb) as environmental variables in the CCA of the pooled bird abundance data.

The first two CCA axes explained 18.4% of the variation in the bird community data among the 33 patches. Axis one explained 10.1% of the variation among patches. The Carolina Chickadee (*Parus carolinensis*), Gray Catbird (*Dumetella carolinensis*), Warbling Vireo (*Vireo gilvus*), Downy Woodpecker (*Picoides pubescens*), and American Crow (*Corvus brachyrhynchos*) had ordination scores in the lowest 25% of axis one and were in the top third of bird abundance scores. The House Sparrow (*Passer domesticus*), Northern Bobwhite (*Colinus virginianus*), Northern

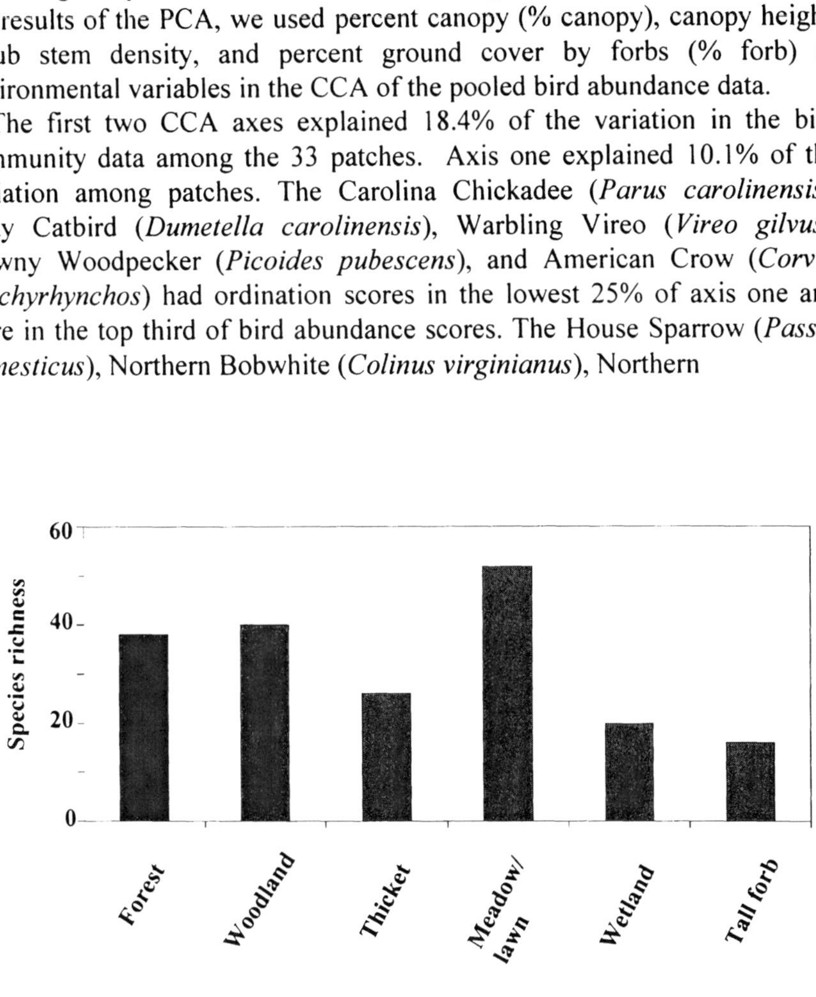

Figure 23.3. Species richness per formation group.

Mockingbird (*Mimus polyglottos*), and House Finch (*Caprodacus mexicanus*) had axis one ordination scores in the top 25% and were in the top third of bird species abundance scores. The selection of these 12 species to characterize the first ordination axis indicates a separation of species associated with forest canopy versus species associated with grassland habitat. Percent canopy closure (r = -0.964), canopy height (r = 0.889), and shrub stem density (r = -0.637) were negatively correlated with axis one scores.

Axis two explained 8.2% of variation in species composition among patches. The Gray Catbird, Northern Cardinal, Indigo Bunting (*Passerina cyanea*), House Finch, American Goldfinch (*Carduelis tristis*) had ordination scores in the lowest 25% of axis two and abundance values in the top third. The Red-headed Woodpecker (*Melanerpes erythrocephalus*) and House Wren (*Troglodytes aedon*) had ordination scores in the top 25% and abundance in the top third. The selection of these seven species indicated a

Table 23.3. Mean (± SE) values of important habitat variables by formation group.

Habitat variable	P-value	Forest	Woodland	Thicket	Meadow/ lawn
% canopy*	$P < 0.001$	89 ± 1.6[a]	44 ± 6.9[b]	60 ± 20.6[b]	5 ± 1.8[c]
Canopy cover*	$P < 0.001$	89 ± 1.6	44 ± 7.0	59 ± 20.6	5 ± 1.8
Canopy height*	$P < 0.001$	19 ± 2.0	20 ± 2.2	9 ± 3.8	3 ± 0.9
Shrub stem density*	$P < 0.001$	28 ± 4.1	0 ± 0.0	31 ± 9.9	1 ± 0.6
% grass*	$P < 0.001$	1 ± 0.4	49 ± 7.9	4 ± 1.5	46 ± 7.7
% forb	$P > 0.8$	29 ± 6.1	30 ± 5.1	30 ± 15.6	37 ± 6.2
% artificial	$P > 0.4$	4 ± 1.7	8 ± 3.2	3 ± 0.8	5 ± 1.5
% leaf litter*	$P < 0.001$	23 ± 4.2	3 ± 2.2	15 ± 12.8	1 ± 0.2
% woody litter*	$P < 0.001$	11 ± 2.3	3 ± 1.1	6 ± 2.5	2 ± 0.7
% rock	$P > 0.5$	1 ± 0.4	1 ± 1.0	2 ± 1.0	3 ± 1.8
% bare soil	$P > 0.4$	5 ± 0.9	5 ± 2.2	8 ± 2.9	6 ± 2.4
% water	$P > 0.9$	0 ± 0.0	0 ± 0.0	0 ± 0.1	0 ± 0.0
% shrub*	$P < 0.001$	25 ± 5.3	0 ± 0.1	31 ± 8.9	1 ± 0.8
% tree*	$P < 0.003$	1 ± 0.4	0 ± 0.1	1 ± 0.4	0 ± 0.0

*$P < 0.05$ for Kruskal-Wallis one-way ANOVA

† Different letters indicate differences ($P < 0.05$) among plant formation groups.

Table 23.4. Loadings for the important habitat variables for first two principal components from PCA analysis of habitat variables.

Variable	Axis 1	Axis 2
% canopy	-0.411	-0.198
% forb	0.001	0.594
No shrub	-0.391	0.068
Canopy height	-0.265	-0.403

separation of species associated with edge habitats and species associated with open woodlands. Shrub stem density (r = - 0.634) and percent forbs (r = -0.510) were negatively correlated with axis two. The CCA ordination of the 33 habitat patches along the first two ordination axes showed that forest and woodland patches were separated from meadow/lawn patches and that thicket patches were separated from forest and woodland patches (Fig. 23.4).

There were differences ($P < 0.05$) in mean detections per count among the four formation groups for seven of the 14 species we selected to characterize the ordination axes: Carolina Chickadees ($P < 0.02$), Red-headed Woodpecker ($P = 0.02$), Northern Cardinal ($P < 0.001$), Warbling Vireo ($P = 0.01$), Downy Woodpecker ($P = 0.03$), American Crow ($P = 0.01$), and Northern Mockingbird ($P = 0.01$) (Appendices 23.2 and 23.3). Downy woodpeckers were most abundant in forest patches, Carolina Chickadees and American Crows were more abundant in forest and woodland patches, Warbling Vireos and Northern Cardinals were more abundant in forest and thicket patches, and Red-headed Woodpeckers more abundant in meadow/lawn patches.

3.3 Similarity and Guild Analyses

The woodland and meadow-lawn plant formation groups were distinct in species composition and abundance from the other formation groups (PD = 0.70) and formed a distinct cluster in the resulting dendrogram (Fig. 23.5). Forest and thickets were similar in their overall species composition (PD = 0.06). The meadow/lawn and woodland formations also were similar in species composition (PD = 0.18).

There were differences among formation groups in mean detections per count for cavity ($P < 0.001$), ground/low ($P < 0.01$), and middle/upper nesters ($P = 0.01$). Cavity nesters and middle/upper canopy nesters were most abundant in the woodland formation group. Ground/low nesters were most abundant in tall perennial forb formation group (Table 23.5).

4. DISCUSSION

Our first objective was to determine if differences in bird species composition and abundance existed among the major habitat types (formations) occurring in greenspaces in the North City Neighborhoods of St. Louis. Although the most abundant species found in our study were a small group of species typical to urban areas (European Starling, Red-winged Blackbird, Common Grackle, American Robin, Northern Cardinal), none of the five species were among the 14 that met our criteria for characterizing the ordination axes that explain differences in species composition among patches. Moreover, seven of the 14 characteristic species differed in abundance among the four dominant formation groups, indicating that these four habitat types supported different bird species and, therefore, distinct bird communities.

Our second objective was to identify important habitat features that can be used to guide greenspace planning. We identified four variables (canopy height, canopy closure, forb cover, and shrub density) that could be used to characterize habitat differences among bird census transects. The four habitat variables showed strong correlations with axes that explained differences in bird species composition. Shrub density and canopy closure also were important for determining richness, diversity, and abundance in Cicero's (1989) study of bird habitats in Golden Gate Park. Although additional research is needed to determine if these variables are associated with abundance of selected bird species and we do not know which bird species are actually breeding and capable of maintaining viable populations,

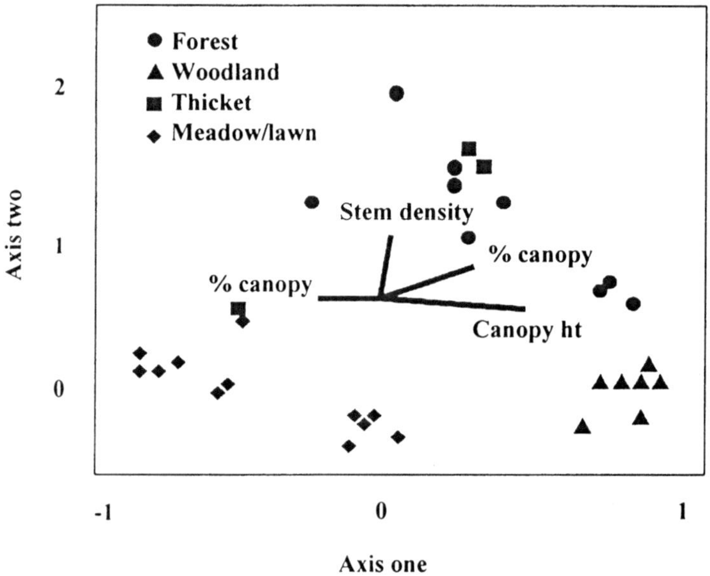

Figure 23.4. Canonical correspondence analysis ordination of forest, woodland, thicket, and meadow/lawn patches.

this information is a logical starting point that challenges the view that urban greenspace is monotypic.

Our final objective was to determine if specific formation groups were important as habitat for unique groups of bird species. We determined from the ordination and classification analyses that woodlands were distinct from other plant formation groups in species composition. We also found that woodlands had a greater abundance of species in the middle/upper nesting guild and canopy nesting guild species and that the tall forb formation group had a greater abundance ground/low nesting guild species. This supports findings from other studies investigating birds in urban environments. Similar to our findings in woodland habitat, research in Amherst, Massachusetts demonstrated that park-like oak-pine forest support an abundance of cavity nesting birds (DeGraaf and Wentworth 1986). Breeding birds are also more abundant in urban woodlands than other urban habitat types (DeGraaf and Wentworth 1986 Tilghman 1987).

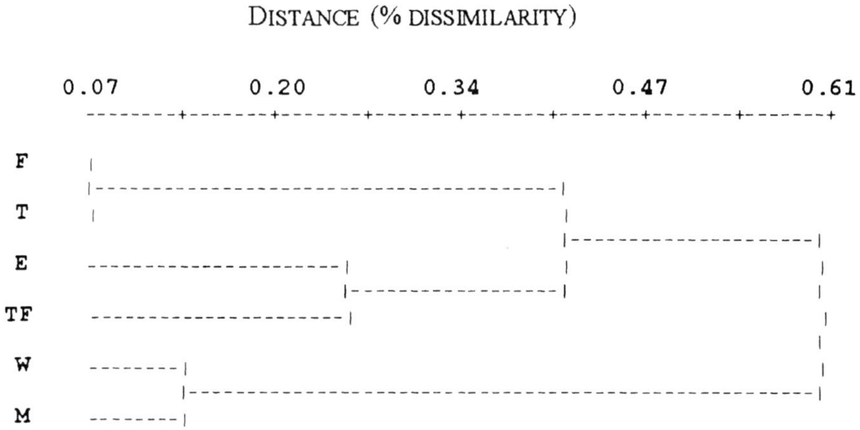

Figure 23.5. Cluster analyses or all species by formation group. Formation groups are abbreviated as: Forest (F), Woodland (W), Thicket (T), Meadow/lawn (M), Emergent wetland (E), and Tall/perennial forb (TF).

5. POLICY AND MANAGEMENT IMPLICATIONS

Hurley (1997) noted that the North City Neighborhoods have a complex history shaped by changing residential patterns, abandonment of commercial and industrial land, and efforts to control flooding along the Missouri River. This history has shaped not just the pattern of open spaces and the habitats within them, but has resulted in two very different settings in which management occurs: the Mississippi River floodplain and upland areas outside the floodplain.

Greenspace on the Mississippi River floodplain contains many habitat types: remnant woodland and forest patches that still experience periodic flooding, thickets developing on forest isolated by construction of a flood control levee in 1959 (Hurley 1997), several relatively large meadow/lawn patches associated with the levee and the recently developed Riverfront Trail, and a few tall forb and emergent wetland patches adjacent to railroad right-of-ways. Most of these floodplain habitats are embedded within a matrix of industry, utility facilities, and housing. By contrast, upland habitat is generally surrounded by housing and some of the habitat patches are quite large (e.g., cemeteries). The bird communities on the floodplain and in the uplands vary with the vegetation. Species diversity is highest in the meadow/lawn habitat. Ground nesting species were most abundant in the tall forb habitat. Species composition was unique in the woodland habitat.

Land managers are more likely to assist bird conservation if we inform them of how habitats are distributed across the landscape and emphasize that key habitat features (e.g., % cover by grasses for ground nesters in tall forbs) can be incorporated into their primary management activities. A conservation plan should also be sensitive to its social context; local residents have their own views on appropriate site management, views that are often tied to their intimate knowledge of the history of a particular site and their feelings about the "popular values of nature" that they associate with a site (Harrison et al. 1987). Improving resident knowledge of the

Table 23.5. Mean detections (± SE) per count of nesting guilds by formation group.

Formation group	Ground/low*	Middle/upper*	Cavity*
Forest	0.12 ± 0.01^{a}	0.15 ± 0.02^{a}	0.08 ± 0.01^{ac}
Woodland	0.01 ± 0.01^{a}	0.53 ± 0.13^{b}	0.60 ± 0.18^{bc}
Thicket	0.13 ± 0.04^{ab}	0.16 ± 0.04^{a}	0.05 ± 0.02^{ad}
Meadow/lawn	0.11 ± 0.02^{a}	0.24 ± 0.05^{a}	0.23 ± 0.05^{de}
Emergent wetland	0.22 ± 0.00^{ab}	0.13 ± 0.00^{ab}	0.04 ± 0.00^{abe}
Tall/perennial forb	0.29 ± 0.05^{b}	0.12 ± 0.10^{ab}	0.13 ± 0.13^{ac}

* $P < 0.05$ for one-way ANOVA

† Different letters indicate differences ($P < 0.05$) among plant formation groups.

ecology of these sites and encouraging their direct participation in management planning, data collection, and site management is both wise and ethical.

ACKNOWLEDGEMENTS

Our research was supported by a U.S. Environmental Protection Agency Sustainable Cities Challenge Grant to Grace Hill Neighborhood Services, and by funds from the Missouri Department of Conservation. We thank Alison Williams and three reviewers for their comments on previous drafts of this paper. This is a contribution from the Missouri Cooperative Fish and Wildlife Research Unit (Biological Resources Division of U.S. Geological Survey, Missouri Department of Conservation, University of Missouri, and the Wildlife Management Institute cooperating).

REFERENCES

Antikainen, E. 1992. The vertical use of a city park by urban birds in Poland. Ornis. Fennica 69:92-96.

Barker, G. M. A., and A. Graf. 1989. Principles for nature conservation in towns and cities. Urban Wildlife Now, No. 3, Nature Conservancy Council, Peterborough, United Kingdom.

Blair, R. B. 1996. Land-use and avian species diversity along an urban gradient. Ecol. Appl. 6:506-519.

Burgess, J., C. M. Harrison, and M. Limb. 1988. People, parks, and urban green: A study of popular meanings and values for open spaces in the city. Urban Studies 25:455-473.

Christman, S. P. 1984. Plot mapping: estimating densities of breeding bird territories by combining spot mapping and transect techniques. Condor 86:237-241.

Cicero, C. 1989. Avian community structure in a large urban park: controls of local richness and diversity. Landscape Urban Plann. 17:221-240.

DeGraaf R. M.., A. D. Geis, and P. A. Healy. 1991. Bird population and habitat surveys in urban areas. Landscape Urban Plann. 21:181-188.

DeGraaf, R. M., and J. M. Wentworth. 1986. Avian guild structure and habitat associations in suburban bird communities. Urban Ecology 9:399-412.

Flores, A., S. T. A. Pickett, W. C. Zipperer, R. V. Pouyat, and R. Pirani. 1998. Adopting a modern ecological view of the metropolitan landscape: the case study of a greenspace system for the New York City region. Landscape Urban Plann. 39:295-308.

Gavareski, C. A. 1976. Relation of park size and vegetation to urban bird populations in Seattle, Washington. Condor 78:375-382.

Geis, A. D. 1974. Effects of urbanization and type of urban development on bird populations, p.97-105. *In* J. H. Noyes and D. R. Progulske [EDS.], Wildlife in an urbanizing environment. Cooperative Extension Service University of Massachusetts, Amherst, MA.

Germaine, S. S., S. S. Rosenstock, R. E. Schweinsburg, and W. S. Richardson. 1998. Relationships among breeding birds, habitat, and residential development in greater Tucson, Arizona. Ecol. Appl. 8:680-691.

Harrison, C., M. Limb, and J. Burgess. 1987. Nature in the city – popular values for a living world. J. of Environ. Manag. 25:347-362.

Hunt, M. 1986. Manual of plant formation entitation. New York City Department of Parks and Recreation Natural Resource Group, New York, NY.

Hurley, A. 1997. Floods, rats, and toxic waste, p.242-261. *In* A. Hurley [ED.], Common fields: an environmental history of St. Louis. Missouri Historical Society Press. St. Louis, MO.

James, F. C., and H. H. Shugart. 1970. A quantitative method of habitat description. Audubon Field Notes 24:727-736.

Lussenhop, J. 1977. Urban cemeteries as bird refuges. Condor 79:456-461.

Matthews, M. J., S. O'Connor, and R. S. Cole. 1988. Database for the New York State urban wildlife habitat inventory. Landscape Urban Plann. 15:23-37.

McCune, B. and M. J. Mefford. 1997. Multivariate analysis of ecological data version 3.0. MJM Software, Glenden Beach, OR.

Mueller-Dombois, D., and H. Ellenberg. 1974. Aims and methods of vegetation ecology. John Wiley and Sons, New York, NY.

Natuhara, Y. and C. Imai. 1996. Spatial structure of avifauna along urban-rural gradients. Ecol. Res. 11:1-9.

Nilon, C. H. and G. S. Huckstep. 1998. Analysis of Chicago River recreation habitats, p.161 - 172. *In* P. H. Gobster and L. M. Westphal [EDS.], People and the river: perception and use of Chicago waterways for recreation. USDI National Park Service, River, Trails and Conservation Assistance Program. Milwaukee, WI.

Nilon, C. H., C. N. Long, and W. C. Zipperer. 1995. Effects of wildland development on forest bird communities. Landscape Urban Plann. 32:81-92.

Peladeau, N. 1996. SIMSTAT for Windows user's guide. Provalis Research, Montreal, QC, Canada.

Reiley, J. O., and S. E. Page. 1995. Survey mapping and evaluation of green space in the federal territory of Kuala Lampur, Malaysia, p.173-184. *In* H. Sukopp, M. Numata, and A. Huber [EDS.], Urban ecology as the basis of urban planning. SPB Academic Publishers, The Hague, Netherlands.

Roberts, D. C. 1994 The design of an urban open-space network for the city of Durban (South Africa). Environmental Conserv. 21:11-17.

Rogers G. F., and R. A. Rowntree, 1988. Intensive surveys of structure and change in urban natural areas. Landscape Urban Plann. 15:59-78.

Sauer, J. R., J. E. Hines, G. Gough, I. Thomas, and B. G. Peterjohn. 1997. The North American Breeding Bird Survey Results and Analysis. Version 96.4. Patuxent Wildlife Research Center, Laurel, MD.

Schroeder, W. 1997. Environmental setting of the St. Louis region, p.13-37. *In* A. Hurley [ED.], Common fields: an environmental history of St. Louis. Missouri Historical Society Press. St. Louis, MO.

Suhonen, J., and J. Jokimaki. 1988. A biogeographical comparison of the breeding bird species assemblages in twenty Finish urban parks. Ornis Fennica 65:76-83.

Sukopp H., and S. Weiler. 1988. Biotope mapping and mature conservation strategies in urban areas of West Germany. Landscape Urban Plann. 15:39-58.

ter Braak, J. F., 1986. Canonical correspondence analysis: a new eigenvector technique for multivariate direct gradient analysis. Ecology 67:1167-117.

Thomas, J. W., and R. A. Dixon. 1974. Cemetery ecology, p.107-110. *In* J. H. Noyes, and D. R. Progulski [EDS.], A symposium of wildlife in an urbanizing environment. University of Massachusetts Cooperative Extension Service, Amherst, MA.

Thom, R. H., and J. H. Wilson. 1980. The natural divisions of Missouri. Transactions of the Missouri Academy of Science 14:9-23

Tilghman, N. G. 1987. Characteristics of urban woodlands affecting breeding bird diversity and abundance. Landscape Urban Plann. 14:481-495.

Wayman, N. L. 1979. History of St. Louis neighborhoods. St. Louis Community Development Agency, St. Louis, MO.

Zipperer, W. C., S. M. Sisinni, and R. V. Pouyat. 1997. Urban tree cover: an ecological perspective. Urban Ecosystems 1:229-246.

Appendix 23.1. Mean detections per count for bird species in North City greenspaces, 1997 and 1998.

Species	Guild	Forest	Woodland	Thicket	Meadow	Wetland	Tall forb
Green-backed Heron		0.00	0.00	0.00	0.01	0.00	0.00
Canada Goose		0.00	0.02	0.00	0.00*	0.00	0.00
Wood Duck	cavity	0.00	0.00*	0.00	0.00	0.00	0.00
Killdeer		0.00	0.00	0.00	0.06	0.00	0.00
Red-tailed Hawk	cavity	0.00	0.06	0.00	0.00*	0.00	0.00
American Kestrel		0.00	0.00	0.00	0.01	0.00	0.00
Northern Bobwhite		0.05	0.00	0.18	0.19	0.00	0.10
Rock Dove		0.00	0.11	0.00	0.08	0.00	0.00
Mourning Dove		0.11	0.37	0.23	0.29	0.64	0.07
Yellow-billed Cuckoo	ground	0.04	0.00	0.02	0.02	0.29	0.00
Chimney Swift)		0.00	0.09	0.00	0.02	0.00	0.00
Belted Kingfisher		0.00	0.00	0.00	0.00	0.00*	0.00
Northern Flicker	cavity	0.04	0.14	0.02	0.01	0.14	0.00
Red-bellied Woodpecker	cavity	0.02	0.00	0.00	0.00*	0.00	0.00
Red-headed Woodpecker	cavity	0.02	0.26	0.00	0.06	0.00	0.00
Downy Woodpecker	cavity	0.15	0.02	0.05	0.01	0.07	0.00
Hairy Woodpecker	cavity	0.01	0.01	0.00	0.00	0.00	0.00
Eastern Kingbird	upper	0.03	0.20	0.00	0.11	0.00	0.00
Western Kingbird	upper	0.00	0.00	0.14	0.01	0.00	0.07
Great crested Flycatcher	cavity,upper	0.04	0.01	0.00	0.01	0.00	0.00
Eastern Wood-pewee	upper	0.12	0.07	0.00*	0.00*	0.00	0.00
Yellow-bellied Flycatcher	ground	0.00	0.02	0.00	0.00	0.00	0.00
Tree Swallow	cavity,upper	0.00	0.00	0.00	0.04	0.00	0.07
Purple Martin	cavity,upper	0.00	0.00	0.00	0.01	0.00	0.00
Barn Swallow	upper	0.00	0.07	0.00	0.09	0.00	0.00
Blue Jay	upper	0.10	0.22	0.00	0.06	0.00	0.00
American Crow	upper	0.37	0.26	0.05	0.09	0.00	0.14

Species	Guild	Forest	Woodland	Thicket	Meadow	Wetland	Tall forb
Tufted Titmouse	cavity,upper	0.05	0.02	0.00	0.00*	0.00	0.00
Black-capped Chickadee	cavity,upper	0.00	0.00	0.11	0.00	0.14	0.00
Carolina Chickadee	cavity,upper	0.25	0.08	0.02	0.00*	0.00	0.00
White-breasted Nuthatch	cavity,upper	0.00	0.01	0.00	0.00	0.00	0.00
House Wren	cavity,upper	0.02	0.07	0.00	0.08	0.00	0.00
Carolina Wren	ground	0.09	0.00*	0.00	0.01	0.00	0.00
Blue-gray Gnatcatcher		0.00	0.00	0.00	0.00*	0.14	0.00
Wood Thrush	upper	0.01	0.00	0.00	0.00	0.00	0.00
Swainson's Thrush	ground	0.00	0.00*	0.00	0.00	0.00	0.00
American Robin	upper	0.83	2.13	0.48	0.70	1.64	0.50
Gray Catbird	ground	0.30	0.07	0.05	0.02	0.00	0.07
Northern Mockingbird	ground	0.01	0.01	0.02	0.12	0.07	0.00
Brown Thrasher	ground	0.01	0.03	0.00	0.03	0.00*	0.00
Cedar Waxwing	upper	0.01	0.01	0.00	0.03	0.00	0.00
European Starling	cavity,upper	0.62	8.93	0.57	3.41	0.36	1.93
Yellow-throated Vireo	upper	0.00	0.00	0.00	0.00	0.00*	0.00
Red-eyed Vireo	upper	0.01	0.00	0.00	0.00*	0.00	0.00
Warbling Vireo	upper	0.18	0.04	0.14	0.01	0.29	0.00
Black-and-white arbler	ground	0.01	0.00	0.00	0.00	0.00	0.00
Chestnut-sided Warbler	ground	0.00	0.00	0.02	0.00*	0.00	0.00
Blackpoll Warbler	upper	0.00	0.03	0.00	0.00	0.00	0.00
Yellow Warbler	upper	0.01	0.01	0.00	0.00*	0.29	0.00
Ovenbird	ground	0.05	0.00	0.00	0.00	0.00	0.00
Common Yellowthroat	ground	0.03	0.00	0.00	0.03	0.00	0.07
Yellow-breasted Chat	ground	0.01	0.00	0.00	0.00	0.00	0.00
Northern Cardinal	ground	1.21	0.11	0.91	0.17	0.79	0.14
Blue Grosbeak	ground	0.00	0.00	0.00	0.00*	0.00	0.00
Indigo Bunting	ground	0.26	0.01	0.21	0.20	0.36	1.28
Song Sparrow	ground	0.00	0.00	0.02	0.06	0.29	0.21
White-throated Sparrow	ground	0.00	0.03	0.00	0.00	0.00	0.00

Species	Guild	Forest	Woodland	Thicket	Meadow	Wetland	Tall forb
Dickcissel	ground	0.00	0.00	0.00	0.07	0.00	0.00
Eastern Meadowlark	ground	0.00	0.00	0.00	0.02	0.00	0.00
Red-winged Blackbird	ground	0.30	0.00		1.44	2.71	4.07
Brown-headed Cowbird		0.04	0.06		0.02	0.00	0.00
Common Grackle	upper	1.45	2.66		0.26	0.64	0.28
Baltimore Oriole	upper	0.00	0.03		0.00	0.00	0.00
Eurasian Tree Sparrow		0.00	0.00		0.04	0.00	0.00
House Sparrow	upper	0.00	0.03		0.18	0.00	0.00
American Goldfinch	upper	0.06	0.00		0.23	0.14	0.28
House Finch	upper	0.08	0.06		1.51	0.00	0.07

Appendix 23.2. Axis 1 CCA scores and mean detections (± SE) per count for bird species that best describe the communities.

Species	*P*-value for year effect	CCA score	Forest	Woodland	Thicket	Meadow/lawn
Carolina Chickadee*	0.02	-2.13	0.25 ± 0.08^{a}	0.08 ± 0.04^{a}	0.02 ± 0.02^{ab}	0.00 ± 0.00^{b}
Gray Catbird	0.44	-2.08	0.30 ± 0.15	0.07 ± 0.07	0.05 ± 0.03	0.02 ± 0.02
Warbling Vireo*	0.61	-2.07	0.18 ± 0.11^{a}	0.04 ± 0.04^{bd}	0.14 ± 0.07^{ab}	0.01 ± 0.01^{cd}
Downy Woodpecker*	0.26	-1.94	0.15 ± 0.05^{a}	0.02 ± 0.01^{ab}	0.05 ± 0.05^{ab}	0.01 ± 0.01^{b}
American Crow*	0.24	-1.40	0.37 ± 0.08^{a}	0.26 ± 0.16^{ab}	0.05 ± 0.03^{b}	0.09 ± 0.04^{b}
House Sparrow	0.03	1.22	0.00 ± 0.00	0.03 ± 0.02	0.27 ± 0.27	0.18 ± 0.08
Northern Bobwhite	0.19	1.23	0.02 ± 0.01	0.00 ± 0.00	0.09 ± 0.09	0.11 ± 0.05
Northern Mockingbird*	0.03	1.24	0.01 ± 0.01^{a}	0.01 ± 0.01^{a}	0.02 ± 0.02^{a}	0.12 ± 0.04^{b}
House Finch	0.45	1.45	0.08 ± 0.05	0.06 ± 0.04	0.80 ± 0.47	1.51 ± 0.83

* $P < 0.05$ for two-way ANOVA with year and habitat effects on rank transformed data.

† Different letters indicate differences ($P < 0.05$) among plant formation groups.

Appendix 23.3. Axis 2 CCA scores and mean detections (± SE) per count for bird species that best describe the communities

Species	*P*-value for year effect	CCA score	Forest	Woodland	Thicket	Meadow/lawn
Gray Catbird	0.44	-2.17	0.30 ± 0.15	0.07 ± 0.07	0.05 ± 0.03	0.02 ± 0.02
Northern Cardinal*	0.65	-1.65	1.21 ± 0.22^{a}	0.11 ± 0.07^{b}	0.91 ± 0.21^{a}	0.17 ± 0.04^{b}
House Finch	0.45	-1.51	0.08 ± 0.05	0.06 ± 0.04	0.80 ± 0.47	1.51 ± 0.83
American Goldfinch	0.06	-1.48	0.06 ± 0.02	0.00 ± 0.00	0.21 ± 0.11	0.23 ± 0.13
Indigo Bunting	0.07	-1.34	0.26 ± 0.06	0.01 ± 0.01	0.21 ± 0.17	0.20 ± 0.10
Red-headed Woodpecker*	0.31	1.76	0.02 ± 0.01^{a}	0.26 ± 0.11^{b}	0.00 ± 0.00^{a}	0.06 ± 0.03^{a}
House Wren	0.52	2.15	0.02 ± 0.02	0.07 ± 0.07	0.00 ± 0.00	0.08 ± 0.08

* $P < 0.05$ for two-way ANOVA with year and habitat effects on rank transformed data.

† Different letters indicate differences ($P < 0.05$) among plant formation groups.

Chapter 24

The importance of the Chicago region and the "Chicago Wilderness" initiative for avian conservation

Jeffrey D. Brawn[1] and Douglas F. Stotz[2]
[1]*Illinois Natural History Survey, 607 Peabody Dr. Champaign, IL 61820 USA, e-mail: j-brawn@uiuc.edu*

[2]*Field Museum of Natural History, Environment and Conservation Programs, Roosevelt Rd. at Lake Shore Dr., Chicago, IL 60605 USA*

Key words: Biodiversity, brood parasitism, Chicago, conservation, habitat restoration, metropolitan, migrant, oak savanna, prairie, public education, urban, wetland

Abstract: The Chicago region is a large, established metropolitan region, that may play a critical role for the conservation of breeding and migratory birds in midwestern North America. The region includes over 80,000 ha of protected lands that include some of the best remaining grassland, oak savanna, and marsh habitat in the state of Illinois. Because of its juxtaposition with Lake Michigan, the region serves as a significant stop-over site for spring and fall migrants that comprise a diverse group of terrestrial and aquatic species. Efforts to conserve and restore habitat for birds in the region are being coordinated by a partnership of over 60 public and private organizations known collectively as the "Chicago Wilderness." These organizations have joined to protect, manage, and restore the habitats and biological communities of the region. Specific activities include ecological inventory, the creation of recovery plans for various types of biologically significant habitats, restoration research, education and outreach, and the identification of region-wide conservation priorities. The benefits of these activities for birds are promising. Studies conducted in lands managed by the forest preserve district of Cook County indicate that the breeding bird communities in the preserves are quite similar to those in the rural areas of Illinois. Moreover, estimated rates of parasitism by Brown-headed Cowbirds (*Molothorus ater*) are less than those observed in the more agricultural landscapes of the upper midwestern United

States. Habitat protection in otherwise developed metropolitan areas may offer significant conservation opportunities in regions where natural habitat is scarce.

1. INTRODUCTION

Wildlife-related issues in large, established metropolitan areas often stem from human-wildlife conflicts rather than conservation opportunities (VanDruff et al. 1996). Although the value of such areas is not generally recognized by conservationists, interest in opportunities and needs are being explored (Beissinger and Osborn 1984, Ilyichev et al. 1986, DeGraff and Wentworth 1991, Marzluff et al. 1998). The City of Chicago and its metropolitan region comprises 13 counties from three states (Illinois, Indiana, and Wisconsin) and encompasses nearly 1.75 million ha (Openlands Project 1999, Fig. 24.1). With over 7 million people, this region is the third largest metropolitan area in North America. However, for several reasons the Chicago region may play an important role for avian conservation in midwestern North America and elsewhere. First, significant protected areas remain in the region and these offer diverse habitats for breeding and migratory birds. For breeding birds, the region offers some of the best remaining grasslands, oak savannas, and wetlands in Illinois (Sullivan 1998). These habitats are now rare in the Midwest in areas where agriculture, especially row-cropping, predominates (Warner 1994). Second, urbanization and, especially, suburbanization are expected to increase significantly in the region over the next 25 years and this will place increasing pressure on the region's natural areas (Openlands Project 1999). Third, because of its proximity to Lake Michigan, the region serves as a critical stopover site for spring and fall migrants that comprise a diverse group of terrestrial and aquatic species. And fourth, efforts to conserve and restore habitat for birds and other components of biodiversity in the region are being coordinated by a partnership of over 60 public and private organizations known collectively as the "Chicago Wilderness." These organizations have joined to protect, manage, and restore the ecosystems and biological communities of the region. The Chicago Wilderness may be an important model for concerted conservation programs in metropolitan regions.

In this paper, we provide an overview of issues in the Chicago region with respect to avian conservation. Our objectives are to: 1) summarize the natural history of the area and outline projected trends in land-use; 2) summarize the ornithological resources of the region with respect to breeding birds and migrants; 3) present an overview of the Chicago Wilderness; and 4) discuss the conservation implications of our findings.

2. METHODS

We estimated migrant abundance at three sites in 1997 and 1998. Thatcher Woods, is a preserve of mostly oak-dominated floodplain forest along the DesPlaines River, a southward flowing river into the Mississippi

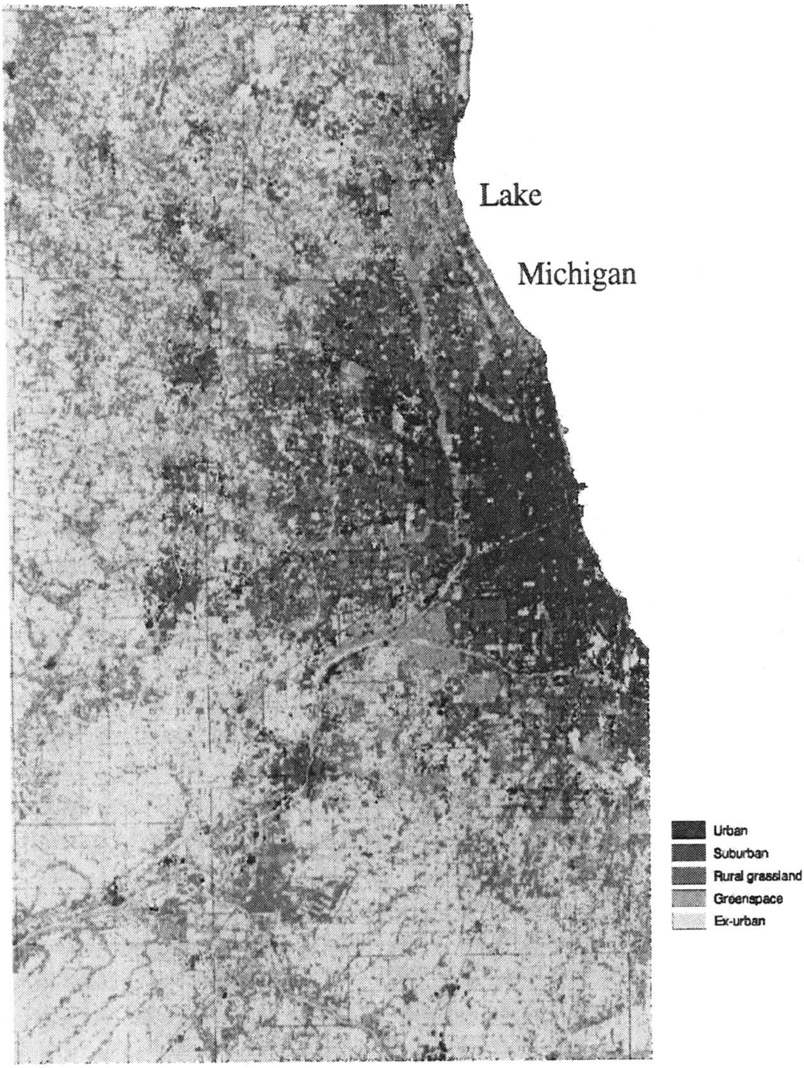

Figure 24.1. Classified image of the Chicago Region showing urban, suburban, and rural areas. Original image was acquired by Landsat in 1997 and covers 1,254,250 ha.

basin that drains about two-thirds of the Chicago region. Swallow Cliff, located in western Cook County in the Palos area, is within the largest area of natural habitat in Northern Illinois and includes upland mesic oak woodland with maples (*Acer* spp.). Grant Park is an urban park on the Lake Michigan lakefront immediately adjacent to downtown Chicago. It is relatively open, with mowed lawns, low shrub rows, and lines of trees, mostly American elms (*Ulmus americana*) and crabapples (*Pyrus cornaria*).

We counted migrants two or three days a week during May of each year along a single transect route at each site. These transects ranged from 2.5 km in Grant Park to 3.5 km at Swallow Cliff. Transects were traversed at a rate of approximately 1 km hr^{-1}.

We estimated rates of nest parasitism in several sites in forest preserves located in western Cook County (near Palos). Located nests were visited every three days until the nest failed or young were fledged. Contents of nests were recorded each visit. Nests were inspected carefully for the presence and number of Brown-headed Cowbird eggs or nestlings.

3. RESULTS

3.1 Ornithological Resources in the Chicago region

The Chicago region is one of the most ecologically diverse areas in Illinois and the Midwest. The major habitats in the region include prairies, shrublands, savannas, woodlands, forests, wetlands, still and moving waters, and sandy dunes along the lakeshore (Sullivan 1998). Many of these natural communities in the region are further classified according to soil texture (fine vs. sandy) and soil moisture. Forest communities (i.e., those with 80-100% canopy cover) can be divided into upland, floodplain, and flatwoods. Savannas (10-50% canopy closure) and woodlands (50 - 80% closure) are divided into fine textured and sandy soil (Taft 1997). Each subcategory is a unique habitat because of characteristic floristic composition and structure and all occur in the Chicago region.

The major breeding bird communities in the Chicago region are found in grasslands or prairies, shrublands, savannas and woodlands, forests, and wetlands (Sullivan 1998). The diversity of birds reflects the habitat diversity and geographic location of the region. Overall, over 400 species have been reported in the region (including counties from Michigan along the lakeshore; Mlodinow 1984). This list includes 125-150 regional breeders and passage migrants not known to breed in this area.

3.2 Status of Breeding Habitat for Birds in the Chicago Region

The overall status of avian habitat is bleak in areas with heavy agricultural development. Overall, Illinois has lost over 98% of the natural habitat that was thought to be present at the time of European settlement (Schwartz 1997). Less than 0.1% of the original native prairie remains and over 98% of the original oak savannas and woodlands have been lost or degraded. Aquatic habitats have fared poorly as well; over 90% of the original wetlands have been lost (Havera et al. 1997).

The remaining terrestrial habitats in the Midwest and the Chicago region tend to be chronically fragmented (Robinson et al. 1995, Brawn and Robinson 1996). The effects of habitat fragmentation on breeding forest and grassland birds of the Midwest are well known; area-sensitive species are generally absent from small fragments and species that breed in fragments experience low breeding success because of high rates of nest predation and brood parasitism (Robinson et al. 1995, Brawn and Robinson 1996, Herkert 1994). Despite chronic fragmentation and habitat loss in the upper midwestern U.S., the Chicago region contains some of the largest intact tracts of tallgrass prairie and oak savannas that remain east of the Mississippi River. The Midewin National Tallgrass Prairie (formerly the site of the Joliet Arsenal in Will Co., IL) will eventually comprise nearly 7,600 ha of natural habitat. Goose Lake Prairie comprises nearly 700 ha and supports some of the region's largest populations of grassland birds, including several species that are locally endangered such as Upland Sandpipers (*Batramia longicauda*) and Northern Harriers (*Circus cyaneus*).

Significantly, the Chicago region contains relatively more natural habitat for breeding birds than areas of Illinois and the Midwest with agricultural development. For example, the Chicago Wilderness counties are 12% (urban) woodland and 4.5% wetlands, compared to 2% and 1%, respectively, for the counties immediately south, and 4.5% wooded and 1% wetlands for counties immediately west (Wang and Moskovits, in press). In the heavily agricultural counties of Illinois, forest-woodland and wetland cover typically comprise less than 1% (Wang and Moskovits, in press).

The formation of county-level forest preserve districts and state parks helped keep the region relatively "green" during the 20th century. First created in 1915 in Cook and DuPage counties, the forest preserve districts were dedicated to preservation of local fauna and flora (Sullivan 1998). Other counties have since followed suit and over 80,000 ha of natural areas are now preserved in these districts. Nearly all the forest, woodland, and savanna habitat in the Chicago region resides in the forest preserve districts (Sullivan 1998). Key state parks have promoted the preservation of sensitive sand dune habitat along the lakeshore and large tracts of grassland habitat. Federal holdings include the Indiana Dunes National Lakeshore and the Midewin National Tallgrass Prairie.

The value of these relatively large forest preserves for breeding birds is evident from results of extensive census work in the forest preserve district of Cook County, where breeding bird communities are similar to those in forest tracts located in more rural parts of Illinois (Robinson et al. 1997, Brawn 1998). Another interesting finding specific to the urban environment is that rates of parasitism by Brown-headed Cowbirds in forest preserves do not appear as high as those observed in other, more rural, parts of Illinois (Table 25.1). Estimated rates of parasitism were lower for five of the most common host species in the Midwest (Robinson et al. 1997). Although present, female cowbirds are not as common in urban woodlots as in more rural landscapes. In rural forest tracts surrounded by row crops, female cowbirds were twice as abundant as in the urban woodlots (Brawn 1998). Differences between urban and rural landscapes in the proximity of woodlots to feeding areas for cowbirds may underlie this pattern.

Loss of wetlands (here referring to fens, sedge meadow, marshes, and floodplains) is nearly complete in the agricultural areas of the Midwest. In the Chicago region, however, losses of wetlands have been less severe. For example, Cook County has more than 6,500 ha of wetlands and ranks in the top five counties in Illinois (Sullivan 1998). Accordingly, marsh bird populations are healthier in the Chicago region than in most other regions of the state. Examples are the American Bittern (*Botaurus lentiginosus*), the Black-crowned Night Heron (*Nycticorax nycticorax*), and the Yellow-headed Blackbird (*Xanthocephalus xanthocephalus*). At present, more than 35 marsh bird nesting colonies are found within the Chicago region (Openlands Project 1999).

Table 24.1. Frequency of cowbird parasitism in urban and rural landscapes in Illinois. Urban sites are located in the forest preserve district of Cook County, Chicago (1997 and 1998 breeding seasons). Rural sites are located in central and southern Illinois' Nature Preserves (Robinson et al. 1997) and woodlots (Brawn and Robinson 1996).

Host species	Percent of nests parasitized	
	Urban	Rural
Wood Thrush (*Hylocichla mustelina*)	56 (24)	97 (206)
Red-eyed Vireo (*Vireo olivaceus*)	50 (20)	77 (13)
Northern Cardinal (*Cardinalis cardinalis*)	7 (28)	39 (77)
Rose-breasted Grosbeak (*Pheucticus ludovicianus*)	5 (22)	28 (25)
Indigo Bunting (*Passerina cyanea*)	33(24)	57 (63)

3.3 The Chicago Region as a Stopover Site for Migrants

The importance of the Chicago region as a stopover site for migrants is potentially critical. During spring migration especially, birds navigate along the edge of, or over, Lake Michigan. In stormy conditions, the fallout of birds, presumably coming in from Lake Michigan, is phenomenal. As a result, even small city parks play an important role and can attract huge numbers of migrants. Along Montrose Point (a small spit jutting out along Chicago's north shore) there is a small hedge now known as the "Magic Hedge." This hedge attracts huge numbers of birds during spring migration and is known as an excellent place to observe migrants. The consistently large numbers of migrant birds attracted to this single hedge surrounded by a city landscape is a strong indication that suitable stopover sites are in short supply and how even modest conservation or restoration efforts can furnish much needed stopover habitat.

Loss of natural habitat along the traditional flyways in the Midwest has been heavy and there is emerging recognition that stopover sites of sufficient quality may play a critical role in limiting populations of many species (Hutto 1998). In the Chicago area, besides the lakefront parks, the forest preserves that have protected land in the Palos region and along rivercourses also attract large numbers of migrants. The three sites where we censused migrants showed similar numbers of migrants overall, with Grant Park having the most individuals per hr in each year. In 1997, the number of migrants per hr ranged from 106 at Grant Park to 86 at Swallow Cliff and 84 at Thatcher Woods. In 1998, the relative abundance at the three sites was similar, but migrants were about one-half as common, ranging from 51 per hr at Grant Park to 41 at Thatcher Woods. Comparative data are unknown to us, but the abundance of migrants in the Chicago region are generally held to be higher than those in other regions of Illinois.

Despite the similarities in migrant numbers at the three sites, marked differences existed in the composition of the migrant avifauna. In particular, we observed a strong tendency for certain species to be much more common on the lakefront vs. inland or vice-versa. Species that are predominantly lakefront migrants include most sparrows (especially the White-throated sparrow [*Zonotrichia albicollis*]), Empidonax flycatchers (*Empidonax spp.*), Ruby-crowned Kinglets (*Regulus calendula*), and some warblers including Common Yellowthroats (*Geothlypis trichas*), Cape May Warblers (*Dendroica tigrina*) and Wilson's Warbler (*Wilsonia pusilla*). Predominantly inland migrants are vireos, Great-crested Flycatchers (*Myiarchus crinitus*), Eastern Wood Pewees (*Contopus virens*), Rose-breasted Grosbeaks (*Pheucticus ludovicianus*), Baltimore Orioles (*Icterus galbula*) and some species of warblers including Chestnut-sided (*Dendroica *pensylvanica*), Black-throated Green (*Dendroica virens*), and Bay-breasted (*Dendroica castanea*) (Table 24.2). The only migrant that is common on the lakefront and on an inland site, is the American Redstart (*Setophaga ruticilla*) which was common both in Grant Park and Thatcher Woods, but extremely rare at Swallow Cliff (Table 24.2).

The three sites also differed dramatically in the magnitude of day to day variation in migrant numbers. The two inland sites were much less variable than Grant Park on the lakefront (Coefficient of Variation (CV) among days in number of birds observed hr^{-1}: Swallow Cliff CV = 19; Thatcher Woods CV = 29; Grant Park CV = 93). Although the inland sites varied from day to day, migrants were common everyday, and the only substantial variation in numbers occurred in the last week of May in 1998 when migrant numbers declined dramatically as migration came to an end. In contrast, numbers of birds in Grant Park fluctuated wildly. Birds arrived in big numbers on days with strong southwesterly winds, especially if a front passed during the night, contributing to the fallout of birds. On days with winds off the lake, there were few or no new migrants and if good flight conditions existed the previous night, overall numbers of migrants could be extremely low. Numbers of birds per hr could vary by nearly two orders of magnitude across a period of three days in Grant Park. Inland, the variance within a week was never more than twofold.

The lakefront parks in Chicago do not have significant breeding populations of birds; only species that tolerate human disturbance and have

Table 24.2. Selected migrant species illustrating variation in abundance between inland and lakefront sites in Chicago Region (no. of individuals hr^{-1} of observation in May 1997).

Species	Swallow Cliff	Thatcher Wood	Grant Park
American Redstart	0.21	5.80	2.85
Lakefront Migrants			
Empidonax spp. (4)	1.19	1.85	6.53
Ruby-crowned Kinglet	0.50	0.33	1.61
White-throated Sparrow	0.64	0.84	35.66
Cape May Warbler	0.00	0.11	1.09
Wilson's Warbler	0.21	0.62	2.31
Common Yellowthroat	0.21	1.35	4.64
Inland Migrant			
Eastern Wood Pewee	1.71	1.38	0.11
Great Crested Flycatcher	1.93	1.16	0.05
Rose-breasted Grosbeak	1.29	1.93	0.48
Bay-breasted Warbler	2.21	2.40	0.48
Chestnut-sided Warbler	2.21	4.00	0.66
Black-throated Green Warbler	5.43	3.60	0.34
Baltimore Oriole	1.54	0.58	0.11

relatively unspecialized habitat requirements are able to regularly breed in these parks. As a result, conservation activities in the lakefront park can concentrate on providing appropriate habitat for migrating birds without having to balance the effects on breeding populations.

In the forest preserves, the value of habitat for stopover migrants may be a function of management practices that affect floristic composition. The oak woodlands and savannas of the Chicago region need active management (i.e., periodic disturbance via prescribed fire) to maintain an open canopy and prevent the encroachment of mesophytic species such as sugar maples (*Acer saccharum*, Fralish et al. 1991). Such management has begun at a number of sites, including Swallow Cliff.

The effect of management activities such as burning and canopy thinning on use of these habitats by migrants has not been studied comprehensively, but preliminary data do suggest a preference for certain tree species as foraging substrates. In 1997 and 1998, we examined the abundance of warblers at Swallow Cliff in areas that had under gone restoration (brush-clearing and prescribed burns) in the spring of 1996, areas that were unrestored (but with oak-dominated canopies), and areas without restoration where maples and other fire-tolerant trees are invading the canopy. We found that the abundance of warblers was much higher in areas dominated by oaks rather than maples (Fig. 24.2). The estimated abundance of migrants appeared relatively similar at managed (i.e., burned) and unmanaged sites; however, continued lack of disturbance will lead to further loss of oaks from the canopy. These data indicate that forest preserves managed for oak may be especially valuable for warbler species that rely more heavily on the inland areas and are sparse on the lakefront.

Much more research on migrants is needed. Research has been initiated using Doppler radar to document the use of stopover sites and the quantity of birds going through the region (R. Larkin and R. Diehl, pers. comm.). A particularly important issue is the foraging behavior and nutritional ecology of birds during stopover (Hutto 1998). Conceivably, certain types of trees and shrubs may offer better feeding conditions than others (Reichard et al. 2001).

3.4 Projected Trends in Land-Use and Potential Threats to Avian Conservation in the Chicago region

Projected trends in population growth and the amount of land that will be developed by urbanization and suburbanization have important implications for avian conservation in the Chicago region. By the year 2020, the population of the Chicago region is expected to increase by about 25% to well over 10 million (Openlands Project 1999). However, the most dramatic growth will not be in the City of Chicago where the increase is expected to be less than 5%. Rather, the outlying counties are expected to experience unprecedented growth rates. For example, the population of Will County (just south of Chicago) is expected to more than double from about 357,000 to over 800,000 in the next 20 years. Between 1970 and 1990, developed land in the Chicago region increased by 47%, while population increased by only 4% (Wang and Moskovits, in press). The rate of development to suburban and urban conditions will likely increase. In Will County, for example, urban and suburban land-covers are projected to increase from 13% to 63% between 1998 and 2028 (Openlands Project 1999).

Sprawl will pose significant threats to habitat for birds, other wildlife, and plants throughout the region. The Illinois Natural Area Inventory (Sulloway and Hubbell 1994) identified 256 significant natural areas in the Illinois portion of the Chicago region; of these, 146 are not protected. The situation is similar in the Wisconsin and Indiana portions of the region and, overall, over 300 natural areas are currently unprotected (Openlands Project

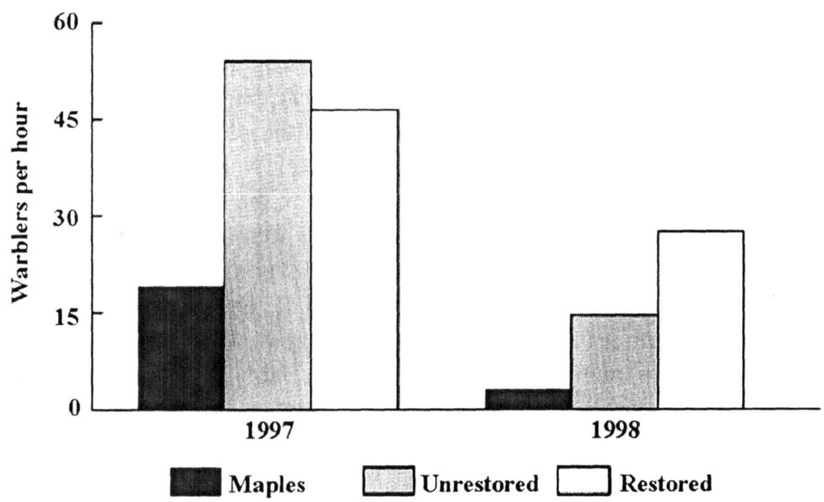

Figure 24.2. Relative abundance of spring migrants in the Chicago region during 1997 and 1998 at sites dominated by maples and oak dominated sites that are undergoing restoration (see text) or have remained undisturbed for at least 60 years.

1999). Much of the land that is likely to be developed is currently under row crops, but without planning the remaining natural terrestrial habitat for birds will be further lost and fragmented. Wetlands will be either lost because of increased water demands or increasingly polluted owing to greater urban and suburban runoff.

3.5 The Chicago Wilderness Biodiversity Initiative

One of the unique aspects of the Chicago region regarding the conservation of birds and biodiversity is the Chicago Wilderness (http://www.chiwild.org). Initiated in 1996, the Chicago Wilderness now consists of over 60 organizations that include: government agencies at the local, state, and federal level; conservation organizations; universities; and research institutions. This partnership was formed to coordinate and plan the protection and management of over 80,000 ha of natural areas in the Chicago region. These areas include the aforementioned forest preserve districts, the Midewin National Tallgrass Prairie, and various state holdings. The Chicago Wilderness has its own staff and an executive steering committee consisting of persons from several key partner institutions. Many of the activities of the Chicago Wilderness are directed or assisted by professional personnel from member institutions and by volunteers.

Major activities and accomplishments of the organization include coordination of conservation activities across political boundaries and disciplines, facilitating an increase in restoration activity (especially oak savanna and prairie habitat), direct support of research on local biodiversity, development of ecological monitoring and inventory programs, development of a Biodiversity Recovery Plan, a marked increase in educational activities directly related to the conservation of local biodiversity, development of new volunteer programs, and involvement of local corporations in biodiversity issues. These activities are coordinated and planned through four teams (science, land management, education and outreach, and policy and strategy). The Chicago Wilderness has attracted national and international attention and is being considered as a model for the coordination and planning of urban conservation activities for cities in Mexico, Brazil, and Hawaii (D. Moskovits, pers. comm.).

Two factors have been fundamental to the effectiveness of the Chicago Wilderness. First, significant resources were secured by the initiative in its formative stage. As of 1998, nearly eight million dollars in support for the Chicago Wilderness had been designated in direct and matching funds from local, state, and federal agencies; private institutions; and non-governmental organizations. The USDA Forest Service and the US Fish and Wildlife Service provided over 40% of the direct support (Chicago Wilderness, Two

Year Progress Report). The early establishment of a funding base was instrumental in attracting interest from member institutions, initiating research and education projects, and maintaining momentum.

Second, the initiative was highly effective in recruiting the participation of agencies that are directly responsible for the management and stewardship of natural habitat. By cutting across political boundaries and being highly inclusive, the Chicago Wilderness assembled teams of experts and personnel that make decisions and directly affect policy on land-use and planning.

The Biodiversity Recovery Plan is especially relevant to avian conservation. This plan includes a thorough inventory of biodiversity in the region, identifies threats to different ecosystems and biological communities, defines recovery/restoration goals for these ecosystems, and recommends specific actions. Birds are considered to be a key element of biodiversity in all major ecosystems and the enhancement of avian habitat has been an important factor in formulating the plan. Without the Chicago Wilderness, the coordination that was needed for this effort would not have been possible. For large metropolitan areas, it is unlikely that any one agency can perform all the tasks that are needed to effectively plan conservation efforts and educate the public.

4. POLICY AND MANAGEMENT IMPLICATIONS

A major implication of our research and review of avian conservation in the Chicago Region is that, with planning, large metropolitan areas can be very useful for birds. In the regions such as Illinois where much land is in row crops, the value of large urban areas with parks or preserves can be immense. The lesson for newer or expanding urban and suburban areas is clear; habitat protection must be established early. Today, Cook County would be hard pressed to set aside land for the forest preserves it now enjoys. Even provision of small hedge rows may provide critical habitat for birds during a stressful stage of their annual cycle. Effective preservation of habitat for birds and other components of biodiversity requires planning and strong public support. We believe that the Chicago Wilderness is an ideal model for meeting these needs. The political and bureaucratic complexity that is typical of urban areas requires an effective organization to coordinate conservation and efforts at a geographic scale relevant to the habitat needs of birds. With a concentrated populace and easy access to the media, outreach and education programs can be particularly effective.

ACKNOWLEDGEMENTS

We thank D. Moskovits for providing insight and unpublished data. The Landsat image was provided by the Chicago Wilderness and NASA. Jocelyn Aycrigg and Liane Suloway kindly provided help with editing the image. We thank the Chicago Wilderness, The Illinois Department of Natural Resources, and the Field Museum of Natural History for supporting various aspects of our research in the Chicago region. We dedicate this paper to the memory of Jerry Sullivan; his deep knowledge and affection for Chicago birds coupled with an extraordinary ability to communicate natural history will be missed.

REFERENCES

Beissinger, S. R., and D. R. Osborne. 1984. Effects of urbanization on avian community organization. Condor 84:75-83.

Brawn, J. D. and S. K. Robinson. 1996. Source-sink population dynamics may complicate the interpretation of long-term census data. Ecology 77:3-12.

Brawn, J. D. 1998. Effects of oak savanna restoration on avian populations and communities in Illinois. Final report, Illinois Natural History Survey, Champaign, IL.

DeGraff, R. M., and Wentworth, J. M. 1981. Urban bird communities and habitats in New England. Transactions of the North American Wildlife and Natural Resources Conference 46:396-413.

Fralish, J. S., F. B. Crooks, J. L. Chambers, and F. M. Harty. 1991. Comparison of presettlement, second-growth and old-growth forest on six site types in the Illinois Shawnee Hills. Amer. Midl. Nat. 125:294-309.

Havera, S. P., L. B. Sulloway, and J. E. Hoffman. 1997. Wetlands in the Midwest with special reference to Illinois, p.88-104. *In* M. W. Schwartz [ED.], Conservation in highly fragmented landscapes. Chapman and Hall, New York, NY.

Herkert, J. R. 1994. The effects of habitat fragmentation on midwestern grassland birds species. Ecol. Appl. 4:461-471.

Hutto, R. L. 1998. On the importance of stopover sites to migrating birds. Auk 115:823-825.

Ilyichev, V. D., V. M. Konstantinov, and B. M. Zvonov. 1986. The urbanized landscape and an arena for mutual relations between man and birds, p.123-130. *In* Urban ecological studies in Central and Eastern Europe. Polish Academy of Sciences, Warsaw, Poland.

Openlands Project. 1999. Under Pressure: Land Consumption in the Chicago Region, 1998-2028. Openland Project, Chicago, IL.

Marzluff, J. M., F. R. Gehlbach, and D. A. Manuwal. 1998. Urban environments: influences on avifauna and challenges for the avian conservationist, p.283-306. *In* J. M. Marzluff and R. Sallabanks [EDS.], Avian conservation. Island Press, Washington, DC.

Mlodinow, S. 1984. Chicago area birds. Chicago Audubon Society, Illinois.

Reichard, S., L. Chalker-Scott, and S. Buchanan. 2001. Interactions among non-native plants and birds, p. 179-223. *In* J. M. Marzluff, R. Bowman, and R. Donnelly [EDS.], Avian ecology and conservation in an urbanizing world. Kluwer Academic, Norwell, MA.

Robinson, S. K., F. R. Thompson, III, T. M. Donovan, D. R. Whitehead, and J. Faaborg. 1995. Regional forest fragmentation and the nesting success of migratory birds. Science 267:1987-1990.

Robinson, S. K., J. D. Brawn, and J. P. Hoover. 1997. Effectiveness of small preserves for breeding birds, p.154-188. *In* M. W. Schwartz [EDS.], Conservation in highly fragmented Landscapes. Chapman and Hall, New York, NY.

Schwartz, M. W. [ED.]. 1997. *Conservation in Highly Fragmented Landscapes.* Chapman and Hall, New York, NY.

Sullivan, J. 1998. An Atlas of Biodiversity. Chicago Wilderness, IL.

Suloway, L., and M. Hubbell. 1994. Wetland resources of Illinois. Illinois Natural History Survey, Special Publication 15. Champaign.

Taft, J. 1997. Savanna and open woodland communities, p.24-54. *In* M. W. Schwartz [EDS.], Conservation in highly fragmented landscapes. Chapman and Hall, New York.

VanDruff, L. W., E. G. Bolen, and G. J. San Julien. 1996. Management of urban wildlife, p.507-530. *In* T. A. Bookhout [EDS.], Research and management techniques for the management of wildlife, 5th edition. The Wildlife Society, Bethesda, MD.

Wang, Y., and D. K. Moskovits. In Press. Tracking fragmentation of natural communities and changes in land-cover: applications of Landsat data for conservation in Chicago. Wilderness. Cons. Biol.

Warner, R. 1994. Agricultural land-use and grassland habitat in Illinois: future shock for Midwestern birds? Cons. Biol. 8:147-156.

Chapter 25

Do temporal trends in Christmas Bird Counts reflect the spatial trends of urbanization in southwest Ohio?

Nancy A. Crosby and Robert B. Blair
Department of Biology, Miami University, Oxford, OH USA 45056, e-mail: crosbyna@muohio.edu

Key words: Christmas Bird Count, conservation, fragmentation, sprawl, urbanization

Abstract: We compared the temporal and spatial effects of urbanization on birds by examining the greater Cincinnati region Christmas Bird Count (CBC) data from 1960-1998, and contrasting it to the current winter distribution and abundance of birds across an urban gradient in Butler County, OH. As urban zones expand, the changes in the regional bird community through time should reflect changes similar to those seen along a gradient of urban land-use. Along the gradient, the pre-development bird species (assumed to be those found at the most undisturbed site) drop out gradually as the land-use becomes more urban. This local extinction is accompanied by the invasion of other species that exploit the changes brought about by urbanization. Consequently, we expected the CBC data to reflect a shift, over time, from predominantly native species to invasive species, as urbanization spread in the region. However, results from the CBC data did not reflect the patterns observed along the spatial urban gradient. CBC circles, because of their size (approx. 384 km^2) may contain enough fragmented habitat to support small populations of some native species, even near rapidly expanding urban centers such as the greater Cincinnati region. The nature of CBCs encourages coverage of as many habitat types as possible in order to observe the greatest number of species. Together, these factors produce data that do not accurately represent spatial population trends. Thus single CBC circles may be at a scale inappropriate for assessing the impacts of expanding human populations on native bird communities.

1. INTRODUCTION

Expansion of human populations and the concomitant development have increasingly altered the natural environment. Human land-use has resulted in fragmentation of natural areas and outright loss of habitat and a growing body of evidence exists concerning the impact of such changes on the distribution and abundance of animal and plant species (Ehrlich and Ehrlich 1981, Wilson 1988, Dale and Hill 1996). Urbanization is an extreme form of human modification of the landscape, with the least amount of original habitat remaining in urban centers. Changes in land-cover, in weather patterns, and accumulation of human wastes alter the distribution of organisms in urban areas (Jokimaki et al. 1996).

Urban sprawl has been defined as low-density development beyond the edge of city services and employment (Vojnovic 1999). Sprawl separates where people live from where they work, shop, and recreate; it increases both dependence on automobiles and the amount of land area devoted to supporting an automobile-dependent culture (Sierra Club 2000). Between 1960 and 1990, the size of metropolitan areas in the United States has more than doubled, while populations have grown by less than one-half (Benfield et al. 1999). Sprawl has changed the landscape in SW Ohio, earning Cincinnati, OH the designation as fourth most sprawl-threatened of large (those containing more than one million people) cities in the United States (Sierra Club 2000). The land area of metropolitan Cincinnati has spread out steadily from 856 km^2 in 1970 to 1310 km^2 in 1990, a 53% increase, while the human population grew less than 10% in the same period (Sierra Club 2000). Sprawl is contributing to lost open space and threatening the agricultural base that generates six billion dollars a year. The sprawl affecting Cincinnati also impacts the surrounding states of Kentucky and Indiana. The southeastern region of Indiana is the fastest growing in the state, a growth rate attributable to the expansion of Cincinnati.

Ohio was originally a densely forested region, with an estimated 95% forest cover in the 1700s (Trautman 1977). The southwest region of the state was dominated by beech-maple forest interspersed with prairie (Beissinger and Osborne 1982). But by 1950, cutting of forests for wood and human settlements had reduced forest coverage by 90% and construction of roads and facilities for human use spread the impact of growing human populations to the remaining natural habitats. In particular, construction of the Interstate-275 beltway around Cincinnati in the 1970s contributed to the geographical expansion of the greater Cincinnati metropolitan area and of human populations beyond the central urban zone.

Urbanization and its corresponding alteration and loss of habitat have effects on bird distribution and abundance. Studies have documented decreases in the numbers and diversity of some species (Robbins et al. 1989, Hagan 1993, Zebehazy and Rossell 1996), sometimes to the point of near extinction (Virkkala et al. 1993). Urbanization causes shifts in the dominant bird species of a community (Jokimaki et al. 1996). This occurs as non-native species adapted to living near humans are able to successfully invade or are introduced accidentally or deliberately by humans (Ehrlich et al. 1988). In a study of avian communities in southwest Ohio, Beissinger and Osborne (1982) found distinct differences between the communities in forested and in urban areas. Urban areas exhibited reduced species diversity and evenness, relative to the forest, and were dominated by a few "weedy" species.

Many studies of the effects of urbanization compare urban and natural areas, with little regard for the gradations that exist between them. To address this, Hedeen and Hedeen (1998) compared bird communities in urban, residential and forested habitats. They found that while the urban habitat showed the highest bird density, it was dominated by three or four non-native species. Composition of the avian communities changed across the habitats, with the urban area having the fewest of the native species found in the forest. Blair (1996 and 2001) studied bird communities along a gradient of habitats increasing in the amount of human-induced change, including a nature preserve, an open-space recreational area, golf course, residential neighborhood, apartment complex and business district. He found that while diversity peaked at the moderately disturbed sites, native species (assumed to be those found at the nature preserve) disappeared as the level of urbanization increased. He classified species according to the site at which they reached peak density. "Urban avoiders" were those species that reached maximum density at the nature preserve, while "urban exploiters" were those with maximum density in the business district. "Suburban adaptable" species were those that reached highest density at one of the sites in between, which exhibited varying levels of disturbance. This trend also has been observed in butterfly communities (Blair and Launer 1997).

Urbanization is a process, not a static condition. Consequently, urban gradients may have a temporal as well as a spatial dimension. As human populations increase through time, areas that once were pristine natural habitat become dominated by human structures. Therefore, temporal trends may reflect spatial trends observed along an urban gradient. If so, it may be possible to predict the level of future impact of expanding urban areas on avian communities by examining the current distribution and abundance of birds along a spatial gradient ranging from natural to urban. This ability to predict changes in distribution and abundance would prove useful in conservation and urban planning.

The Audubon Christmas Bird Count (CBC) provides one means of testing the idea that spatial trends in urban bird distribution and abundance reflect temporal trends. The CBC provides data on bird abundance over long periods of time in specific areas and consequently may be used to address this issue. The CBC is an annual event in which volunteer observers count and identify all birds they can find in a circle of 24-km (15 mile) diameter, on a single calendar day within two weeks of Christmas Day (Butcher 1990). CBC data have been used in a number of studies to address trends in bird populations (Atkinson 1995, Cade and Woods 1997, McCrimmon et al. 1997, Marzluff et al. 2001). Because of the nature of CBC data, there has been concern regarding its use in scientific research (Bock and Root 1981, Butcher et al. 1990). But a number of studies have found concordance in trends detected by CBCs and those detected using more standardized data sets (Pyle and Desante 1994, Viverette et al. 1996, Dunn and Sauer 1997). Drennan (1981) stated that the high levels of concern expressed regarding the non-random location of CBC count locations were unwarranted, as bird populations are not randomly distributed over a geographic range. CBC data can be made more reliable by applying simple corrective measures, such as accounting for the amount of observer effort (Peterson 1995), and are considered appropriate for use under a number of conditions (Butcher 1990). In particular, problems with non-standardized data are less likely to be significant for comparisons over long periods of time or across broad regions.

To address the relationship between temporal and spatial urban gradients, we used CBC data from the greater Cincinnati area to compare temporal trends in bird communities to the spatial patterns of winter bird communities along an urban gradient in nearby Butler County, Ohio. We asked whether 1) "urban exploiter" species tend to increase in time, reflecting their dominance at urban sites, as urbanization has increased in the greater Cincinnati area; 2) "urban avoider" species tend to decrease in abundance over time, reflecting their loss from more urban sites; and 3) there is a discernable pattern in the spatial and temporal trends of suburban adaptable species.

2. METHODS

2.1 Urban Gradient Study Sites

The six sites along the urban gradient were located within a 10 km radius of Miami University, Oxford, OH (39°30' N, 84°45' W, typical elevation

~290 m). The six sites all consisted of climax beech-maple forest prior to European settlement and include from least urban to most urban:

(1) a nature preserve -- Hueston Woods Nature Preserve, a 67-ha woodlot containing primarily mature beech-maple forest. Moderate-use dirt footpaths, several small parking areas and a narrow service area are maintained. The preserve is surrounded by a 975-ha state park which consists of young and mature deciduous forest, old fields, infrequent stands of conifers, and a 250-ha lake.

(2) a recreational area -- Peffer Memorial Park, an 80-ha recreational park, is owned by Miami University. Large portions of it are managed as a multiple-use trail system for residents of Oxford, OH and the surrounding community. Purchased between 1955 and 1966 as two separate parcels of pastureland and agricultural field, the vegetation currently is predominantly secondary growth of low stature.

(3) a golf course -- The 3,600-m Hueston Woods Golf Course was established in 1969 along the border of the nature preserve. During primary months of use (approximately March - November), the course accommodates 10,000 players. Rough consists mostly of native trees such as beech and maple, and grasses.

(4) a residential district -- This residential neighborhood is composed of detached, single-family houses built between 1960 and 1974. The houses are densely surrounded by lawns, and also frequently by gardens, shade trees, and native and ornamental flora. Lots range in size from approximately 1600 to 2500 square meters, with an average lot size of 1988 square meters.

(5) an area of apartment developments -- This consists of multi-level apartment complexes and intervening parking lots with interspersed lawns of small-scale landscaping and isolated trees. The development of apartment complexes occurred between 1960 and 1980. The building height, density, layout, parking, and plantings are comparable to that of an office park.

(6) a business district -- This site includes a 4 x 2 city block area that consists primarily of low-rise office buildings, parking lots, and a one-half block landscaped urban park. Paved roads range from access alleys to a well-traveled state highway. Initial construction of these properties began in the early to mid-1800s.

For each study site, we calculated the proportion of the area covered by buildings, pavement, lawn, grassland, and trees or shrubs within an approximately 50-m radius (depending on the particular scale of the map or photograph) of each survey point. We used ortho-digitally corrected land-use maps available from the City of Oxford and aerial photographs available from the Department of Geography at Miami University to these proportions. We used this information to rank the sites from most natural to most urban (for detailed metrics and photos see Blair 2001).

2.2 Urban Gradient Data Collection and Calculations

We visited each site on a total of 10 occasions in January and February of 1997, 1998, and 1999. Within each year, visits were spaced at least two to three days apart. We conducted bird surveys at 16 points within each site using the variable circular plots method as described in Reynolds et al. (1980). We calculated the density of each species using the area of a circle with radius equal to an estimated cutoff distance determined for each species in each site. To calculate these distances, we plotted a histogram of abundance of individuals in concentric bands of specified size. The distance at which observations show a decline of greater than 50% (without increases at further distances) was taken as the cutoff distance for accurate detection and any observations beyond these cutoff distances were eliminated from the data. We also eliminated from the analysis all species that were observed at less than 1% of the total observations within a site. We calculated average daily densities for each species at each site.

2.3 Greater Cincinnati Christmas Bird Count Data

The site we chose to represent the greater Cincinnati area was the Western Hamilton County circle of the Christmas Bird Count, on the western side of the metropolitan Cincinnati area (Fig. 25.1). Until 1977, this site was centered at $39^{0}14'$ N, $84^{0}38'$ W, at Springdale and Blue Rock Roads. In 1977, the center was changed to at $39^{0}12'$ N, $84^{0}38'$ W, centered at Gaines and Jessup Roads. The two points are 3.0 km apart, and both are approximately the same distance from the center of Cincinnati. Difference in landscape coverage of the CBC circle before and after the change was treated as negligible, due to the overall size of the circle relative to the distance moved and the positions of each center point relative to the urban center, interstate highways and other major roads. Elevation ranges between 146 and 274 meters (480 and 900 feet) for the 24-kilometer diameter circle. In a search of the available literature and government documents, we were unable to find quantified information regarding the change in land-use in the western Hamilton County area for the time period covered. However, the Hamilton County Regional Planning Commission (2000) reported that land devoted to residential use has expanded while land devoted to agriculture has decreased steadily since 1940. In 1999, county land devoted to residential use accounted for 39% of the total land-use, while commercial/industrial/institutional uses accounted for 26.4%. Agriculture accounted for 10.3% and only 3.7% was devoted to parks and recreational areas.

We obtained bird counts and observer party information for years 1961-1998 from the Christmas Bird Count (Audubon Society 2000) and from the National Audubon Society Field Notes (1961-1998), which contain the annual results of all Christmas Bird Counts. Corrections to count numbers were made using observer party hour information by dividing number of birds observed by number of party hours spent observing, after Peterson (1995). We performed linear regressions for each species over the specified time period using StatView v.4.5. We then compared results to trends along the spatial urban gradient.

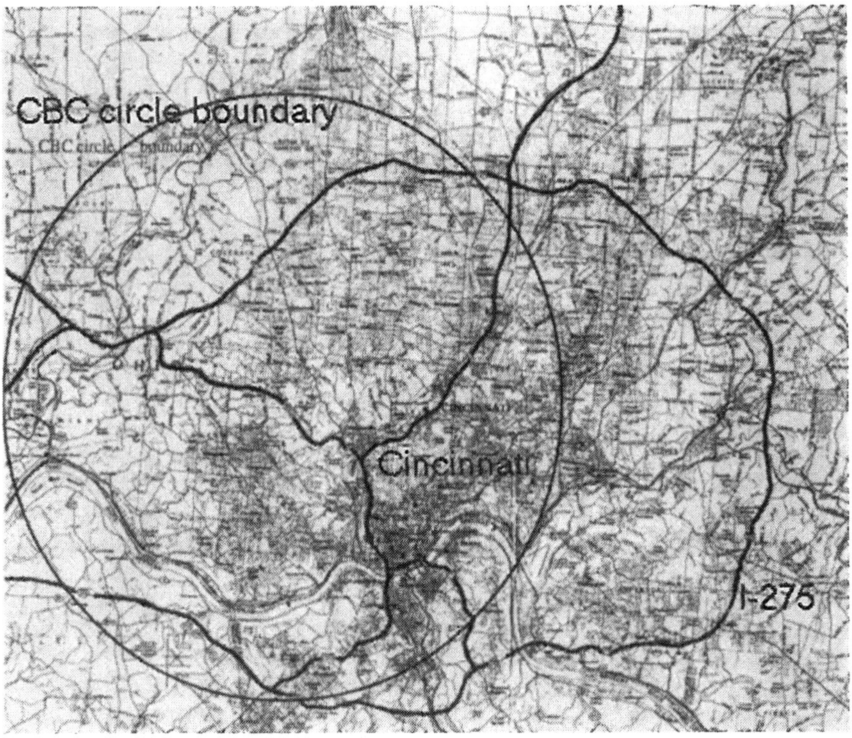

Figure 25.1. Location of western Hamilton County Christmas Bird Count circle. Prior to construction of the Interstate-275 beltway, there was little development beyond the urban core. Since construction in the 1970s, development has proceeded out to and beyond the beltway.

3. RESULTS

We observed 21 species across the urban gradient in Butler County. The densities of all species varied across the gradient to some extent (Appendix 25.1). Distributions across the gradient sites also varied, with two species (American Robins and Northern Cardinals) appearing at all six sites of the gradient, and five appearing at only one site of the gradient (Northern Flicker, Pileated Woodpecker, Golden-crowned Kinglet, Eastern Bluebird, and Song Sparrow).

Two "urban exploiters" reached maximum density in the business district: European Starling and House Sparrow. Three more reached maximum density in the apartment complex, with either no appearance at any of the less developed sites (Rock Dove and Song Sparrow) or significantly lower density at the less developed sites (House Finch) (Fig. 25.2).

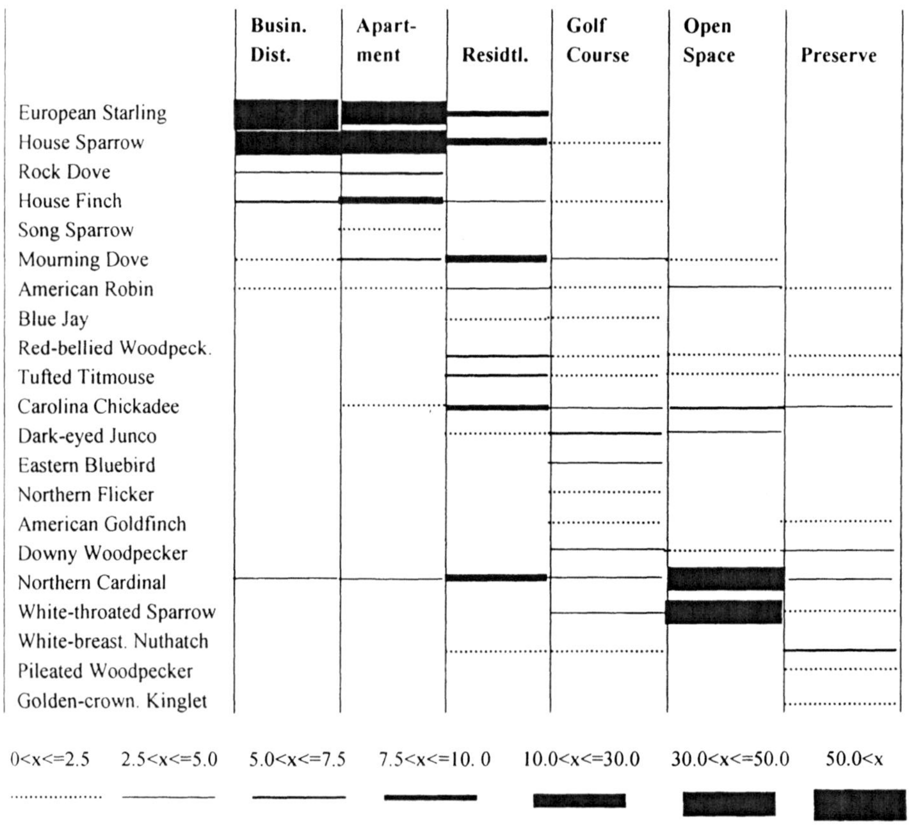

Figure 25.2. Ranking of 21 species observed at Oxford, OH gradient, based on site of maximum density. This ranking illustrates the pattern of distribution of species along the urban gradient. Thickness of each line represents density (in birds ha^{-1}).

Comparing these spatial patterns to the temporal data from the CBCs, four of these five urban exploiting species showed significant increases over time in numbers observed per party hour (European Starling, $P < 0.0001$; Rock Dove, $P < 0.0001$; House Finch, $P < 0.0001$; Song Sparrow, $P = 0.0003$), while the other (House Sparrow) exhibited no significant trend ($P = 0.7404$) (Table 25.1).Three of "urban avoiders" reached maximum density at the preserve (Pileated Woodpecker, White-breasted Nuthatch, and Golden-crowned Kinglet). Two more reached maximum density at the open-space recreational area (White-throated Sparrow and Northern Cardinal). The White-throated Sparrow was found only at the three least developed sites (Fig. 25.2).

In contrast to these spatial trends, the temporal data from the CBCs showed that four of the urban avoiding species significantly increased in numbers of birds observed per party hour (Golden-crowned Kinglet, $P = 0.0098$; Pileated Woodpecker and White-breasted Nuthatch, $P <= 0.001$;White-throated Sparrow, $P < 0.0001$). The remaining urban avoider, Northern Cardinal, exhibited no significant trend ($P = 0.2042$) (Table 25.1)

Eleven species were characterized as "suburban adaptable," with maximum density at the residential neighborhood (Mourning Dove, Red-bellied Woodpecker, Blue Jay, Tufted Titmouse, Carolina Chickadee and American Robin), or the golf course (Northern Flicker, Downy Woodpecker, Eastern Bluebird, Dark-eyed Junco, and American Goldfinch). (Fig. 25.2).

In comparison, four of the suburban adaptable species showed significant increases over time (Red-bellied Woodpecker; $P = 0.0006$; Eastern Bluebird, $P = 0.0458$; American Robin, $P = 0.0012$; American Goldfinch, $P = 0.0036$). Five exhibited no significant temporal trend (Mourning Dove, $P = 0.0658$; Northern Flicker, $P = 0.07824$; Downy Woodpecker, $P = 0.2591$; Blue Jay, $P = 0.8347$; Tufted Titmouse, $P = 0.2287$; Carolina Chickadee, $P = 0.6264$), and one declined significantly over time (Dark-eyed Junco, $P < 0.0001$) (Table 25.1).

4. DISCUSSION

The temporal trends represented by the CBC data did not necessarily reflect the spatial trends found for winter birds along the urban gradient. Urban exploiting species did show increases in time, as urbanization has expanded in the greater Cincinnati area. Four of the five species peaking at the two most urban sites in the spatial gradient showed significant temporal increases. In the one case of decline (House Sparrow), the explanation may lie, at least in part, with one of the other urban species, the House Finch. Ehrlich et al. (1988) noted that in some places where their ranges overlap, competition between House Finches and House Sparrows has resulted in declines in House Sparrows. This may be the case in southwest Ohio because CBC data show declines in House Sparrows beginning in the few years following the first appearances and subsequent increases of House Finches in the area. Consequently, competition may be influencing the temporal trend observed here. Thus, we conclude the temporal trends observed for most of these urban adapted species did mirror their distribution patterns along the spatial urban gradient.

Table 25.1. Results of regressions of number of individuals seen per party hour based on CBC data from western Hamilton county 1961-1998. Listing of species is in the order as they appear in Figure 25.2.

Species	Slope	Intercept	R^2	P
Urban Exploiters				
European Starling	8.856	-17450.844	0.401	<0.0001
House Sparrow	-0.058	133.458	0.004	0.7404
Rock Dove	0.260	-512.799	0.729	<0.0001
House Finch	0.238	-469.905	0.504	<0.0001
Song Sparrow	0.181	-353.825	0.342	0.0003
Suburban Adaptable				
Mourning Dove	0.128	-247.104	0.105	0.0658
American Robin	0.898	-1769.986	0.291	0.0012
Blue Jay	0.003	-3.656	0.001	0.8347
Red-bellied Woodpecker	0.017	-32.908	0.328	0.0006
Tufted Titmouse	-0.024	50.177	0.041	0.2287
Carolina Chickadee	0.133	-252.948	0.008	0.6264
Dark-eyed Junco	-0.337	674.978	0.511	<0.0001
Eastern Bluebird	0.013	-25.999	0.127	0.0458
Northern Flicker	-0.001	2.687	0.002	0.7824
American Goldfinch	0.074	-144.059	0.247	0.0036
Downy Woodpecker	-0.009	19.936	0.041	0.2591
Urban Avoiders				
Northern Cardinal	0.066	-121.655	0.051	0.2042
White-throated Sparrow	0.175	-343.89	0.631	<0.0001
White-breasted Nuthatch	0.019	-37.368	0.65	<0.0001
Pileated Woodpecker	0.004	-8.457	0.665	<0.0001
Golden-crowned Kinglet	0.012	-24.253	0.196	0.0098

In contrast to the predictions generated from their spatial patterns, urban avoider species did not decrease significantly over the time period considered. In fact, four of the five species that peaked at the two least urban sites showed significant increases in abundance, despite the increasing urbanization of the study area. None of the urban avoiders decreased significantly in the time period covered. For urban avoiding species, temporal trends did not mirror their distribution patterns along the spatial urban gradient.

There were no clear patterns in the temporal changes of the suburban adaptable species, with four species showing significant increases, one a significant decrease, and five showing no significant temporal trend.

With reference to our original hypotheses, our results show that 1) exotic urban avoiders increased in time, reflecting their dominance at urban sites along the spatial gradient; 2) native species did not necessarily decrease in time and thus did not reflect their loss from more urban sites across the spatial gradient; and 3) no clear pattern could be discerned in the temporal trends of suburban adaptable species.

A number of factors could explain the lack of congruence between temporal and spatial trends for urban avoiding species. Bird feeding by humans has increased significantly since the 1950s and has resulted in range expansion for a number of species (Jokimaki et al. 1996). Due to the availability of this artificial food supply during winter, winter populations of many of these species may have increased.

Dunn (1995) found that feeder watching as part of the Christmas Bird Count could introduce a bias into the data collected. This effect may be greater in northern regions than elsewhere, because of the harshness of northern winters and the increased importance to birds of winter food in feeders. CBC totals have shown inflation of 1 to 7% in the estimates of species that commonly visit feeders (Dunn 1995). Urban avoiders or suburban adapted species in this analysis that are known to commonly visit feeders are White-breasted Nuthatch, Downy Woodpecker, American Goldfinch, Northern Cardinal, Northern Flicker, Dark-eyed Junco, Carolina Chickadee, Tufted Titmouse, Red-bellied Woodpecker and Blue Jay. Therefore, the trends observed for these species might reflect a bias in CBC data related to the method by which feeder observations and effort are included in the data. Trends may be artificially inflated, masking actual declines of some of these species. Berry (1992) noted, in an analysis of CBC data from one location, that some increasing trends have been influenced by increases in observer effort at feeders as well as by increasing inclusion of feeder-watching data in CBCs, producing erroneous results. He also noted that because of this potential inflation any declines that are observed should be of special concern, since the actual decline may be even more precipitous. This is of particular concern for the Dark-eyed Junco, which exhibited a significant decline even with the use of potentially inflated data.

The Christmas Bird Count originally began as a recreational event in response to an earlier Christmas Day tradition of hunting and killing songbirds (Dunn and Sauer 1997). At that time, contests were held to amass the largest number of bird carcasses. While the killing tradition has ceased, an aspect of the "trophy hunting" remains. Participants in CBCs generally endeavor to count as many species as possible, to obtain the largest species total for their circle, in competition both with other circle locations, as well as their own from one year to the next. This can bias coverage of the circle landscape, with participants seeking as many different habitat types as possible to increase their likelihood of seeing different species, regardless of the percent of land-cover represented by any given habitat type. This "trophy" aspect of the CBC counts are not necessarily representative of the location, and thus will influence trends estimated from the data.

Another possible factor in the failure to detect declines in the region relates to the dynamics of population change. Hayward and Iverson (1998) noted that often time lags exist between demographic patterns that work to reduce population sizes, and the actual numbers of birds observed on a survey. Tilman et al. (1994) noted that habitat fragmentation can result in species extinction generations after the original fragmentation occurred.

A further observation can be noted by looking not simply at overall trends, but at absolute numbers of birds observed in the CBC data. While urban exploiters and urban avoiders may both show increasing trends, the overall abundances of each species are quite different. The urban avoider species are consistently observed at much lower numbers than urban exploiters. Due to this, and the uncertainty regarding artificial inflation in CBC data, the impact of future development will be correspondingly larger on these less abundant species.

Perhaps the most significant factors in explaining the temporal trends observed relate to the size of a CBC circle and the level of urbanization in the area. CBC circles are large (380 km^2 [177 mi^2]) and it is unlikely that an area of this size will have become completely urbanized. The position of CBC circles also may influence the results. The western Hamilton County circle includes only portions of the central, most heavily urbanized area of Cincinnati, and hence data from this specific location may not be entirely representative of patterns that would be observed in a CBC circle centered on the urban core.

Because of its location relative to the metropolitan Cincinnati area and I-275, western Hamilton County has been undergoing significant development. But land-use in the CBC circle has not been systematically characterized. The level of development in some parts of the circle, particularly those furthest removed from the urban center, may not yet have reached some critical point at which native species drop out. The topography of the Cincinnati metropolitan area prevents development out from the central urban core in smooth, continuous rings. The terrain, with sometimes steep changes in elevation precludes development in some areas. As a result, some isolated patches of native habitat remain, even as urbanization expands outward. Enough minimally sufficient natural habitat may remain to support some populations of native species. Further, as urbanization proceeds and natural habitat becomes increasingly fragmented, some remaining native species may become temporarily concentrated in the isolated patches of habitat that do remain (Bierregaard et al. 1992, Hagan et al. 1996). This may actually increase the likelihood of sighting individuals and thereby inflate counts, even while overall numbers are declining.

Even without data quantifying land-use change for the time period involved, these two factors -- size and level of development within a CBC circle -- suggest that CBC circles may be an inappropriate spatial scale for assessing temporal trends of avian species in urbanizing areas. The results here suggest that the change in urbanization along the temporal gradient is not comparable to range of urbanization observed along the spatial gradient. The problem is exacerbated by a lack of control over the positioning of CBC circles, although this could be overcome to some extent where multiple circles in the same area exist, by combining data from the multiple data sets. Even in areas of rapid development, the change in land-use across the temporal gradient is often less than that over the spatial gradient, on the scale of a CBC circle.

4.1 Policy and Management Implications

Single CBC circles may not be the best scale for assessing the impact of temporal changes in patterns of urbanization on winter bird populations, but CBC data can still be useful in detecting trends and these trends should be further investigated. Additionally, where available, data from multiple circles should be combined, even where some overlap exists. This could provide a more representative view of on-going trends by providing more thorough landscape coverage. And while CBC data may not show temporal trends comparable to trends observed along spatial gradients, they may be useful for examining trends at specific areas along the spatial urbanization gradient. For example, CBC data may be useful for examining population trends associated with the conversion of natural habitat to moderately disturbed habitat associated with residential development. CBC data also may be useful as warning flags, as urban growth becomes widespread enough to affect native bird species.

Further improvements in methods for standardizing CBC data may make them more reliable for detecting such trends (Butcher and McCulloch 1990). Changes in the recording of feeder-watcher bird counts and effort in CBC data also may make it possible to eliminate potential biases.

These results further make apparent the need for studies quantifying land-use practices to reliably assess how expanding development impacts avian species. It would be particularly useful to collect this information for CBC circles, to reliably determine the scale of urbanization within them and allow for more accurate assessments of trends in avian populations. Even in areas where urbanization is occurring rapidly there may be remnant habitat and as a result, trends for the entire CBC circle could be skewed. Organizers of future CBC events may consider the feasibility of including data on location of sightings within the circle.

CBC data do indicate that temporal trends of urban adapted species, the species often considered to be weedy, undesirable species, do reflect their distribution along the spatial urban gradient. However, this does not appear to be true for urban avoiding species, that decline along a spatial urban gradient. The lack of concordance between temporal and spatial trends for these species should not justify elimination or reduction in the level of concern about the effects of urbanization on these species or avian communities in general. Determining the most suitable spatial scale for temporal analyses will improve our ability to estimate population trends with increasing urbanization (Miller et al. 2001, Hostetler 2001). Additional research is needed to provide data about temporal patterns of land-use change. Careful analysis of temporal patterns in land-use and changes in avian community structure and composition will ultimately provide a useful tool in urban planning and land-use impact studies. Given the limitations discussed, CBC data can be useful to policy makers, planners and residents interested in initiatives for controlling sprawl, in designing residential developments that minimize the loss of natural habitat, and helping to maintain native avian species. Urbanization will continue in the greater Cincinnati area and elsewhere but more people are becoming aware of the problems it causes for human and avian populations alike. Any analysis that has the potential to be useful in addressing these concerns is worth pursuing.

REFERENCES

Atkinson, E. C. 1995. Northern Shrikes (*Lanius excubitor*) wintering in North America: a Christmas Bird Count analysis of trends, cycles, and interspecific interactions. Proc. West. Found. Vert. Zoo. 6:39-44.

Beissinger, S. R., and D. R. Osborne. 1982. Effects of urbanization on avian community organization. Condor 84:75-83.
Benfield, F. K., M.D. Raimi, and D.D.T. Chen. 1999. Once there were greenfields: How urban sprawl is undermining America's environment, economy and social fabric. National Resources Defense Council, New York.
Berry, J. 1992. A selective analysis of data from the Newburyport Christmas Bird Count. Bird Observ. 20:314-323.
Bierregaard, R. O., T. E. Lovejoy, V. Kapos, A. A. dos Santos, and R. W. Hutchings. 1992. The biological dynamics of tropical rainforest fragments. BioScience 42:859-866.
Blair, R. B. 1996. Land use and avian species diversity along an urban gradient. Ecol. Appl. 6:506-519.
Blair, R. B. 2001. Creating a homogeneous avifauna, p. 461-488. *In* J. M. Marzluff, R. Bowman, and R. Donnelly [EDS.], Avian ecology and conservation in an urbanizing world. Kluwer Academic Norwell, MA.
Blair, R. B., and A. E. Launer. 1997. Butterfly diversity and human land use: species assemblages along an urban gradient. Biol. Conserv. 80:113-125.
Bock, C. E., and T. L. Root. 1981. The Christmas Bird Count and avian ecology. Stud. Avian Biol. 6:17-23.
Butcher, G. S. 1990. Audubon Christmas Bird Counts, p. 5-13. *In* J.R. Sauer and S.W. Droege (EDS.), Survey designs and statistical methods for the estimation of avian population trends. U. S. Fish and Wildlife Service Biological Report 90.
Butcher, G. S., and C. E. McCulloch. 1990. Influence of observer effort on the number of individual birds recorded on Christmas Bird Counts, p. 120-129. *In* J.R. Sauer and S.W. Droege (EDS.), Survey designs and statistical methods for the estimation of avian population trends. U. S. Fish and Wildlife Service Biological Report 90.
Butcher, G. S., M. R. Fuller, L. S. McAllister, and P. H. Geissler. 1990. An evaluation of the Christmas Bird Count for monitoring population trends of selected species. Wild. Soc. Bull. 18:129-134.
Cade, T. J., and C. P. Woods. 1997. Changes in distribution and abundance of the Loggerhead Shrike. Conserv. Biol. 11:21-31.
Cornell Laboratory of Ornithology. 1999. National Audubon Society's Christmas Bird Count. [Online]. Available: http://birdsource.tc.cornell.edu/cbcdata. [28 Sept 1999]
Dale, A., and S. B. Hill. 1996. Biodiversity conservation: a decision-making context, p. 97-118. *In* A. Dale and J. B. Robinson (EDS.), Achieving sustainable development. UBC Press, Vancouver.
Drennan, S. R. 1981. The Christmas Bird Count: an overlooked and underused sample. Stud. Avian Biol. 6:24-29.
Dunn, E. H. 1995. Bias in Christmas Bird Counts for species that visit feeders. Wilson Bull. 107:122-130.
Dunn, E. H., and J. R. Sauer. 1997. Monitoring Canadian bird populations with winter counts, p. 49-55. *In* E. H. Dunn, M. D. Cadman, and J. B. Falls (EDS.), Monitoring bird populations: the Canadian experience. Can. Wildl. Serv. Occasional Paper No. 95.
Ehrlich, P. R., D. S. Bodkin, and D. Wheye. 1988. The birder's handbook: a field guide to the natural history of North American birds. Simon and Schuster, New York.
Ehrlich, P., and A. Ehrlich. 1981. Extinction: the causes and consequences of the disappearance of species. Random House, New York.
Hagan, J. M. III. 1993. Decline of the rufous-sided towhee in the eastern United States. Auk. 110:863-874.
Hagan, J. M., W. M. Vader Haegen, and P. S. McKinley. 1996. The early development of forest fragmentation effects on birds. Conserv. Biol. 10:188-202.

Hamilton County Regional Planning Commission. 2000. Land use distribution by political jurisdiction, p. 14-16. Decision Support Manual Vol. 5. Cincinnati, OH.

Hayward, G. D., and G. C. Iverson. 1998. Long-term trend in Marbled Murrelets in southeast Alaska based on Christmas Bird Counts. Northwest Science. 72:170-179.

Hedeen D. L., and S. E. Hedeen. 1998. Winter-bird communities in a western mesophytic forest and two proximate urban habitats. J. Ky. Acad. Sci. 59:168-173.

Hostetler, M. 2001. The importance of multi-scale analyses in avian habitat selection studies in urban environments, p. 139-154. *In* J. M. Marzluff, R. Bowman, and R. Donnelly [EDS.], Avian ecology and conservation in an urbanizing world. Kluwer Academic, Norwell, MA.

Jokimaki J., J. Suhonen, K. Inki, and S. Jokinen. 1996. Biogeographical comparison of winter bird assemblages in urban environments in Finland. J. Biogeogr. 23:379-386.

Marzluff, J. M., K. J. McGowan, R. Donnelly, and R. L. Knight. 2001. Causes and consequences of expanding American Crow populations, p. 333-365. *In* J. M. Marzluff. R. Bowman, and R. Donnelly [EDS.], Avian ecology and conservation in an urbanizing world. Kluwer Academic, Norwell, MA.

McCrimmon, D. A. Jr., S. T. Fryska, J. C. Ogden, and G. S. Butcher. 1997. Nonlinear population dynamics of six species of Florida Ciconiiformes assessed by Christmas Bird Counts. Ecol. Appl. 7:581-592.

Miller, J. R., J. M. Fraterrigo, N. T. Hobbs, D. M. Theobald, and J. A. Wiens. 2001. Urbanization, avian communities, and landscape ecology, p. 117-137. *In* J. M. Marzluff, R. Bowman, and R. Donnelly [EDS.], Avian ecology and conservation in an urbanizing world. Kluwer Academic, Norwell, MA.

Peterson, A. P. 1995. Erroneous party-hour data and a proposed method of correcting observer effort in Christmas Bird Counts. J. Field Ornithol. 66:385-390.

Pyle, P., and D. F. Desante. 1994. Trends in waterbirds and raptors at Southeast Farallon Island, California, 1974-1993. Bird Populations 2:33-43.

Reynolds, R. T., J. M. Scott, and R. A. Nussbaum. 1980. A variable circular-plot method for estimating bird numbers. Condor 82:309-313.

Robbins C. S., J. R. Sauer, R. S. Greenberg, and S. Droege. 1989. Population declines in North American birds that migrate to the Neotropics. Proc. Nat. Acad. Sci. USA 86:7658-7662.

The Sierra Club. 2000. Sierra Club Sprawl Report; Sprawl Report. [Online]. Available http://www.sierraclub.org/sprawl/report98/cincinnati.html [1 Aug 2000].

Tilman, D., R. M. May, C. L. Lehman, and M. A. Nowak. 1994. Habitat destruction and the extinction debt. Nature. 371:65-66.

Trautman, M. B. 1977. The Ohio country from 1750 to 1977: a naturalist's view. Ohio Biological Survey Biological Notes No. 10. Columbus, OH.

Virkkala, R., T. Alanko, T. Laine, and J. Tiainen. 1993. Population contraction of the White-backed Woodpecker, *Dendrocopos leucotos*, in Finland as a consequence of habitat alteration. Biol. Conserv. 66:47-53.

Viverette, C. B., S. Struve, L. J. Goodrich, and K. L. Bildstein. 1996. Decreases in migrating Sharp-shinned Hawks (*Accipter striatus*) at traditional raptor-migration watch sites in eastern North America. Auk. 113:32-40.

Vojnovic, I. 1999. The environmental costs of modernism: an assessment of Canadian cities. Cities 16:301-315.

Wilson, E. O. 1988. The current state of biological diversity, p. 3-18. *In* E. O. Wilson and F. M. Peters (EDS.), Biodiversity. National Academy Press, Washington, D.C.

Zebehazy, L. A., and C. R. Rossell Jr. 1996. Winter bird use of Eastern White Pine (*Pinus strobus*) stands with different structural characteristics. J. Elisha Mitchell Scientif. Soc. 112:94-97.

Appendix 25.1. The mean daily density (MDD; birds/ha and SE) of 21 species of birds observed in January and February 1997-1999, at six sites along an urban gradient in SW Ohio.

	Business Dist.		Apartments		Residential		Golf Course		Open Space		Preserve	
	MDD	SE	MDD	S E	MDD	SE	MDD	SE	MDD	SE	MDD	S.E
Rock Dove (*Columba livia*)	3.18	1.70	5.60	1.41	0.00	0.00	0.00	0.00	0.00	0.00	0.00	0.00
Mourning Dove (*Zenaida macroura*)	0.60	0.60	5.60	2.07	22.64	7.08	3.69	0.86	1.39	0.52	0.00	0.00
Red-bellied Woodpecker (*Centurus carolinus*)	0.00	0.00	0.00	0.00	3.06	2.04	0.76	0.34	0.64	0.28	0.80	0.34
Northern Flicker (*Colaptes auratus*)	0.00	0.00	0.00	0.00	0.00	0.00	1.66	1.11	0.00	0.00	0.00	0.00
Downy Woodpecker (*Picoides pubescens*)	0.00	0.00	0.00	0.00	0.00	0.00	3.89	1.78	1.41	0.78	3.18	0.51
Pileated Woodpecker (*Dryocopus pileatus*)	0.00	0.00	0.00	0.00	0.00	0.00	0.00	0.00	0.00	0.00	1.79	0.46
Blue Jay (*Cyanocitta cristata*)	0.00	0.00	0.00	0.00	1.40	0.52	0.38	0.19	0.00	0.00	0.00	0.00
Tufted Titmouse (*Parus bicolor*)	0.00	0.00	0.00	0.00	3.06	0.93	1.15	0.61	2.42	0.86	1.40	0.67
Carolina Chickadee (*Parus carolinensis*)	0.00	0.00	0.13	0.13	9.55	2.50	3.06	0.54	5.09	0.12	3.06	0.79
White-breasted Nuthatch (*Sitta carolinensis*)	0.00	0.00	0.00	0.00	0.76	0.34	2.29	0.42	0.00	0.00	7.00	0.79
Golden-crowned Kinglet (*Regulus satrapa*)	0.00	0.00	0.00	0.00	0.00	0.00	0.00	0.00	0.00	0.00	0.80	0.61
Eastern Bluebird (*Sialia sialis*)	0.00	0.00	0.00	0.00	0.00	0.00	3.06	2.54	0.00	0.00	0.00	0.00
American Robin (*Turdus migratorius*)	0.76	0.28	1.39	0.52	4.84	2.09	0.64	0.51	4.07	0.17	0.38	0.38
European Starling (*Sturnus vulgaris*)	76.14	7.20	36.92	8.28	9.42	2.13	0.00	0.00	0.00	0.00	0.00	0.00

	Business Dist.		Apartments		Residential		Golf Course		Open Space		Preserve	
	MDD	SE	MDD	S E	MDD	SE	MDD	SE	MDD	SE	MDD	S.E
Song Sparrow (*Melospiza melodia*)	0.00	0.00	0.99	0.80	0.00	0.00	0.00	0.00	0.00	0.00	0.00	0.00
White-throated Sparrow (*Zonotrichia albicollis*)	0.00	0.00	0.00	0.00	0.00	0.00	4.60	3.90	43.80	5.89	0.64	0.64
Dark-eyed Junco (*Junco hyemalis*)	0.00	0.00	0.00	0.00	2.48	1.50	5.09	4.08	3.06	0.14	0.00	0.00
Northern Cardinal (*Cardinalis cardinalis*)	2.93	0.47	4.71	0.81	12.22	1.38	4.07	0.73	38.55	4.20	2.67	0.55
House Finch (*Carpodacus mexicanus*)	5.66	3.26	16.27	7.37	3.69	1.50	1.15	0.82	0.00	0.00	0.00	0.00
American Goldfinch (*Carduelis tristis*)	0.00	0.00	0.00	0.00	0.00	0.00	0.64	0.34	0.00	0.00	0.60	0.43
House Sparrow (*Passer domesticus*)	41.89	6.34	40.36	6.22	11.84	2.00	0.64	0.28	0.00	0.00	0.00	0.00

Chapter 26

Survey techniques and habitat relationships of breeding birds in residential areas of Toronto, Canada

Jean-Pierre L. Savard[1] and J. Bruce Falls[2]

[1] *Canadian Wildlife Service, 1141 route de l'Église, P.O. Box 10100, Sainte-Foy, QUE Canada G1V 4H5, e-mail: Jean-Pierre.Savard@ec.gc.ca*

[2] *Department of Zoology, University of Toronto, Toronto, ONT Canada M5S 3G5*

Key words: American Robin, European Starling, habitat selection, House Sparrow, residential areas, survey techniques, urban birds

Abstract: We measured bird abundance and diversity in seven types of residential areas in Toronto, Canada, during the breeding season. We used transects and a combination of nest searching and spot-mapping. Transects underestimated the size for the breeding population for most species. Relationships between the two methods varied between species and residential areas indicating that for several species relative abundance was not directly proportional to breeding density. Several bird species responded similarly to habitat features. Rock Dove (*Columbia livia*), European Starling (*Sturnus vulgaris*) and Chimney Swift (*Chaetura pelagica*) were positively associated with house density and negatively with lawn area. House Sparrow was somewhat different, possibly because they built nests in conifers whenever cavities were absent thus they may not be so dependent on buildings. Another group, including Mourning Dove (*Zenaida macroura*), American Robin (*Turdus migratorius*), Common Grackle (*Quiscalus quiscula*) and Northern Cardinal (*Cardinalis cardinalis*), all were positively associated with low foliage and lawn area and negatively associated with house density. Other native species (Blue Jay [*Cyanocitta cristata*], Chipping Sparrow [*Spizella passerina*], Song Sparrow [*Melospiza melodia*] and Brown-headed Cowbird [*Molothrus ater*] shared the same habitat associations as the latter group to a greater or lesser extent. Conifers and shrubs appear to be the two key components linked to species diversity in temperate cities. Our results highlight the duality of the urban bird community structure: 1) the presence of typical urban birds (House Sparrow [*Passer domesticus*], European Starling, Rock Dove and Chimney Swift) that respond to architectural features first and to natural features second, and 2) the presence of native birds associated with remnant natural features, especially vegetation structure and volume.

1. INTRODUCTION

The last few decades have witnessed an extensive growth in the number and size of cities to the detriment of natural areas worldwide. More and more people now live in cities. It is increasingly common that most human-wildlife contact occurs within city limits. Hence, the importance of understanding the ecological requirements of urban wildlife (Savard et al. 2000). The explosion of suburban developments and the creation of urban parks reflect human desires for open space and wildlife.

Urban areas constitute unique and highly dynamic ecosystems that support a variety of plants and wildlife (Gilbert 1989, Adams 1994). These are organized along various gradients of urbanization from wildlands (natural areas) to town centers (Natuhara and Imai 1996). In general, as the level of urbanization increases, species richness decreases (Batten 1972, Bezzel 1985, Mills et al. 1989) and this is especially true when urbanization replaces natural, productive and heterogeneous habitats. Urban ecosystems are quite dynamic, experiencing species addition and removal through time (Morneau et al. 1999). Within a given town, residential neighborhoods vary in terms of architectural features, building density, and vegetation composition and volume, and support a variety of bird species (Hadidian et al. 1997, Clergeau et al.1998). These neighborhoods should be the focal point of urban wildlife management aimed at increasing positive interactions between people and wildlife. Therefore, it is important to have an understanding of habitat-wildlife relationships in urban residential areas. Estimation of bird abundance is a key component of this understanding and it is important to understand the strengths, weaknesses, and biases of the techniques used.

We examined some of these relationships in Toronto, Canada, a city large variation in the composition of residential neighborhoods (see Fig. 26.1 for a representative image of one study area). Our study focused on three main objectives: first to compare estimates of breeding bird abundance from transects with those derived from a combination of nest searching and the spot-mapping methods in residential areas; second, to determine the breeding densities of birds in seven residential areas; and third, to identify habitat features responsible for variation in breeding densities among these neighborhoods.

2. METHODS

2.1 Study Area

Seven residential areas were selected for study (Table 26.1), representing different degrees of urbanization (Table 26.2). In each residential area, we established a large, approximately square plot in which we conducted our bird survey transects and habitat analyses. Plots ranged in size from 13-17 ha (Table 26.1) and included a mixture of developed and undeveloped areas that was representative of the larger neighborhood/landscape. We used aerial photographs and city maps to select plots surrounded by similar degrees of urbanization. All areas were at least several kilometers from the edge of town. The landscape surrounding all the selected areas can be classified suburban, as building density always exceeded five units per ha. Buildings were typically single family dwellings and vegetation covered 30-60% of the ground (Table 26.2).

Landsdowne was the most urban plot with nearly 70% of the ground covered by buildings and streets. It had dense semi-detached two-story dwellings, a few tall trees, small lawns (<0.02 ha), few shrubs and few conifers. Rosemary and Rosedale had large two-story houses and the streets were lined with tall deciduous trees shading the streets. Lawns were <0.03 ha. Etobicoke and O'Connor were characterized by a high density of bungalows (one-story houses) with small trees which did not shade the streets. They had medium size lawns (0.02 - 0.04 ha) and shrubs were abundant. Don Mills had larger bungalows and larger lawns (>0.04 ha) than Etobicoke and O'Connor. It had numerous shrubs and medium and small sized trees that did not shade the streets. Bayview was the least urbanized of the residential areas. It had scattered large houses, a few tall trees, but mostly medium and small trees. It had large lawns (>0.09 ha) and numerous shrubs and conifers.

2.2 Census Techniques

Between May and June 1975, we used three methods concurrently to assess densities of breeding birds in the seven residential areas. First, an intensive nest search was carried out in each plot to estimate densities of non-territorial species such as the House Sparrow, European Starling, Rock Dove and Common Grackle. Nest searches occurred throughout the day. Second, nest searching was used in combination with the spot-mapping method (Bibby et al. 1992) to survey territorial species. Each census lasted a minimum of three hours, during which the location and activity of each bird encountered were recorded on a map of the area and an effort was made to locate nests. Each foraging bird was observed to determine if it was gathering nest material or food for its young. In these cases, the bird was followed until it returned to its nest. Most found nests were located in this way. Third, a single transect was located along the streets within the boundaries of each plot. Each transect was centered in the middle of the streets and divided into 92 x 92 m sections (Table 26.1). This width is approximately equal to the width of the street plus the depth of the lots on either side of the street and was similar to the size used by Geis (1974) and

Figure 26.1. Spin-2 satellite image of a residential portion of Toronto, Canada, including the study neighborhood Etobicoke. The image encompasses approximately 21 km^2 and was taken in 1989.

Lucid (1974) in urban areas. The observer walked in the center of the street but did not attempt to visually cover every backyard. We recorded all birds seen or heard within each section for a four-minute period. Three observers were involved in nest searching and spot mapping activities and their efforts were distributed equally among plots. We believed the combination of nest searching and spot-mapping method provided accurate estimates of the size of the breeding bird population, especially in the suburban environment where nests may be easier to locate than in natural areas. Two of the three observers also were involved in transect surveys and again their efforts were distributed equally among plots. Transects provided estimates for non-territorial species whose nests were difficult to locate and for non-breeding species using the area.

2.3 Measurement Of Habitat Features

Natural habitat features were measured in July of the same year, after the breeding season. Because of the general uniformity of suburban lots within a given residential sector, we considered sampling of 25-30% of each plot adequate to characterize the habitat structure and composition of the area. In Etobicoke and Rosedale, all habitat metrics were measured in a single, smaller plot (4.41 ha and 3.40 ha, respectively, located in the center of the plot within those neighborhoods. In the other neighborhoods, habitat characteristics were measured on several randomly selected residential lots within the larger plot. In both cases, a similar proportion of the larger plot was sampled. Because of the uniformity of the residential areas selected, both methods seemed to provide adequate estimates of the features of the different neighborhoods. We measured the area of lawn, garden, earth, and artificial ground (rocks, tennis courts, brick patios, and driveways, etc.) in

Table 26.1. Sampling efforts for spot-mapping and transect surveys. Effort is measured as the number of hours nest searching and number of spot-mapping and transect surveys.

	Nest search and spot-mapping		Transects	
Residential area	Plot area (ha)	Census effort (h) (No. of spot-mapping surveys)	Number of 92x92 m sections	Number of surveys
Bayview	17	117 (12)	16	7
Don Mills	14	97 (10)	9	9
O'Connor	15	61 (11)	16	6
Etobicoke	17	93 (10)	14	9
Rosemary	16	93 (10)	12	8
Rosedale	13	61 (12)	9	10
Lansdowne	16	72 (12)	14	4

Table 26.2. Measurements of foliage volume, area of natural habitat, and area of man-modified habitat in seven residential areas.

Residential area	# of stories	# houses/ha (% ground covered)	Artificial substrate cover [1]	Natural area cover [2]	Deciduous foliage volume (m^3 ha^{-1})	Coniferous foliage volume (m^3 ha^{-1})	% foliage < 4.5m	% foliage < 9m	Stratum with most foliage volume
Bayview	1	6.2 (11)	41	59	24,894	3140	24	59	4.5-9 m
Don Mills	1	11.1 (11)	49	51	20,568	1488	23	54	4.5-9 m
O'Connor	1	20.0 (15)	60	40	8353	1108	22	63	4.5-9 m
Etobicoke	1	13.5 (15)	55	45	10,284	407	22	62	4.5-9 m
Rosemary	2	17.4 (25)	57	43	37,181	1696	10	32	9.0-13.5 m
Rosedale	2	13.4 (18)	62	38	49,687	729	7	17	13.5-18 m
Lansdowne	2	27.3 (29)	69	31	18,688	224	7	21	13.5-18 m

[1] % ground covered by artificial substrate. Includes roads, houses, paved areas and other hard surfaces.
[2] % ground covered by natural area. Includes lawns, gardens, forested shrubby areas.

each of the selected lots. We used the technique described by Thomas (1973) to determine the volume of vegetation in different height strata. This technique classifies foliage of shrubs and trees by their appropriate geometrical form to calculate their volume. This technique also calculates vegetation volume by height strata. This provides an excellent way of characterizing the vertical structure as well as composition of the vegetation. We used only a portion of the height strata for the analysis as adjacent strata were highly correlated.

The area of the ground covered by streets and sidewalks was measured directly from city maps using a scale of 1:1200 and 1:2400. To estimate the area covered by houses, we randomly selected a third of the houses in each neighborhood and calculated their average area and then multiplied this area by the total number of houses in each plot.

2.4 Statistical Analysis

We used Spearman's rank correlation to identify habitat features most strongly associated with the abundance of a given bird species. No correction factors were applied to account for multiple comparisons (Rice 1989), thus significance values should be used only for comparisons among variables. Bird diversity and evenness were measured with the Shannon-Wiener function (Krebs 1989).

3. RESULTS

3.1 Characteristics Of Residential Areas

Residential areas differed in housing type and density, in the amount of unpaved and natural areas and in vegetation volume and structure (Table 26.2). House density ranged from 6.2 houses ha^{-1} in the least urbanized neighborhood (Bayview) to 27.3 houses ha^{-1} in the most urbanized (Lansdowne). Two urbanization gradients occurred among our plots, one for neighborhoods with one story buildings (Bayview, Don Mills, O'Connor and Etobicoke) and one for neighborhoods with two-story buildings (Rosemary, Rosedale and Landsdowne). In the least developed neighborhood, vegetated areas comprised 59% of the total area. In the most urban neighborhood, vegetated areas comprised only 31% of the total area. The volume of deciduous vegetation ranged from 8,400 to 49,700 m^3 per ha and that of coniferous vegetation from 224 to 3,100 m^3 per ha. The vertical distribution of foliage differed greatly between neighborhoods (Fig. 26.2, 26.3). In relatively open canopy neighborhoods, where tall trees did not entirely shade the streets (Bayview, Don Mills, Etobicoke, O'Connor) most of the volume of deciduous foliage was below 13.5 m. However, Bayview and Don Mills had nearly twice the volume of deciduous foliage of Etobicoke and O'Connor. In closed canopy residential areas, where tall trees shade the streets, the relative density of deciduous foliage varied among neighborhoods (Rosemary, Rosedale, Lansdowne) because trees differed in height. Rosedale, followed by Rosemary, had tall trees and high foliage densities. In Lansdowne, trees were smaller and foliage density much lower. In these closed canopy neighborhoods most of the volume of deciduous foliage was over 13.5 m in height.

In all neighborhoods, the greatest volume of coniferous foliage was below 9 m; in no neighborhoods did coniferous foliage occur above 18 m. Bayview had nearly twice as much coniferous foliage as other neighborhoods. Rosemary had nearly twice as much coniferous foliage as Rosedale. Conifers were quite rare in Lansdowne, the most urbanized residential area.

3.2 Comparison Of Survey Techniques

Transect counts underestimated breeding densities of nearly all species (Table 26.3). Native territorial species such as the American Robin, Northern Cardinal, Chipping Sparrow and Song Sparrow were underestimated more than introduced species (House Sparrow, European Starling, Rock Dove). Breeding densities of Common Grackle and European Starling, Blue Jay and Red-winged Blackbird were overestimated in a few plots.

Although transects underestimated breeding densities of American Robin, results were proportional to the size of the breeding population i.e. the larger the breeding population, the larger the transect estimate (Table 26.4). A similar trend was observed for House Sparrow counts. However, counts of Common Grackle, European Starling, Mourning Dove, Blue Jay, Northern Cardinal, Chipping Sparrow, and Song Sparrow did not consistently reflect the size of the breeding population.

We also tested whether the number of birds recorded in a 92 x 92 m section of a transect was proportional to the number of nests and/or territories present in that section. Only the four most numerous species were tested, (House Sparrow, European Starling, American Robin and Common Grackle). Although significant correlations were found between the number of birds observed and the number of nests present in these sections, the degree of association varied greatly among species and residential areas.

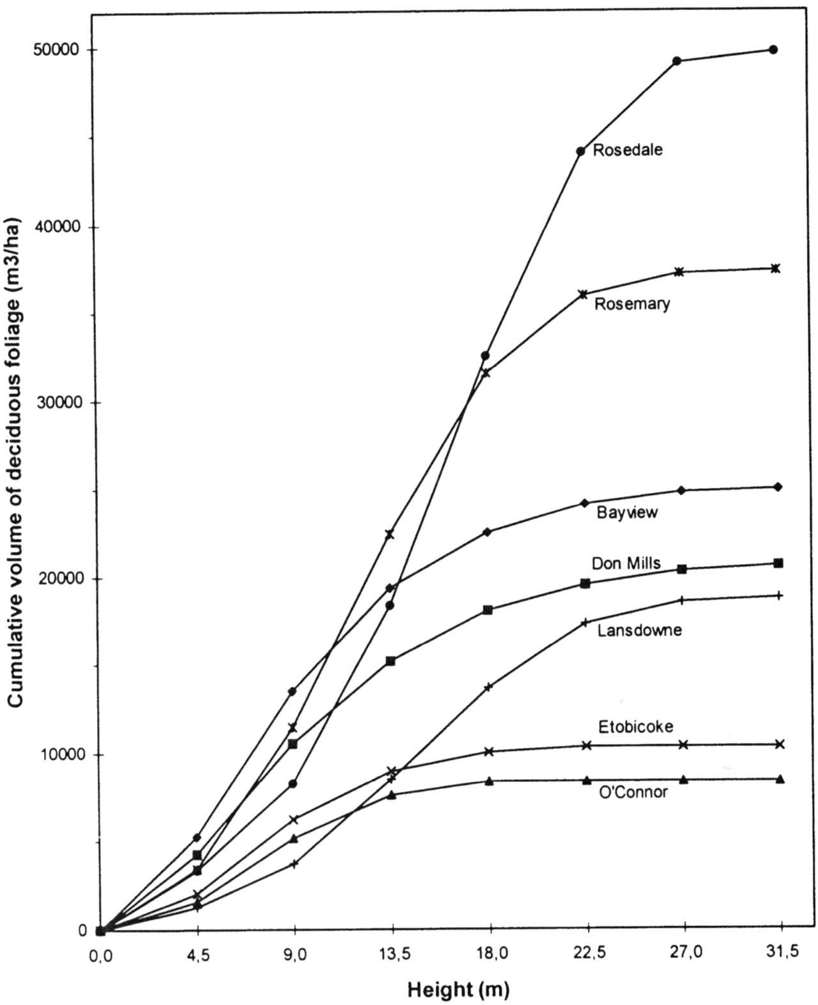

Figure 26.2. Comparison of the vertical distribution of deciduous foliage among plots. The cumulative volume of deciduous foliage (m^3 ha^{-1}) is given by 4.5 m height strata.

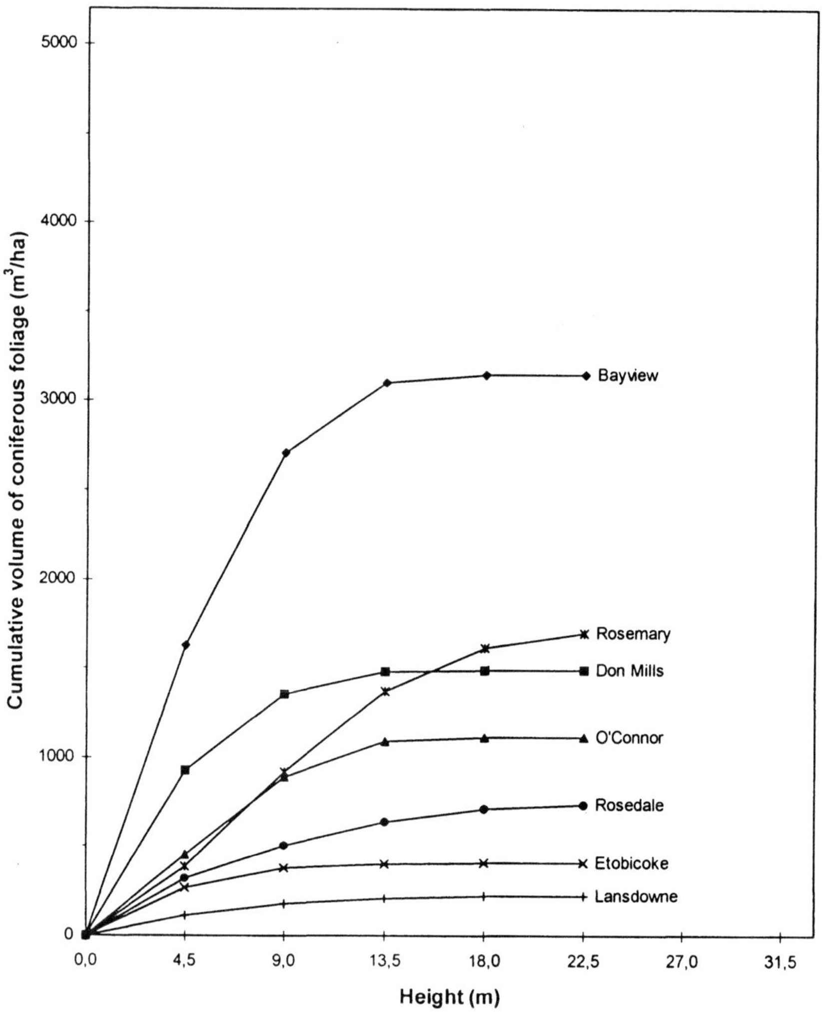

Figure 26.3. Comparison of the vertical distribution of coniferous foliage among plots. The cumulative volume of coniferous foliage ($m^3 ha^{-1}$) is given by 4.5 m height strata.

Table 26.3. Percentage of the breeding population as determined by a combination of nest search and spot-mapping that was estimated by transect counts.

	Bayview	Don Mills	O'Connor	Etobicoke	Rosemary	Rosedale	Lansdowne	Mean
Rock Dove	–	–	–	–	50	52	150	84
Mourning Dove	34	3	29	–	55	–	–	30
Blue Jay	42	–	29	–	25	207	–	76
Gray Catbird	–	0	–	–	–	–	–	0
American Robin	51	56	45	45	58	50	55	51
European Starling F[1]	128	50	161	132	61	49	63	92
Red-eyed Vireo	50	–	–	–	–	–	–	50
House Sparrow	71	72	79	99	46	59	91	74
Red-winged Blackbird	140	–	–	–	–	–	–	140
Northern Oriole	33	–	–	–	–	40	–	37
Common Grackle F	59	23	139	170	45	–	54	82
Northern Cardinal	53	18	14	29	20	0	–	27
Chipping Sparrow	10	39	52	50	25	–	–	35
Song Sparrow	32	16	–	32	–	–	–	27

[1] F = flocking species.

– indicates a species was not found by any of the method.

3.3 Breeding Bird Density From Nest Searching And Spot-mapping Data

Breeding bird densities ranged between neighborhoods from 595 to 824 pairs per km^2 (Table 26.5). Introduced species (House Sparrow, European Starling and Rock Dove) averaged nearly twice as many pairs as native species and were more abundant than native species in all neighborhoods but one (Bayview). Densities of introduced species were less variable (CV = 28%) among neighborhoods than densities of native species (CV = 57%). Diversity measures were more variable (CV = 18%) than evenness measures (CV = 8%). Native species were more abundant in the least urbanized neighborhoods. Total breeding bird density was not strongly related to the level of urbanization. Species richness and diversity tended to decrease with increasing urbanization in neighborhoods dominated by both single- and two-story buildings. No clear pattern was obvious in evenness measures, although a trend existed for the least urbanized neighborhoods to have higher evenness than the more urbanized neighborhoods (Table 26.5). House Sparrow dominated the avifauna of all neighborhoods and were the least variable species between neighborhoods in terms of abundance. The American Robin was the second most abundant species in open canopy neighborhoods, but were replaced by European Starling in closed canopy areas (Table 26.5). Common Grackle was the third most numerous species in five of the neighborhoods, reaching a density of 119 pairs per km^2 in Bayview. All other native species had breeding densities below 30 pairs per km^2.

3.4 Relative Bird Abundance From Transect Data

The relative abundance of birds counted on transects ranged between 688 and 1079 birds per km^2 (Table 26.6). Introduced species were nearly three times more abundant than native species. As expected, several species used residential areas for purposes other than nesting. Transects provided estimates for Chimney Swift and Brown-headed Cowbird, two species that breed in the area but were not adequately surveyed with nest searching and spot-mapping. Common Grackle and European Starling were observed in greater number in Etobicoke and O'Connor, two open canopy neighborhoods, than the number actually breeding there. The relative abundance of introduced species was positively associated with the level of urbanization in neighborhoods with one- and two-story buildings. The relative abundance of native species, species richness, and species diversity tended to be negatively associated with level of urbanization.

Table 26.4. Spearman's rank correlations between the number of nests and/or territories in a transect section and the average number of birds recorded in a section (transect survey). The unit of comparison here is each 92 x 92 m section.

Plot name	Bayview	Don Mills	O'Connor	Etobicoke	Rosemary	Rosedale	Lansdowne	All plots
Degrees of freedom	14	7	14	12	10	7	12	88
House Sparrow	0.85*	0.74	0.22	0.44	0.84*	0.70	0.56	0.64*
Starling	0.33	0.81*	0.32	0.22	0.38	0.27	0.63*	0.56*
Robin	0.50	0.00	0.69*	0.37	0.28	0.38		0.50*
Grackle	0.65*	0.37	0.47	0.32	0.51			0.42*
All four species	0.67*	0.64*	0.61*	0.48*	0.68*	0.40*	0.59*	0.60*
Degrees of freedom	62	26	62	46	38	26	46	318

* $P < 0.05$

Table 26.5. Density of breeding birds (pairs per km^2) based on nest searching and spot-mapping in seven residential areas of Toronto.

	Bayview	Don Mills	O'Connor	Etobicoke	Rosemary	Rosedale	Lansdowne	Mean density ± S.E.	CV (%)
INTRODUCED SPECIES									
House Sparrow	227	329	353	299	203	405	282	299.7 ± 26.6	24
European Starling	47	138	42	77	166	168	242	125.7 ± 27.9	59
Rock Dove	–	–	–	–	37	77	32	20.9 ± 11.2	142
NATIVE SPECIES									
American Robin	215	153	119	143	148	91	40	129.9 ± 20.7	42
Common Grackle	119	116	35	30	37	–	12	49.9 ± 18.2	96
Chipping Sparrow	5	22	27	17	12	–	–	11.9 ± 4.0	90
Northern Cardinal	17	22	7	12	25	5	–	12.6 ± 3.5	73
Blue Jay	12	–	7	–	12	7	–	5.4 ± 2.1	101
Mourning Dove	25	15	7	–	20	–	–	9.6 ± 3.6	109
Song Sparrow	17	22	–	17	–	–	–	8.0 ± 3.8	126
Northern Oriole	12	–	–	–	–	5	–	2.4 ± 1.7	190
Red-winged Blackbird	5	–	–	–	–	–	–	0.7 ± 0.7	265
Red-eyed Vireo	5	–	–	–	–	–	–	0.7 ± 0.7	265
Gray Catbird	–	7	–	–	–	–	–	1.0 ± 1.0	265
Introduced species	274	467	395	376	406	650	556	446.3 ± 47.0	28
Native species	432	357	202	219	254	113	47	232.0 ± 50.1	57
Total	706	824	597	595	660	763	603	6783 ± 90.1	13
N. of species	12	9	8	7	9	7	5	8.1 ± 0.8	26
Diversity (H')	1.82	1.67	1.30	1.42	1.75	1.27	1.13	1.48 ± 0.10	18
Evenness (J')	0.71	0.73	0.59	0.68	0.76	0.65	0.70	0.69 ± 0.02	8

Table 26.6. Relative abundance (birds per km^2) of birds based on 53 transect surveys in residential areas of Toronto.

	Bayview	Don Mills	O'Connor	Etobicoke	Rosemary	Rosedale	Lansdowne	Mean	CV (%)
INTRODUCED SPECIES									
House Sparrow	324	471	561	593	185	476	515	446.4±54.4	32
European Starling	120	139	135	203	203	166	304	181.4±23.9	35
Rock Dove	6	8	8		37	80	96	33.6±14.8	117
NATIVE SPECIES									
American Robin	219	170	108	128	172	91	44	133.1±22.2	44
Common Grackle	140	53	97	102	33	–	13	62.6±19.6	83
Chipping Sparrow	1	17	28	17	6	–	–	9.9±4.1	111
Brown-headed Cowbird	33	6	6	8	4	–	–	8.1±4.3	140
Northern Cardinal	18	8	2	7	10	–	–	6.4±2.4	101
Blue Jay	10	–	4	–	6	29	–	7.0±3.9	149
Mourning Dove	17	1	4	–	22	–	–	6.3±3.5	147
Chimney Swift	–	–	1	1	4	4	24	4.9±3.3	177
Barn Swallow	4	11	–	2	1	–	–	2.6±1.5	155
Song Sparrow	11	7	–	11	–	–	–	4.1±2.0	129
Northern Oriole	8	5	–	1	1	4	–	2.7±1.1	112
Red-winged Blackbird	14	1	–	2	–	–	–	2.4±2.0	212
Northern Flicker	–	–	–	2	2	1	–	0.7±0.4	133
Eastern-wood Pewee	–	–	–	–	1	4	–	0.7±0.6	209
Red-eyed Vireo	5	–	–	1	–	–	–	0.9±0.7	218
Great-crested Flycatcher	–	–	–	–	1	–	–	0.1±0.1	265
Black-capped Chickadee	–	–	–	–	–	1	–	0.1±0.1	265
Gray Catbird	1	–	–	–	–	–	–	0.1±0.1	265
Cedar Waxwing	1	–	–	1	–	1	–	0.43±0.2	125
Warbling Vireo	–	–	1	–	–	–	–	0.1±0.1	265

American Goldfinch	1	–	2	–	–	–	–	0.43±0.3	184

Table 26.6 (cont'd)

	Bayview	Don Mills	O'Connor	Etobicoke	Rosemary	Rosedale	Lansdowne	Mean	CV (%)
All Introduced Species	450	618	704	796	125	722	915	661.4±67.3	27
All Native Species	483	279	250	283	263	135	81	253.4±48.3	51
ALL SPECIES	933	897	954	1079	688	957	996	914.9±46.4	13
Number of species	18	15	11	15	16	11	6	13.1±1.5	31
Diversity (H')	1.84	1.46	1.31	1.37	1.73	1.31	1.21	1.46±0.09	16
Evenness (J')	0.64	0.54	0.55	0.51	0.63	0.55	0.68	0.59±0.02	11

3.5 Bird-habitat Relationships

In general, bird-habitat relationships based on breeding birds densities and on bird relative abundance were similar. Breeding densities of introduced species were negatively correlated with lawn area and positively correlated with deciduous foliage over 27 m (Table 26.7) whereas their relative densities were negatively associated with coniferous and deciduous foliage below 9 m (Table 26.8). Breeding densities of native species were positively correlated woth lawn area as well as with volume of coniferous and deciduous foliage between 4.5 and 9 m and negatively correlated with area covered by houses. Their relative densities were negatively correlated with house area and positively with lawn area and with deciduous vegetation below 9 m. Total bird diversity based on breeding densities was positively correlated with volume of coniferous foliage between 4.5 and 9 m, total coniferous foliage and lawn area. Bird diversity based on relative bird abundance was positively correlated with total coniferous foliage and the volume of deciduous foliage between 4.5 and 9 m.

Breeding and relative densities of Rock Dove were positively correlated with house area and with volume of deciduous foliage over 13.5 m, and negatively correlated with lawn area. House Sparrow breeding densities were not strongly correlated with any measured features but their relative abundance was negatively correlated with the volume of deciduous and coniferous foliage between 4.5 and 9 m and with the total volume of coniferous and deciduous foliage (Table 26.8). Blue Jay breeding densities were positively correlated with volume of coniferous foliage and their relative abundance was correlated with the volume of coniferous foliage between 13.5 and 18 m and with the total volume of deciduous foliage. Breeding and relative densities of Chipping Sparrow were negatively correlated with the volume of relatively high deciduous foliage and those of Song Sparrow were correlated positively with lawn area. Breeding and relative densities of Mourning Dove were positively associated with the volume of deciduous and coniferous foliage below 9 m and with total foliage volume of conifers.

Breeding densities of American Robin and Common Grackle were positively correlated with the volume of deciduous and coniferous foliage below 9 m, lawn area, and negatively with house area. However, relative densities of Common Grackle were most strongly positively correlated with lawn area and negatively with deciduous foliage between 13.5 and 18 m. Northern Cardinal breeding and relative densities were positively correlated with the volume of deciduous and coniferous foliage below 9 m, and their relative density with lawn area. Chimney Swift abundance was positively

Table 26.7. Relationships between breeding bird densities, based on nest searching and spot mapping, and habitat features (Spearman's rank correlation). Underlined correlation coefficients are significant at 90% (∝ = 0.10) and those with an asterisk (*) at 95% (∝ = 0.05) (not corrected for multiple comparisons).

	Coniferous foliage			Deciduous foliage						
	4.5-9.0m	13.5-18.0m	Total Volume	4.5-9.0m	13.5-18.0m	27.0 31.5m	Total Volume	Lawn area	House area	Paved area
House Sparrow	-.357	-.214	-.393	-.429	-.179	+.288	-.143	-.143	-.250	+.500
European Starling	-.571	+.179	-.464	-.214	<u>+.786*</u>	+.649	+.500	<u>-.750</u>	<u>+.714</u>	-.036
Rock Dove	-.197	<u>+.670</u>	-.197	-.020	<u>+.906*</u>	+.497	<u>+.709</u>	<u>-.788*</u>	<u>+.749</u>	+.394
American Robin	<u>+.750</u>	+.036	<u>+.857*</u>	<u>+.857</u>	-.214	-.126	+.179	<u>+.893*</u>	<u>-.714</u>	-.571
Common Grackle	<u>+.786*</u>	+.036	<u>+.857*</u>	<u>+.714</u>	-.321	-.198	+.000	<u>+.821*</u>	<u>-.679</u>	-.571
Chipping Sparrow	+.288	-.378	+.252	-.018	<u>-.829</u>	-.582	-.613	+.523	-.613	+.198
Northern Cardinal	<u>+.679</u>	+.179	<u>+.786*</u>	<u>+.786</u>	-.071	-.198	+.250	+.607	-.428	-.286
Blue Jay	<u>+.850*</u>	<u>+.869*</u>	<u>+.756*</u>	+.661	+.340	-.124	+.510	+.151	-.000	+.132
Mourning Dove	<u>+.927*</u>	+.408	<u>+.964</u>	<u>+.815*</u>	-.037	-.112	+.259	+.630	-.482	-.408
Song Sparrow	+.120	-.598	+.299	+.418	-.418	+0.60	-.120	<u>+.837*</u>	<u>-.777*</u>	<u>-.717</u>
Introduced species[1]	-.536	+.179	-.464	-.357	+.607	<u>+.739</u>	+.393	<u>-.750</u>	+.464	+.286
Native species [2]	<u>+.750</u>	+.036	<u>+.857*</u>	<u>+.857*</u>	-.214	-.126	+.179	<u>+.893*</u>	<u>-.714</u>	-.571
Total density	+.286	+.321	+.464	+.571	+.500	<u>+.847*</u>	+.714	+.179	-.357	-.357
Diversity (H')	<u>+.821*</u>	+.214	<u>+.893*</u>	<u>+.892</u>	-.107	-.234	+.250	<u>+.786*</u>	-.536	-.464

1 = House Sparrow, European Starling and Rock Dove.
2 = Other species

Table 26.8. Relationships between relative bird density, based on transect results, and habitat features (Spearman's rank correlation). Underlined coefficients are significant at 90% (∝ = 0.10) and those with an asterisk (*) at 95% (∝ = 0.05) (not corrected for multiple comparisons).

	Coniferous foliage			Deciduous foliage						
	4.5-9.0m	13.5-18.0m	Total volume	4.5-9.0m	13.5-18.0m	27.0-31.5m	Total volume	Lawn area	House area	Paved area
House Sparrow	-.679	-.679	-.786*	-.821*	-.571	-.378	-.750	-.179	+.071	+.357
European Starling	-.721	-.144	-.685	-.487	+.360	+.027	+.000	-.721	+.829*	+.126
Rock Dove	-.342	+.432	-.342	-.342	+.685	+.573	+.378	-.865	+.685	+.270
American Robin	+.821*	+.214	+.893*	+.893*	-.107	-.234	+.250	+.786*	-.536	-.464
Common Grackle	+.464	-.357	+.429	+.321	-.714	-.631	-.464	+.857*	-.643	-.393
Chipping Sparrow	+.255	-.400	+.200	-.055	-.855*	-.670	-.655	+.509	-.564	-.236
Northern Cardinal	+.793*	+.162	+.865*	+.847*	-.144	-.273	+.180	+.775*	-.505	-.523
Blue Jay	+.556	+.852*	+.482	+.519	+.556	+.262	+.704	-.037	+.037	+.259
Mourning Dove	+.927*	+.556	+.890*	+.667	-.037	-.318	+.185	+.408	-.259	-.074
Song Sparrow	+.159	-.518	+.259	+.418	-.418	-.201	-.159	+.837*	-.618	-.677
Chimney Swift	-.532	+.312	-.606	-.532	+.569	+.130	+.165	-.973*	+.973*	+.459
Brown-headed Cowbird	+.491	-.346	+.491	+.436	-.655	-.523	-.346	+.927*	-.727	-.436
Introduced species[1]	-.929*	-.571	-.964*	-.857*	-.107	+.072	-.429	-.500	+.393	+.143
Native species[2]	+.536	-.214	+.607	+.678	-.393	-.342	-.036	+.929*	-.679	-.536
Total density	-.536	-.786*	+.607	-.643	-.643	-.469	-.821*	+.071	+.000	-.143
Diversity (H')	+.750	+.286	+.857*	+.964*	.107	-.036	+.464	+.714	-.464	-.500

1 = House sparrow, European starling and Rock Dove.
2 = Other species

associated with house area and negatively associated with lawn area, whereas Brown-headed Cowbird had exactly the opposite relationships.

4. DISCUSSION

4.1 Efficiency Of Transects

It is not surprising that transects underestimated the size of the breeding population because during the breeding season some proportion of birds are incubating and not detectable. Woolfenden and Rohwer (1969) found that transects estimated only a third of the breeding bird population in suburban St. Petersburg, Florida. Transects also tend to underestimate breeding densities in natural habitats (Enemar 1959, Haukiöja 1968, Franzreb 1976). In our study, the relationship between transect results and the size of the breeding population varied with species and habitat types. Transects provided a reliable density index for American Robin and House Sparrow but not for flocking species like European Starling and Common Grackle.

Within a given suburban habitat, the accuracy of transects varied with species. This implies that calculations of diversity indices will be biased because each species is not sampled with the same accuracy and/or precision. Calculations of relative abundance also will be affected. Any comparisons between species will be difficult to interpret because of the problem of defining what proportion of the breeding population of each species was actually sampled. Several efforts have been made to modify transects so that actual densities may be derived (Emlen 1971, Franzreb 1976, Burnham et al. 1980). However, most proposed techniques assume habitats are uniform and/or birds are randomly distributed. Urbanizing habitats often are not uniform but highly heterogeneous, and birds are far from being uniformly or randomly distributed. Therefore, these techniques cannot be used adequately in most urbanizing habitats.

In our study, the four most abundant breeding species also were the four most abundant species as determined by transects. Woolfenden and Rohwer (1969) found the same in their study. Although both methods rank species similarly, the estimated relative abundance often differs. For example, in Don Mills, House Sparrow was the most abundant species representing 59% of breeding birds. Based on transects, House Sparrow was still the most abundant species, but represented only 40% of the total number of birds observed.

In our study, transects did detect a few species not found breeding thus giving a more complete list of birds present in the plots. However a similar species list can be derived from nest searching and spot-mapping surveys since non-breeding birds are still detected.

4.2 Bird Community Structure

The structure of urban bird communities is characterized by low species richness, high bird abundance, and low species evenness (Emlen 1974, Lancaster and Rees 1979, Jokimaki and Suhonen 1993, Jokimaki and Huhta 1996). The urban bird community is dominated by a few abundant species, usually the House Sparrow or the European Starling. In our residential areas, the ratio of the number of pairs per species observed per 20 ha averaged 17.5 ± 1.6 (n = 7, range: 11.7 - 24.1). In natural forest habitats in Ontario this ratio was much lower: 1) Mature poplar forest: 3.1 (Erskine 1971); 2) Mature maple-hemlock forest: 1.9 (Taylor 1983); 3) Mature (85-95 years old) pine stand: 5.1 (Oelke 1967a); 4) Mature (100-110 years old) pine stands: 6.0 (Oelke 1967b). In our study, species richness seemed to be more strongly related to the level of urbanization than was evenness. However, the range of variation among the residential areas we surveyed may not have been sufficient to clearly identify the relationship of urban patterns to species evenness.

4.3 Bird-habitat Relationships

Urban neighborhoods have a mix of natural and human elements that may influence avian communities. Our results highlight the duality; in a single community you find species such as Chimney Swift, Rock Dove and European Starling that respond to architectural features first and to natural features second, and other, mostly native birds associated with remnant natural features, especially vegetation structure and volume.

The absence of any significant correlation between breeding densities of House Sparrows and measured habitat features may reflect their great behavioral plasticity in urban habitats or their occurrence in most different urban habitats. In some neighborhoods where cavities in buildings were scarce, House Sparrows nested in conifers. Several studies have found a positive correlation between building abundance and House Sparrow densities (Thomas 1973, Lucid 1974, Williamson 1974, Geis 1974, Schinner 1974).

Building abundance also was positively correlated with European Starling densities in several studies (Lucid 1974, Williamson 1974, Geis 1974, Schinner 1974). We found a similar correlation; however, we also found a strong correlation between starling density and volume of deciduous foliage. In part, this might reflect an association of tall trees in neighborhoods with two-story buildings. Starlings prefer to nest high (Verheyen 1969) and may find more nesting sites in two-story than in one-story houses. The third introduced species, Rock Dove, was found only in neighborhoods with two-story houses and was positively correlated with house area and paved area, similar to previous studies (Weber 1972, Erskine 1974, Williamson 1974).

Consistent with other studies (Hooper et al. 1975, Williamson and DeGraaf 1980), we found that native species abundance was related to vegetative features. Common Grackle was closely associated with the presence of coniferous vegetation as was noted by Lucid (1974) and Maxwell et al. (1976). In our study area grackles nested exclusively in conifers. Robins were more numerous than grackles and more closely associated with both coniferous and deciduous vegetation. Most urban studies have found that the abundance of robins was positively associated with vegetation volume, especially below 9 m, and negatively associated with building density (Lucid 1974, Schinner 1974, Williamson 1974). Lawns also are a major habitat feature associated with robin abundance (Williamson and DeGraaf 1980). The preference of dense, shrubby vegetation by Northern Cardinal (Shinner 1974, Goetz 1975) explains its correlation with vegetation below 9 m. The avoidance of tall deciduous vegetation by Chipping Sparrow is consistent with other studies that show a preference for open areas and avoidance of closed canopy (Smith 1956, Tate 1973, Goetz 1975). In general, Song Sparrow is associated with open areas and low vegetation in urban areas (Suthers 1960, Thomas 1973, Lucid 1974, Williamson 1974, Goetz 1975) consistent with the positive association we found with lawn area and the negative association with house area. We found that Mourning Dove abundance was positively associated with deciduous and coniferous foliage below 9m and with area of natural ground; however, relationships between Mourning Doves and vegetative features were less consistent in other studies, from a positive relationship with coniferous foliage (Lucid 1974), to a positive relationship with deciduous foliage (Shinner 1974, Goetz 1975).

In other suburban neighborhoods, Lucid (1974) found a positive correlation between total bird species diversity and the total volume of vegetation. Hooper et al. (1975) found the number of breeding species related to the amount of foliage below 4 m. All such studies have found a negative relationship between building volume or density and bird diversity (Lucid 1974, Schinner 1974, Hooper et al. 1975). The composition and structure of urban bird communities can be manipulated by varying these various components. Thus, in highly urbanized residential areas bird diversity can be increased by diversifying and increasing the native vegetative components of these areas. Native vegetation may support a greater density of insects than exotic vegetation (Mills et al. 1989). Abundance of introduced bird species can be increased, if this is a desired goal, by adding artificial structures (Luniak 1992). In temperate areas like Toronto, conifers also may play an important role in diversifying the urban avifauna. Several species were associated with the presence of coniferous trees and shrubs which often were the preferred nesting substrate especially early in the breeding season before the growth of deciduous leaves. The importance of conifers to birds, especially in winter, has been noted in other studies of temperate cities (Jokimaki et al. 1996, Jokimaki and Suhonen 1998)

To date, most studies including ours have used linear relationships to identify bird-habitat relationships. It is likely that many of these relationships are nonlinear. Therefore, we should be cautious in interpreting the non-significant relationships found in these studies (Marzluff 1986). Our relatively small sample size of only seven neighborhood plots should be considered when comparing our results to those of others. However, even with our limited sample size, we were successful in identifying several strong relationships between habitat and individual species that are consistent with the results of other studies.

We are beginning to understand how habitat features relate to bird abundance and diversity in urbanizing areas -(Jokimaki and Suhonen 1993, Jokimaki and Huhta 1996, Jokimaki 1999). However, little research has been done on the population dynamics of most native species of birds or on their microhabitat requirements in the urban environment. Advances in these areas are required if we are to foster self-sustaining bird communities with a greater proportion of native species and contribute to the quality of urban life for humans.

4.4 Policy And Management Implications

Estimating the structure of avian communities in urbanizing areas relies on several methods which, as we have shown, may not always provide similar estimates. For several species, breeding densities estimated from nest searching and spot mapping were lower than relative densities estimated from transect counts which include breeding as well as non-breeding birds. This has implications for understanding bird-habitat relationships in urban areas. However, the biases of different techniques are species-specific and managers and researchers should be aware of the strengths and weaknesses of the techniques they choose and which species are likely to be under or overestimated. Nesting and foraging requirements differ for several species in the urban environment and both requirements have to be met to maximize breeding densities. One of the most important features for increasing breeding bird diversity in temperate cities is the presence of conifers. Several species were closely linked with the presence of conifers. Also, shrubs were important, especially for native species. Bird diversity in urban residential areas could be increased by the addition of native shrubs and coniferous trees.

ACKNOWLEDGEMENTS

We thank our field assistants, M. Stasiulis, R. Harris, H. Kirsch, M. Cadman for their help in the field. We also express our thanks to D. A. Aborn and J. Jokimaki for their thorough and helpful review of the manuscript.

REFERENCES

Adams, L. W. 1994. Urban wildlife habitats: a landscape perspective. Univ. Minnesota Press, Minneapolis, MN.

Batten, A. L. 1972. Breeding bird species diversity in relation to increasing urbanization. Bird Study 19:157-166.

Bibby, C. J., N. D. Burgess, and D. A. Hill, 1992. Bird Census Techniques. British Trust for Ornithology and Royal Society for the Protection of Birds, Academic Press, London.

Bezzel, E. 1985. Birdlife in intensively used rural and urban environments. Ornis Fennica 62:990-95.

Burnham, K. P., D. R. Anderson, and J. L Laake, 1980. Estimation of density from line transect sampling of biological populations. Wildl. Monogr. 72:1-202.

Clergeau, P., J.-P. L. Savard, G. Mennechez, and G. Falardeau. 1998. Bird abundance and diversity along an urban-rural gradient: a comparative study between two cities on different continents. Condor 100:413-425.

Enemar, A. 1959. On the determination of the size and composition of a passerine bird population during the breeding season. Var Fagelvarld, Suppl. 2.

Emlen, J. T. 1971. Population densities of birds derived from transect counts. Auk 88:323-342.

Emlen J. T. 1974. An urban bird community in Tucson, AZ: derivation, structure, regulation. Condor 76:184-197.

Erskine, A. J. 1971. Mature poplar forest. American Birds 25:977-978.

Erskine, A. J. 1974. Small town residential area with wooded vacant lots. American Birds 28:1053-1054.

Franzreb, K. E. 1976. Comparison of variable strip transect and spot-map methods for censusing avian populations in a mixed coniferous forest. Condor 78:260-262.

Geis, A. D. 1974. Effect of urbanization and type of urban development on bird populations, p. 97-105. *In* J. H. Noyes and D.R. Progulski [EDS.], Wildlife in an urbanizing environment. Planning and Resource Development Series No. 28. Noldsworth Natural Resources Center. Cooperative Extension Service, Univ. of Massachussetts, Amherst, MA.

Gilbert, O. L. 1989. The ecology of urban habitats. Chapman and Hall, New York.

Goetz, E. 1975. Habitat variables and urban songbird population. M.S. Thesis, West Virginia Univ., Morgantown, WV.

Hadidian, J., J. Sauer, C. Swarth, P. Hanly, S. Droege, C. Willilams, J. Huff, and G. Didden. 1997. A citywide breeding bird survey for Washington, D.C. Urban Ecosystems 1:87-102.

Haukiöja, E. 1968. Reliability of the line transect survey method in bird census with reference to reed bunting and sedge warbler. Ornis Fenn. 45:105-113.

Hooper, R. G., E. F. Smith, H. S. Crawford, B. S. Mcginnes, and V. J. Walker. 1975. Nesting bird populations in a new town. Wildl. Soc. Bull. 3:111-118.

Jokimaki, J. 1999. Occurrence of breeding bird species in urban parks: effects of park structure and broad-scale variables. Urban Ecosystems 3:21-34.

Jokimaki, J., and E. Huhta. 1996. Effects of landscape matrix and habitat structure on a bird community in northern Finland: A multi-scale approach. Ornis Fennica 73:97-113.

Jokimaki, J., and J. Suhonen. 1993. Effects of urbanization on the breeding bird species richness in Finland: a biogeographical comparison. Ornis Fennica 70:71-77.

Jokimaki, J., and J. Suhonen. 1998. Distribution and habitat selection of wintering birds in urban environments. Landscape and Urban Planning 39:253-263.

Jokimaki, J., J. Suhonen, K. Inski, and S. Jokinen. 1996. Biogeographical comparison of winter bird assemblages in urban environments in Finland. J. Biogeogr. 23:379-386.

Krebs, C. J. 1989. Ecological methodology. Harper and Row, New York.

Lancaster, R. K., and W. E. Rees. 1979. Bird communities and the structure of urban habitats. Can. J. Zool. 57:2358-2368.

Lucid, V. J. 1974. Bird utilization of habitat in residential areas. Ph.D. Diss., Va. Polytech. Inst. and State Univ., Blacksburg, VA.

Luniak, M. 1992. The use of nest-boxes for the management of breeding avifauna in urban parks - studies in Warsaw and Poynan (Poland). Acta Ornithologica.

Marzluff, J. M. 1986. Assumptions and design of regression experiments: The importance of lack-of-fit testing, p165-170. In J. Verner, M. L. Morrison, and C. J. Ralph [EDS.], Modelling habitat relationships of terrestrial vertebrates, University of Wisconsin Press, Madison.

Maxwell, G. R. II, J. M. Nocilly, and R. I. Shearer. 1976. Observations at a cavity nest of the common grackle and an analysis of grackle nest sites. Wilson Bull. 88:505-507.

Mills, G. S., B. Dunning Jr., and J. M. Bates 1989. Effects of urbanization on breeding bird community structure in south-western desert habitats. Condor 91:416-428.

Morneau, F., R. Décarie, R. Pelletier, D. Lambert, J.-L. DesGranges, and J.-P. L. Savard. 1999. Changes in breeding bird richness and abundance in Montreal parks over a period of 15 years. Landscape and Urban Planning 44:111-121.

Natuhara, Y. and C. Imai. 1996. Spatial structure of avifauna along urban-rural gradients. Ecol. Res. 11:1-9.

Oelke, H. 1967a. Mature pine stands (85-95 years old) of the southern Canadian shield. Audubon Field-Notes 21:617.

Oelke, H. 1967b. Mature pine stands (100-110 years old) of the southern Canadian shield. Audubon Field-Notes 21:617-618.

Rice, W. R. 1989. Analyzing statistical tests. Evolution 43:223-225.

Savard, J.-P. L., P. Clergeau, and G. Mennechez. 2000. Biodiversity concepts and urban ecosystems. Landscape and Urban Planning. 48:131-142.

Schinner, J. R. 1974. An analysis of the interrelationship of habitat and avifauna in Metropolitan Detroit. Ph.D. Diss., Michigan State Univ., Lansing, MI.

Smith, R. L. 1956. An evaluation of conifer plantations as wildlife habitat. Ph.D. Diss., Cornell Univ., Ithaca, NY.

Suthers, R. A. 1960. Measurement of some lake-shore territories of the song sparrow. Wilson Bull. 72:232-237.

Tate, D. 1973. Habitat usage by the chipping sparrow (Spizella passerina) in Northern lower Michigan. Ph.D. Diss., University of Nebraska-Lincoln, NE.

Taylor, P. 1983. Mature maple-hemlock forest. American Birds 37:70.

Thomas, J. 1973. The determination of habitat requirements for birds in suburban areas. A pilot study. Ph.D. Diss., University of Massachusetts., Amherst, MA.

Verheyen, R. F. 1969. Le choix du nichoir chez l'étourneau (Sturnus v. vulgaris). Le Gerfaut 59:239-259.

Weber, W. C. 1972. Birds in cities: a study of populations, foraging ecology, and nest-sites of urban birds. Can. Field Nat. 89:457-460.

Williamson, R. D. 1974. Birds in Washington, D.C., p. 131-135. *In* J. H. Noyes and D. R. Progulske [EDS.], Wildlife in an urbanizing environment. Planning and Resource Development Series No. 28. Noldsworth Natural Resources Center. Cooperative Extension Service, Univ. of Massachussetts, Amherst, MA.

Williamson, R. D and R. M. DeGraaf. 1980. Habitat associations of 10 bird species in Washington, DC. Urban Ecology 5:125-136.

Woolfenden, G. E. and S. A. Rohwer. 1969. Breeding birds in a Florida suburb. Bull. Florida State Museum 13:1-83.

Chapter 27

Integrating avian ecology into emerging paradigms in urban ecology

Reed Bowman[1] and John M. Marzluff[2]
[1]*Archbold Biological Station, P.O. Box 2057, Lake Placid, FL 33862 USA, e-mail:rbowman@archbold-station.org.*

[2]*College of Forest Resources, University of Washington, Box 352100, Seattle, WA 98195 USA.*

Key words: Avian ecology, experimental design, spatial scale, synthesis, urban ecology

Abstract: We discuss emerging paradigms in the conceptual and methodological approaches to the study of urban ecosystems. We review methods used in urban studies of avian ecology, both historically and in this book, and discuss how the results of these studies can be incorporated into these paradigms. We explore the mechanistic links between the urban environment and between different levels of biological organization, from physiology to individuals to populations to communities and we suggest future approaches to increase the applicability of urban avian ecology to understanding the functioning of cities as ecosystems. These approaches require and will benefit from interdisciplinary research and education.

1. INTRODUCTION

Few could doubt the pervasive influence of humans on virtually every ecosystem on this planet. Some ecosystems have been wholly transformed while in others the effects of humans have been more subtly expressed. The cumulative effects have been enormous and these effects are on-going and accelerating (Vitousek et al. 1997). A complete understanding of ecosystem dynamics in any of these impacted systems depends on incorporating these effects; yet the response of ecologists has been relatively slow (Collins et al. 2000). As young graduate students, we were steered away from ecological studies in human environments with admonishments of the difficulty of interpreting our results biased by the artifacts of humans on the environment. However, these artifacts are the very effects that influence ecological pattern and process in urban environments that subsequently influence the structure and composition of bird communities. The challenge to urban avian ecologists and to urban ecologists in general is to describe the urban ecosystem and then measure the appropriate attributes that influence the patterns and processes of interest (Collins et al. 2000).

A new paradigm of urban ecology is emerging that seeks to disentangle the patterns and processes of ecosystems driven by biogeophysical phenomena from patterns of human activity driven by socioeconomic phenomena (Grimm et al. 2000). This paradigm includes a dichotomy in research approaches, described by Nancy Grimm and her colleagues (2000) as "ecology in the city", an effort to understand how ecological patterns and processes differ in the city from less urbanized habitats or wildlands and, "ecology of the city", an effort to understand the dynamics of these processes within the city, especially as they are influenced by human activity in its many forms. Virtually all urban avian ecological studies have taken the form of "ecology in the city" in that they contrast urban bird communities or populations with areas at different stages of development during the process of urbanization (e.g. suburban, exurban, etc., see Marzluff et al. 2001) either using temporal comparisons or simultaneous spatial comparisons along an urban-rural gradient (McDonnell et al. 1993). Obviously these two approaches to understanding urban ecosystems are complementary and studies of ecology in the city can be interpreted in the larger conceptual framework of the city as an ecosystem.

A rigorous approach to urban ecology demands rigorous methodology, including observational studies to define patterns and experimental studies to understand how and which processes regulate those patterns. Development of realistic urban models that incorporate the socio-cultural human axis will be necessary to determine how these patterns and process are driven by human values. Our perceived societal gains and losses are shaped by these values, and these gains and losses shape public policy, thus forming a feedback loop by which human-driven change can influence the biotic and abiotic environment (McDonnell and Pickett 1990). If one of the goals of urban ecology is to inform management of cities to mitigate the deleterious effects of human progress (Niemelä 1999), then the integration of these two methodological approaches is critical. Here we review the methodological framework of the more traditional "ecology in cities" approach, with specific examples from the studies in this book, but emphasize how these studies may fit into an "ecology of cities" paradigm.

2. METHDOLOGY IN URBAN AVIAN ECOLOGY AND THE INTEGRATION OF TWO URBAN PARADIGMS

Because cities are enormously complex, including a built environment and anthropogenic alterations to the biotic and abiotic environments, the first challenge is to develop metrics that reflect variation likely to influence the ecological patterns and processes under study. In part, this requires identifying specific ecological processes affected by specific urban patterns, and then selecting the appropriate metric (Niemelä 1999, Grimm et al. 2000). The studies in this book have used a wide variety of metrics, often to define an urban gradient. These have included human housing density (Bowman and Woolfenden 2001; Haskell et al. 2001; McGowan 2001), the type of built environment (i.e. single-story houses versus multi-family dwellings and the percent of green space) (Savard and Falls 2001; Jerzak 2001), and percent area vegetated (Mennechez and Clergeau 2001). Although distance from urban centers has been used as a metric in the past, because urban areas are spatially heterogeneous, this metric is not particularly useful (Alberti et al. 2001). Other metrics that might be useful include percent impervious surface area, ratios of commercial to residential development, ratios of native to non-native species, and even socioeconomic metrics such as property values. Alberti et al. (2001) proposes a series of pattern metrics designed to link urban development to ecological conditions and expands the utility of urban ecological gradient analysis. Miller et al. (2001) suggest that these metrics can be nested at a variety of spatial scales. Spatial patterns of heterogeneity require their inclusion in defining the urban landscape, but they also provide the link to the "ecology of cities" paradigm. These spatial patterns may be influenced by both the natural and the human ecosystem, such as the environment, both biotic and abiotic, in which the city occurs, the age of the city, the growth of the city, the demography of its occupants, etc. (Grimm et al. 2000). Understanding these relationships may help us predict how and why these spatial patterns exist, how they change over time and space, and how these patterns influence ecological processes, which then enter into the feedback loop of public policy through society's perceived gains or losses.

This spatial heterogeneity of the urban ecosystem also may require its own set of metrics. Landscape ecology provides a series of metrics such as patch size and shape, patch isolation and dispersion, corridors, matrices and networks and the configuration of these elements in the landscape (Forman and Godron 1986) that may be a useful starting point for defining the urban landscape (Alberti et al. 2001). Bolger (2001) defined patches in an urban environment by size and context; the interiors of large unfragmented habitat blocks, the edge areas of blocks adjacent to the urban matrix, and fragments embedded in the urban matrix, but relatively few other studies are explicit about landscape patterns. Coppedge et al. (2001) used a large spatial scale to examine patterns of urbanization, juniper encroachment, and changes in bird communities, but limited their analysis to a single physiographic province. Several studies attempt to address spatial heterogeneity by including multiple study sites with different urban attributes (Azzerad and Nilon 2001; Blair 2001; Clergeau et al. 2001; Mennechez and Clergeau 2001; Savard and Falls 2001). Undoubtedly, human social and economic factors play a role in the distribution of neighborhood types within cities.

Incorporating spatial heterogeneity requires large-scale and multi-scale studies (Hostetler 2001) which often have enormous logistical and economical constraints. Rarely can these studies be designed with effective replicates and controls. Without this rigorous experimental framework, the validity of inferential statistics may be questioned (Oksanen in press). Miller et al. (2001) suggest a method of nested spatial sampling that lends itself to traditional statistical tools and geostatistical methods. Methods such as canonical correspondence analysis and principal components analysis, such as used by Azzerad and Nilon (2001) and others (Jokimäki and Suhonen 1998), might be used to reduce many related variables into composites that explain variation in the dependent variables. Sometimes large-scale patterns might be reproduced at smaller-scales where confounding variables might be controlled in a more rigorous experimental design. Such microcosm experiments (Okasanen in press) may allow complex interactions in the urban environment to be simplified to test the effect of just a single variable. Incorporating spatial heterogeneity at large-scales and developing experimental designs from which causality can be induced may be one of the biggest challenges to urban avian ecology. Deductive studies that lead to experimental inductive designs may be the best approach.

To assess urban effects on birds, the impacts of urbanization on the biotic and abiotic components of the ecosystem that are likely to affect birds also must be measured. Abiotic impacts might include changes in local climate, nutrient cycling, hydrology, air and water pollution, and disturbance regimes. For example, the urban infrastructure may have profound effects on fire regimes; habitat fragmentation caused by road construction is likely to reduce the frequency and extent of fire. Simultaneously, protection of private property results in fire suppression, further altering fire regimes. Coppedge et al. (2001) examine the interactions between alterations in fire regime, vegetation community composition, and bird communities. Recovery from disturbance also is likely to be modified in the urban environment. Impacts to the biotic community are likely to influence post-disturbance recovery. The structure and composition of the native vegetation is likely to vary in relation to human land use (Kent et al. 1999, Reichard et al. 2001) and anthropogenic alterations to this structure are likely to result in novel responses to disturbance. Interactive effects such as phenological changes resulting from local climatic shifts, may influence post-disturbance recovery, but also may affect other resources such as food. Urban effects on climate and even global climate change may alter the phenological patterns of resource abundance; birds that cannot shift their timing of breeding to coincide with peak food abundance may suffer demographic impacts (Thomas et al. 2001). Urbanization is likely to directly affect arthropods, the primary foods for many birds, especially during reproduction, yet studies of the effects of urbanization on arthropods are few and often poorly designed to elucidate the mechanism by which the distribution and abundance patterns may be affected (McIntyre 2000). Introduction of novel predators such as domestic cats may have profound affects on birds (Churcher and Lawton 1987), but the distribution of native predators also may be influenced by the urban environment. Haskell et al. (2001) examined how terrestrial predator abundance varied with human housing density and the impact that variation had on predation rates on bird nests. However, they also extended their questions to the interaction of laws designed to slow human development, property tax rates, and human attitudes about both conservation and taxation. Ineffective planning may not achieve conservation goals and reduce overall tax revenues that inevitable increase the tax burden on others. Bowman and Woolfenden (2001) examined variation in nest success across and urban gradient and concluded that patterns were a result of a complex interaction between predators and the timing of breeding, both of which were influenced by the urban environment. However, providing food for birds and letting domestic cats roam freely are both human social traits, illustrating that some demographic variation in urban bird populations may be driven as much by the human social and cultural parameters as local ecological variables.

Goldstein et al. (1980/1981) described conflicts between the desire of humans for particular structural patterns of vegetation around their homes and the needs of wildlife for a more complex structural pattern. Clearly, human values can influence the biotic and abiotic environment. These human preferences might have a historical or cultural origin (Orians 1998), but if different structural patterns favor wildlife, then understanding the human perceptual basis is necessary if the goal is to shift public behavior towards one favoring wildlife rather than people.

3. MECHANISTIC LINKS IN ECOLOGICAL ORGANIZATION

Although the changes in bird community structure and composition in response to urbanization seem to be consistent enough to make broad generalizations (Blair 2001, Marzluff 2001), relatively few studies have examined the response of birds at underlying levels of biological organization, such as populations and individuals. These levels are often hierarchical; effects on individuals may often be manifested through changes in survival and/or fecundity, which in turn effect demographic patterns and population trajectories. Species-specific responses then alter community structure through additional complex interactions such as changes in competitive interactions. Yet growing evidence exists that urbanization does have impacts at all levels of organization. Lead from car exhaust may lower hemoglobin concentration in urban populations of Tree Sparrows (*Passer montanus*) (Kostelecka-Myrcha et al. 1997). Access to protein and lipid-rich anthropogenic foods results in higher levels of plasma cholesterol and blood urea nitrogen in urban House Sparrows (*P. domesticus*) (Gavett and Wakely 1986). Schoech and Bowman (2001) attempted to relate phenological differences in the timing of breeding between suburban and wildland populations of Florida Scrub-Jays (*Aphelocoma coerulescens*) by examining blood constituents. A vast suite of physiological indices, such as blood characteristics, nutritional condition, and growth rates could be influenced by the urban environment that might affect demographic patterns. In addition, the behavior of birds might be directly and indirectly affected. Human disturbance can influence time and activity patterns of birds and this may vary depending on the urban context, as shown by Fernández-Juricic et al. (2001) Time and energy budgets, dominance hierarchies, and interspecific interactions are all likely to be influenced in an urban environment.

Developing mechanistic links between individual and population level patterns requires detailed studies of demography, ideally longitudinal comparisons of marked populations across an urban gradient. Bowman and Woolfenden (2001), Marzluff et al. (2001), and McGowan (2001) use such studied to examine how urbanization influences demography and how these demographic patterns may affect community structure. In contrast, Bolger (2001) compares top-down and bottom-up regulation of population processes and how these may vary depending on urban context. Changes in both regulatory processes and demographic patterns themselves can alter population structure and life history parameters such as age at first breeding. Urban populations of Florida Scrub-Jays have higher adult mortality rates, but lower age of first reproduction (Breininger 1999) and fewer young remain as helpers on their natal territory. In part, this may reduce the resilience of the population to stochastic demographic variation.

Changes in the urban landscape also are likely to have large-scale effects on dispersal and immigration and emigration rates. These alterations are likely to have effects beyond the scale of individual populations, impacting metapopulation dynamics and regional patterns of genetic diversity. Relatively few studies have addresses such issues. Urban populations of Florida Scrub-Jays disperse farther than those in wildlands and the movement patterns are unidirectional; jays move from suburbs to wildlands, but not vice versa (Thaxton and Hingtgen 1996), but is difficult to determine if this is an effect of the urban environment or habitat fragmentation or an interaction between both. Alterations in dispersal patterns and changes in population size and structure could lead to changes in genetic diversity and the loss of local adaptations. However, the urban environment may have more direct effects on genetic patterns. Exposure to anthropogenic toxins may increase rates of heritable mutation (Yauk and Quinn 1996). Indeed, rapid changes in avian population and community structure with urbanization and the concomitant changes in niche space may result in rapid evolution of urban avifaunas (Diamond 1986, Johnston 2001).

Developing mechanistic links between population and community level patterns requires understanding the demographic trends and underlying mechanisms for a large variety of species. The links between the environment, demography, and community structure may be expressed in community measures seldom examined in urban avian ecological studies. Although community measures such as density, species richness and evenness are relatively common, measures such as resource partitioning, guild representation and trophic relations are relatively rare. Blair (2001) demonstrates a process where urban avifaunas in different regions may converge through processes of local extinction and local invasion, providing the best example of how population-level processes can influence community structure.

4. CONCLUSIONS

The research approach outlined here and evident in the body of urban bird literature is clearly that of the ecology of birds in cities. A rigorous ecological and experimental approach to understanding avian ecology in an urban context, such as that outlined by Bolger (2001), is necessary to understand how process influences patterns, but we have largely ignored how human values influence these processes. If humans are part of their environment rather than separate and distinct from it, then our social and cultural values must be shaped and, in turn, shape the environment. Understanding that feedback loop and how human values influence variation in urban patterns and processes is necessary to ultimately manage the urban environment to improve the quality of life for birds and for humans. Only by understanding the ecology of cities and integrating the vast patterns of human activities into the conceptual framework can we hope to develop urban environments that lessen their ecological footprint (Wackernagel and Rees 1996; Collins et al. 2000) and reduce their impact on avian biodiversity. Studies of the ecology in cities are fine, but we also need to be able to integrate our finding into the larger conceptual framework of the ecology of cities.

Understanding the ecology of cities will require broad interdisciplinary partnerships. Avian ecologists will increase the applicability and relevance of their research by working with urban planners, landscape architects, geographers, and policy scientists. Productive teams also will need a diversity of social scientists, economists, and demographers to better understand the socio-economic and psychological drivers of human decisions that shape urban patterns. Integrated, interdisciplinary research on urban ecosystems is challenging, but possible and productive as evidenced by recent NSF LTER sites in the urban settings of Phoenix, Arizona and Baltimore, Maryland.

Sustaining interdisciplinary research teams requires novel graduate education programs. Students from the natural and social sciences need to be immersed in interdisciplinary teamwork, including team-authored graduate theses. Programs at the University of Washington and Arizona State University are training urban ecologists using these methods with success. The result will be a new generation of graduate students that seek the challenges of teamwork in urban ecosystems, rather than steer away from ecological studies in urbanizing settings.

REFERENCES

Alberti, M., E. Botsford, and A. Cohen. 2001. Quantifying the urban gradient: linking urban planning and ecology, p. 89-115. *In* J. M. Marzluff, R. Bowman, and R. Donnelly [EDS.], Avian ecology and conservation in an urbanizing world. Kluwer Academic, Norwell, MA.

Azerrad, J. M. and C. H. Nilon. 2001. Avian community characteristics of urban greenspaces in St. Louis, Missouri, p. 489-509. *In* J. M. Marzluff, R. Bowman, and R. E. Donnelly [EDS.], Avian ecology and conservation in an urbanizing world. Kluwer Academic, Norwell, MA.

Blair, R. B. 2001. Creating a homogeneous avifauna, p. 461-488. *In* J. M. Marzluff, R. Bowman, and R. E. Donnelly [EDS.], Avian ecology and conservation in an urbanizing world. Kluwer Academic, Norwell, MA.

Bolger, D. T. 2001. Urban birds: population, community, and landscape approaches, p. 155-177. *In* J. M. Marzluff, R. Bowman, and R. E. Donnelly [EDS.], Avian ecology and conservation in an urbanizing world. Kluwer Academic, Norwell, MA.

Bowman, R. and G. E. Woolfenden. 2001. Nest success and the timing of nest failure of Florida Scrub-Jays in suburban and wildland habitats, p. 385-404. *In* J. M. Marzluff, R. Bowman, and R. E. Donnelly [EDS.], Avian ecology and conservation in an urbanizing world. Kluwer Academic, Norwell, MA.

Breininger, D. R. 1999. Florida Scrub-Jay demography and dispersal in a fragmented landscape. Auk 116:520-527.

Churcher, P. B. and J. H. Lawton. 1987. Predation by domestic cats in an English village. J. Zool. Lond. 212:439-455.

Clergeau, P., G. Mennechez, A. Sauvage, and A. Lemoine. 2001. Human perception and appreciation of birds: A motivation for wildlife conservation in urban environments of France, p. 69-88. *In* J. M. Marzluff, R. Bowman, and R. E. Donnelly [EDS.], Avian ecology and conservation in an urbanizing world. Kluwer Academic, Norwell, MA.

Collins, J. P., A. Kinzig, N. B. Grimm, W. F. Fagan, D. Hope, J. Wu, and W. T. Borer. 2000. A new urban ecology. American Sci. 88:416-425.

Coppedge, B. R., D. M. Engle, S. D. Fuhlendorf, R. E. Masters, and M. S. Gregory. 2001. Urban sprawl and juniper encroachment effects on the abundance of wintering passerines in Oklahoma, p. 225-243. *In* J. M. Marzluff, R. Bowman, and R. E. Donnelly [EDS.], Avian ecology and conservation in an urbanizing world. Kluwer Academic, Norwell, MA.

Diamond, J. M. 1986. Rapid evolution of urban birds. nature 324:107-108.

Fernández-Juricic, E., M. D. Jimenez, and E. Lucas. 2001. Bird tolerance to human disturbance in urban parks of Madrid (Spain): Management implications, p. 261-275. *In* J. M. Marzluff, R. Bowman, and R. E. Donnelly [EDS.], Avian ecology and conservation in an urbanizing world. Kluwer Academic, Norwell, MA.

Forman, R. T. T. and M. Godrun. 1986. Landscape ecology. John Wiley and Sons, New York.

Gavett, A. P. and J. S. Wakely. 1986. Blood constituents and their relation to diet in urban and rural House Sparrows. Condor 88:279-284.

Goldstein, E. L., M. Gross, and R. M. DeGraaf. 1980/1981. Explorations in bird-land geometry. Urban Ecology 5:113-124.

Grimm, N. B., J. M. Grove, S. T. A. Pickett, and C. L. Redman. 2000. Integrated approaches to long-term studies of urban ecological systems. BioScience 50:571-584.

Haskell, D. G., A. M. Knupp, and M. C. Schneider. 2001. Nest predator abundance and urbanization, p. 245-260. *In* J. M. Marzluff, R. Bowman, and R. E. Donnelly [EDS.], Avian ecology and conservation in an urbanizing world. Kluwer Academic, Norwell, MA.

Hostetler, M. 2001. The importance of multi-scale analyses in avian habitat selection studies in urban environments, p. 139-154. *In* J. M. Marzluff, R. Bowman, and R. E. Donnelly [EDS.], Avian ecology and conservation in an urbanizing world. Kluwer Academic, Norwell, MA.

Jerzak, L. 2001. Synurbanization of the magpie in the Palearctic, p. 405-427. *In* J. M. Marzluff, R. Bowman, and R. E. Donnelly [EDS.], Avian ecology and conservation in an urbanizing world. Kluwer Academic, Norwell, MA.

Johnston, R. F. 2001. Synanthropic birds of North America, p. 49-67. *In* J. M. Marzluff, R. Bowman, and R. E. Donnelly [EDS.], Avian ecology and conservation in an urbanizing world. Kluwer Academic, Norwell, MA.

Jukimäki, J. and J. Suhonen. 1998. Distribution and habitat selection of wintering birds in urban environments. Landscape and Urban Planning 39:253-263.

Kent, M., R. A. Stevens, and L. Zhang. 1999. Urban plant ecology patterns and processes: a case study of the flora of the City of Plymouth, Devon, U.K. J. Biogeogr. 26:1281-1298.

Kostelecka-Myrcha, A., J. Zukowski, and E. Oksiejczuk. 1997. Changes in the red blood indices during nestling development of the Tree Sparrow *Passer montanus* in an urban environment. Ibis 139:92-96.

McGowan, K. J. 2001. Demographic and behavioral comparisons of suburban and rural American Crows, p. 367-383. *In* J. M. Marzluff, R. Bowman, and R. E. Donnelly [EDS.], Avian ecology and conservation in an urbanizing world. Kluwer Academic, Norwell, MA.
Marzluff, J. M. 2001. Worldwide increase in urbanization and its effects on birds. *In* J. M. Marzluff, R. Bowman, and R. E. Donnelly [EDS.], Avian ecology and conservation in an urbanizing world, p. 19-47. Kluwer Academic, Norwell, MA.
Marzluff, J. M., R. Bowman, and R. Donnelly. 2001. A historical perspective on urban bird research: trends, terms, and approaches, p. 1-17. *In* J. M. Marzluff, R. Bowman, and R. E. Donnelly [EDS.], Avian ecology and conservation in an urbanizing world. Kluwer Academic, Norwell, MA.
McDonnell, M. J. and S. T. A. Pickett. 1990. Ecosystem structure and function along urban-rural gradients: an unexploited opportunity for ecology. Ecology 71:1232-1237.
McDonnell, M. J., S. T. A. Pickett, and R. V. Pouyat. 1993. The application of the ecological gradient paradigm to the study of urban effects, p. 175-189. *In* M. J. McDonnell and S. T. A. Pickett [EDS.], Humans as components of ecosystems: The ecology of subtle human effects and populated areas. Springer-Verlag, New York.
McIntyre, N. E. 2000. Ecology of urban arthropods: a review and a call to action. Ann. Entomol. Soc. Amer. 93:825-835.
Miller, J. R., J. M. Fraterrigo, N. T. Hobbs, D. M. Theobald, and J. A. Wiens. 2001. Urbanization, avian communities, and landscape ecology, p. 117-137. *In* J. M. Marzluff, R. Bowman, and R. E. Donnelly [EDS.], Avian ecology and conservation in an urbanizing world. Kluwer Academic, Norwell, MA.
Niemelä, J. 1999. Ecology and urban planning. Biodiversity and Conservation 8:119-131.
Oksanen, L. In press. Logic of experiments in ecology: is 'pseudoreplication' a pseudoissue? Oikos.
Orians, G. 1998. Human behavioral ecology: 140 years without Darwin is too long. Bulletin of the Ecological Society of America 79:15-28
Reichard, S. H., L. Chalker-Scott, and S. Buchanan. 2001. Interactions among non-native plants and birds, p. 179-223. *In* J. M. Marzluff, R. Bowman, and R. E. Donnelly [EDS.], Avian ecology and conservation in an urbanizing world. Kluwer Academic, Norwell, MA.
Savard, J-P. L. and J. B. Falls. 2001. Survey techniques and habitat relationships of breeding birds in residential areas of Toronto, Canada, p. 543-568. *In* J. M. Marzluff, R. Bowman, and R. E. Donnelly [EDS.], Avian ecology and conservation in an urbanizing world. Kluwer Academic, Norwell, MA.
Schoech, S. J. and R. Bowman. 2001. Variation in the timing of breeding between suburban and wildland Florida Scrub-Jays: Do physiologic measures reflect different environments?, p. 291-308. *In* J. M. Marzluff, R. Bowman, and R. E. Donnelly [EDS.], Avian ecology and conservation in an urbanizing world. Kluwer Academic, Norwell, MA.
Thaxton, J. E. and T. M. Hingtgen. 1996. Effects of suburbanization and habitat fragmentation on Florida Scrub-Jay dispersal. Fla. Field Nat. 24:25-37.
Thomas, D. W., J. Blondel, P. Perret, M. M. Lambrechts, and J. R. Speakman. 2001. Energetic and fitness costs of mismatching resource supply and demand in seasonally breeding birds. Science 291:2598-2600.
Vitousek, P. M., H. A. Mooney, J. Lubchenco, and J. M. Melillo. 1997. Human domination of Earth's ecosystems. Science 277:494-499.
Wackernagel, M. and W. E. Rees. 1996. Our ecological footprint: reducing human impact on the earth. New Society Publishers, Philadelphia, PA.
Yauk, C. L. and J. S. Quinn. 1996. Multilocus DNA fingerprinting reveals high rate of heritable genetic mutation in herring gulls nesting in an industrialized urban site. Proc. Natl. Acad. Sci. USA 93:12137-12141.

INDEX

Printed in the United States
54428LVS00001B/15

9 780792 374589